Handbook of Rigging

Other McGraw-Hill Handbooks of Interest

Alpern • HANDBOOK OF SPECIALTY ELEMENTS IN ARCHITECTURE
Avallone and Baumeister • MARKS' STANDARD HANDBOOK FOR MECHANICAL ENGINEERS
Brady and Clauser • MATERIALS HANDBOOK
Brater • HANDBOOK OF HYDRAULICS
Callender • TIME-SAVER STANDARDS FOR ARCHITECTURAL DESIGN DATA
Church • EXCAVATION HANDBOOK
Conover • GROUNDS MAINTENANCE HANDBOOK
Crocker and King • PIPING HANDBOOK
Croft and Summers • AMERICAN ELECTRICIANS' HANDBOOK
Davis and Sorensen • HANDBOOK OF APPLIED HYDRAULICS
DeChiara and Callender • TIME-SAVER STANDARDS FOR BUILDING TYPES
Fink and Beaty • STANDARD HANDBOOK FOR ELECTRICAL ENGINEERS
Gaylord and Gaylord • STRUCTURAL ENGINEERING HANDBOOK
Gieck • ENGINEERING FORMULAS
Harris • HANDBOOK OF NOISE CONTROL
Harris and Crede • SHOCK AND VIBRATION HANDBOOK
Havers and Stubbs • HANDBOOK OF HEAVY CONSTRUCTION
Heyel • THE FOREMAN'S HANDBOOK
Hicks • STANDARD HANDBOOK OF ENGINEERING CALCULATIONS
Higgins • HANDBOOK OF CONSTRUCTION EQUIPMENT MAINTENANCE
Higgins and Morrow • MAINTENANCE ENGINEERING HANDBOOK
King and Brater • HANDBOOK OF HYDRAULICS
Manas • NATIONAL PLUMBING CODE HANDBOOK
Mantell • ENGINEERING MATERIALS HANDBOOK
McPartland • MCGRAW-HILL'S NATIONAL ELECTRICAL CODE HANDBOOK
Merritt • BUILDING DESIGN AND CONSTRUCTION HANDBOOK
Merritt • STRUCTURAL STEEL DESIGNERS' HANDBOOK
O'Brien • CONTRACTOR'S MANAGEMENT HANDBOOK
Parmley • FIELD ENGINEER'S MANUAL
Rosaler and Rice • STANDARD HANDBOOK OF PLANT ENGINEERING
Rossnagel • HANDBOOK OF RIGGING
Smeaton • SWITCHGEAR AND CONTROL HANDBOOK
Tuma • ENGINEERING MATHEMATICS HANDBOOK
Tuma • TECHNOLOGY MATHEMATICS HANDBOOK

Handbook of Rigging

For Construction and Industrial Operations

W. E. Rossnagel (deceased)

Lindley R. Higgins, P.E.

Joseph A. MacDonald

Fourth Edition

McGraw-Hill Book Company

New York St. Louis San Francisco Auckland
Bogotá Hamburg London Madrid Mexico
Milan Montreal New Delhi Panama
Paris São Paulo Singapore
Sydney Tokyo Toronto

Library of Congress Cataloging-in-Publication Data

Rossnagel, W. E.
Handbook of rigging.

Includes index.
1. Hoisting machinery—Rigging. 2. Scaffolding.
I. Higgins, Lindley R. II. MacDonald, Joseph A.
III. Title.
TJ1367.R6 1988 621.8′62 87-22641
ISBN 0-07-053941-3

1234567890 DOC/DOC 8932109

ISBN 0-07-053941-3

The editors for this book were Harold B. Crawford and Ingeborg M. Stochmal, the designer was Naomi Auerbach, and the production supervisor was Dianne Walber. It was set in Century Schoolbook. It was composed by the McGraw-Hill Book Company Professional & Reference Division composition unit. Printed and bound by R. R. Donnelley & Sons, Inc.

Contents

Section 5. Scaffolding and Ladders

Section 6. Procedures and Precautions

Appendix American National Standards for Safety and Health Construction (1987)

Preface

This fourth edition of the acknowledged standard rigging reference—*The Handbook of Rigging*—has been revised, expanded, and updated to incorporate not only the advances in methods and technologies of rigging and scaffolding that have occurred since publication of the first edition in 1957, but also the regulations governing the safe practices and procedures for using rigging equipment and materials on projects that have been prescribed and enforced by OSHA since 1970.

The handbook is intended to be a ready reference and guide for expert riggers engaged in full-time rigging operations; erectors of buildings and other structures; maintenance mechanics in industrial plants and electric generating stations having less frequent rigging jobs to perform; operators of all types of hoists, derricks, and cranes; and other workers engaged in the erection of scaffolds or signage, or in climbing tall structures for repair or maintenance work.

Proper training can significantly improve accident prevention. The handbook attempts to identify the elements of safe rigging practice and expand upon the minimum safety regulations promulgated by the various standards organizations and by OSHA. It should be used as a guide in conjunction with the applicable safety regulations by contractors, supervisors, riggers, equipment operators, and managers concerned with or responsible for the safety of employees at work on a project, as well as for the general public and surrounding structures.

The information contained in specific chapters such as "Derricks and Cranes" or "Scaffolds and Ladders," and so forth can provide the basis for developing instructional material used in the training of personnel engaged in rigging operations. It can also be included in standard instructions issued to employees for the safe use of rigging equipment and materials.

Safe practices recommendations in the handbook, of necessity, are framed in general terms to accommodate the many variations in rigging practices and the different ways in which rigging is used. Because these recommendations are only advisory in nature, they must be supplemented by strict observance of specific relevant regulations as well as manufacturers' recommendations and requirements.

The authors are grateful to the many technical and trade associations, government agencies, equipment manufacturers, and materials producers cited in the text and illustrations, that assisted in the preparation of this handbook. Special thanks go to: Catherine M. Barth for her conscientious assistance in collecting and cataloging reference material solicited from equipment manufacturers, technical associations, and government agencies; Eugenie L. Gray for her critical review of editorial material to ensure grammatical consistency and technical accuracy of both manuscript and page proofs; Ingeborg M. Stochmal and Rita Margolies for their patient and thorough review of the finished manuscript and guidance through the editing and production stages.

J. A. MacDonald, C.E.
L. R. Higgins, P.E.

Introduction

The art of rigging may be traced to prehistoric times. Levers were used then, as now, to pry stones, roll logs, and move objects that were too heavy to be moved by hand. The inclined plane, or natural ramp, was in use even then to help move heavy objects up to higher elevations.

The first major rigging job, of which there is not only a record but also indisputable evidence, was in the construction of the three pyramids at Gizeh, near Cairo, Egypt, about 2700 B.C. It's estimated that preparation work must have taken almost 10 years; and construction about 20 years.

As it stands today, the large pyramid—built to contain the remains of Pharaoh Cheops (Khufu)—is 746 ft square at the base and 451 ft high. Originally, the structure was encased in a fine grain limestone. But at some unknown time during the past 4,600 years this sheath was removed.

The large pyramid contains about 2.3 million stones weighing from 2 to 30 tons each; a total of about 5.75 million tons (nearly 20 times the weight of the masonry in the 102-story Empire State Building in New York City). These huge blocks of stone had to be moved from the quarries to the bank of the Nile River, ferried across during the annual three-month period when the river was at its flood stage, and then dragged to the construction site.

Records indicate that a sand ramp, requiring nearly one million tons of sand transported from the desert, was built up one side as the pyramid rose in height. Another million tons of sand were then required to backfill the interior of the pyramid. And when the job was completed, the ramp had to be removed.

The construction crews had no mechanical equipment. Instead they used levers, rollers, crude ropes, sledges, plumb lines, and string sightings to get the massive job done. The huge stones were hauled up the ramp on rollers, for an average lift of 100 ft, by the brute strength of 100,000 slaves in teams of 50 men each, driven by the slave master's whips.

The pyramid remains today as indisputable proof of the ingenuity of the Egyptians. To fully appreciate the enormity of this undertaking, consider the power one man can develop:

> Turning a crank, such as on a winch, a man may exert 15 to 18 lb force continuously. Intermittently, he may exert 25 to 40 lb.
>
> Pulling downward on a rope or on the hand chain of a chain hoist, he may be able to exert a 40-lb pull for a long time; but for a short period, his pull may approach his weight.
>
> In lifting a few inches off the ground (assuming that he does it in a proper manner), he may lift up to 300 lb.
>
> Pushing or pulling an object, such as a vehicle, he may (with good footing) exert a force of 100 to 300 lb.

Now, considering the amount of work that a man may be expected to accomplish in an 8-h day, compare his effort with the cost of doing the job electrically.

Assume that electric power costs about $0.10 per kilowatt-hour and that electrically driven machinery is 50% efficient. Further, for continuous work, a man may be expected to deliver about 0.10 hp, while for a very short time he may exert from 0.4 to 0.5 hp. Then the cost equivalent of a man's labor may approximate the following calculations:

> A strong man can lift 86 tons (such as bags of cement) from the ground to a height of 4 ft in an 8-h work day, averaging 0.045 hp. The cost of doing this with electric power would be about $27 per day.
>
> A man can carry 22.3 tons up a ramp or stair to a height of 12 ft in an 8-h workday, averaging 0.034 hp. The equivalent electric power would cost about $20 per day.
>
> Pushing a wheelbarrow, he can move 40.7 tons up a 3-ft ramp in 8 h and average 0.015 hp. With electric power this would cost approximately $9.00 each working day. Shoveling loose earth, he can raise 20 tons to shoulder height in 8 h, exerting 0.013 hp. Using electric power, he can do this for about $7.00.

Thus, it can be shown that man is a very inefficient machine, and it is easy to understand why 100,000 men were required to transport material for the Great Pyramid.

The art of rigging has developed to the degree that today manufacturers build 200- to 400-ton traveling cranes for power plants; hammerhead cranes of even greater capacity for shipyards; and mobile cranes and derricks capable of handling trusses and girders weighing up to 200 tons for buildings or bridges.

However, this book is not intended to deal with rigging operations of this magnitude. Rather it covers conventional rigging operations in industrial plants, factories, and power plants; in transporting and handling heavy machinery; in mines and port facilities; and on construction sites for erection and demolition of structures.

Chapters include rigging equipment, materials, accessories, procedures, and precautions used in the practice of rigging. Included also is a section on the erection of temporary scaffolding for painting, repairing, construction or demolition of structures, and supporting heavy loads.

Section

1

Engineering Principles

Chapter

1

Basic Machines

The first and most important step in any rigging operation is to determine the forces that will affect the job, and then to select and arrange the equipment that will move the loads safely, both horizontally and vertically.

Too often, though, the load-bearing parts of a rigging system are stressed to a point dangerously close to the breaking point without the rigger realizing it. This element of chance can be reduced to a minimum by a simple knowledge of how to determine the loads that must be moved and the capacity of the equipment being used.

The forces involved in rigging will vary with the method of connection, the direction of support in reference to the load, and the effects of motion. Thus the rigger must know something about mechanical laws, the determination of stresses, the effect of motion, the weight of loads, the centers of gravity, and factors of safety. Also essential is a basic knowledge about the strength of materials, which is covered in succeeding chapters.

Mechanical Laws

Machinery offers a mechanical advantage in moving loads. The elementary machines from which all machinery is constructed are the inclined plane, the lever, the wheel and axle (gear or pulley and shaft), the screw (derived from the inclined plane), and the block and fall.

Riggers frequently make use of the inclined plane when hauling a load on rollers up a ramp or on skids onto a truck.

To estimate roughly the pull required to haul a load of 15,000 lb on rollers up an incline of 4 ft in 20 ft, draw a diagram *ABC* representing

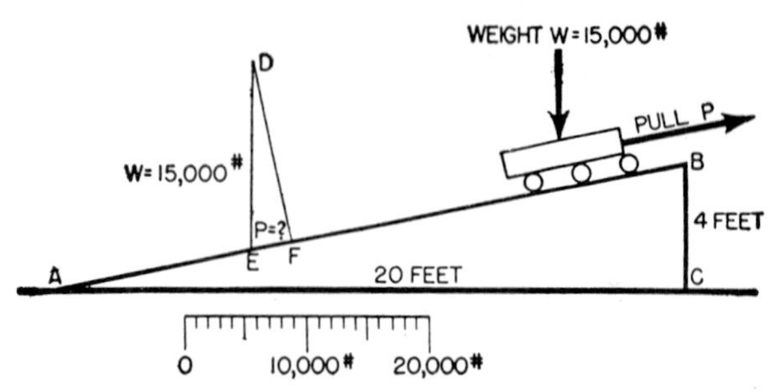

Figure 1.1 Calculating the force required to move an object up an inclined plane or ramp.

the incline. Then draw to any suitable scale a vertical line *DE* representing the weight of the load (see Fig. 1.1).

Assume a scale of 1 in. = 10,000 lb so that a line 1½ in. long will represent 1½ × 10,000 lb, or 15,000 lb. From *D* draw a line *DF* at a right angle to the slope of the incline *AB*.

Using the same scale of 1 in. = 10,000 lb, measure the distance *EF*, which will be the theoretical pull required. This scales to about 0.3 in., or 3,000 lb, which is the pull if frictionless rollers were used. To this value the resistance to friction must be added to determine the actual pulling force.

Another simple application of the mechanical advantage is through leverage, in which one point (the fulcrum) of a rigid bar is fixed, another point is connected with the force (the load) to be acted upon, and a third point is connected with the force (the power) applied. The crowbar is a typical lever.

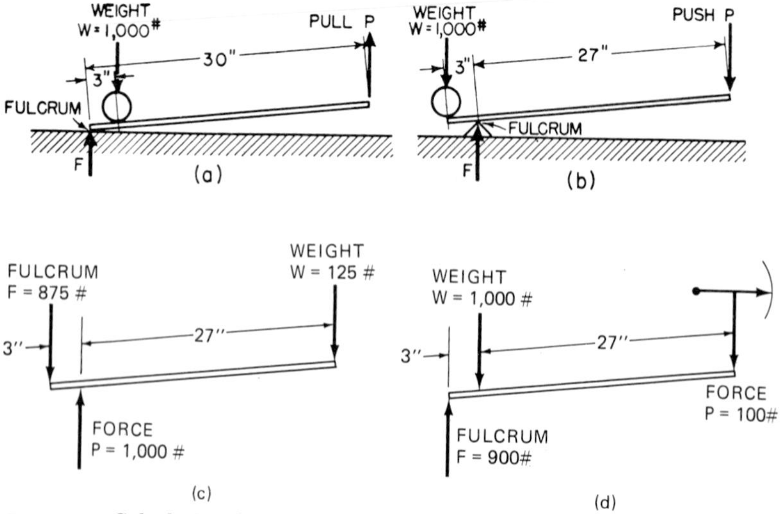

Figure 1.2 Calculating the force required to lift or move a load by means of levers.

As commonly used, there are two arrangements of the fulcrum. Figure 1.2*a* shows how an upward pull P on the handle lifts the weight W because gravity acts as a downward force. The toe of the crowbar pivots about the fulcrum F and, in effect, the floor or ground exerts an upward force to resist the downward pressure.

Assume a weight W of 1,000 lb acting at a distance of 3 in. from the fulcrum F and a person lifting up on the lever at a distance 30 in. from the fulcrum. The force P times the force distance (30 in.) always equals the weight W (1,000 lb) times the weight distance (3 in.). Thus $P \times 30$ in. = 1,000 lb × 3 in., or the force required is $P = 100$ lb. In other words, if the force distance is 10 times the weight distance, then the force is one-tenth the weight, and the lever is said to provide a mechanical advantage of 10. This rule holds true for other ratios of distances also.

However, if the crowbar is used as shown in Fig. 1.2*b*, where the fulcrum is between the force and the load, the force then is a push P in the downward direction. Using the same 30 in., the force P times the force distance (27 in.) equals the weight W (1,000 lb) times the weight distance (3 in.). Thus $P \times 27$ in. = 1,000 lb × 3 in., or the force required is $P = 111$ lb.

By observation, then, the crowbar is most efficient, or requires less force, when used as shown in Fig. 1.2*a* although sometimes it may be necessary to limit the pressure on the fulcrum, as in Fig. 1.2*b*. In this latter application, the pressure is always the sum of the weight W and the force P.

A third form of leverage, with fulcrum and load at the ends of a rigid member and the applied force at an intermediate point, produces a mechanical disadvantage, as illustrated in Fig. 1.2*c*. However, this arrangement is important in a fishing rod, where a relatively light weight at the end of the rod must be lifted quickly by a relatively heavy pull of small travel above the butt end.

In construction work, a heavy load of a very small travel, on a scale used for weighing loads on trucks, is balanced by a small force with considerable travel on the indicator end, as shown in Fig. 1.2*d*.

Yet another application of the basic lever is in the use of a pulley, where the block-and-tackle arrangement acts as the lever, with the fulcrum at the side of the block and the rope fixed to the ceiling (see Fig. 1.3). The mechanical advantage obtained with the block-and-fall arrangement is explained in detail in Chapter 10.

Just as the block and tackle, or hoisting derrick, can be considered as a form of lever, so almost every other tool becomes some form of the same simple device. Even the simple twisting of two lines with a pipe to pull a load is merely the application of a lever with a long driving arm and a short resisting arm.

An important adaptation of the lever principle is the wheel and axle. Although the movement of the lever is limited, the motion of the wheel may continue indefinitely. For the wheel-and-axle arrangement with belts and pulleys, the belt pulls, or tensions, are inversely proportional

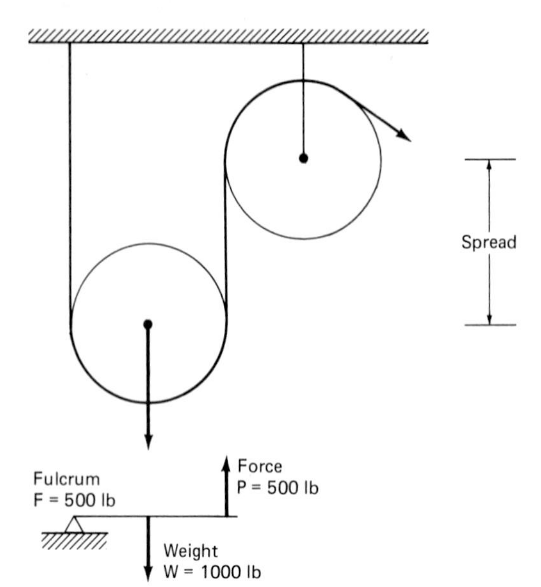

Figure 1.3 How a pulley system increases mechanical advantage.

to the pulley diameters, that is, when one pulley is four times as large as the other, the belt pull on the large pulley is one-fourth the belt pull on the small pulley (see Fig. 1.4).

In this example a force P is exerted on the radius R (12 in.) of the large pulley, while the weight W (1,000 lb) is supported on the radius r (3 in.) of the small pulley. Thus $P \times 12$ in. $= 1{,}000$ lb $\times 3$ in., or $P =$ 250 lb.

For gears on the same shaft, the mechanical advantage of the driven pulley relative to the driver is directly proportional, and the speed is inversely proportional, to the lengths of the respective radii (see Fig. 1.5).

If there is a train of gears, such as on a hand winch consisting of several gear reductions, the theoretical pull that can be exerted on the hoist rope may be readily estimated:

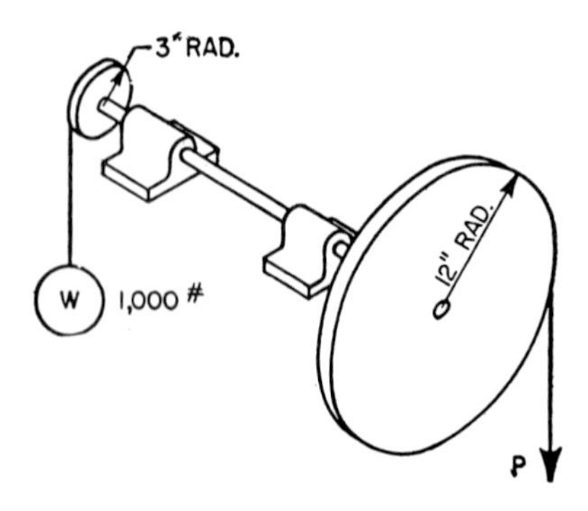

Figure 1.4 Calculating the force required to lift a load by means of wheels and a shaft.

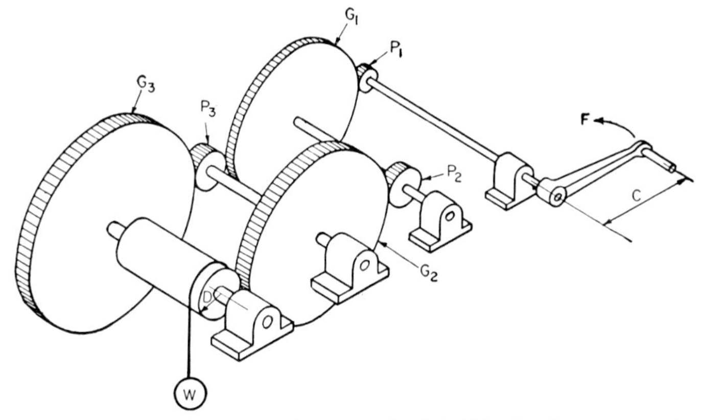

Figure 1.5 Calculating the force required to lift a load by means of a train of gears.

$$W = \frac{F \times C \times G_1 \times G_2 \times G_3 \times \ldots}{D \times P_1 \times P_2 \times P_3 \times \ldots}$$

where

W = pull on hoist rope, pounds
F = force on hand crank, about 17 lb at 2½ ft/s
C = radius of hand crank, inches
$G_1, G_2, G_3, \ldots$ = number of teeth in gear $G_1, G_2, G_3, \ldots$
D = radius of drum, to center of rope, inches
$P_1, P_2, P_3, \ldots$ = number of teeth in pinion $P_1, P_2, P_3, \ldots$

(Of course, a deduction must be made for friction.)

Determination of Stresses

To figure the stress developed by a load, it is important to know that the forces acting in one direction must be equal and opposite to the forces acting in the other direction; that is, all forces must be in equilibrium.

When an ordinary outrigger is used on a building to support a swinging scaffold, the reactions at the supporting and hold-down points are easily calculated, once the load is known (see Fig. 1.6). The load force multiplied by the length of its lever arm must equal the hold-down force multiplied by the length of its lever arm. Thus the fulcrum reaction is equal and opposite to the two forces.

But for a derrick additional factors complicate the determination of these reactions since the supporting members act at numerous angles and in various combinations. To find the resultant stresses, a simple

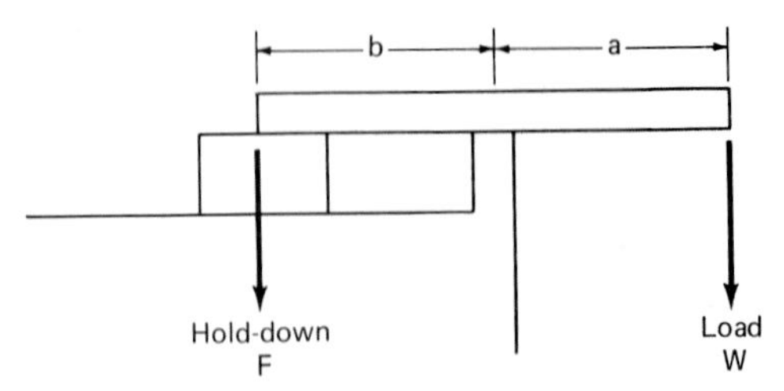

Figure 1.6 Load distribution for an outrigger when used to support a scaffold.

graphic method provides adequate results for all members. This picture, called a triangle of forces, shows how all the forces at one point must be in equilibrium (see Fig. 1.7).

Note: Conventionally, vector arrows heading toward a point under consideration indicate that the member is under compression; arrows heading away from the point indicate tension.

Starting at point 1 (see Fig. 1.7*b*), the derrick's load-line pull must be equally balanced by the boom resistance and the topping-lift pull. Thus three forces act at this point. Draw line *AB* at a convenient scale parallel to the load line. Next, draw lines *BC* and *AC* parallel to the derrick boom and to the topping lift, respectively. These lines will intersect at a point *C*. Measuring the lengths of these two lines will provide directly the forces induced in the respective members.

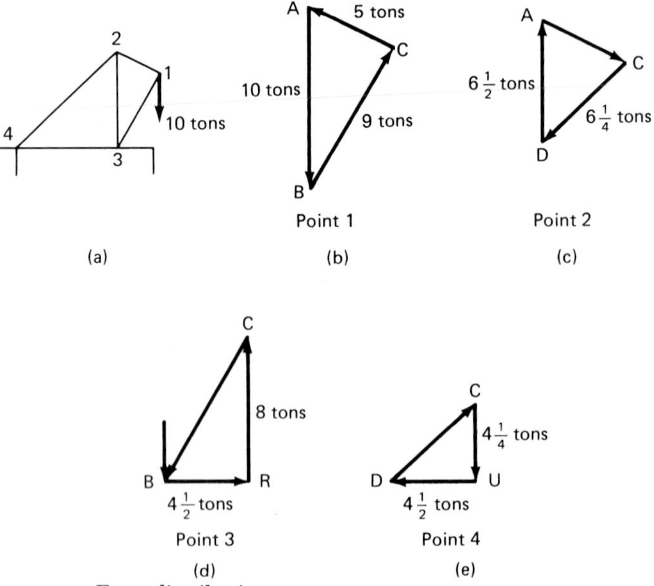

Figure 1.7 Force distribution vectors.

Assume the derrick load is 10 tons. By drawing line AB 5 in. long, the diagram will have a scale of 2 tons = 1 in. Then if line BC is 4.5 in. long, the stress in the derrick boom is 9 tons; if line AC is 2.5 in. long, then the stress in the topping lift is 5 tons. Any one force in this triangle is the resultant that balances the other two forces.

To analyze the forces at point 2 (see Fig. 1.7*c*), draw a new force triangle using the same stress and length of the line for the topping lift as at point 1, but reverse the direction of the arrow. Draw line CD to represent the guy, and line DA to represent the mast. Again, the completed diagram constitutes a triangle of forces with the respective lengths a direct measure of the stresses in the various parts.

Next, to determine the stress at the foot of the mast, note that the boom stress BC is made up of a vertical and a horizontal reaction (see Fig. 1.7*d*). CR is the vertical reaction, and it goes down into the foot block along with the vertical reaction found previously in the mast reaction AD. Since CR equals 8 tons and AD equals 6.5 tons, the total vertical load on the foot block, point 3, is 14.5 tons. BR is the horizontal force of 4.5 tons that must be resisted to keep this point from slipping.

Similarly, the guy reaction can be found by drawing DC equal and opposite to the force noted on the drawing for point 2 (see Fig. 1.7*e*). The horizontal and vertical reactions are necessary to form a closed triangle of forces.

To check the reactions of the derrick upon the supporting points, eliminate all of the members that are used to make the mechanical device workable; then simply consider the problem as a rigid lever.

In Fig. 1.8 project the load line down to meet the horizontal line of the

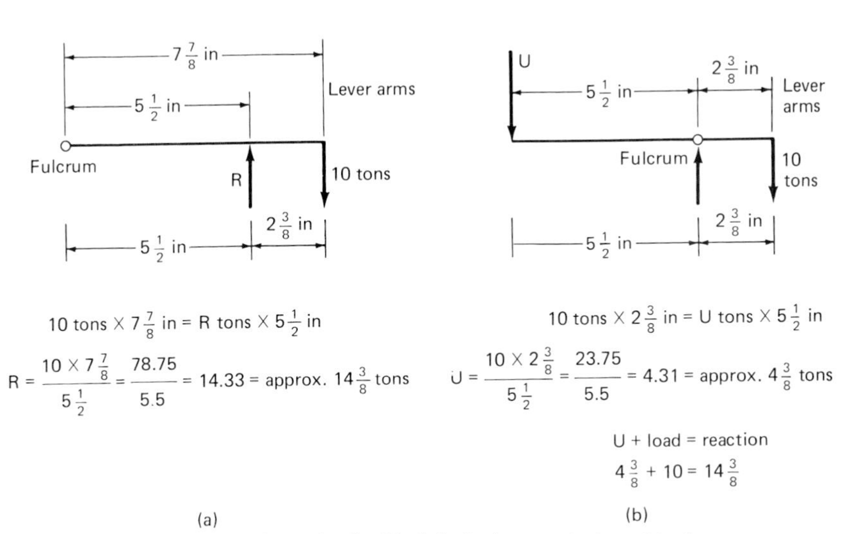

Figure 1.8 (*a*) Calculating lever loads. (*b*) Calculating vertical positioning.

lever, because the effective load lever arm and the load are perpendicular to it. Next consider point 4 as the fulcrum of a lever and point 3 (Fig. 1.7) as a resisting load. Then, using simple arithmetic, calculate the resting load as 14.33 tons (Fig. 1.8*a*). This checks close enough with the 14.5-ton reaction previously determined. By the same method, calculate the vertical reaction at *U* as 4.31 tons, which is close enough to the 4.5-ton force determined to keep point 3 from slipping (Fig. 1.8*b*). (The differences in the values obtained by the check method result from graphic discrepancies in the triangle of forces method, and are small enough to be acceptable.)

When a sling suspension is being analyzed, the use of a triangle of forces clearly shows the effect of decreasing the angles of lift. Assume a 10-ton load with the sling arrangements shown in Fig. 1.9. Then:

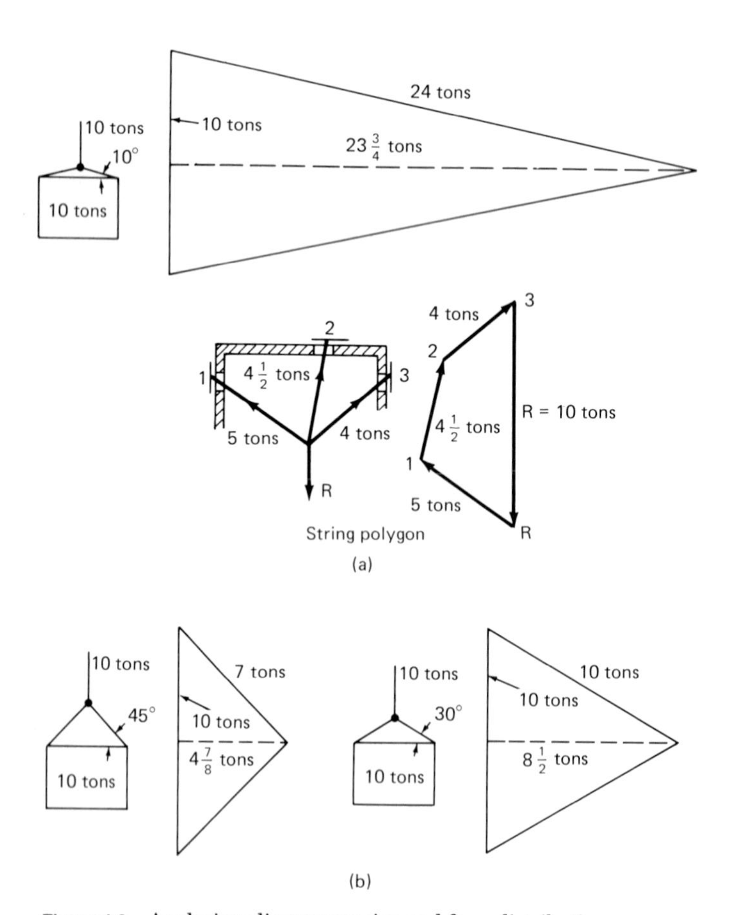

Figure 1.9 Analyzing sling suspension and force distribution.

With the lift angle at 10°, each arm is stressed to 24 tons, with the resistance reaction reaching 23.3 tons.

When the sling's lift angle is 45°, the load puts a 7-ton stress in each sling arm that must be resisted by a 4.875-ton dividing reaction.

At a lift angle of 30°, each arm is stressed 10 tons, and the dividing reaction is 8.5 tons.

Thus, it is apparent that when a choker hitch is jammed down for hoisting, the resultant very small lift angle produces sling-arm stresses considerably greater than the load. Also the stress is very high at the point where the hitch bends over the upper corner of the load, and can cause quick damage at this point. Such hitches require the use of a supplementary compression member (such as a spreader bar) to provide the cross resistance necessary for the weaker load member.

When three or more known forces act at one point, a string polygon can be used to determine the resultant force. Simply draw a set of lines (to scale) parallel to the known forces and place the lines end to end. The beginning and end points are connected with a line that represents the resultant of all the forces. The resultant line is opposite in direction to the known forces and of a magnitude equal to the scaled length.

Three or more unknown forces acting on a point constitute an indeterminate problem, and a triangle of forces cannot be drawn for the solution. Such a problem requires a design engineer and must be solved using calculus.

Effect of Motion

A knowledge of the principles of basic machines described thus far would be quite adequate if the rigger were to deal only with stationary forces. However, most rigging operations involve loads in motion. The rigger therefore should also be aware of the effect that motion has on loads and the stress placed on a rope, chain, beam, scaffold, ladder, or other load-bearing member because of a suddenly applied load, such as a jerk or an impact.

For example, a beam may safely support a concentrated load of 1,000 lb at its center. However, if a wheel carrying a 1,000-lb load is rolled at high speed over this beam, it will produce a stress in the beam twice as great as that produced by the static (stationary) load. Similarly, if the 1,000-lb load (assumed to be incompressible) is dropped onto the beam from a height of 2 ft, the energy developed at the instant of impact would be 1,000 lb × 2 ft, or 2,000 ft-lb. This energy must be absorbed by the beam.

If the beam could be deflected 2 ft, then the impact would be (2,000 ft-lb)/(2 ft) = 1,000 lb, to which must be added the static load of 1,000 lb that would be applied if the load were at rest on the beam.

If the beam deflects only 1 in., then the impact would be (2,000 ft-lb)/($\frac{1}{12}$ ft) = 24,000 lb, plus 1,000 lb, or a total of 25,000 lb.

If the beam deflects only ¼ in., then the impact would be (2,000 ft-lb)/($\frac{1}{48}$ ft) = 96,000 lb, plus 1,000 lb, or a total of 97,000 lb, under which loading the beam would fail.

(Actually, the load will compress slightly upon impact or the sling will stretch, or both, and therefore the stopping distance of the load's center of gravity will be somewhat greater.)

When the movement of the load is horizontal, or nearly so (such as when a heavy slab on a pair of rollers gets out of control on a slight downgrade), and the load strikes a fixed object (such as a building column), the same rule applies, except that the static weight is not added. The impact on a column can readily be estimated if the vertical distance, the distance the load drops (probably only a few inches or so), is known and the deflection of the column is assumed.

If, on the other hand, the speed is known, then it is possible to calculate the theoretical vertical drop by using the formula

$$H = \frac{V^2}{2g} \quad \text{or} \quad H = \frac{V^2}{64.32}$$

where H = theoretical drop, feet
V = velocity, feet per second
g = acceleration due to gravity, ft/s

For example, if a 3,600-lb automobile traveling 60 mi/h (88 ft/s) runs head-on into a massive concrete bridge abutment backed by an earth embankment, the movement of the abutment will be negligible. But when the car comes to rest, if the rear axle is 3 ft nearer to the wall than when the front bumper first touched it, then the stopping distance of the car's center of gravity is 3 ft. Since $H = V^2/64.32 = 88 \times 88/64.32 = 120$ ft (the equivalent drop to produce this speed), the energy developed would be 3,600 lb × 120 ft = 432,000 ft-lb at the instant of contact. Thus, the impact force would be (432,000 ft-lb)/(3-ft stopping distance) = 143,000 lb, or nearly 72 tons.

The inertia of a rigged object in motion that takes place with a decrease or increase in speed tends to increase the stresses the rigging must bear. As a hoist load line starts to move, the load must accelerate from zero to the normal speed of ascent. This requires an additional force over and above the weight of the load, which will vary with the rate of change of the velocity.

A load lifted very slowly by a hand winch induces little additional stress in the load line. However, a load accelerated quickly by mechanical power may put twice as much stress on the rope as the weight to be lifted.

Friction, too, can impose severe stresses in rigging. Where sheave maintenance is neglected, additional friction in each sheave may be as much as 10% of the rope tension for every sheave used. Thus, if a hoist line is reeved up, friction will add considerably to load and acceleration requirements. In addition, the rope may have been weakened by continuous use, involving small sheaves with worn grooves, reverse bending, or generally harsh construction applications.

Remember: It is imperative to apply a load or take up slack on a rope slowly.

Weight of Loads

The most important step in any rigging operation is to determine the load's weight. If this information cannot be obtained from shipping papers, design plans, catalog data, or other dependable sources, it may be necessary to calculate such weights.

The weights and properties for all sizes of standard and special structural metal shapes and members, and methods of finding weights of built-up or special members, are contained in three handbooks:

1. *Manual of Steel Construction.* (American Institute of Steel Construction)
2. *Cold Formed Steel Design Manual.* (American Iron and Steel Institute)
3. *Aluminum Construction Manual.* (American Aluminum Association)

However, because these books may not be available on all job sites, it is well to remember that the weight of a square foot of steel, 1 in. thick, is 40.8 lb, and that of aluminum is 14.4 lb. Since it is customary to figure weights based on ⅛-in. increments of material, a square foot of

⅛-in.-thick steel weighs 5.1 lb, and any additional thickness is some multiple of that. A square foot of ⅛-in.-thick aluminum weighs 1.8 lb.

Structural beam sizes are usually supplied on erection plans together with their weight per foot of length and the length of a member. Thus, it is easy to compute the weight of any member to be lifted. However, the weights of angle, plate, or built-up members will have to be calculated.

Weights of any structural shape can be computed by separating the parts or by flattening them into rectangles, which in turn become parts or multiples of a square foot of metal, 1 in. thick.

For example, a 2-ft-wide steel plate, ½ in. thick and 20 ft long, would weigh $2 \times 20 \times 4 \times 5.1 = 816$ lb. If this were a cover plate on the top and bottom flange of a wide-flange beam, W 30×132, which is 24 ft long, the composite weight of the member, not including fastenings, would be $816 \times 2 + 132 \times 24 = 4{,}800$ lb.

Angle weights can be approximated closely enough for safe use by considering the angle as flattened out to form a plate. A steel angle, L $5 \times 3 \times$ ¼ in., would flatten out to approximately an $8 \times$ ¼-in. plate (see Fig. 1.10). Similarly, a $7 \times$ ⅝-in. plate would have its weight calculated, approximately, as $\frac{7}{12} \times 5 \times 5.1 = 14.7$ lb. Since the actual weight of this angle is 13.6 lb (from tables in the *Manual of Steel Construction*), the calculated weight is close enough for figuring rigging requirements.

Likewise, tanks or other cylindrical shapes can readily be considered as rectangular plates for calculating the weight of the load to be lifted. For this, simply multiply the circumference, that is, the distance around the cylinder ($3.14 \times$ diameter), by the length, or height, of the cylinder and add to this value the closed end areas of the cylinder ($3.14 \times \frac{1}{2}$ diameter $\times \frac{1}{2}$ diameter).

Thus a steel boiler 5 ft in diameter has a circumference of $3.14 \times 5 = 15.7$ ft. If the boiler is 8 ft long and made of ⅜-in. plate, it would flatten out to a rectangle $15.7 \times 8 = 125.6$ ft^2, which would weigh $3 \times 5.1 \times 125.6 = 1{,}922$ lb. Since the end area of the boiler is $3.14 \times \frac{5}{2} \times \frac{5}{2} = 19.62$ ft^2, if the ends were capped with ½-in. plate, these flattened pieces would weigh $2 \times 19.62 \times 4 \times 5.1 = 800.5$ lb. The boiler's total weight would then be $1{,}922 + 800.5 = 2{,}722.5$ lb.

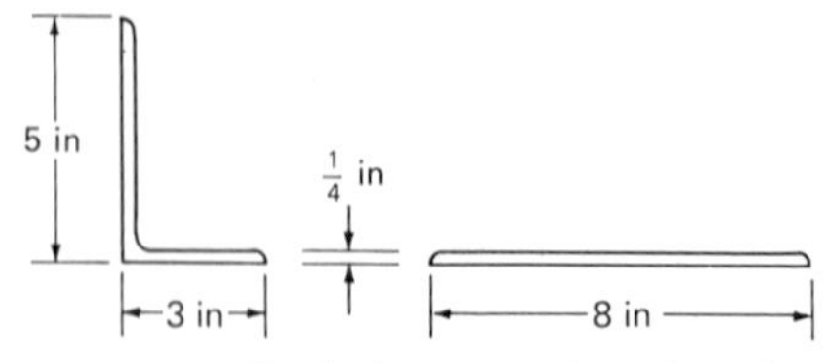

Figure 1.10 Typical structural steel angle.

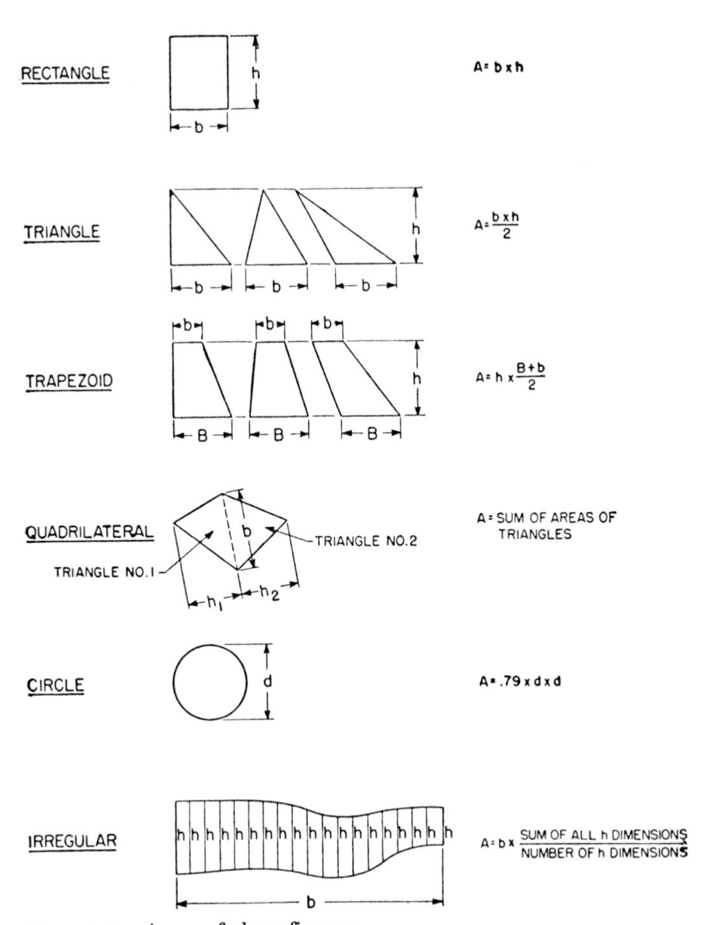

Figure 1.11 Areas of plane figures.

When the weight of a load cannot be obtained from readily available sources, it may be necessary for the rigger to roughly calculate the weight of such objects. If the object is of a simple shape, consider it cut up into a number of regular geometric shapes, the cubic contents (and the weight) of which then can be calculated with a fair degree of accuracy (see Fig. 1.11). The sum of these values will give the total weight.

For example, to find the area of a flat plate of irregular shape (see Fig. 1.12), draw chalk lines on the plate from any corners desired so as to subdivide the shape into a number of triangles. Take the necessary measurements and apply the formula for triangles given in Fig. 1.11. The sum of the triangles equals the area of the plate.

The shape is cut into three triangles *A*, *B*, and *C*. Triangle *A* has a base of 30 in. and a height of 9 in. Thus its area is $bh/2 = 30 \times 9/2 =$ 135 in.^{2} Triangle *B* has an area of $28 \times 8\frac{1}{2}/2 = 119$ in.^{2}, and triangle

C has an area of $15 \times 3\frac{1}{2} / 2 = 26\frac{1}{4}$ in.^{2}. Thus the area of the plate is the sum of the triangular areas: $135 + 119 + 26\frac{1}{4} = 280\frac{1}{4}$ in.^{2}. If the thickness of the plate is $\frac{3}{4}$ in., the volume is $280\frac{1}{4} \times \frac{3}{4} = 210$ in.^{3}. Using a table of material weights, steel weighs 0.28 lb/in.^{3} and the plate weighs $210 \times 0.28 = 59$ lb.

Likewise, if it is necessary to calculate the weight of a complex solid shaped object, such as a concrete member, first divide the object into smaller simple geometric shapes and then find the volumes of the individual parts (see Fig. 1.13).

For example, with the irregularly shaped concrete block shown in Fig. 1.14, divide the shape into a rectangular block A and a frustrum of a pyramid B. The volume of part A is $4 \times 5 \times 3\frac{1}{2}$ ft $= 70$ ft^3. The volume of part B is the volume of the entire pyramid (6 ft high) minus the volume of the upper pyramid (3 ft high).

The volume of the large pyramid is the area of the base times one-third the height $4 \times 5 \times 6/3 = 40$ ft^3. From this, subtract the volume of the small pyramid $2 \times 2\frac{1}{2} \times 3/3 = 5$ ft^3. Thus the volume of the frustrum is $40 - 5 = 35$ ft^3. The total volume of the concrete structure is $70 + 35 = 105$ ft^3. Since concrete weighs about 144 lb/ft^3 the block weighs $105 \times 144 = 15{,}120$ lb, or about $7\frac{1}{2}$ tons.

Since the weight of most materials to be lifted is based on their weight per cubic foot, it is first necessary to determine how many cubic feet of material (its volume) must be hoisted before the weight can be estimated. Table 1.1, which lists the weights of common materials, should enable any rigger to compute the approximate weight of a given load. But, when in doubt, do not hesitate to seek advice from an engineer or a foreman on the job.

Remember: The time taken to calculate the approximate weight of any object is time well spent and may prevent a serious accident through overloading of lifting gear.

Center of Gravity

Frequently the rigger must determine the center of gravity of an irregularly shaped load before safely slinging it for lifting. The center of gravity is the location where the center of the object's entire weight is theoretically concentrated and where the object will balance when lifted. When the object is suspended freely from a hook, this point will always be directly below the hook. Thus, a load that is slung above and through the center of gravity will be in equilibrium and will not tend to slide out of the hitch or become unstable in any manner.

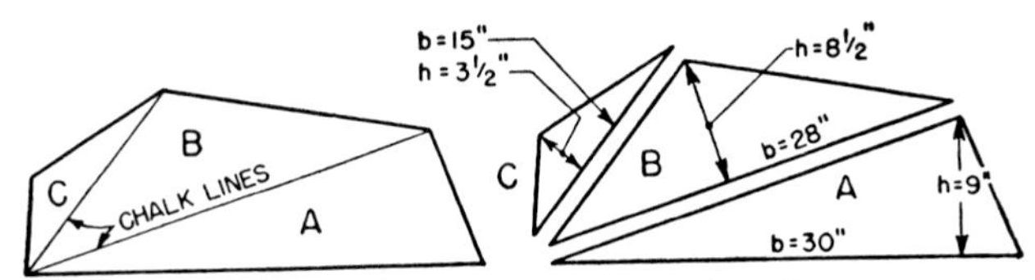

Figure 1.12 Calculating the area of a complex plane figure.

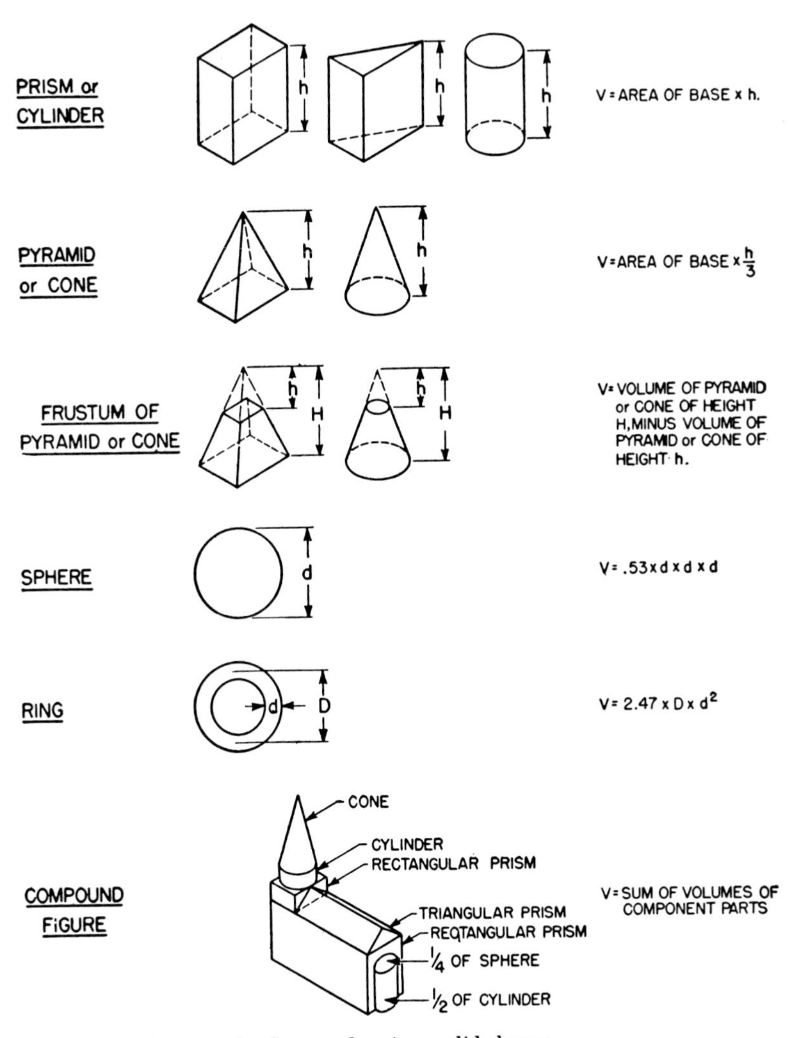

Figure 1.13 Areas and volumes of various solid shapes.

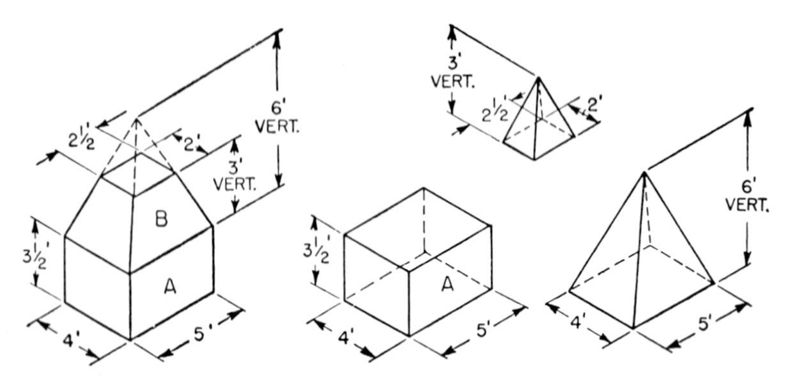

Figure 1.14 Calculating the volume of a complex solid figure.

TABLE 1.1 Weights of Common Materials

Material	Weight, lb/ft³
Earth (Excavated)	
Clay	
Dry	63
Damp, plastic	110
Clay and gravel, dry	100
Earth	
Dry, loose	76
Dry, packed	95
Mud, packed	115
Riprap	
Limestone	80–85
Sandstone	90
Shale	105
Sand, gravel	
Dry, loose	90–105
Dry, packed	100–120
Wet	118–120
Excavations in Water	
Sand or gravel	60
Sand or gravel and clay	65
Clay	80
River mud	90
Soil	70
Stone riprap	65
Timber (at 15 to 20% MC)	
Ash, white, red	33.5–44.5
Birch	41.9–47.2
Cedar, white-red	21.0–25.5
Cypress	32.1–33.7

TABLE 1.1 Weights of Common Materials (*Continued*)

Material	Weight, lb/ft^3
Fir	
Douglas spruce	32.1–35.2
Eastern	23.1–26.2
Hemlock	29.3–33.0
Hickory	47.5–50.9
Maple	39.8–44.9
Oak	
Red	44.0–44.9
White	41.9–47.2
Pine	
Red	29.3–31.4
White	25.9–27.0
Yellow	36.3–37.4
Redwood, California	26.6–27.7
Spruce, white, black	23.8–29.2
Masonry	
Brick	
Soft	100
Medium	115
Hard	130
Cement	
Portland, loose	90
Portland, set	183
Concrete, plain	
Cinder	108
Expanded slag	100
Stone, gravel	144
Light aggregate, load-bearing	70–105
Concrete, reinforced	
Cinder	111
Slag	138
Stone	150
Lime, gypsum, loose	53–64
Mortar, set	103
Stone, quarried, piled	
Granite, basalt, gneiss	96
Limestone, marble, quartz	95
Sandstone	82
Shale	92

TABLE 1.1 Weights of Common Materials (*Continued*)

Material	Weight, lb/ft³
Metals and Alloys	
Aluminum, cast, hammered	165
Brass, cast, rolled	526
Copper, cast, rolled	556
Iron	
Cast, pig	450
Wrought	480
Lead	710
Steel	
Cold-drawn	489
Rolled	490
Tin, cast, hammered	459
Zinc, rolled, sheet	449
Other Materials	
Asphaltum	81
Coal, bituminuous	40–54
Glass	
Common	156
Plate or crown	161
Slags	
Bank slag	67–72
Bank screenings	96–117
Tar, bituminous	75
Liquids	
Gasoline	42
Petroleum	54
Water	
Fresh	62.4
Ice	56
Snow, fresh fallen	8
Sea	64

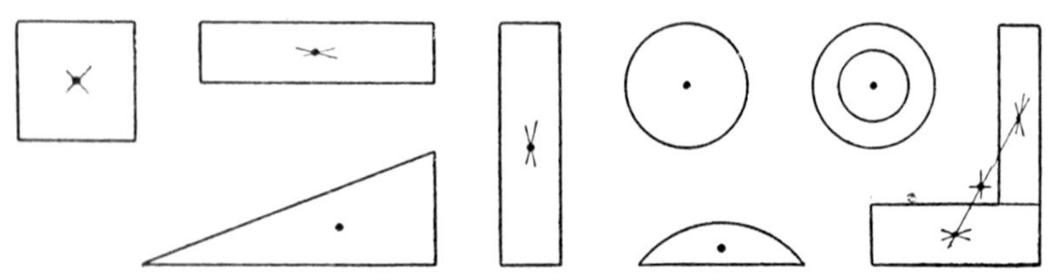

Figure 1.15 Approximating the location of the center of gravity of plane figures.

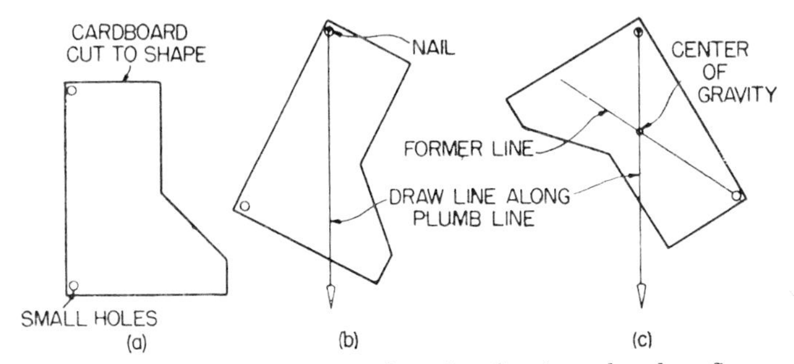

Figure 1.16 Locating the center of gravity of an irregular plane figure.

Of course, the center of gravity of a solid figure has to be located in three planes or directions: lengthwise, crosswise, and vertically. Finding the exact location of the center of gravity of a solid figure requires mathematical calculations. However, for the average rigging job, the center of gravity can be estimated closely enough. Figure 1.15 contains some familiar and some irregularly shaped plane figures, representing the length or cross section of an object, and the approximate locations of their centers of gravity. Most of the centers of gravity lie within the object; a few are located outside.

A simple but exact method of locating the center of gravity of a plane figure or an irregular shape is to cut it out, at any convenient scale, from a piece of cardboard. Punch pinholes near two adjacent corners of the cardboard shape and then suspend it freely from a pin stuck into a wall or other vertical surface (see Fig. 1.16). Also suspend from the pin a small plumb bob or other weight attached to a string, and draw a line on the cardboard along the string. Repeat, using the other hole. The object's center of gravity will lie at the intersection of the two pencil lines, and its location can be transferred to the shape to be lifted.

Another way to determine the center of gravity of an odd-shaped object is to break up the shape into simple masses, then determine the resultant balancing load and its location at a point where the weights multiplied by their respective lever arms will balance (see Fig. 1.17).

Thus, in Fig. 1.17*b*, if load 1 is estimated to be 10 tons and load 2 is 5 tons, the resultant balancing load will be 15 tons, acting in the opposite direction at a point where $10 \times 1 = 5 \times 2$. The center of gravity would be located at a point one-third of the way between the loads, measured from the larger of the loads.

The location of such a point can be computed for any quantity by using 1 for the lever arm of the larger weight LW. To be in equilibrium, LW $\times$ 1 must equal the small weight SW times the unknown arm length, and the unknown arm is then obtained from the equation LW $\times$ 1 = SW $\times$ unknown arm.

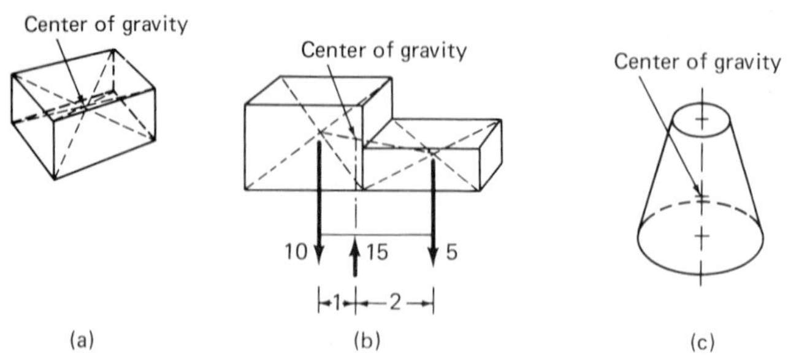

Figure 1.17 Location of the center of gravity for various irregular solids.

In odd-shaped bodies, where the center of gravity cannot be calculated readily, the rigger must assume a center of gravity, then correct it by trial and error, moving the hoist, load, and sling suspension a little at a time until a satisfactory result is obtained (see Fig. 1.18).

If any load tilts more than 5% after it is lifted clear off the ground, it should be landed and rerigged. A hitch should be arranged to support the load above the center of gravity at all times during the rotation for correct positioning so that the load will never tend to turn without being under control. The center of gravity must always be below the effective point of lift.

The need for determining the location of the center of gravity of a load to be hoisted is explained in detail in Chapter 8.

Factor of Safety

Although the foregoing factors must be considered before the type and size of rope to be used for moving any particular load can be selected, it is virtually impossible for the rigger to evaluate all the variables that can affect rigging and lifting equipment. Therefore, to compensate for

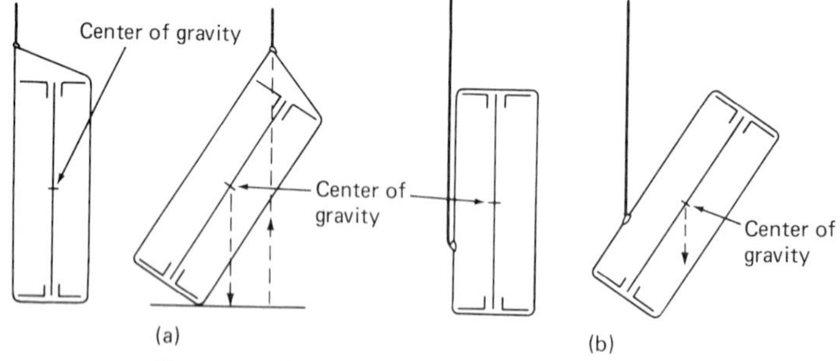

Figure 1.18 Correct and incorrect methods of making a hitch arrangement for turning a load onto its side. (*a*) Correct; can be turned over under control at all times. (*b*) Incorrect; will fall out of control as soon as center of gravity passes outside of lower corners.

any unforeseen influences, a factor of safety is usually applied to the material being used.

This factor is defined as the usual breaking strength of a material divided by the allowable load weight. In the case of plow-steel cable, the factor is 10; for manila rope it is 5.

However, factors of safety can vary with a manufacturer's experience of what is safe for the use and material under consideration. For example, if a particular ¾-in.-diameter plow-steel cable has a breaking strength of 50,000 lb, a rigger might use 10,000 lb as the maximum load weight that such a cable should lift; whereas for a ¾-in.-diameter manila rope having a breaking strength of 5,000 lb, he might restrict the loading to 500 lb.

In hoisting operations where many cumulative effects are often encountered, the actual stress in a rope may come close to the actual breaking strength, even if a factor of safety is applied. Before determining the safe working load of a particular rope, check the rope manufacturer's ratings. If the manufacturer provides rope capacity in terms of breaking strength, be sure to divide that value by the factor of safety to get the safe working load.

Remember: Never consider the factor of safety as a reserve strength. It is not available for additional capacity. Moreover, it must never be lowered under any circumstances.

Chapter

2

Wood Technology

In scaffolding, wood has been all but superseded today by tubular steel structures. Yet wood is still commonly used by riggers in many applications, such as for structural members, in rigging and shoring timbers, as scaffold planks, and in ladders.

It is essential, then, that the rigger be able to ensure that the correct species and quality of wood are used and also to check depreciation and loss of strength of wood elements. This means that the rigger must have at least a basic knowledge of the structure of wood and its physical properties.

Most riggers can readily identify spruce, yellow pine, Douglas fir, and a few other kinds of wood, while the average carpenter has a much broader knowledge of the various species. But identification is more or less by general appearance, rather than by any technical knowledge.

When it comes to determining the actual condition of a piece of wood that has been in use for some time and exposed to the elements, only a competent inspector examining a specimen with extreme care can even roughly provide such information. To help the project engineer, rigger, or job foreman to make inspections with somewhat greater accuracy, this chapter includes basic information relative to the structure of wood and its inherent defects, uses, and identification. A glossary at the end of this chapter will enable the reader to identify the various parts of wood, its characteristics, and its susceptibility to deterioration.

Structure of Wood

Wood essentially is composed of millions of tiny cells, which run lengthwise in the tree or log. These longitudinal or axial cells primarily store food and transport sap. Sandwiched between them at various places and extending in a radial direction are long slender groups of cells known as pith rays, which vary greatly in size in different species of trees, and which are most easily seen on quartersawed surfaces. They serve to bond the wood together radially and to conduct the sap radially across the grain (see Fig. 2.1).

In the wood of growing trees, only sapwood, the portion up to about 3 in. thick beneath the bark, contains living cells. Maples, hickories, ashes, some of the southern pines, and ponderosa pine may have sapwood from 3 to 6 in. or more in thickness, especially in second-growth trees. Unless treated, all sapwood is nondurable when exposed to conditions that favor decay (see Fig. 2.2).

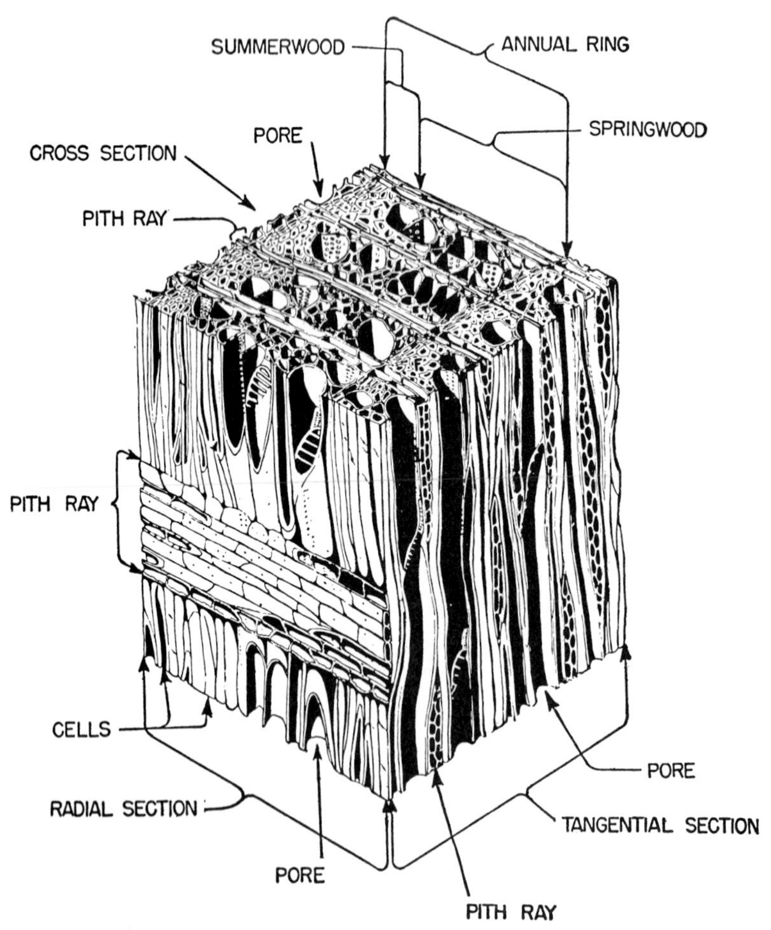

Figure 2.1 A 1/32 -in. cube of hardwood greatly magnified. *(Courtesy of Forest Products Laboratory.)*

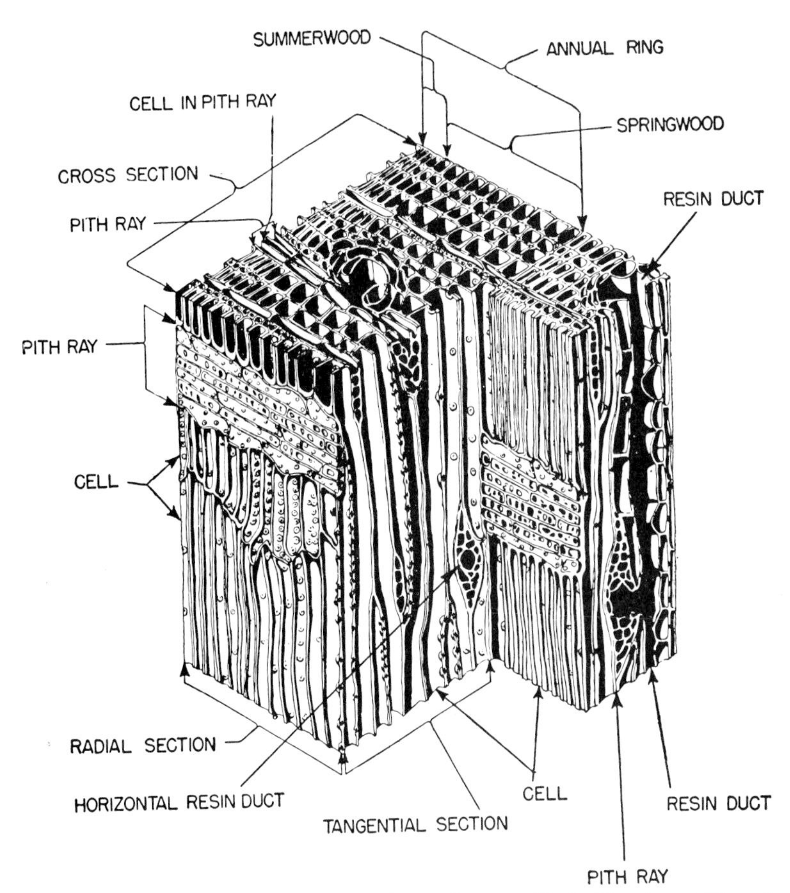

Figure 2.2 A 1/32 -in. cube of softwood greatly magnified. *(Courtesy of Forest Products Laboratory.)*

The darker, central portion of a living tree consists of lifeless cells and is dead. This central portion is known as heartwood, and its inactive cells have been slightly changed, both chemically and physically, from the cells of the inner sapwood rings. In addition, heartwood cell cavities may contain deposits of various materials, which frequently give the wood a dark color and usually make heartwood more durable when used in exposed conditions.

A tree's growth cycle begins when moisture taken in by the roots ascends through the cells that constitute the band of sapwood and travels out through the branches of the leaves. There it combines with carbon from the air to form food material that descends through a layer of thin-walled living cells between the bark and the wood. This layer then divides itself, forming new wood cells on the inside and new bark cells on the outside.

It should be noted that no growth in either diameter or length takes place in wood already formed—new growth is simply the addition of new cells, not the further development of old ones. As the diameter of the woody trunk increases, the bark is pushed outward, its outer layers stretching and cracking in ridged patterns characteristic of individual species.

In temperate climates, sap stops flowing during the winter and the tree ceases to grow. When spring rains arrive, tree growth resumes with large but thin-walled wood cells being produced quite rapidly to supply the growing tree with sufficient moisture. During the hot summer months, the cells produced are smaller and have thicker walls, which make this portion of wood heavier and stronger than that developed during spring growth. Thus each year an additional pair of concentric rings is added to the tree, the lighter part being known as springwood or earlywood, and the denser part as summerwood or latewood. Collectively, they are known as one annual ring.

Earlywood is lighter in weight, softer, and weaker than latewood; it shrinks less across and more lengthwise along the grain of the wood. Because of its greater density, the proportion of latewood is sometimes used to judge the quality or strength of wood.

If the growth of a tree's diameter is interrupted by drought or defoliation by insects, more than one ring may be formed in the same growing season. These inner rings usually do not have sharply defined boundaries and are termed false rings.

Domestic woods are divided into two classes, hardwoods and softwoods. But because domestic woods are classified botanically, rather than according to their physical and mechanical properties, not all hardwoods are hard; nor are all softwoods soft.

Wood from all coniferous (cone-bearing) trees is known as softwood and is nonporous in structure (see Fig. 2.3). Hardwood, which comes from trees having broad leaves (deciduous), is subdivided into two types: ring-porous, such as oak, ash, and hickory, in which the pores are arranged in definite rows of rings along the layer of springwood; and diffuse-porous, such as gum, walnut, and maple, in which the pores are more uniform in size and scattered throughout the annual ring.

Growth rings are most readily seen in species having sharp contrast between earlywood and latewood, such as ring-porous hardwoods and most softwoods, except the soft pines (see Figs. 2.4 and 2.5).

Physical Properties of Wood

Some physical properties of wood are influenced by species as well as by variables such as moisture content; other properties tend to be more independent of species.

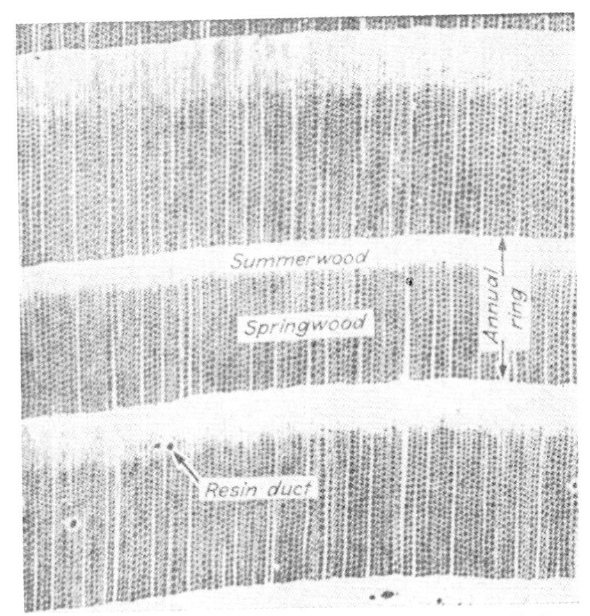

Figure 2.3 Cross section of nonporous wood, magnified.

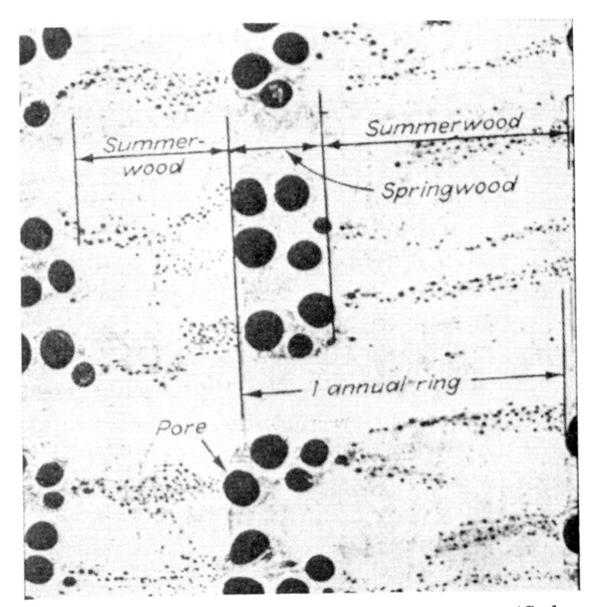

Figure 2.4 Cross section of ring-porous wood, magnified.

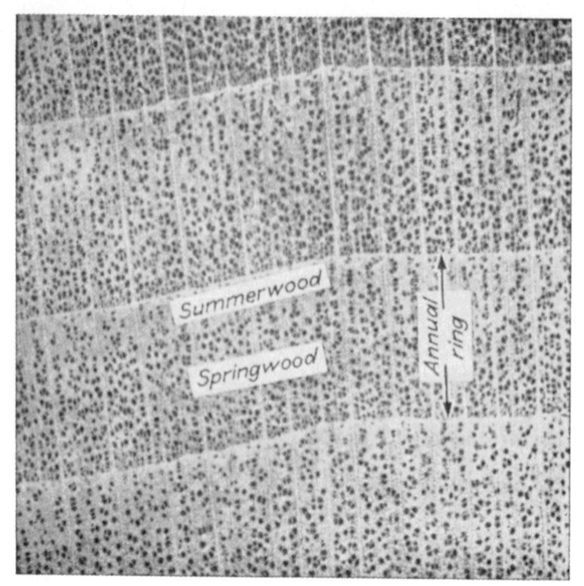

Figure 2.5 Cross section of diffuse-porous wood, magnified.

The weight, shrinkage, and strength of wood depend on the moisture content, which is expressed as a fraction or percentage of the weight of oven-dry wood.

The tree moisture content may range from about 30 to over 200 of the weight of the wood substance, with the sapwood portion usually having the highest moisture content. While the heartwood moisture content is usually much less than that of sapwood, in some species it can be greater.

Although the moisture content values given in Table 2.1 are considered typical for various species, there is considerable variation within and between trees. This variability often exists even within individual boards cut from the same tree. Moisture can exist in wood as water or water vapor in cell cavities, and as water bound chemically within cell walls. Thus *green* wood is often defined as wood in which the cell walls are completely saturated with water vapor in cell cavities.

The moisture content at which the cell walls are completely saturated without water being present in the cell cavities is called the fiber saturation point. This averages about 30 for most wood. It is often considered that the moisture content below which the physical and mechanical properties of wood begin to change is a function of both the relative humidity and the temperature of the surrounding air.

Wood in service usually is exposed to both long-term (seasonal) and short-term (daily) changes in the relative humidity and temperature of the surrounding air. Thus, wood is always undergoing at least slight

TABLE 2.1 Moisture Content of Wood

Species	Moisture content, percent	
	Heartwood	Sapwood
Hardwoods		
Elm		
American	95	92
Cedar	66	61
Rock	44	57
Maple		
Silver	58	97
Sugar	65	72
Oak		
California black	76	75
Northern red	80	69
Southern red	83	75
White	64	78
Softwoods		
Cedar		
Alaska	32	166
Eastern red	33	249
Western red	58	249
Douglas fir, coast-type	37	115
Fir		
Grand	91	136
Noble	34	115
Pacific silver	55	164
White	98	160
Hemlock		
Eastern	97	119
Western	85	170
Pine		
Longleaf	31	106
Ponderosa	40	148
Red	32	134
Shortleaf	32	122
Sugar	98	219
Western white	62	148
Redwood	86	210
Spruce		
Eastern	34	128
Sitka	41	

changes in moisture content. The long-term changes are usually gradual, while the short-term fluctuations tend to influence only the wood surface.

The practical objective of all wood seasoning, handling, and storing methods should be to minimize moisture content changes in wood during use. Protective coatings, such as varnish, lacquer, or paint, help retard these changes.

Shrinkage

Wood tends to shrink mainly tangentially, that is, in the direction of the annual growth rings; about one-half as much radially, across the rings; and only slightly longitudinally, across the grain. However, the combined effects of radial and tangential shrinkage can distort the shape of wood pieces because of the difference in shrinkage and the curvature of the annual rings.

Wood shrinkage is affected by a number of variables. In general, though, wood having a greater density will shrink more. The size and shape of a piece of wood may also affect shrinkage, and in some species, the temperature and the rate of drying may influence shrinkage. Transverse and volumetric shrinkage can vary by as much as 15, while longitudinal shrinkage is generally quite small. Average values for green to oven-dry shrinkage are between 0.1 and 0.2 for most species of wood.

Certain types of wood, however, exhibit excessive longitudinal shrinkage and should be avoided in use where longitudinal stability is important:

Reaction wood (see p. 47), whether compression wood in softwoods or tension wood in hardwoods, tends to shrink excessively along the grain, and this longitudinal shrinkage can cause serious warping, such as bow, crook, or twist, and cross breaks may develop in the zones of high shrinkage.

Wood from near the center of the tree of some species also exhibits lengthwise excessive shrinkage.

Wood with cross grain is subject to increased shrinkage along the longitudinal axis of the piece.

Density, specific gravity, and weight

Two primary sources of variation affect the weight of wood products: the wood's density and its variable moisture content.

The density of wood, exclusive of water, varies greatly, both within and between species. For most it falls between about 20 and 45 lb of mass per cubic foot, and can be as high as 65 lb of mass per cubic foot.

Although calculated density values should always be considered as approximations, because of the variations in wood structure, moisture content, and the ratio of heartwood to sapwood found in the different species, these values are sufficiently accurate to permit estimation of structural loads and calculations of lifting weights.

To compare species or products and to estimate weights, specific gravity is used as a standard reference basis rather than density. The specific gravity of wood is usually based on the oven-dry weight and the volume at some specified moisture content. The specific gravity of oven-dry wood does not change at moisture contents above approximately 30, the wood fiber saturation point.

If the specific gravity of wood is known, based on its oven-dry weight, its volume at a specified moisture content between 0 and 30 can be approximated.

Weathering

Freshly cut wood exposed to weathering, without first being treated for protection, undergoes certain physical changes, such as change in color, warping, loss of surface fibers, surface roughening, and checking (see Fig. 2.6).

The most severe chemical degradation of untreated woods occurs from exposure to ultraviolet light—the sun's rays. In most woods, cycles of wetting and drying produce checks or cracks that are easily visible. Moderate- to low-density woods develop fewer checks than do high-density woods, and vertical-grain woods check less than flat-grain boards.

As a result of weathering, boards tend to warp (particularly cup) and pull out of their fastenings. To resist this action, it is best to use boards having widths no greater than eight times their thickness.

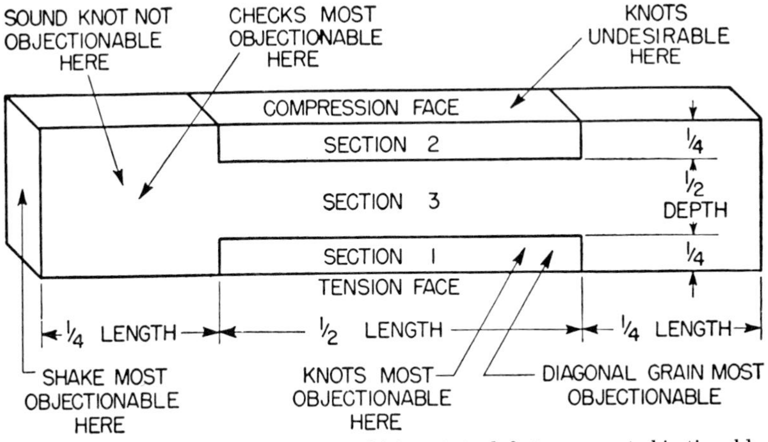

Figure 2.6 Areas in a wood beam in which certain defects are most objectionable.

Warping

When wood dries, it shrinks more tangentially than radially, and an almost insignificant amount longitudinally. Unless influenced by outside factors, dimensional changes caused by shrinkage follow a definite pattern (see Fig. 2.7).

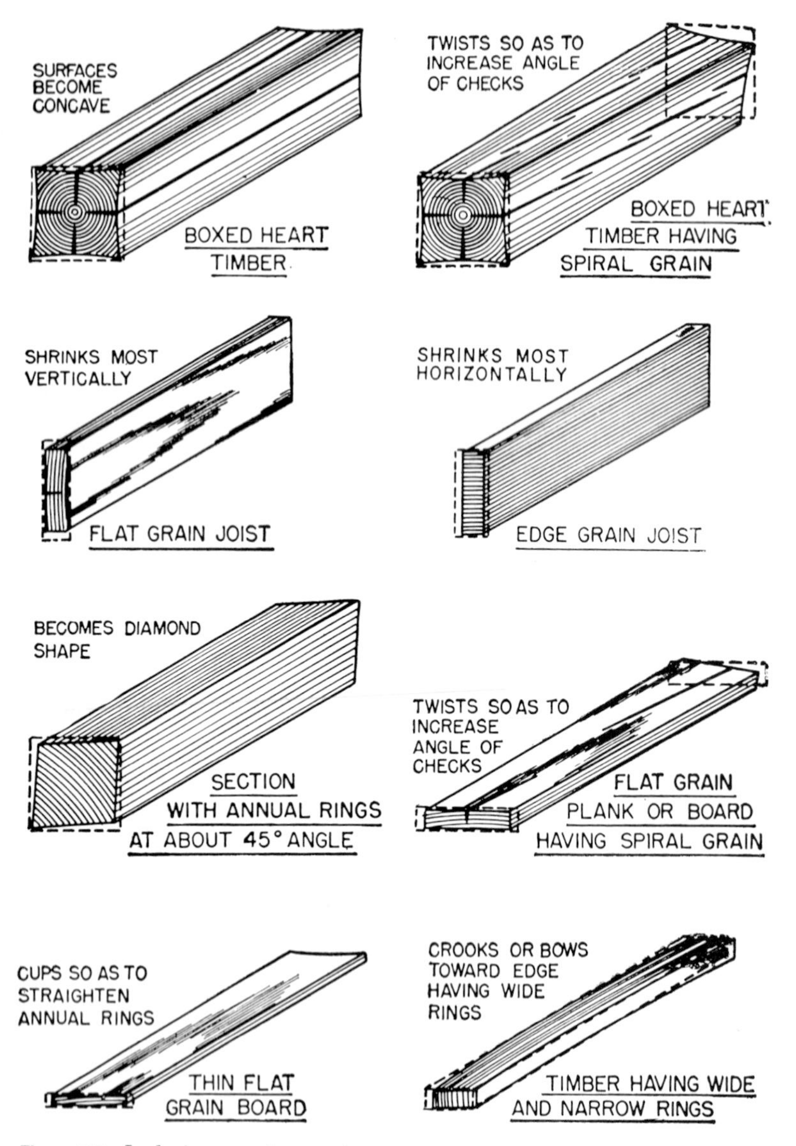

Figure 2.7 In drying, wood warps in a certain predetermined manner unless influenced by outside factors.

Edge-grain planks shrink more on the thickness.

Flat-grain planks shrink more on the width.

Boxed heart timbers cause all faces to become slightly convex. If large checks develop, the faces will become somewhat concave.

Mechanical Properties of Wood

The mechanical properties most commonly measured to determine wood strengths include the modulus of rupture in bending, the maximum stress in compression parallel to the grain, the compression strength perpendicular to the grain, and the shear strength parallel to the grain. Additional measurements often include work-to-maximum-load in bending, impact bending strength, tensile strength perpendicular to the grain, and hardness.

Notes

1. The modulus of rupture in bending reflects the maximum load-carrying capacity of the member and is proportional to the maximum moment borne by the wood member.
2. The work-to-maximum-load in bending is a measure of the energy absorbed by the member as it is slowly loaded to failure.
3. The impact bending height of drop is related to the energy absorption due to a rapid or falling load.
4. Hardness is the load required to embed a 0.444-in.-diameter ball to one-half its diameter in a direction perpendicular to the grain.

The mechanical properties of wood listed in Table 2.2 were derived from extensive sampling and analysis procedures. These properties often are represented as the average mechanical property of the species and are used to derive allowable properties for design.

Although only clear straight-grained wood is used to determine basic mechanical properties, wood products vary considerably in their physical properties because of a tree's growth characteristics. Variables such as specific gravity, cross grain, knots, and localized slopes of grain, plus natural defects, such as pitch pockets, must be considered when assessing the mechanical properties or estimating the actual performance of wood products.

Knots

The most common defect in wood is the knot—that portion of a branch that becomes incorporated in the bole of the tree. Its influence on the

TABLE 2.2 Strength of Straight-Grained Wood as a Function of Grain Slope

Slope of grain	Strength of straight-grained* wood, percent: Beams†	Strength of straight-grained* wood, percent: Posts and columns‡
1:40	100	100
1:20	85–100	100
1:18	80– 85	up to 100
1:16	76– 80	up to 100
1:15	74– 76	87–100
1:14	69– 74	82– 87
1:12	61– 69	74– 82
1:10	53– 61	66– 74
1: 8	50– 53	56– 66
1: 6	50– 53	50– 56

*Cross grain causes a noticeable loss of strength in beams and columns.
†Applies only to middle half of length.
‡Applies to entire length.

mechanical properties of wood depends on the size, location, shape, soundness, accompanying local slope of grain, and type of stress to which the knot is subjected.

Knot shapes vary and are determined by the angle at which a piece of wood is cut from the tree in relation to its branches or limbs. When a branch or limb extends at about a right angle to the face of the piece containing the knot, the knot will appear round; when it extends at an angle to the face, it will be an oval; and when it extends more nearly parallel to the surface, a spike will be formed. When estimating the strength of a piece of wood that contains a knot, remember that the axis of the knot extends to the pith of the tree and forms a cone with the knot's apex at the pith.

Knots are further classified as intergrown and encased. The intergrown kind is formed where a limb is alive and growth is continuous at the junction of limb and trunk (see Fig. 2.8). The encased knot, however, is formed after the branch dies, with new growth of the trunk enclosing the dead limb; thus the fibers of the trunk no longer connect continuously with fibers of the knot (see Figs. 2.9 and 2.10).

There is a pronounced decrease in most mechanical properties of wood where a knot exists in a member. Knots displace clearwood; cross grain is formed by the distorted fibers around the knot; discontinuity in the fibers leads to stress concentrations; and checking often occurs around knots during the drying operation.

Conversely, knots increase the hardness and strength of a member when they are in compression perpendicular to the grain, although they cause nonuniform wear or stress distribution at contact surfaces. Wood

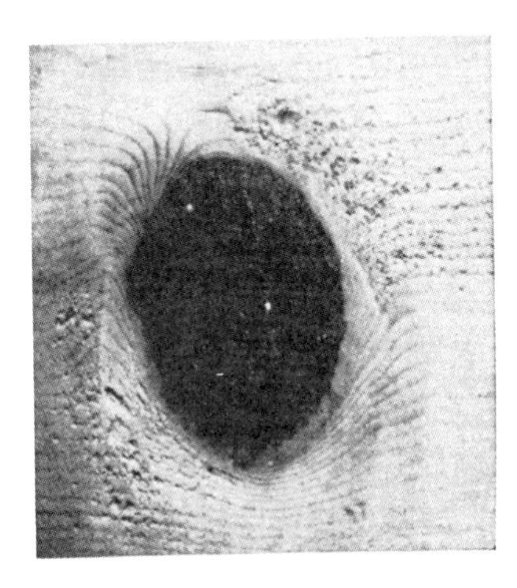

Figure 2.8 Sound or intergrown knot.

members loaded uniformly in tension are usually more seriously affected by knots than if loaded in other ways.

The strength of some structural members is affected not only by the size, but also by the location of a knot. For example, in a simply supported beam, knots on the lower side are in tension; those on the upper side, in compression; and those at or near the neutral axis, in horizontal shear. Thus, a knot has a marked effect on the maximum load a beam will sustain when it is located on the tension side at the point of maximum stress.

In long columns, where stiffness rather than direct crushing strength is the controlling factor, the loss of strength due to knots or other defects is relatively small. However, in short or intermediate columns, the reduction in strength caused by knots is approximately proportional to the size of the knot. Knots in round timbers (poles and piles) have a lesser effect on strength than knots in sawed timbers. Knots also reduce the fatigue strength of clear wood.

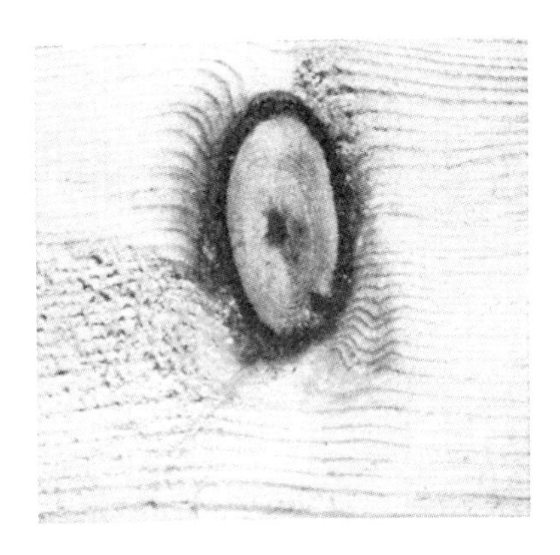

Figure 2.9 Encased knot surrounded by a layer of dead bark.

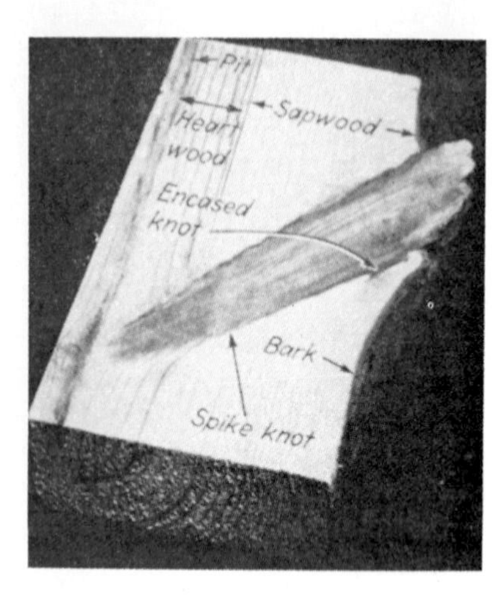

Figure 2.10 A spike knot is a longitudinal (or nearly longitudinal) section through a limb or branch of the tree.

Checks

The next most common wood defect is the check. Checks develop because of uneven shrinkage when the wood is dried. This results in the pulling apart or separating of cell rows in a radial split, usually starting at the bark or outside the timber. A check of almost imperceptible width is as detrimental to the wood member as a wider one (see Fig. 2.11).

Checks are most harmful if located near the ends of a beam's vertical faces, at about the middle of the beam's height. At this point, the horizontal shearing stress is at its maximum, and the beam tends to split in half. Checks at this location tend to reduce resistance to this

Figure 2.11 A check is a radial split in a piece of wood resulting from unequal shrinkage.

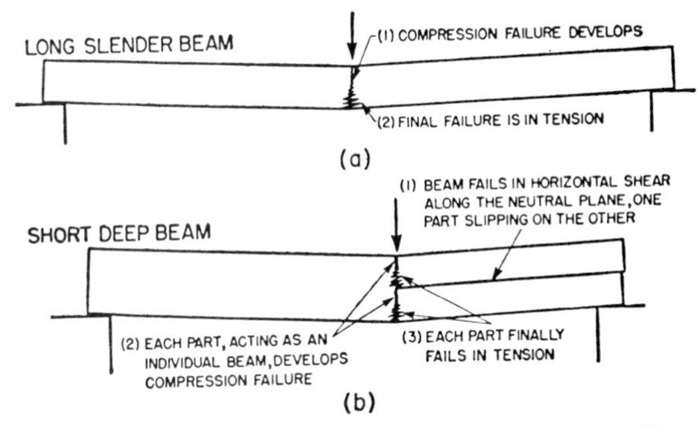

Figure 2.12 Typical failures. (*a*) Long slender beam. (*b*) Short deep beam.

horizontal shear, with resultant beam failure under small loads (see Fig. 2.12). This is particularly true of beams having a depth greater than one-twelfth of their span.

Shakes

Like checks, shakes are most objectionable when occurring at the points of maximum horizontal shear. Shakes are the lengthwise separations of wood that occur between and parallel to the annual rings (see Fig. 2.13).

Occasionally, in an old timber, splits may be observed running longitudinally on one face and diagonally on the adjacent face. Inserting a knife blade into the split and then running it lengthwise should show if it has a tendency to rotate around the pith. If it does, the split is a check; if not it probably is a shake, assuming that the rings cannot be observed at the end of the timber.

Figure 2.13 A shake is a separation between successive annual rings.

Pitch pockets

This type of defect in large timber is not too serious, provided the dip grain that accompanies it is not excessive. However, a large number of pitch pockets indicate a lack of bond of the wood and the probable presence of a shake.

A pitch pocket is a well-defined opening that contains free resin, extends parallel to the annual rings, and is almost flat on the pith side and curved on the bark side. They are characteristic of pines, spruces, and Douglas firs (see Fig. 2.14). The effect of pitch pockets on strength depends on their number, size, and location in the piece of wood.

Cross grain

Cross grain, one of the most treacherous defects in a piece of wood, is a general classification that includes various grain patterns such as diagonal, spiral, interlocked, dip, wavy, and curly.

Diagonal grain results from sawing a straight piece of wood from a crooked log, or from not sawing parallel to the bark in a straight but tapered log (the wood fibers do not run parallel to the edges of the piece). Such a grain may occur on either the edge-grain or the flat-grain face of the piece. More generally it occurs on the former, in which it is usually readily discernible by the angular direction of the annual ring markings. If it occurs on the flat-grain face, it will be considered in the same class as spiral grain.

Spiral grain is not as easy to detect (see Fig. 2.15). Wood fibers in a log generally run in a longitudinal direction. However, in some logs the fibers run at a slight angle, thus taking a spiral path around the log. Spiral grain may frequently be observed on telephone poles and flagpoles, where it is definitely indicated by checks. On the flat-grain face of a piece of wood, the markings of the annual rings do not indicate the direction of the grain. If there are a number of herringbone V's on this face, they indicate diagonal grain, but the angle must be measured on the edge grain or radial face. If the points of these V's are

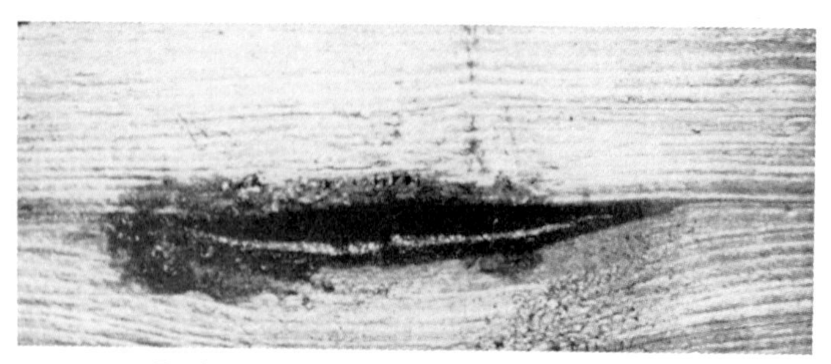

Figure 2.14 Pitch pockets occur only in certain softwoods.

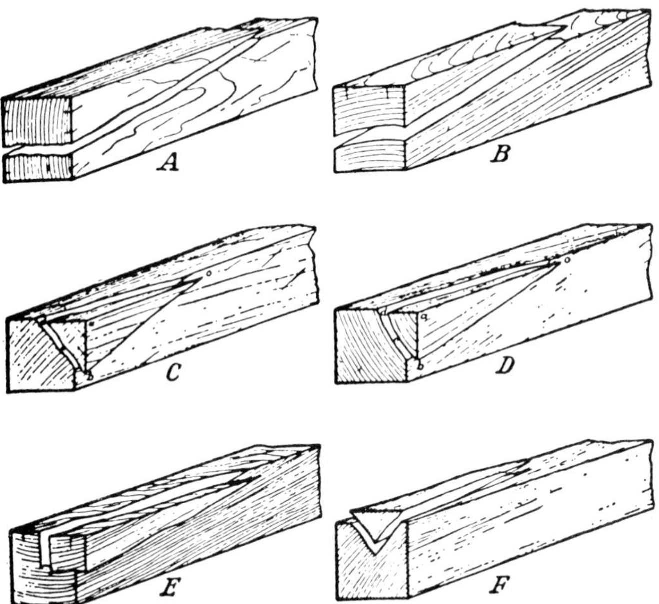

Figure 2.15 Indications of spiral and diagonal grain and combinations thereof. Annual rings parallel to edge of piece: *A*—spiral grain; *B*—diagonal grain. Annual rings oblique to edge of piece: *C*—spiral grain; *D*—diagonal grain. Spiral and diagonal grain in combination: *E*—rings parallel to edge; *F*—rings oblique to edge.

not in a line parallel to the edge of the piece, then spiral grain exists.

To determine the direction of the true grain in a piece of wood, first observe the markings or visible grain on the radial face. If these do not run parallel to the edge of the piece, the grain is diagonal, the angle being measured by the length required to produce a deviation of 1 in. Next, observe the tangential or flat-grain surface. Checks, if present, indicate the direction of the grain (see Fig. 2.16).

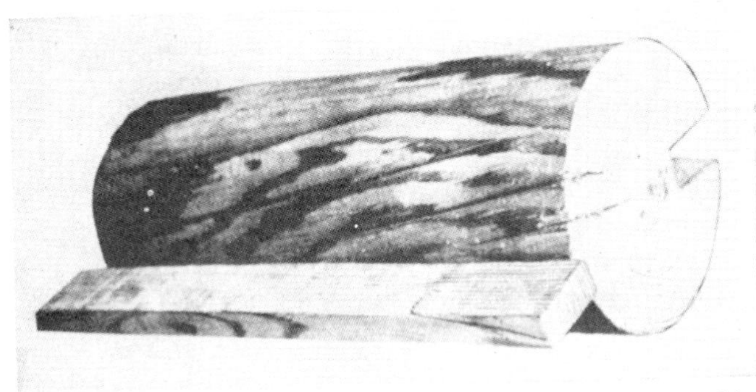

Figure 2.16 Beam cut from a large log. Spiral grain is indicated by checks and split.

Also, in hardwoods the direction of the rays, or of the large pores, indicates the grain. In certain softwoods (those that may contain pitch pockets) the resin ducts are positive indicators of the grain. These ducts appear as fine brownish lines similar to tiny pin scratches.

If none of these indicators are observed, apply the point of a fountain pen to the wood and note the direction in which the ink runs in the wood cells. Another test is to jab the point of a small sharp knife blade into the tangential face at an angle of about 15 to the surface, with the knife set crossways to the length of the piece. Slowly rotate the knife blade with the cutting edge toward the piece of wood. This will lift up a splinter, and the direction in which the fibers tend to pull out indicates the direction of the grain.

Torn or chipped grain on the radial face also indicates spiral grain. Spiral grain should always be measured on the flat-grain surface farthest from the pith.

If the annual rings on a cross section run diagonally across the piece, or if there are not true flat-grain or edge-grain faces, determining the angle of cross grain becomes more involved. To measure the angle of diagonal grain on such a piece, locate on the end (the cross section of the piece) that corner which is farthest from the pith. From this corner draw a line radially, or at right angle to the rings. On this line, if the size of the piece will permit, measure 1 in. from the corner. Then follow the annual ring markings until they intersect that edge farthest from the pith. The ratio B:A indicates the angle of diagonal grain (see Fig. 2.17).

To measure the angle of spiral grain in a piece of wood whose faces are neither radial nor tangential, mark a point some distance from the end on one of the edges that is neither nearest to nor farthest from the pith and follow along the fibers of a check, or parallel to a check, to the end of the piece. Then draw a line from the end of this line radially toward the pith. The minimum distance of this radial line from the starting edge D, used in conjunction with the distance of the starting point from the end of the member C, will give the angle of the spiral grain.

Some wood may contain both diagonal grain and spiral grain. To obtain the true angle of the grain in such a piece, first obtain the angles of the spiral grain and the diagonal grain, both being expressed decimally. Thus an angle of 1:15 is $\frac{1}{15}$ or 0.066. Square the angle of the diagonal grain and add it to the square of the spiral grain. Then extract the square root of the sum to obtain the true angle of the grain.

Small pieces of wood are generally weakened by the presence of dip grain or burls, which are usually the result of defects that may or may not exist in the piece (see Fig. 2.18). The cross grain in the dip usually runs at a severe angle and greatly reduces the strength, in particular when this defect is on the tension side of a beam. To determine the

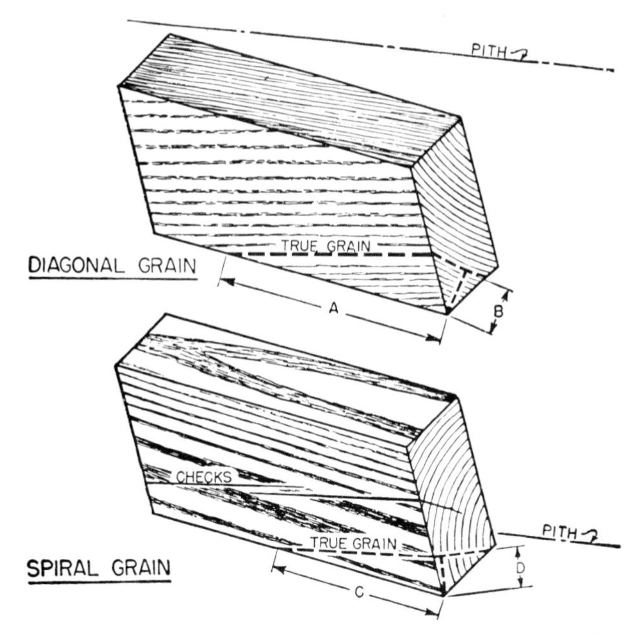

Figure 2.17 Method of measuring the angle of diagonal and spiral grain when the rings are oblique to the edge of the piece.

strength of a beam of small cross section, such as a ladder side rail containing dip grain at the tension face, assume that the dip cannot take tensile stress and therefore the depth of the section should be deducted from the depth of the beam.

Brashness

This is an abnormal condition that causes wood in bending to break suddenly and completely across the grain when deflected by only a relatively small amount. Brashness is usually associated with slow grown hardwoods, very fast or very slow grown softwoods, and wood with pre-existing

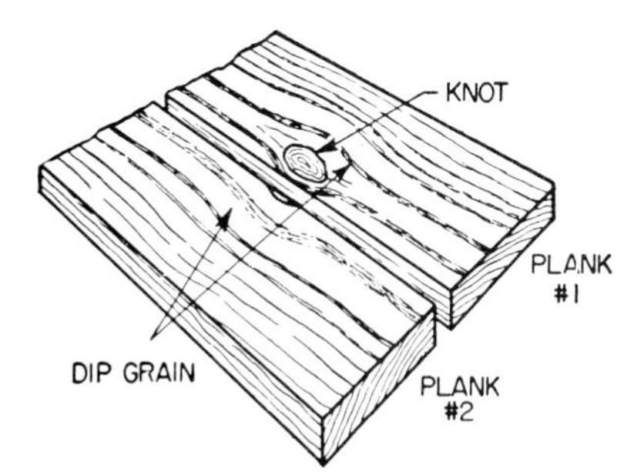

Figure 2.18 A knot in one piece of wood may cause dip grain in the adjacent piece.

compression failures. Wood that is exceptionally lightweight for its species is usually brash, as is wood that has been exposed to a high temperature for a long time, such as wood ladders used in boiler rooms.

A piece of tough wood and a piece of brash wood may have identical strengths in static bending, but under impact, the brash piece will fail at a much lower stress. Under ordinary conditions, the strength of wood in tension is from two to four times its compressive strength, but this ratio is greatly reduced in brash wood.

The length of the splinters produced on the tension side of a beam that has failed in bending is proportional (approximately) to the span of the beam. Thus in beams of very short span, the splinters produced are very short. Dry wood usually breaks more suddenly than moist wood, but still the fracture contains splinters (see Fig. 2.19). On the other hand, brash wood usually fails with a splinterless fracture, which gives a suggestion of crystallization, with the open ends of the cells or pores conspicuous, especially under a magnifying glass (see Fig. 2.20).

Because summerwood is more dense, it has greater strength than springwood. In softwoods, wide rings may be predominantly springwood, which is one of the reasons why wide-ring softwoods are brash.

On the other hand, the difference between wide- and narrow-ring, ring-porous hardwoods occurs mostly in the summerwood. This means that in very narrow rings the wood is mostly porous springwood and consequently low in strength and brashy.

Splintering occurs only in the tension half of a fracture, even in tough wood. Hence, abrupt failure on the compression half of a member should not in itself be taken as an indication of brashness, pre-existing compression failure, or other defect.

Abrupt failures are not usually caused by exposure to excessive heat, unless the temperature has been high enough to darken the wood throughout. Prolonged moderately high temperatures, however, may make the wood brash. Heating wet wood may set up damaging stresses

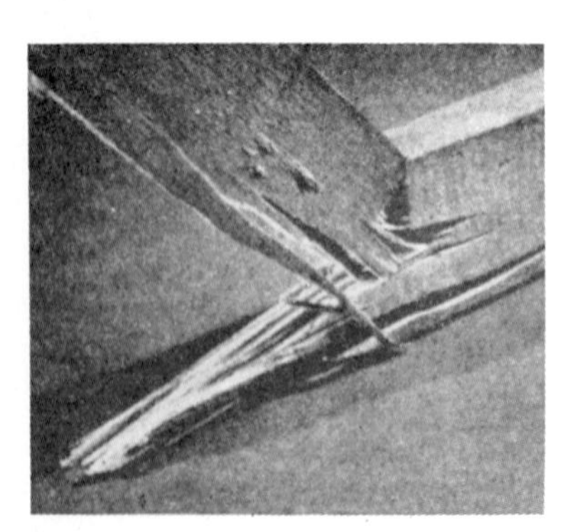

Figure 2.19 Typical fracture in tough wood. Note the relatively long splinters.

Figure 2.20 Typical fracture in brash wood. Note the absence of any splinters. In making the splinter test with the knife small chips of wood may fall out.

within the piece, especially if it contains the pith and if it is heated to about 160°F.

Another test for the presence of brash wood is to make a knife test (as described under cross grain). If the wood is tough, a splinter will be raised up but will not tear out unless the knife blade is rotated too far. On the other hand, if the wood is brash, the piece will fly out with a faint snap, leaving a recess of punky wood. Of course, there are various degrees of brashness. Therefore, there is no definite line of demarcation between tough and brash wood.

Compression failures

Next to cross grain and brashness, a tendency for compression failure is perhaps the most hazardous defect in structural wood. Although this defect is not too common, its danger is that it is sometimes extremely difficult, if not actually impossible, to detect such a defect with the unaided eye (see Fig. 2.21).

Compression failures should not be confused with compression wood. Such failures result from excessive bending of standing trees from wind or snow, or rough handling of logs and lumber. If a piece of wood is subject to excessive bending stress, the fibers on the compression side of the neutral plane will be compressed to such a degree that they are buckled.

Members containing visible compression failures can have seriously low strength and shock resistance. Although most species of wood are from two to four times as strong in tension as in compression, the tensile strength of wood containing compression failures can be as low as one-third the strength of matched clear wood. Even small compression failures, visible only under the microscope, reduce the strength seriously and cause brittle failures.

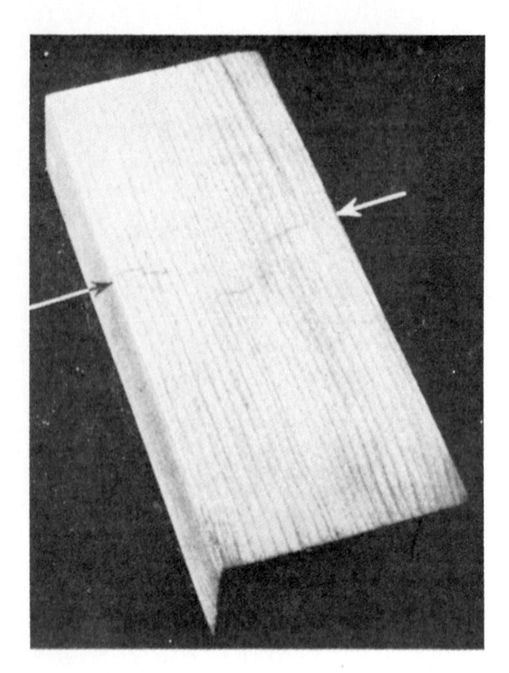

Figure 2.21 Compression failure is one of the most difficult defects to detect.

Note: Because of this low strength associated with compression failure, many safety codes require certain structural members, such as ladder rails and scaffold planks, to be entirely free of such conditions.

If the load on a wood member is relieved after the compression failure occurs, the piece will appear to be in perfect condition. In fact, even though the average carpenter or rigger is told that a compression failure exists in a piece of finished lumber being examined, the chances are 100 to 1 that he will be able to locate it. If the piece being inspected is rough sawed, it is impossible to find the point of failure, even though the piece may have failed halfway through. If the beam is reversed so that the compression failure is on the tension side, it may collapse without warning under a very small load. Such a break will appear without splinters and may be confused with a fracture in brash wood.

To inspect a piece of finished lumber for a suspected compression failure, move the wood in relation to a light source (or vice versa) until there is a glare reflected on the surface of the area being examined. A compression failure then can be detected as a very faint irregular line running crossways to the piece at an angle of 70 to 90° to the edge. A liberal application of carbon tetrachloride or tung oil to the surface of wood often makes compression failures more easily visible.

Compression failures may occur in several places, their spacing depending on the length of the beam. Usually they occur only on one face and halfway across the two adjacent faces, assuming of course that the injury occurred to the piece in its existing dimensions. If, however, the compression failure caused damage halfway through the log, then perhaps the beam being inspected has been cut from the defective side of the log. In this case, the compression failure should be visible on all four finished faces of the piece.

Wood containing a compression failure is extremely weak under impact. Therefore, if the piece suspected of containing a compression failure is not too heavy, hold one end of it with the questionable face downward and allow the other end of the piece to drop onto the concrete floor or pavement. The ease with which a piece containing a compression failure is broken may come as a real surprise.

Examination of a timber that has failed under load and has an abrupt splinterless fracture may raise the question of whether it failed because of a compression failure or because it was brash wood. To determine the cause of failure, examine the width of the annual rings, feel the heft of the piece, and make the knife test for brash wood. If no indication of brashness is observed, then it can be assumed that compression failure existed prior to final failure.

Then carefully examine the wood for several feet each side of the break to locate any secondary compression failures, as described previously. Their presence is a definite indication that the break was due to a compression failure.

Reaction wood

Abnormal woody tissue, frequently associated with leaning trunks and crooked limbs of both conifers and hardwoods, is termed reaction wood.

In softwoods this tissue is found on the lower side of an inclined member and is called compression wood. In hardwoods it is located on the upper side of the inclined member and is known as tension wood.

Both types of reaction wood undergo excessive longitudinal shrinkage when subjected to moisture loss below the fiber saturation point—up to 10 times normal in compression wood and five or more times normal in tension wood. When either of these woods is present in the same board with normal wood, unequal longitudinal shrinkage causes internal stresses that result in warping.

Pronounced compression wood can be detected by ordinary visual examination. It is characterized by eccentric growth of annual rings, which are of a dense nature and are predominantly in summerwood (see Fig. 2.22).

In finished lumber, compression wood is usually somewhat darker because of the greater proportion of summerwood. Compression wood on the flat-grain and edge-grain faces is dull and lifeless in appearance. It is the one exception to the rule that the strength of any piece of wood, regardless of its species, can be judged by its weight. Com-

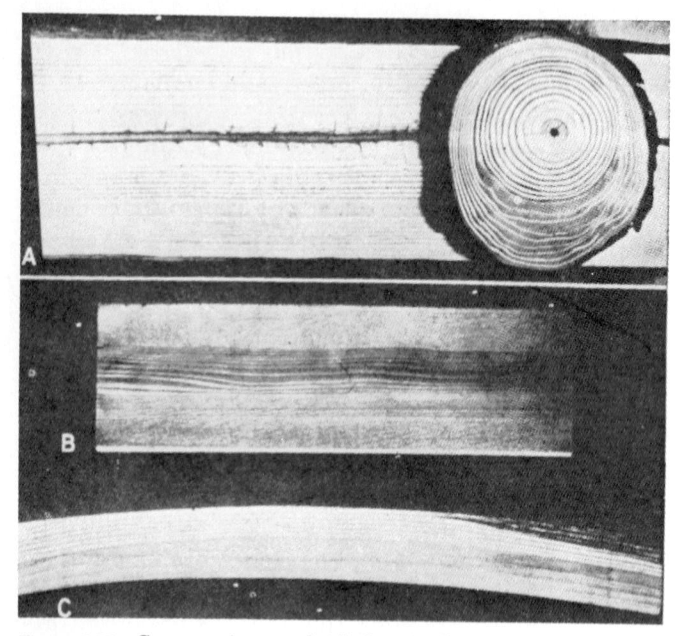

Figure 2.22 Compression wood. (*A*) Longitudinal and cross sections through part of a tree trunk with compression wood on lower side. (*B*) Cross break in compression wood and split between compression wood and normal wood due to greater longitudinal shrinkage of compression wood. (*C*) Crook caused by longitudinal shrinkage of the compression wood on the lower side of the piece. *(Courtesy of Forest Products Laboratory.)*

pression wood, by its very nature, should be excluded from all uses where strength under shock or impact is essential.

The great longitudinal shrinkage of compression wood (2½ to 20 times normal wood), if located near the center of the piece, may cause this part of the wood to fail in tension, thereby producing cross cracks in the compression wood. If located near the edge of a small piece, the compression wood, in shrinking, will cause crooking or bowing.

Also, because of abnormal shrinkage, spike knots are frequently twisted from their positions and tend to protrude above the face of the material. Poles containing compression wood are hazardous for linemen to climb, for if cross checks develop at the surface, the shell of the outer wood may peel off, with the result that the points of their climbers will lose their hold on the pole.

Tension wood is somewhat more difficult to detect. However, eccentric growth, as seen on the transverse section, frequently indicates its presence. Also, the tough tension wood fibers resist being cut cleanly and result in a woolly condition on the surface of sawn boards. In some species, tension wood may show up on a smooth surface as areas of contrasting colors.

Reaction wood, compared to normal wood of comparable specific gravity, is definitely weaker.

Chemicals

Wood generally is highly resistant to many chemicals, mild acids, and solutions of acidic salts. However, various chemical solutions will have different effects on the mechanical properties of wood. Two general types of actions result from contact with chemical solutions.

The first is an almost completely reversible effect, involving swelling of the wood member. Liquids such as water, alcohols, and some other organic liquids swell the wood and thereby lower its mechanical properties. But these liquids do not degrade the wood structure chemically. Removal of the swelling liquid allows the wood to return to its original condition. Petroleum oils and creosote do not swell wood and have no appreciable effect on its properties.

The second type of action is irreversible and involves permanent changes in the wood's structure because of alteration of one or more of its chemical constituents. These changes result in the hydrolysis of celluloses and hemicelluloses by acids or acidic salts, oxidation of wood substances by oxidizing agents, or delignification and solution of hemicelluloses by alkalies or alkaline salt solutions. An acidic condition tends to make the wood brittle.

In general, heartwood of such species as Douglas fir, southern pine, and white oak is quite resistant to attack by dilute mineral and organic acids. Oxidizing acids such as nitric acid have a greater degradative action than nonoxidizing acids. Alkaline solutions are more destructive than acidic solutions. Finally, hardwoods are more susceptible to attacks by both acids and alkalies than softwoods.

Wood products are sometimes treated with decay preservatives or fire-retarding salts, usually in water solution, and then are kiln-dried. While wood properties are affected to some extent by the combined effects of such water-soluble chemicals, treatment methods, and kiln drying, overall the mechanical properties remain essentially unchanged.

Decay or rot

Nearly all domestic wood is susceptible, in varying degrees, to common decay fungi. This is not an inorganic oxidizing process, such as the rusting of steel or the crumbling of stone, but rather a disease of the wood—wood-destroying fungi that seriously reduce the strength of wood members. For wood to be susceptible to decay, four conditions must be present:

1. The wood must provide food for the fungus. Unless chemically treated, most woods are able to do this to some extent.

2. There must be a sufficient amount of moisture present. When wood is kept constantly dry, or when it contains less than 20% moisture, it will not decay; and when submerged continuously in water, wood shows very little decaying.
3. There must be air present, a moist stagnant air being the most effective.
4. There must be warmth, as wood will not decay in extremely cold climates.

Everyone is familiar with the fruiting bodies produced by decay fungi on fallen trees and tree stumps, including toadstools, shelves, and crusts. On the underside of these fruiting bodies, millions of tiny spores or germs are produced which, when matured, become free and are blown by the wind. When deposited on susceptible wood, they will germinate and form minute fibrous strands. As the strands grow, they puncture the cell walls and not only feed on the contents of the cells, but also actually devour the cell walls.

Even apparently sound wood adjacent to obviously decayed parts may contain hard-to-detect early (incipient) decay that decidedly weakens wood, especially in shock resistance.

The term dry rot is frequently used, but this is a misnomer, for no wood that is dry can rot. It usually refers to decay where there is no visible evidence of contact with moisture, such as in portable ladders. However, moisture had to be present at some time, though not necessarily continuously. Wood that has been painted in the green, or unseasoned, state may decay under the coating of paint.

In its incipient stage, decay may not be too advanced for remedial action to be taken, but the symptoms must first be recognized. Among the surface indications of incipient decay are the following:

Small bleached or otherwise discolored areas on the surface wood

Zigzag zone lines not far from the ends of structural timbers

Oozing of extractive liquids from the joints between wood structural members

Dark zones in the wood separated by zones of lighter-color tissue caused by certain fungi

Persistently moist appearance of freshly cut sections

Further decay of such wood can be arrested by keeping the wood dry. Under no circumstances should wood that shows evidence of incipient decay be used where members might be subject to shock or impact loads.

In its advanced stage, decay is more readily recognizable, but the only remedy then is to remove and replace the rotted members. Advanced decay in the interior of wood structural members can be detected by:

Inspecting structural members for sagging

Jabbing a knife blade into the wood and noting the resistance to penetration; rotted wood is very soft

Drilling small holes into the timber and observing the resistance to drilling, and the color of the chips removed; dark wood, powder, or paste is an indication of internal decay

Striking the timber with the round end of a ball-peen hammer and noting if the sound is dead and hollow

Striking the wood member against a concrete floor to determine loss of resonance

In addition to structural weakness, decayed wood is much more combustible than sound wood and therefore presents a greater fire hazard.

It is poor practice to leave the end grain of any wood exposed, since it readily absorbs moisture from the air. The ends of lumber should be heavily painted or tarred.

However, ordinary paints, varnishes, and similar protective coatings cannot be relied upon to preserve wood against decay. These contain no substances that are poisonous to fungi, and they may themselves support the growth of fungi in the presence of dampness. Although protective coatings retard the absorption of moisture by wood, they also retard the drying out of wood that has taken up moisture through an uncoated surface or joint.

Note: End grain of oak in contact with galvanized metal will cause local rotting of the wood, while similar contact with steel beams or plates will cause rapid local corrosion of the steel.

Termites

The termite is perhaps the most troublesome insect to contend with in structural timbers. Superficially it resembles an ant in size, general appearance, and habit of living in colonies. Although there are numerous species of termites in this country, they can be grouped into two general categories: ground-inhabiting or subterranean and wood-inhabiting or nonsubterranean termites.

Most of the termite damage in this country results from the subterranean variety, which is more prevalent in the southern than in the northern states where colder weather does not favor development. These termites develop their colonies and live in the ground, building their tunnels through earth and around obstructions to get at the wood they need for food. They also must have a constant source of moisture.

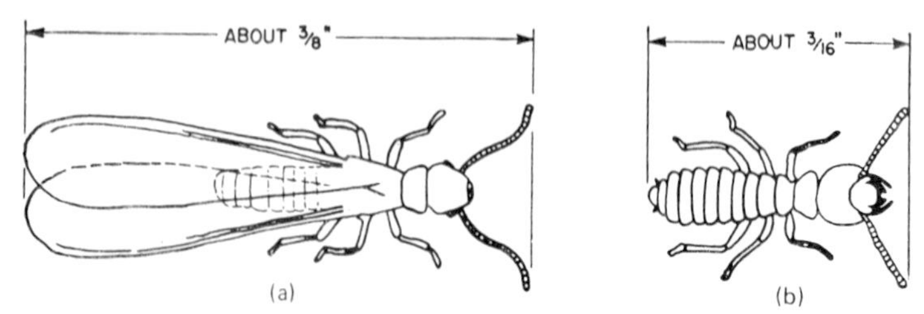

Figure 2.23 Two principal forms of the subterranean termite. (*a*) Winged sexual adult commonly observed when migrating. (*b*) Tiny but destructive worker, which never comes out into the light and hence is seldom seen.

The most frequently observed termite is the winged adult, but it is the grayish-white worker termite that is the most destructive (see Fig. 2.23). Winged adult termites are brownish or blackish in color, with elongated body and long white wings extending beyond it at the rear. At certain seasons, usually in the spring and fall, the winged sexual adults migrate in large numbers and at such times may be observed for a short period of several hours. They then lose their wings, enter crevices between timbers, and breed new colonies.

Worker termites are blind, shun the light, and are seldom seen—except when a structure or building is demolished or altered, and the timber in which they are living is suddenly cut into or the soil is excavated.

Termites frequently eat away the entire inside of a timber, leaving nothing but a thin shell of wood (and possibly paint) and with no exterior evidence of attack until the member ultimately fails structurally. Signs of infestation include the earthlike shelter tubes bypassing masonry to provide communications between ground and wood; migrating termites; the wings that the termites have shed just before reentering the wood; pellets of fine, digested, excreted wood similar to sawdust on the floor below joints in the wood; holes in the surface of the wood about the size of a BB shot, and the sagging or collapse of structural members. (Often termite damage is confused with decay, which is quite different and is caused by fungi.)

Combatting termites is nearly a hopeless task, unless all the infested wood is removed and burned. When contact with moisture from the earth, a leaking water pipe, or a roof leak is cut off, the termites, which depend on such a supply, will die. Among the poisons used in combatting termites are orthodichlorobenzene, Paris green, sodium fluorosilicate, and carbon tetrachloride.

Glossary of Wood Technology

To help the rigger understand references made to the various parts of wood and wood products, its characteristics, defects, infection, and infestation, the following key definitions are provided.

Annual Rings Concentric, but often irregular, recurring rings that appear in whole or in part upon the cross section of any wood, whether in the form of tree, log, timber, plank, or finished wood part. Each complete annual ring consists of an inner band of low-density wood (springwood) and an outer band of higher-density wood (summerwood).

Bark Pocket Opening between annual growth rings that contains bark. These pockets appear as dark streaks on radial surfaces and as rounded areas on tangential surfaces.

Bastard Sawn Lumber so cut that the annual rings make an angle of between 30 and 60° with the faces of the piece.

Board Foot Unit of measurement of lumber represented by a board 1 ft long, 12 in. wide, and 1 in. thick, or its cubic equivalent. In practice, the board-foot calculation for lumber 1 in. thick or thicker is based on its nominal thickness, width, and length. Lumber with a nominal thickness of less than 1 in. is calculated as 1 in.

Bole Main stem of a tree of substantial diameter. Saplings and small-diameter trees have stems, not boles.

Boxed Heart Timber where the pith is located within the piece.

Brashness Brittle condition of wood characterized by a more or less abrupt failure across the grain, instead of a tendency to splinter when broken.

Brown Rot In wood, any decay in which the attack concentrates on the cellulose and associated carbohydrates, rather than on the lignin, producing a light to dark brown friable residue, loosely termed dry rot.

Bruise Injury to the wood at or near the surface, caused by its having been struck by a hammer or other object. Such spots are vulnerable to decay.

Check Lengthwise separation or split in the wood that occurs radially, or normal to the annual rings, as seen on the cross section. This is the result of uneven shrinkage of the wood.

Compression Failure Buckling of fibers due to excessive compression along the grain as a result of flexure or end compression of the piece. It may occur as a result of severe bending of a tree by wind, ice, or snow or by stressing wood structural members beyond the proportional limit.

Compression Wood Abnormal growth occurring in conifers and characterized by relatively wide annual rings, usually eccentric, and a comparatively large proportion of summerwood (usually 50% or more) that merges into the springwood without a marked contrast in color.

Conifer Tree bearing seed cones, usually an evergreen. Its wood is known as softwood.

Cross Break Separation of wood cells across the grain due to tension resulting from unusual shrinkage or mechanical stress.

Cross Section Section of a tree, at right angle to the pith or axis, as observed on the end view of a log or piece of wood cut from it.

Decay, Rot Disintegration of the wood substance due to the action of wood-destroying fungi.

Earlywood *See* **Springwood.**

Grain Direction, size, arrangement, appearance, or quality of the fibers in wood or lumber. To have specific meaning, the term must be qualified.

Burl: Local disturbance in the grain, usually associated with knots, or undeveloped bands produced by the healing of wounds during the life of the tree.

Cross grain: Grain or direction of the cells in a piece of wood, which does not run parallel with the axis of the piece.

Curly grain: Fibers being irregularly distorted, as in maple, birch, and other species. This may be several inches long as observed on the tangential surface.

Diagonal grain: Fibers running diagonally across the piece as a result of the latter having been sawed at an angle across the annual rings. It may be observed on the radial and occasionally on the tangential face.

Dip grain: Wave or undulation such as occurs around a knot, pitch pocket, or other defect.

Edge grain, vertical grain, rift grain, comb grain, or quartersawed: Lumber that has been sawed so that the annual rings run at an angle of 45° or more with the face of the piece.

Flat grain: Lumber in which the annual rings form an angle of less than 45° with the wide face of the piece; also called slash grain, plain sawed, or tangential cut.

Green Freshly sawed or undried wood.

Hardwood One of the botanical groups of trees that have broad leaves, in contrast to the conifers or softwoods. The term has no reference to the actual hardness of the wood.

Heartwood Inner and usually darker portion of the cross section of a tree, all cells of which are lifeless and serve only the mechanical function of keeping the tree from breaking under its own weight and from the force of the wind.

Hollow Heart Cavity in the heart of a log, resulting from decay.

Honeycombing Open checks in the interior of a piece of lumber, often not visible on the surface.

Knot Portion of a branch or limb that has been surrounded by subsequent growth of the stem. The shape of the knot as it appears on a cut surface depends on the angle of the cut relative to the long axis of the knot. Knots may be:

Encased: Knot whose rings of annual growth are not intergrown with those of the surrounding wood.

Intergrown: Knot whose rings of annual growth are completely intergrown with those of the surrounding wood.

Loose: Knot that is not held firmly in place by growth or position and that cannot be relied upon to remain in place.

Pin: Knot that is not more than ½ in. in diameter.

Sound: Knot that is solid across its face, at least as hard as the surrounding wood, and shows no indication of decay.

Spike: Knot cut approximately parallel to its long axis so that the exposed section is definitely elongated.

Latewood *See* **Summerwood.**

Live Timber Timber cut from a tree that was living at the time of cutting.

Low-Density Wood Wood that is exceptionally light in weight for its species, due usually to abnormal growth conditions. It is frequently referred to as brash wood and breaks with a splinterless fracture.

Lumber Product of the saw and planing mill not manufactured other than by sawing, resawing, passing lengthwise through a standard planing machine, cross-cutting to length, or matching. Lumber is further designated as:

Board: Lumber that is nominally less than 2 in. thick and 2 in. or more wide.

Dimension: Lumber with a nominal thickness of from 2 in. up to, but not including, 5 in. and a nominal width of 2 in. or more.

Dressed size: Dimensions of lumber after being surfaced with a planing machine. The dressed size is usually ½ to ¾ in. less than the nominal or rough size.

Nominal size: As applied to timber or lumber, size by which it is known and sold in the market, often differing from the actual size.

Rough lumber: Lumber that has not been dressed, but has been sawed, edged, and trimmed.

Side lumber: Board from the outer portion of the log, ordinarily one produced when squaring off a log for a tie or timber.

Structural lumber: Lumber that is intended for use where allowable properties are required. Grading of structural lumber is based on the strength of the piece as related to its anticipated uses.

Timbers: Lumber that is nominally 5 in. or more in least dimension; used for beams, stringers, posts, girders, and other heavy support members.

Pitch Poorly defined accumulation of resin in the wood cells in a more or less irregular patch.

Pitch Pocket Well-defined opening between annual rings, containing more or less pitch.

Pith Soft core at the center of the trunk or limb of a tree or log.

Pores Hollow cells or vessels of which the wood is composed, usually larger and thin-walled in the springwood, and smaller and heavy-walled in the summerwood, as observed on the cross section under a microscope. These cells, tubes, pores, or tracheids run vertically in the tree and conduct the sap up and down the tree.

Radial Face Face of a piece of wood that extends in a generally radial direction in the tree. This surface cuts the annual rings at nearly a right angle and usually presents edge grain.

Rays Tiers of cells extending radially in the tree. They are seen as bands on the radial face of a piece of wood and as dashes on the tangential face. The entire radial face will be seen almost covered with these tiny structures, which appear as fine to conspicuous short lines.

Red Heart Incipient stage of a destructive heart rot in certain coniferous trees caused by fomes pini. It is characterized by an abnormal pink to purplish red or brownish color in the heartwood.

Resin Ducts Very small irregular openings. They are visible on all three sections of a tree, but sometimes appear as fine pin scratches on the tangential section. They occur normally only in pines, spruce, Douglas fir, larch, and tamarack.

Ring-Porous Woods Group of hardwoods in which the pores are comparatively large at the beginning of each annual ring and decrease in size more or less abruptly toward the outer portion of the ring. This forms a distinct inner zone of pores, known as earlywood, and an outer zone with smaller pores, known as latewood.

Sap All the fluids in a tree.

Sapwood Zone of light-colored wood near the bark on the cross section of a tree, about 1 to 3 in. wide and containing 5 to 50 annual rings. The cells in the sapwood are active and store up starch and otherwise assist in the life processes of the tree, although only the last or outer layer of cells forms the growing part and true life of the tree.

Second Growth Timber that has grown after the removal of all or a large portion of the previous stand, whether by cutting, fire, wind, or other agency. Second-growth material is frequently of rapid growth during its early life.

Shake Lengthwise separation of wood that occurs usually between and parallel to the annual rings. A through shake is one extending between two opposite or adjacent faces of a piece of wood.

Sound Wood Wood that has not been affected by decay.

Split Lengthwise separation of wood, extending from one surface through the piece to the opposite or adjacent surface.

Springwood Inner portion of the annual ring that is grown in the spring season. This portion of the ring is less dense than the summerwood.

Summerwood Dense outer portion of the annual ring that is grown in the summer season.

Tension Wood Form of wood found in leaning trees of some hardwood species and characterized by the presence of gelatinous fibers and excessive longitudinal shrinkage. Tension wood fibers hold together tenaciously, so that sawed surfaces usually have projecting fibers, and planed surfaces often are torn or have raised grain. Tension wood may cause warping.

Tangential Surface Face of a piece of wood that is tangential to the annual rings in the piece, usually showing flat grain.

Wane Bark or lack of wood on the edge or corner of a piece.

Warp Any variation from a true or plane surface, including:

Bow: Distortion of lumber in which there is a deviation, in a direction perpendicular to the flat face, from a straight line from end to end of the piece.

Crook: Distortion of lumber in which there is a deviation, in a direction perpendicular to the edge, from a straight line from end to end of the piece.

Cup: Distortion of a board in which there is a deviation flatwise from a straight line across the width of the board.

Twist: Distortion caused by the turning or winding of the edges of a board so that the four corners of any face are no longer in the same plane.

Chapter

3

Wood Structural Members

Wood Structural Members

Wood structural members—beams, columns, joists, and planks—are usually rectangular in cross section and have constant depth and width throughout their span. They should be designed large enough so that loading will not stress the member beyond allowable values for tension, compression, shear, and bending. Under certain conditions, such as to maintain specified clearances, beams, joists, and planks should also be large enough to limit deflection under loading.

A wood member's strength differs in each axis, with the member being strongest when the load is applied parallel to its grain. When properly designed, a wood member may also be safely loaded perpendicular to the grain. Table 3.1 illustrates the effects of stress for loads parallel and perpendicular to the grain.

Properties of a Wood Structural Member

Cross section: Section taken through the member, perpendicular to the member's longitudinal axis. Typical cross sections are shown in Fig. 3.1.

Cross-sectional area: Product of a section's base by its depth,

$$A = bd \qquad (\text{in.}^2)$$

Neutral axis: Line in the cross section of a member on which there is neither tension nor compression.

TABLE 3.1 Stress Effects for Load Conditions Parallel and Perpendicular to Grain

Stress parallel to grain	Stress perpendicular to grain
Tension	
Tension parallel to grain	
Wood fibers tend to stretch along the grain.	Wood fibers tend to separate from the grain.
A wood member's highest strength property is its natural resistance to tension parallel to the grain.	A wood member's least strength is when the fibers are under tension perpendicular to the grain. As a result, design values are seldom given.
Knots or holes reduce a member's cross section, thereby substantially reducing fiber resistance to the applied force.	Not applicable.
A force applied at an angle to the grain reduces the fiber resistance.	Not applicable.
Compression	
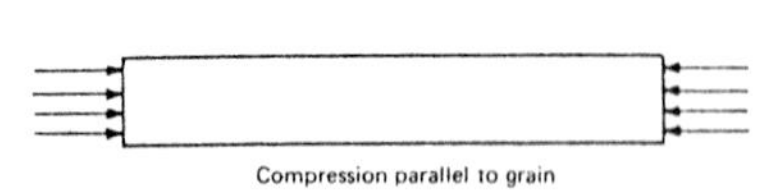 Compression parallel to grain	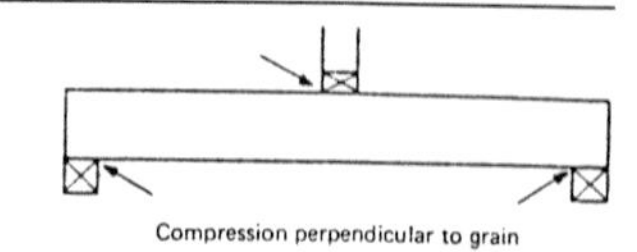 Compression perpendicular to grain
Wood fibers tend to compress lengthwise.	Wood member surfaces tend to compress, resulting in a displacement of the member.
Resistance to compression parallel to the grain is affected by the presence of knots or holes, as well as by the angle of loading.	Not applicable.
Shear	
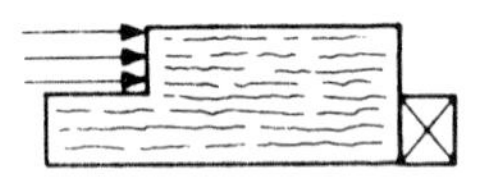 Shear parallel to grain	
Where loading induces a compression stress on one side of a wood member and tension on the other side, shear stress tends to occur parallel to the grain, the largest occurring along the neutral axis on the plane at which the induced stress changes from compression to tension.	This factor need not be considered, since limits on design stresses in shear parallel to the grain, and in compression or bearing perpendicular to the grain, control the design of a wood member.
Shear resistance is reduced by excessive checks and splits in the wood.	Not applicable.

TABLE 3.1 Stress Effects for Load Conditions Parallel and Perpendicular to Grain *(Continued)*

Stress parallel to grain	Stress perpendicular to grain
Bending	
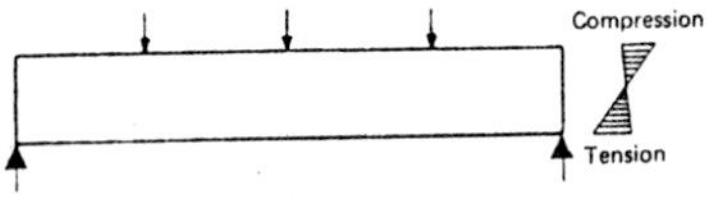 Compression Tension Fiber stress in bending	
Perpendicular loading induces compression stresses in the extreme fibers on the loaded side of a wood member,and tension stresses on the opposite side.	Not applicable.
These stresses diminish in intensity from the outside face fibers to zero at the center of the member's neutral axis.	Not applicable.
Resistance to bending in the extreme fibers is reduced by knots or holes, as well as by deviations in the grain slope.	Not applicable.

Moment of inertia: Sum of each member's basic cross-sectional area, multiplied by the square of its distance from the neutral axis of that particular area,

$$I = \frac{bd^3}{12} \qquad (\text{in.}^4)$$

Section modulus: Moment of inertia of a section divided by the distance from the neutral axis of the cross section,

$$S = \frac{I}{c} = \frac{bd^2}{6} \qquad (\text{in.}^3)$$

Radius of gyration: Square root of a section's moment of inertia divided by its area,

$$r = \sqrt{\frac{I}{A}} \qquad (\text{in.}^2)$$

Design load: Total uniform load, equal to dead load (weight of structural member) plus live load (workers and materials supported by the member), W in pounds.

Effective span: Distance from face to face of supports plus one-half the required length of bearing at each end L in feet.

Linear loading: Uniform load over the length of the member, wL in pound-feet.

Notation

A = cross-sectional area, in.^{2}
E = modulus of elasticity, lb/in.^{2}
F_b = design value for extreme fiber in bending, lb/in.^{2}
f_b = unit stress, lb/in.^{2}
F_c = design value in compression parallel to grain, lb/in.^{2}
f_c = unit stress, lb/in.^{2}
F_t = design value in tension parallel to grain, lb/in.^{2}
F_v = design value in horizontal shear, lb/in.^{2}
f_v = unit stress, lb/in.^{2}
I = moment of inertia, in.^{4}
L = length of member, ft
M = bending of resisting moment, in.-lb or ft-lb
P = total concentrated load, pounds
Q = static moment of area about neutral axis, in.^{3}
r = radius of gyration, in.^{2}
S = section modulus, in.^{2}
V = vertical design shear, lb
W = total weight, lb
w = weight per foot, lb

Actual, not nominal, sizes of sections must be used when determining stresses and deflection of a wood structural member. The properties of structural lumber are listed in Table 3.2.

Beams

The proper design of timber beams involves three steps:

1. Compute the member's section modulus ($S = Mf$) and select an appropriate sized beam from the tables.

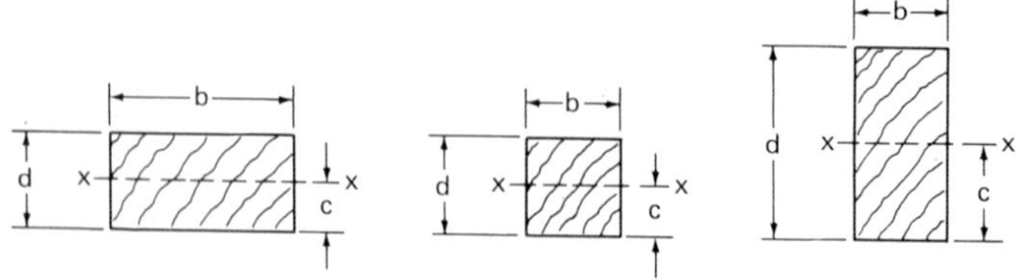

Figure 3.1 Typical cross sections.

2. Test the beam selected for horizontal shear; increase dimensions if necessary.
3. Test the beam selected for deflection.

Most designers simply perform the first computation (section modulus) and then select the proper beam size that they know from experience will meet the requirements for steps 2 and 3. In determining the strength and stiffness of a wood member under transverse loading, it is necessary to consider:

1. Bending moment induced by a load
2. Horizontal shear at beam supports
3. Bearing on supporting members
4. Deflection or deformation caused by load

Any of these four factors may control the design of a member.

Bending

For structural safety, a beam's bending moment, induced by the combination of live and dead loads on the beam, may not be greater than the resisting moment of the wood member. For example, assume a beam having a span L uniformly loaded with w pounds per foot (see Fig. 3.2).

1. Compute the maximum bending moment M, which occurs at the beam's midpoint $L/2$, using the formula

$$M = \frac{wL}{8} \qquad \text{lb-ft}$$

Converting to pound-inches,

$$M = \frac{wL^2 \times 12}{8} = \frac{3wL^2}{2}$$

2. Divide the resisting moment M by the allowable fiber stress in bending F_b for the particular type of wood and grade of lumber, to determine the section modulus S of the member,

$$S = \frac{M}{F_b}$$

TABLE 3.2 Properties of Structural Lumber

Average Specific Gravity and Average Weight in Pounds per Cubic Foot for Commercially Important Species or Species Combinations.

(Specific gravity is based on oven-dry weight and oven-dry volume. Weight per cubic foot is based on weight and volume at a moisture content of 15 percent.)

Species	Specific Gravity	Weight per Cubic Foot
Ash (commercial white)	0.62	40.5
Aspen	0.40	27.2
Balsam Fir	0.38	25.5
Beech	0.68	46.1
Birch, Sweet and Yellow	0.66	44.0
Black Cottonwood	0.33	22.2
California Redwood (Close grain)	0.42	29.2
California Redwood (Open grain)	0.37	24.8
Coast Sitka Spruce	0.39	26.6
Coast Species	0.39	26.6-34.3
Cottonwood, Eastern	0.41	28.5
Douglas Fir - Larch	0.51	34.3
Douglas Fir South	0.48	32.7
Eastern Hemlock - Tamarack	0.45	30.2
Eastern Spruce	0.43	28.9
Eastern White Pine	0.38	25.5
Eastern Woods	0.38	25.5-31.2
Engelmann Spruce - Alpine Fir	0.36	24.3
Hem-Fir	0.42	28.1
Hickory and Pecan	0.75	48.2
Idaho White Pine	0.40	26.0
Lodgepole Pine	0.44	29.8
Maple, Black & Sugar	0.66	44.5
Mountain Hemlock	0.47	32.8
Northern Aspen	0.42	28.7
Northern Pine	0.46	31.2
Northern Species	0.35	24.2-34.3
Northern White Cedar	0.31	22.2
Oak, Red and White	0.67	47.3
Ponderosa Pine (North)	0.49	33.0
Ponderosa Pine - Sugar Pine	0.42	28.6
Red Pine (North)	0.42	28.7
Sitka Spruce	0.43	29.1
Southern Cypress	0.48	33.5
Southern Pine	0.55	37.3
Spruce - Pine - Fir	0.42	26.9
Sweetgum & Tupelo	0.54	35.6
West Coast Woods (Mixed Species)	0.35	24.2-34.3
Western Cedars	0.35	24.2
Western Hemlock	0.48	31.8
Western White Pine	0.40	27.3
White Woods (Western Woods)	0.35	24.2-34.3
Yellow Poplar	0.46	29.4

For species combinations, values are weighted averages. Where two values are shown, they indicate the range of average weights for the species or species combinations permitted under that commercial grouping.

TABLE 3.2 Properties of Structural Lumber (Continued)

NOMINAL SIZE b(inches)d	STANDARD DRESSED SIZE (S4S) b(inches)d	AREA OF SECTION A	MOMENT OF INERTIA I	SECTION MODULUS S	Weight in pounds per linear foot of piece when weight of wood per cubic foot equals:					
					25 lb.	30 lb.	35 lb.	40 lb.	45 lb.	50 lb.
1 x 3	3/4 x 2 1/2	1.875	0.977	0.781	0.326	0.391	0.456	0.521	0.586	0.651
1 x 4	3/4 x 3 1/2	2.625	2.680	1.531	0.456	0.547	0.638	0.729	0.820	0.911
1 x 6	3/4 x 5 1/2	4.125	10.398	3.781	0.716	0.859	1.003	1.146	1.289	1.432
1 x 8	3/4 x 7 1/4	5.438	23.817	6.570	0.944	1.133	1.322	1.510	1.699	1.888
1 x 10	3/4 x 9 1/4	6.938	49.466	10.695	1.204	1.445	1.686	1.927	2.168	2.409
1 x 12	3/4 x 11 1/4	8.438	88.989	15.820	1.465	1.758	2.051	2.344	2.637	2.930
2 x 3	1 1/2 x 2 1/2	3.750	1.953	1.563	0.651	0.781	0.911	1.042	1.172	1.302
2 x 4	1 1/2 x 3 1/2	5.250	5.359	3.063	0.911	1.094	1.276	1.458	1.641	1.823
2 x 6	1 1/2 x 5 1/2	8.250	20.797	7.563	1.432	1.719	2.005	2.292	2.578	2.865
2 x 8	1 1/2 x 7 1/4	10.875	47.635	13.141	1.888	2.266	2.643	3.021	3.398	3.776
2 x 10	1 1/2 x 9 1/4	13.875	98.932	21.391	2.409	2.891	3.372	3.854	4.336	4.818
2 x 12	1 1/2 x 11 1/4	16.875	177.979	31.641	2.930	3.516	4.102	4.688	5.273	5.859
2 x 14	1 1/2 x 13 1/4	19.875	290.775	43.891	3.451	4.141	4.831	5.521	6.211	6.901
3 x 1	2 1/2 x 3/4	1.875	0.088	0.234	0.326	0.391	0.456	0.521	0.586	0.651
3 x 2	2 1/2 x 1 1/2	3.750	0.703	0.938	0.651	0.781	0.911	1.042	1.172	1.302
3 x 4	2 1/2 x 3 1/2	8.750	8.932	5.104	1.519	1.823	2.127	2.431	2.734	3.038
3 x 6	2 1/2 x 5 1/2	13.750	34.661	12.604	2.387	2.865	3.342	3.819	4.297	4.774
3 x 8	2 1/2 x 7 1/4	18.125	79.391	21.901	3.147	3.776	4.405	5.035	5.664	6.293
3 x 10	2 1/2 x 9 1/4	23.125	164.886	35.651	4.015	4.818	5.621	6.424	7.227	8.030
3 x 12	2 1/2 x 11 1/4	28.125	296.631	52.734	4.883	5.859	6.836	7.813	8.789	9.766
3 x 14	2 1/2 x 13 1/4	33.125	484.625	73.151	5.751	6.901	8.051	9.201	10.352	11.502
3 x 16	2 1/2 x 15 1/4	38.125	738.870	96.901	6.619	7.943	9.266	10.590	11.914	13.238
4 x 1	3 1/2 x 3/4	2.625	0.123	0.328	0.456	0.547	0.638	0.729	0.820	0.911
4 x 2	3 1/2 x 1 1/2	5.250	0.984	1.313	0.911	1.094	1.276	1.458	1.641	1.823
4 x 3	3 1/2 x 2 1/2	8.750	4.557	3.646	1.519	1.823	2.127	2.431	2.734	3.038
4 x 4	3 1/2 x 3 1/2	12.250	12.505	7.146	2.127	2.552	2.977	3.403	3.828	4.253
4 x 6	3 1/2 x 5 1/2	19.250	48.526	17.646	3.342	4.010	4.679	5.347	6.016	6.684
4 x 8	3 1/2 x 7 1/4	25.375	111.148	30.661	4.405	5.286	6.168	7.049	7.930	8.811
4 x 10	3 1/2 x 9 1/4	32.375	230.840	49.911	5.621	6.745	7.869	8.933	10.117	11.241
4 x 12	3 1/2 x 11 1/4	39.375	415.283	73.828	6.836	8.203	9.570	10.938	12.305	13.672
4 x 14	3 1/2 x 13 1/2	47.250	717.609	106.313	8.203	9.844	11.484	13.125	14.766	16.406
4 x 16	3 1/2 x 15 1/2	54.250	1086.130	140.146	9.418	11.302	13.186	15.069	16.953	18.837
6 x 1	5 1/2 x 3/4	4.125	0.193	0.516	0.716	0.859	1.003	1.146	1.289	1.432
6 x 2	5 1/2 x 1 1/2	8.250	1.547	2.063	1.432	1.719	2.005	2.292	2.578	2.865
6 x 3	5 1/2 x 2 1/2	13.750	7.161	5.729	2.387	2.865	3.342	3.819	4.297	4.774
6 x 4	5 1/2 x 3 1/2	19.250	19.651	11.229	3.342	4.010	4.679	5.347	6.016	6.684
6 x 6	5 1/2 x 5 1/2	30.250	76.255	27.729	5.252	6.302	7.352	8.403	9.453	10.503
6 x 8	5 1/2 x 7 1/2	41.250	193.359	51.563	7.161	8.594	10.026	11.458	12.891	14.323
6 x 10	5 1/2 x 9 1/2	52.250	392.963	82.729	9.071	10.885	12.700	14.514	16.328	18.142
6 x 12	5 1/2 x 11 1/2	63.250	697.068	121.229	10.981	13.177	15.373	17.569	19.766	21.962
6 x 14	5 1/2 x 13 1/2	74.250	1127.672	167.063	12.891	15.469	18.047	20.625	23.203	25.781
6 x 16	5 1/2 x 15 1/2	85.250	1706.776	220.229	14.800	17.760	20.720	23.681	26.641	29.601
6 x 18	5 1/2 x 17 1/2	96.250	2456.380	280.729	16.710	20.052	23.394	26.736	30.078	33.420
6 x 20	5 1/2 x 19 1/2	107.250	3398.484	348.563	18.620	22.344	26.068	29.792	33.516	37.240
6 x 22	5 1/2 x 21 1/2	118.250	4555.086	423.729	20.530	24.635	28.741	32.847	36.953	41.059
6 x 24	5 1/2 x 23 1/2	129.250	5948.191	506.229	22.439	26.927	31.415	35.903	40.391	44.878
8 x 1	7 1/4 x 3/4	5.438	0.255	0.680	0.944	1.133	1.322	1.510	1.699	1.888
8 x 2	7 1/4 x 1 1/2	10.875	2.039	2.719	1.888	2.266	2.643	3.021	3.398	3.776
8 x 3	7 1/4 x 2 1/2	18.125	9.440	7.552	3.147	3.776	4.405	5.035	5.664	6.293
8 x 4	7 1/4 x 3 1/2	25.375	25.904	14.802	4.405	5.286	6.168	7.049	7.930	8.811
8 x 6	7 1/2 x 5 1/2	41.250	103.984	37.813	7.161	8.594	10.026	11.458	12.891	14.323
8 x 8	7 1/2 x 7 1/2	56.250	263.672	70.313	9.766	11.719	13.672	15.625	17.578	19.531
8 x 10	7 1/2 x 9 1/2	71.250	535.859	112.813	12.370	14.844	17.318	19.792	22.266	24.740
8 x 12	7 1/2 x 11 1/2	86.250	950.547	165.313	14.974	17.969	20.964	23.958	26.953	29.948
8 x 14	7 1/2 x 13 1/2	101.250	1537.734	227.813	17.578	21.094	24.609	28.125	31.641	35.156
8 x 16	7 1/2 x 15 1/2	116.250	2327.422	300.313	20.182	24.219	28.255	32.292	36.328	40.365
8 x 18	7 1/2 x 17 1/2	131.250	3349.609	382.813	22.786	27.344	31.901	36.458	41.016	45.573
8 x 20	7 1/2 x 19 1/2	146.250	4634.297	475.313	25.391	30.469	35.547	40.625	45.703	50.781
8 x 22	7 1/2 x 21 1/2	161.250	6211.484	577.813	27.995	33.594	39.193	44.792	50.391	55.990
8 x 24	7 1/2 x 23 1/2	176.250	8111.172	690.313	30.599	36.719	42.839	48.958	55.078	61.198

(Courtesy of National Forest Products Association.)

TABLE 3.2 Properties of Structural Lumber (Continued)

NOMINAL SIZE b(inches)d	STANDARD DRESSED SIZE (S4S) b(inches)d	AREA OF SECTION A	MOMENT OF INERTIA I	SECTION MODULUS S	Weight in pounds per linear foot of piece when weight of wood per cubic foot equals: 25 lb.	30 lb.	35 lb.	40 lb.	45 lb.	50 lb.
10 x 1	9 1/4 x 3/4	6.938	0.325	0.867	1.204	1.445	1.686	1.927	2.168	2.409
10 x 2	9 1/4 x 1 1/2	13.875	2.602	3.469	2.409	2.891	3.372	3.854	4.336	4.818
10 x 3	9 1/4 x 2 1/2	23.125	12.044	9.635	4.015	4.818	5.621	6.424	7.227	8.030
10 x 4	9 1/4 x 3 1/2	32.375	33.049	18.885	5.621	6.745	7.869	8.993	10.117	11.241
10 x 6	9 1/2 x 5 1/2	52.250	131.714	47.896	9.071	10.885	12.700	14.514	16.328	18.142
10 x 8	9 1/2 x 7 1/2	71.250	333.984	89.063	12.370	14.844	17.318	19.792	22.266	24.740
10 x 10	9 1/2 x 9 1/2	90.250	678.755	142.896	15.668	18.802	21.936	25.069	28.203	31.337
10 x 12	9 1/2 x 11 1/2	109.250	1204.026	209.396	18.967	22.760	26.554	30.347	34.141	37.934
10 x 14	9 1/2 x 13 1/2	128.250	1947.797	288.563	22.266	26.719	31.172	35.625	40.078	44.531
10 x 16	9 1/2 x 15 1/2	147.250	2948.068	380.396	25.564	30.677	35.790	40.903	46.016	51.128
10 x 18	9 1/2 x 17 1/2	166.250	4242.836	484.896	28.863	34.635	40.408	46.181	51.953	57.726
10 x 20	9 1/2 x 19 1/2	185.250	5870.109	602.063	32.161	38.594	45.026	51.458	57.891	64.323
10 x 22	9 1/2 x 21 1/2	204.250	7867.879	731.896	35.460	42.552	49.644	56.736	63.828	70.920
10 x 24	9 1/2 x 23 1/2	223.250	10274.148	874.396	38.759	46.510	54.262	62.014	69.766	77.517
12 x 1	11 1/4 x 3/4	8.438	0.396	1.055	1.465	1.758	2.051	2.344	2.637	2.930
12 x 2	11 1/4 x 1 1/2	16.875	3.164	4.219	2.930	3.516	4.102	4.688	5.273	5.859
12 x 3	11 1/4 x 2 1/2	28.125	14.648	11.719	4.883	5.859	6.836	7.813	8.789	9.766
12 x 4	11 1/4 x 3 1/2	39.375	40.195	22.969	6.836	8.203	9.570	10.938	12.305	13.672
12 x 6	11 1/2 x 5 1/2	63.250	159.443	57.979	10.981	13.177	15.373	17.569	19.766	21.962
12 x 8	11 1/2 x 7 1/2	86.250	404.297	107.813	14.974	17.969	20.964	23.958	26.953	29.948
12 x 10	11 1/2 x 9 1/2	109.250	821.651	172.979	18.967	22.760	26.554	30.347	34.141	37.934
12 x 12	11 1/2 x 11 1/2	132.250	1457.505	253.479	22.960	27.552	32.144	36.736	41.328	45.920
12 x 14	11 1/2 x 13 1/2	155.250	2357.859	349.313	26.953	32.344	37.734	43.125	48.516	53.906
12 x 16	11 1/2 x 15 1/2	178.250	3568.713	460.479	30.946	37.135	43.325	49.514	55.703	61.892
12 x 18	11 1/2 x 17 1/2	201.250	5136.066	586.979	34.939	41.927	48.915	55.903	62.891	69.878
12 x 20	11 1/2 x 19 1/2	224.250	7105.922	728.813	38.932	46.719	54.505	62.292	70.078	77.865
12 x 22	11 1/2 x 21 1/2	247.250	9524.273	885.979	42.925	51.510	60.095	68.681	77.266	85.851
12 x 24	11 1/2 x 23 1/2	270.250	12437.129	1058.479	46.918	56.302	65.686	75.069	84.453	93.837
14 x 2	13 1/4 x 1 1/2	19.875	3.727	4.969	3.451	4.141	4.831	5.521	6.211	6.901
14 x 3	13 1/4 x 2 1/2	33.125	17.253	13.802	5.751	6.901	8.051	9.201	10.352	11.502
14 x 4	13 1/2 x 3 1/2	47.250	48.234	27.563	8.203	9.844	11.484	13.125	14.766	16.406
14 x 6	13 1/2 x 5 1/2	74.250	187.172	68.063	12.891	15.469	18.047	20.625	23.203	25.781
14 x 8	13 1/2 x 7 1/2	101.250	474.609	126.563	17.578	21.094	24.609	28.125	31.641	35.156
14 x 10	13 1/2 x 9 1/2	128.250	964.547	203.063	22.266	26.719	31.172	35.625	40.078	44.531
14 x 12	13 1/2 x 11 1/2	155.250	1710.984	297.563	26.953	32.344	37.734	43.125	48.516	53.906
14 x 16	13 1/2 x 15 1/2	209.250	4189.359	540.563	36.328	43.594	50.859	58.125	65.391	72.656
14 x 18	13 1/2 x 17 1/2	236.250	6029.297	689.063	41.016	49.219	57.422	65.625	73.828	82.031
14 x 20	13 1/2 x 19 1/2	263.250	8341.734	855.563	45.703	54.844	63.984	73.125	82.266	91.406
14 x 22	13 1/2 x 21 1/2	290.250	11180.672	1040.063	50.391	60.469	70.547	80.625	90.703	100.781
14 x 24	13 1/2 x 23 1/2	317.250	14600.109	1242.563	55.078	66.094	77.109	88.125	99.141	110.156
16 x 3	15 1/2 x 2 1/2	38.750	20.182	16.146	6.727	8.073	9.418	10.764	12.109	13.455
16 x 4	15 1/2 x 3 1/2	54.250	55.380	31.646	9.418	11.302	13.186	15.069	16.953	18.837
16 x 6	15 1/2 x 5 1/2	85.250	214.901	78.146	14.800	17.760	20.720	23.681	26.641	29.601
16 x 8	15 1/2 x 7 1/2	116.250	544.922	145.313	20.182	24.219	28.255	32.292	36.328	40.365
16 x 10	15 1/2 x 9 1/2	147.250	1107.443	233.146	25.564	30.677	35.790	40.903	46.016	51.128
16 x 12	15 1/2 x 11 1/2	178.250	1964.463	341.646	30.946	37.135	43.325	49.514	55.703	61.892
16 x 14	15 1/2 x 13 1/2	209.250	3177.984	470.813	36.328	43.594	50.859	58.125	65.391	72.656
16 x 16	15 1/2 x 15 1/2	240.250	4810.004	620.646	41.710	50.052	58.394	66.736	75.078	83.420
16 x 18	15 1/2 x 17 1/2	271.250	6922.523	791.146	47.092	56.510	65.929	75.347	84.766	94.184
16 x 20	15 1/2 x 19 1/2	302.250	9577.547	982.313	52.474	62.969	73.464	83.958	94.453	104.948
16 x 22	15 1/2 x 21 1/2	333.250	12837.066	1194.146	57.856	69.427	80.998	92.569	104.141	115.712
16 x 24	15 1/2 x 23 1/2	364.250	16763.086	1426.646	63.238	75.885	88.533	101.181	113.828	126.476
18 x 6	17 1/2 x 5 1/2	96.250	242.630	88.229	16.710	20.052	23.394	26.736	30.078	33.420
18 x 8	17 1/2 x 7 1/2	131.250	615.234	164.063	22.786	27.344	31.901	36.458	41.016	45.573
18 x 10	17 1/2 x 9 1/2	166.250	1250.338	263.229	28.863	34.635	40.408	46.181	51.953	57.726
18 x 12	17 1/2 x 11 1/2	201.250	2217.943	385.729	34.939	41.927	48.915	55.903	62.891	69.878
18 x 14	17 1/2 x 13 1/2	236.250	3588.047	531.563	41.016	49.219	57.422	65.625	73.828	82.031
18 x 16	17 1/2 x 15 1/2	271.250	5430.648	700.729	47.092	56.510	65.929	75.347	84.766	94.184
18 x 18	17 1/2 x 17 1/2	306.250	7815.754	893.229	53.168	63.802	74.436	85.069	95.703	106.337
18 x 20	17 1/2 x 19 1/2	341.250	10813.359	1109.063	59.245	71.094	82.943	94.792	106.641	118.490
18 x 22	17 1/2 x 21 1/2	376.250	14493.461	1348.229	65.321	78.385	91.450	104.514	117.578	130.642
18 x 24	17 1/2 x 23 1/2	411.250	18926.066	1610.729	71.398	85.677	99.957	114.236	128.516	142.795

(Courtesy of National Forest Products Associations.)

TABLE 3.2 Properties of Structural Lumber (Continued)

NOMINAL SIZE b(inches)d	STANDARD DRESSED SIZE (S4S) b(inches)d	AREA OF SECTION A	MOMENT OF INERTIA I	SECTION MODULUS S	Weight in pounds per linear foot of piece when weight of wood per cubic foot equals:					
					25 lb.	30 lb.	35 lb.	40 lb.	45 lb.	50 lb.
20 x 6	19 1/2 x 5 1/2	107.250	270.359	98.313	18.620	22.344	26.068	29.792	33.516	37.240
20 x 8	19 1/2 x 7 1/2	146.250	685.547	182.813	25.391	30.469	35.547	40.625	45.703	50.781
20 x 10	19 1/2 x 9 1/2	185.250	1393.234	293.313	32.161	38.594	45.026	51.458	57.891	64.323
20 x 12	19 1/2 x 11 1/2	224.250	2471.422	429.813	38.932	46.719	54.505	62.292	70.078	77.865
20 x 14	19 1/2 x 13 1/2	263.250	3998.109	592.313	45.703	54.844	63.984	73.125	82.266	91.406
20 x 16	19 1/2 x 15 1/2	302.250	6051.297	780.813	52.474	62.969	73.464	83.958	94.453	104.948
20 x 18	19 1/2 x 17 1/2	341.250	8708.984	995.313	59.245	71.094	82.943	94.792	106.641	118.490
20 x 20	19 1/2 x 19 1/2	380.250	12049.172	1235.813	66.016	79.219	92.422	105.625	118.828	132.031
20 x 22	19 1/2 x 21 1/2	419.250	16149.859	1502.313	72.786	87.344	101.901	116.458	131.016	145.573
20 x 24	19 1/2 x 23 1/2	458.250	21089.047	1794.813	79.557	95.469	111.380	127.292	243.203	159.115
22 x 6	21 1/2 x 5 1/2	118.250	298.088	108.396	20.530	24.635	28.741	32.847	36.953	41.059
22 x 8	21 1/2 x 7 1/2	161.250	755.859	201.563	27.995	33.594	39.193	44.792	50.391	55.990
22 x 10	21 1/2 x 9 1/2	204.250	1536.130	323.396	35.460	42.552	49.644	56.736	63.828	70.920
22 x 12	21 1/2 x 11 1/2	247.250	2724.901	473.896	42.925	51.510	60.095	68.681	77.266	85.851
22 x 14	21 1/2 x 13 1/2	290.250	4408.172	653.063	50.391	60.469	70.547	80.625	90.703	100.781
22 x 16	21 1/2 x 15 1/2	333.250	6671.941	860.896	57.856	69.427	80.998	92.569	104.141	115.712
22 x 18	21 1/2 x 17 1/2	376.250	9602.211	1097.396	65.321	78.385	91.450	104.514	117.578	130.642
22 x 20	21 1/2 x 19 1/2	419.250	13284.984	1362.563	72.786	87.344	101.901	116.458	131.016	145.573
22 x 22	21 1/2 x 21 1/2	462.250	17806.254	1656.396	80.252	96.302	112.352	128.403	144.453	160.503
22 x 24	21 1/2 x 23 1/2	505.250	23252.023	1978.896	87.717	105.260	122.804	140.347	157.891	175.434
24 x 6	23 1/2 x 5 1/2	129.250	325.818	118.479	22.439	26.927	31.415	35.903	40.391	44.878
24 x 8	23 1/2 x 7 1/2	176.250	826.172	220.313	30.599	36.719	42.839	48.958	55.078	61.198
24 x 10	23 1/2 x 9 1/2	223.250	1679.026	353.479	38.759	46.510	54.262	62.014	69.766	77.517
24 x 12	23 1/2 x 11 1/2	270.250	2978.380	517.979	46.918	56.302	65.686	75.069	84.453	93.837
24 x 14	23 1/2 x 13 1/2	317.250	4818.234	713.813	55.078	66.094	77.109	88.125	99.141	110.156
24 x 16	23 1/2 x 15 1/2	364.250	7292.586	940.979	63.238	75.885	88.533	101.181	113.828	126.476
24 x 18	23 1/2 x 17 1/2	411.250	10495.441	1199.479	71.398	85.677	99.957	114.236	128.516	142.795
24 x 20	23 1/2 x 19 1/2	458.250	14520.797	1489.313	79.557	95.469	111.380	127.292	143.203	159.115
24 x 22	23 1/2 x 21 1/2	505.250	19462.648	1810.479	87.717	105.260	122.804	140.347	157.891	175.434
24 x 24	23 1/2 x 23 1/2	552.250	25415.004	2162.979	95.877	115.052	134.227	153.403	172.578	191.753

(Courtesy of National Forest Products Association.)

Since the bending moment may not be greater than the resisting moment, the two formulas can be equated,

$$F_b S = \frac{3wL^2}{2}$$

3. Determine one of the following:

   a. Size of beam required (when span and load are known),

$$S = \frac{3wL^2}{2F_b}$$

   b. Allowable span (when size and load per linear foot are known),

$$L = \sqrt{\frac{2F_b S}{3w}}$$

   c. Allowable load per linear foot (when size of beam and span are known),

$$w = \frac{2F_b S}{3L^2}$$

Alternatively, once the section modulus of a wood structural member is determined, the rigger can select a properly sized beam directly from

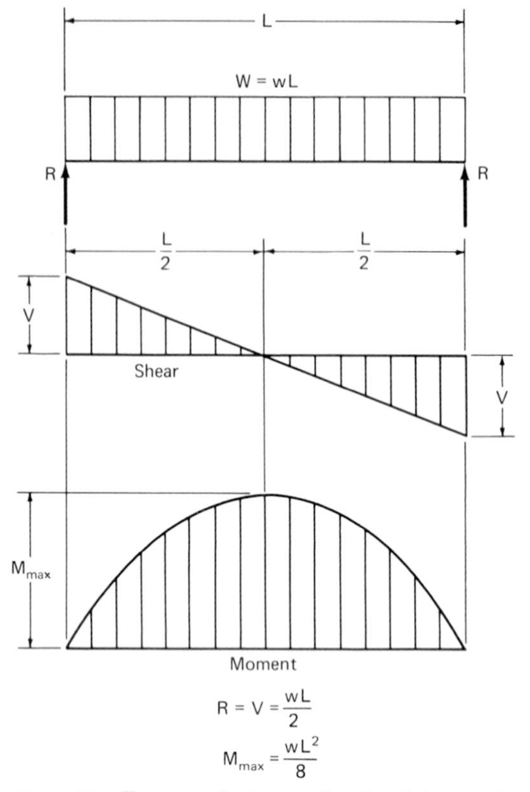

Figure 3.2 Forces and stresses developed in a uniformly distributed load on a simple beam.

precomputed tables. When using tables, make sure to select a wood section that has a section modulus equal to or greater than the computed value. (The section modulus of a wood member is I/c, which for a rectangular member is $bd^2/6$.) Although many beam sizes will meet these requirements, the most practical sizes are those whose breadth ranges from one-half to one-third the beam's depth. Table 3.3 gives safe-load data for wood beams.

Remember: Beams that are relatively deep compared to their width tend to bend sideways under loading, unless properly braced.

Rules of thumb: For beams having the following ratios of depth to width or thickness:

2:1	No lateral support is needed
3:1 or 4:1	Ends should be held in position

5:1	One edge should be held in line for entire length of beam
6:1	Lateral supports are required at intervals of 8 ft

Although lumber is customarily specified in terms of nominal sizes, always use the actual sizes (net dimensions) for your design computations.

Horizontal shear

Horizontal or longitudinal shearing forces in a member must be considered in designing a beam subject to vertical shear (see Fig. 3.3). When a beam is loaded vertically, the upper part of the beam tends to slide along its lower part. To prevent this sliding action, the shear resistance of a wood member must equal or exceed the horizontal shear that the vertical loading induces. Maximum horizontal shear occurs in a rectangular beam at the neutral axis of the section and depends on the magnitude of vertical shear in the member,

$$f_v = \frac{VQ}{Ib}$$

For rectangular beams,

$$f_v = \frac{3V}{2bd}$$

However, f_v may not be greater than the design value in horizontal shear F_v for the particular type and grade of wood used.

Because notching a beam will affect the shear strength of the member, notching should be avoided whenever possible, especially on a beam's tension side. *Never* notch the tension side of a beam that is 4 in. deep or deeper, except at its ends.

To reduce stress concentration where notching is necessary, use a gradually tapered notch instead of a square cornered one. Notches

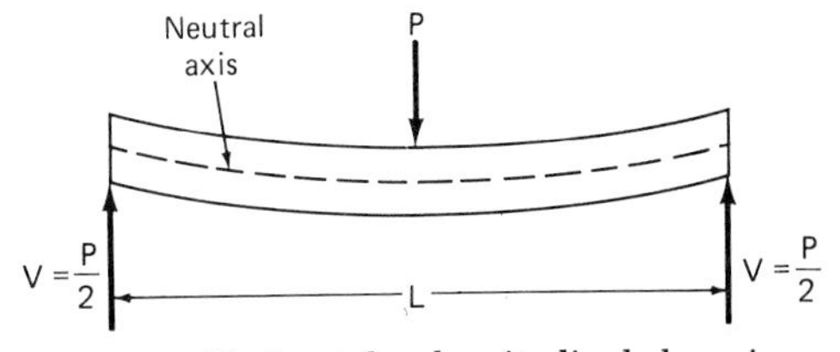

Figure 3.3 Horizontal or longitudinal shear in a simple beam.

TABLE 3.3 Safe-Load Tables for Wood Beams

Use of Tables

To use the tables, locate the appropriate bending design value, F_b, and refer to the span length involved. If the total load W is known read down the column under the appropriate F_b heading to find a matching design load W and then read across the page to see the required beam size. If the beam size is known read across the page to the column under the appropriate F_b heading to find the design load W.

Before selecting a size of beam it is advisable to check the board measure, bm, in several sizes which qualify in order to find the one which has the least amount of lumber and thus is the most efficient.

After determining the required beam size, or design load, W, in the manner just described, it is necessary to check the horizontal shear, F_v, and the modulus of elasticity, E, to make sure that the induced or required values do not exceed the respective values allowed for the species and grade of lumber to be used.

Symbols used in the tables are as follows:

- F_b = Allowable unit stress in extreme fiber in bending, psi.
- W = Total uniformly distributed load, pounds
- w = Load per linear foot of beam, pounds
- F_v = Horizontal shear stress, psi, induced by load W
- E = Modulus of elasticity, 1000 psi, induced by load W for $l/360$ limit

Beam sizes are expressed as nominal sizes, inches, but calculations are based on net dimensions of S4S sizes.

SIZE OF BEAM		F_b									
		900	1000	1100	1200	1300	1400	1500	1600	1800	2000
4'- 0" SPAN											
2 x 4	W	459	510	561	613	664	715	766	817	919	1021
	w	115	128	140	153	166	179	191	204	230	255
	F_v	66	73	80	88	95	102	109	117	131	146
	E	926	1029	1131	1234	1337	1440	1543	1646	1851	2057
3 x 4	W	766	851	936	1021	1106	1191	1276	1361	1531	1701
	w	191	213	234	255	276	298	319	340	383	425
	F_v	66	73	80	88	95	102	109	117	131	146
	E	926	1029	1131	1234	1337	1440	1543	1646	1851	2057

(Courtesy of National Forest Products Association.)

TABLE 3.3 Safe-Load Tables for Wood Beams (Continued)

SIZE OF BEAM		F_b									
		900	1000	1100	1200	1300	1400	1500	1600	1800	2000
4'- 0" SPAN CONT'D											
4 x 4	W	1072	1191	1310	1429	1548	1667	1786	1906	2144	2382
	w	268	298	328	357	387	417	447	476	536	595
	F_v	66	73	80	88	95	102	109	117	131	146
	E	926	1029	1131	1234	1337	1440	1543	1646	1851	2057
2 x 6	W	1134	1260	1386	1513	1639	1765	1891	2017	2269	2521
	w	284	315	347	378	410	441	473	504	567	630
	F_v	103	115	126	138	149	160	172	183	206	229
	E	589	655	720	785	851	916	982	1047	1178	1309
3 x 6	W	1891	2101	2311	2521	2731	2941	3151	3361	3781	4201
	w	473	525	578	630	683	735	788	840	945	1050
	F_v	103	115	126	138	149	160	172	183	206	229
	E	589	655	720	785	851	916	982	1047	1178	1309
2 x 8	W	1971	2190	2409	2628	2847	3066	3285	3504	3942	4380
	w	493	548	602	657	712	767	821	876	986	1095
	F_v	136	151	166	181	196	211	227	242	272	302
	E	447	497	546	596	646	695	745	794	894	993
4 x 6	W	2647	2941	3235	3529	3823	4117	4411	4706	5294	5882
	w	662	735	809	882	956	1029	1103	1176	1323	1470
	F_v	103	115	126	138	149	160	172	183	206	229
	E	589	655	720	785	251	916	982	1047	1178	1309
3 x 8	W	3285	3650	4015	4380	4745	5110	5475	5840	6570	7300
	w	821	913	1004	1095	1186	1278	1369	1460	1643	1825
	F_v	136	151	166	181	196	211	227	242	272	302
	E	447	497	546	596	646	695	745	794	894	993
6 x 6	W	4159	4622	5084	5546	6008	6470	6932	7394	8319	9243
	w	1040	1155	1271	1386	1502	1618	1733	1849	2080	2311
	F_v	103	115	126	138	149	160	172	183	206	229
	E	589	655	720	785	851	916	982	1047	1178	1309
4 x 8	W	4599	5110	5621	6132	6643	7154	7665	8176	9198	10220
	w	1150	1278	1405	1533	1661	1789	1916	2044	2300	2555
	F_v	136	151	166	181	196	211	227	242	272	302
	E	447	497	546	596	646	695	745	794	894	993
6 x 8	W	7734	8594	9453	10313	11172	12031	12891	13750	15469	17188
	w	1934	2148	2363	2578	2793	3008	3223	3438	3867	4297
	F_v	141	156	172	188	203	219	234	250	281	313
	E	432	480	528	576	624	672	720	768	864	960
5'- 0" SPAN											
2 x 4	W	368	408	449	490	531	572	613	653	735	817
	w	74	82	90	98	106	114	123	131	147	163
	F_v	53	58	64	70	76	82	88	93	105	117
	E	1157	1286	1414	1543	1671	1800	1929	2057	2314	2571
3 x 4	W	612	681	749	817	885	953	1021	1089	1225	1361
	w	123	136	149	163	177	191	204	218	245	272
	F_v	53	58	64	70	76	82	88	93	105	117
	E	1157	1286	1414	1543	1671	1800	1929	2057	2314	2571
4 x 4	W	857	953	1048	1143	1239	1334	1429	1524	1715	1906
	w	172	191	210	229	243	267	286	305	343	381
	F_v	53	58	64	70	76	82	88	93	105	117
	E	1157	1286	1414	1543	1671	1800	1929	2057	2314	2571

(Courtesy of National Forest Products Association.)

TABLE 3.3 Safe-Load Tables for Wood Beams (Continued)

Symbols used in the tables are as follows:

F_b = Allowable unit stress in extreme fiber in bending, psi.
W = Total uniformly distributed load, pounds
w = Load per linear foot of beam, pounds
F_v = Horizontal shear stress, psi, induced by load W
E = Modulus of elasticity, 1000 psi, induced by load W for $l/360$ limit

Beam sizes are expressed as nominal sizes, inches, but calculations are based on net dimensions of S4S sizes.

SIZE OF BEAM		F_b									
		900	1000	1100	1200	1300	1400	1500	1600	1800	2000
5'- 0" SPAN CONT'D											
2 x 6	W	908	1008	1109	1210	1311	1412	1513	1613	1815	2017
	w	182	202	222	242	262	282	303	323	363	403
	F_v	83	92	101	110	119	128	138	147	165	183
	E	736	818	900	982	1064	1145	1227	1309	1473	1636
3 x 6	W	1512	1681	1849	2017	2185	2353	2521	2689	3025	3361
	w	303	336	370	403	437	471	504	538	605	672
	F_v	83	92	101	110	119	128	138	147	165	183
	E	736	818	900	982	1064	1145	1227	1309	1473	1636
2 x 8	W	1577	1752	1927	2103	2278	2453	2628	2803	3154	3504
	w	315	350	385	421	456	491	526	561	631	701
	F_v	109	121	133	145	157	169	181	193	218	242
	E	559	621	683	745	807	869	931	993	1117	1241
4 x 6	W	2117	2353	2588	2823	3059	3294	3529	3764	4235	4706
	w	424	471	518	565	612	659	706	753	847	941
	F_v	83	92	101	110	119	128	138	147	165	183
	E	736	818	900	982	1064	1145	1227	1309	1473	1636
2 x 10	W	2567	2852	3137	3423	3708	3993	4278	4563	5134	5704
	w	513	570	627	685	742	799	856	913	1027	1141
	F_v	139	154	170	185	200	215	231	247	278	308
	E	438	486	535	584	632	681	730	778	876	973
3 x 8	W	2628	2920	3212	3504	3796	4088	4380	4672	5256	5840
	w	526	584	642	701	759	818	876	934	1051	1168
	F_v	109	121	133	145	157	169	181	193	218	242
	E	559	621	683	745	807	869	931	993	1117	1241
6 x 6	W	3327	3697	4067	4437	4806	5176	5546	5916	6655	7394
	w	666	739	813	887	961	1035	1109	1183	1331	1479
	F_v	83	92	101	110	119	128	138	147	165	183
	E	736	818	900	982	1064	1145	1227	1309	1473	1636
4 x 8	W	3679	4088	4497	4906	5315	5723	6132	6541	7359	8176
	w	736	818	899	981	1063	1145	1226	1308	1472	1635
	F_v	109	121	133	145	157	169	181	193	218	242
	E	559	621	683	745	807	869	931	993	1117	1241
3 x 10	W	4278	4752	5229	5704	6180	6655	7130	7606	8556	9507
	w	856	951	1046	1141	1236	1331	1426	1521	1711	1901
	F_v	139	154	170	185	200	216	231	247	278	308
	E	438	486	535	584	632	681	730	778	876	973
6 x 8	W	6188	6875	7563	8250	8938	9625	10313	11000	12375	13750
	w	1238	1375	1513	1650	1788	1925	2063	2200	2475	2750
	F_v	113	125	138	150	163	175	188	200	225	250
	E	540	600	660	720	780	840	900	960	1080	1200

(Courtesy of National Forest Products Association.)

in wood beams should not be deeper than one-sixth the depth of the wood member. *Never* put a notch in the middle one-third of a span. Notches at the ends of a beam should not be greater than one-fourth the depth of the wood member (see Fig. 3.4).

Always check the desired bending load of a beam having square cornered notches at its ends against the load obtained by the formula

$$V = \left(\frac{d'}{d}\right)\left(\frac{2F_v bd'}{3}\right)$$

Bearing on supports

When a wood beam is loaded, its fibers tend to compress at the beam support points. Make sure there is sufficient end-bearing area for the beam to transfer the load without damage to its fibers. Determine the bearing area by dividing the beam's end reaction by the design value of compression perpendicular to the grain F_c for the particular type and grade of wood being used. If the bearing area, from the end of a wood beam, is less than 6 in. long, then it is safe to use higher stresses in compression perpendicular to the grain. For such bearing areas, if located 3 in. or more from the ends of a beam, increase the compression values as indicated in Table 3.4.

Deflection

A beam's deflection Δ is the amount of deformation that results from a load applied to the beam. As long as the induced bending stress does not exceed the design value, deflection does not seriously affect the safety of the beam, and deflection can be ignored, with the design of the beam based on strength alone. However, deflection becomes a critical factor for the rigger when a specified clearance is required under the beam.

Joists

Small beams spaced closely together to support floor (or roof) loads are called joists. They are designed in the same manner as wood beams. Table 3.5 has been compiled to facilitate the proper selection of wood joists. It gives maximum spans for various sizes of floor loads and

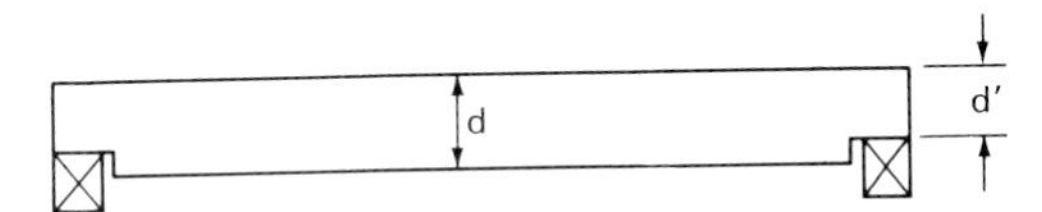

Figure 3.4 Horizontal shear in a notched beam.

TABLE 3.4 Factors by Which to Increase Compression Values Based on Bearing Length

Length of bearing, in.	Factor
½	1.75
1	1.38
1½	1.25
2	1.19
3	1.13
4	1.10
6 or more	1.00

TABLE 3.5 Maximum Spans for Floor Joists

Span

While the effective span length for an isolated beam is customarily taken as the distance from face to face of supports plus one-half the required length of bearing at each end, it is the practice in designing joists spaced not over 24 inches apart to consider the span as the clear distance between supports.

Net Sizes of Lumber

Joists are customarily specified in terms of nominal sizes but calculations to determine the allowable span and required modulus of elasticity are based on dressed sizes.

Design Stresses

Unit design values for design of wood joists are given in the National Design Specification for Wood Construction, available from the National Forest Products Association.

Adjustment of Modulus of Elasticity

The modulus of elasticity values listed in the span tables for joists are those required for the tabulated spans if deflection under the live load is limited to $\ell/360$. Where other deflection limits are acceptable, the tabular E values may be adjusted by multiplying them by the following factors:

For limit of $\ell/300$ -------- 0.833

For limit of $\ell/240$ -------- 0.667

For limit of $\ell/180$ -------- 0.500

(Courtesy of National Forest Products Association.)

TABLE 3.5 Maximum Spans for Floor Joists (Continued)

Symbols used in the tables are as follows:

F_b = Unit design stress at extreme fiber in bending, psi.
L = Clear distance between supports, feet.
E = Minimum required modulus of elasticity, 1000 psi, if deflection under the live load is limited to ℓ/360.

Joist sizes are expressed as nominal but calculations are based on dressed sizes

Size of Joist	Spacing C. to C.		F_b									
			900	1000	1100	1200	1300	1400	1500	1600	1800	2000
2 x 6	12	L	8'8"	9'2"	9'7"	10'0"	10'5"	10'10"	11'3"	11'7"	12'3"	12'11"
		E	1063	1246	1437	1637	1846	2063	2289	2521	3007	3522
	16	L	7'6"	7'11"	8'4"	8'8"	9'1"	9'5"	9'9"	10'0"	10'7"	11'2"
		E	924	1083	1249	1423	1605	1794	1989	2191	2614	3062
	24	L	6'1"	6'5"	6'9"	7'1"	7'4"	7'7"	7'11"	8'2"	8'7"	9'1"
		E	744	871	1005	1144	1291	1443	1600	1762	2103	2463
2 x 8	12	L	11'5"	12'1"	12'7"	13'3"	13'9"	14'3"	14'9"	15'3"	16'2"	17'1"
		E	1063	1246	1437	1631	1846	2063	2289	2521	3007	3522
	16	L	9'11"	10'5"	11'0"	11'6"	11'11"	12'5"	12'10"	13'3"	14'0"	14'10"
		E	924	1083	1249	1423	1605	1794	1989	2191	2614	3062
	24	L	8'1"	8'6"	8'11"	9'4"	9'8"	10'1"	10'5"	10'9"	11'5"	12'0"
		E	744	871	1005	1144	1291	1443	1600	1762	2103	2463
2 x 10	12	L	14'7"	15'5"	16'2"	16'10"	17'6"	18'2"	18'10"	19'5"	20'7"	21'9"
		E	1063	1246	1437	1637	1846	2063	2289	2521	3007	3522
	16	L	12'7"	13'4"	14'0"	14'7"	15'3"	15'10"	16'4"	16'10"	17'11"	18'11"
		E	924	1083	1249	1423	1605	1794	1989	2191	2614	3062
	24	L	10'3"	10'10"	11'4"	11'10"	12'4"	12'10"	13'3"	13'9"	14'7"	15'4"
		E	744	871	1005	1144	1291	1443	1600	1762	2103	2463
2 x 12	12	L	17'9"	18'9"	19'7"	20'6"	21'4"	22'2"	22'11"	23'8"	25'1"	26'6"
		E	1063	1246	1437	1637	1846	2063	2289	2521	3007	3522
	16	L	15'5"	16'3"	17'1"	17'10"	18'6"	19'2"	19'10"	20'6"	21'9"	23'0"
		E	924	1083	1249	1423	1605	1794	1989	2191	2614	3062
	24	L	12'6"	13'2"	13'10"	14'5"	15'0"	15'7"	16'2"	16'7"	17'8"	18'10"
		E	744	871	1005	1144	1291	1443	1600	1762	2103	2463
2 x 14	12	L	20'11"	22'1"	23'2"	24'2"	25'2"	26'1"	27'0"	27'11"	29'7"	31'2"
		E	1063	1246	1437	1637	1846	2063	2289	2521	3007	3522
	16	L	18'2"	19'2"	20'1"	20'11"	21'9"	22'7"	23'5"	24'2"	25'7"	27'0"
		E	924	1083	1249	1423	1605	1794	1989	2191	2614	3062
	24	L	14'9"	15'6"	16'3"	17'0"	17'8"	18'4"	19'0"	19'7"	20'10"	22'0"
		E	744	871	1005	1144	1291	1443	1600	1762	2103	2463
3 x 6	12	L	11'2"	11'10"	12'5"	12'11"	13'6"	14'0"	14'6"	14'11"	15'10"	16'9"
		E	1373	1608	1855	2113	2383	2663	2953	3254	3882	4547
	16	L	9'9"	10'3"	10'9"	11'3"	11'8"	12'2"	12'7"	12'11"	13'9"	14'6"
		E	1193	1397	1612	1836	2071	2314	2567	2827	3374	3952
	24	L	7'11"	8'4"	8'9"	9'2"	9'6"	9'10"	10'2"	10'6"	11'2"	11'9"
		E	960	1124	1297	1478	1666	1862	2065	2275	2714	3179
3 x 8	12	L	14'9"	15'7"	16'4"	17'1"	17'9"	18'5"	19'1"	19'9"	20'11"	22'1"
		E	1373	1608	1855	2113	2383	2663	2953	3254	3882	4547
	16	L	12'10"	13'6"	14'2"	14'10"	15'5"	16'0"	16'7"	17'1"	18'1"	19'1"
		E	1193	1397	1612	1836	2071	2314	2567	2827	3374	3952
	24	L	10'5"	11'0"	11'6"	12'0"	12'6"	13'0"	13'5"	13'10"	14'8"	15'6"
		E	960	1124	1297	1478	1666	1862	2065	2275	2714	3179
3 x 10	12	L	18'10"	19'10"	20'10"	21'9"	22'7"	23'6"	24'4"	25'1"	26'7"	28'1"
		E	1373	1608	1855	2113	2383	2663	2953	3254	3882	4547
	16	L	16'4"	17'3"	18'1"	18'10"	19'7"	20'5"	21'1"	21'10"	23'2"	24'5"
		E	1193	1397	1612	1836	2071	2314	2567	2827	3374	3952
	24	L	13'3"	14'0"	14'8"	15'4"	16'0"	16'7"	17'2"	17'8"	18'9"	19'10"
		E	960	1124	1297	1478	1666	1862	2065	2275	2714	3179
3 x 12	12	L	22'11"	24'2"	25'4"	26'5"	27'6"	28'7"	29'7"	30'7"	32'5"	34'2"
		E	1373	1608	1855	2113	2383	2663	2953	3254	3882	4547
	16	L	19'11"	20'11"	21'11"	22'11"	23'11"	24'10"	25'8"	26'6"	28'1"	29'7"
		E	1193	1397	1612	1836	2071	2314	2567	2827	3374	3952
	24	L	16'2"	17'0"	17'10"	18'8"	19'5"	20'2"	20'10"	21'6"	22'10"	24'1"
		E	960	1124	1297	1478	1666	1862	2065	2275	2714	3179
3 x 14	12	L	27'0"	28'5"	29'10"	31'2"	32'5"	33'8"	34'10"	36'0"	38'2"	40'3"
		E	1373	1608	1855	2113	2383	2663	2953	3254	3882	4547
	16	L	23'5"	24'8"	25'11"	27'1"	28'2"	29'3"	30'3"	31'3"	33'1"	34'11"
		E	1193	1397	1612	1836	2071	2314	2567	2827	3374	3952
	24	L	19'0"	20'0"	21'0"	22'0"	22'11"	23'9"	24'7"	25'5"	26'11"	28'4"
		E	960	1124	1297	1478	1666	1862	2065	2275	2714	3179

(Courtesy of National Forest Products Association.)

TABLE 3.5 Maximum Spans for Floor Joists (Continued)

Symbols used in the tables are as follows:

F_b = Unit design stress at extreme fiber in bending, psi.

L = Clear distance between supports, feet.

E = Minimum required modulus of elasticity, 1000 psi, if deflection under the live load is limited to ℓ/360.

Joist sizes are expressed as nominal but calculations are based on dressed sizes

Size of Joist	Spacing C. to C.		F_b									
			900	1000	1100	1200	1300	1400	1500	1600	1800	2000
2 x 6	12	L	8'1"	8'6"	8'11"	9'3"	9'8"	10'0"	10'5"	10'9"	11'5"	12'0"
		E	1012	1186	1368	1558	1757	1964	2179	2400	2863	3353
	16	L	7'0"	7'4"	7'9"	8'1"	8'5"	8'8"	9'0"	9'4"	9'10"	10'5"
		E	880	1031	1189	1355	1528	1708	1894	2191	2489	2915
	24	L	5'8"	6'0"	6'4"	6'7"	6'10"	7'1"	7'4"	7'7"	8'0"	8'5"
		E	708	829	957	1089	1229	1374	1523	1677	2002	2345
2 x 8	12	L	10'7"	11'2"	11'9"	12'3"	12'9"	13'3"	13'8"	14'1"	15'0"	15'10"
		E	1012	1186	1368	1558	1757	1964	2179	2400	2863	3353
	16	L	9'2"	9'8"	10'2"	10'7"	11'0"	11'5"	11'10"	12'3"	13'0"	13'8"
		E	880	1031	1189	1355	1528	1708	1894	2191	2489	2915
	24	L	7'6"	7'11"	8'3"	8'7"	9'0"	9'4"	9'7"	9'11"	10'7"	11'2"
		E	708	829	957	1089	1229	1374	1523	1677	2002	2345
2 x 10	12	L	13'6"	14'3"	14'11"	15'7"	16'3"	16'10"	17'5"	18'0"	19'1"	20'2"
		E	1012	1186	1368	1558	1757	1964	2179	2400	2863	3353
	16	L	11'9"	12'3"	13'0"	13'6"	14'0"	14'6"	15'1"	15'7"	16'7"	17'6"
		E	880	1031	1189	1355	1528	1708	1894	2191	2489	2915
	24	L	9'6"	10'0"	10'6"	11'0"	11'6"	11'11"	12'4"	12'9"	13'6"	14'3"
		E	708	829	957	1089	1229	1374	1523	1677	2002	2345
2 x 12	12	L	16'6"	17'4"	18'2"	19'0"	19'9"	20'6"	21'3"	21'11"	23'3"	24'6"
		E	1012	1186	1368	1558	1757	1964	2179	2400	2863	3353
	16	L	14'3"	15'0"	15'9"	16'6"	17'2"	17'10"	18'5"	19'0"	20'2"	21'3"
		E	880	1031	1189	1355	1528	1708	1894	2191	2489	2915
	24	L	11'7"	12'3"	12'10	13'5"	13'11"	14'5"	14'11"	15'5"	16'5"	17'5"
		E	708	829	957	1089	1229	1374	1523	1677	2002	2345
2 x 14	12	L	19'5"	20'5"	21'5"	22'4"	23'3"	24'2"	25'0"	25'10"	27'5"	28'11"
		E	1012	1186	1368	1558	1757	1964	2179	2400	2863	3353
	16	L	16'10"	17'8"	18'6"	19'4"	20'2"	20'11"	21'8"	22'5"	23'9"	25'1"
		E	880	1031	1189	1355	1528	1708	1894	2191	2489	2915
	24	L	13'8"	14'5"	15'1"	15'9"	16'5"	17'0"	17'7"	18'2"	19'3"	20'4"
		E	708	829	957	1089	1229	1374	1523	1677	2002	2345
3 x 6	12	L	10'4"	10'11"	11'6"	12'0"	12'6"	13'0"	13'5"	13'10"	14'8"	15'6"
		E	1307	1531	1766	2012	2269	2535	2811	3098	3696	4329
	16	L	9'0"	9'6"	10'0"	10'5"	10'10"	11'3"	11'8"	12'0"	12'9"	13'5"
		E	1136	1330	1535	1748	1972	2203	2444	2691	3212	3762
	24	L	7'4"	7'9"	8'1"	8'5"	8'9"	9'1"	9'5"	9'9"	10'4"	10'11"
		E	914	1070	1235	1406	1586	1773	1966	2166	2584	3026
3 x 8	12	L	13'8"	14'5"	15'2"	15'10"	16'6"	17'1"	17'8"	18'3"	19'4"	20'5"
		E	1307	1531	1766	2012	2269	2535	2811	3098	3696	4329
	16	L	11'10"	12'6"	13'1"	13'8"	14'3"	14'10"	15'4"	15'10"	16'9"	17'8"
		E	1136	1330	1535	1748	1972	2203	2444	2691	3212	3762
	24	L	9'7"	10'1"	10'7"	11'1"	11'7"	12'0"	12'5"	12'10"	13'7"	14'4"
		E	914	1070	1235	1406	1586	1773	1966	2166	2584	3026
3 x 10	12	L	17'5"	18'5"	19'4"	20'2"	21'0"	21'9"	22'7"	23'4"	24'9"	26'1"
		E	1307	1531	1766	2012	2269	2535	2811	3098	3696	4329
	16	L	15'2"	16'0"	16'9"	17'6"	18'2"	18'10"	19'6"	20'2"	21'5"	22'7"
		E	1136	1330	1535	1748	1972	2203	2444	2691	3212	3762
	24	L	12'4"	13'0"	13'7"	14'2"	14'9"	15'4"	15'10"	16'4"	17'5"	18'4"
		E	914	1070	1235	1406	1586	1773	1966	2166	2584	3026
3 x 12	12	L	21'2"	22'4"	23'5"	24'6"	25'6"	26'6"	27'5"	28'4"	30'0"	31'7"
		E	1307	1531	1766	2012	1269	2535	2811	3098	3696	4329
	16	L	18'5"	19'5"	20'4"	21'3"	22'2"	23'0"	23'9"	24'6"	26'0"	27'5"
		E	1136	1330	1535	1748	1972	2203	2444	2691	3212	3762
	24	L	15'0"	15'9"	16'6"	17'3"	18'0"	18'8"	19'4"	20'0"	21'2"	22'4"
		E	914	1070	1235	1406	1586	1773	1966	2166	2584	3036
3 x 14	12	L	25'0"	26'4"	27'7"	28'10"	30'1"	31'3"	32'4"	33'4"	35'4"	37'4"
		E	1307	1531	1766	2012	1769	2535	2811	3098	3696	4329
	16	L	21'8"	22'10"	24'0"	25'1"	26'1"	27'1"	28'0"	28'11"	30'8"	32'4"
		E	1136	1330	1535	1748	1972	2203	2444	2691	3212	3762
	24	L	17'7"	18'7"	19'6"	20'4"	21'2"	22'0"	22'9"	23'6"	24'11"	26'3"
		E	914	1070	1235	1406	1586	1773	1966	2166	2584	3026

(*Courtesy of National Forest Products Association.*)

TABLE 3.5 Maximum Spans for Floor Joists (Continued)

Symbols used in the tables are as follows:

F_b = Unit design stress at extreme fiber in bending, psi.
L = Clear distance between supports, feet.
E = Minimum required modulus of elasticity, 1000 psi, if deflection under the live load is limited to $\ell/360$.

Joist sizes are expressed as nominal but calculations are based on dressed sizes

Size of Joist	Spacing C. to C.		F_b 900	1000	1100	1200	1300	1400	1500	1600	1800	2000
2 x 10	12	L	12'8"	13'4"	14'0"	14'7"	15'2"	15'9"	16'4"	16'10"	17'11"	18'10"
		E	963	1133	1306	1488	1678	1875	2081	2292	2733	3201
	16	L	11'1"	11'7"	12'1"	12'7"	13'2"	13'8"	14'2"	14'7"	15'6"	16'4"
		E	840	984	1135	1294	1459	1631	1808	1992	2376	2783
	24	L	8'11"	9'5"	9'10"	10'3"	10'8"	11'1"	11'6"	11'11"	12'7"	13'3"
		E	676	792	914	1040	1174	1312	1454	1602	1912	2239
2 x 12	12	L	15'5"	16'3"	17'0"	17'9"	18'6"	19 2"	19'10"	20'6"	21'9"	22'11"
		E	963	1133	1306	1488	1678	1875	2081	2292	2733	3201
	16	L	13'4"	14'1"	14'9"	15'5"	16'0"	16'7"	17'3"	17'10"	18'10"	19'11"
		E	840	984	1135	1294	1459	1631	1808	1992	2376	2783
	24	L	10'10"	11'5"	12'0"	12'6"	13'0"	13'6"	14'0"	14'5"	15'4"	16'4"
		E	676	792	914	1040	1174	1312	1454	1602	1912	2239
2 x 14	12	L	18'2"	19'1"	20'0"	20'11"	21'9"	22'7"	23'5"	24'2"	25'7"	27'0"
		E	963	1133	1306	1488	1678	1875	2081	2292	2733	3'01
	16	L	15'9"	16'7"	17'5"	18'2"	18'11"	19'7"	20'3"	20'11"	22'3"	23'5"
		E	840	984	1135	1294	1459	1631	1808	1992	2376	2783
	24	L	12'9"	13'6"	14'2"	14'9"	15'4"	15'11"	16'6"	17'0"	18'01"	19'01"
		E	676	792	914	1040	1174	1312	1454	1602	1912	2239
3 x 8	12	L	12'10"	13'6"	14'2"	14'9"	15'4"	15'11"	16'6"	17'1"	18'1"	19'1"
		E	1248	1462	1686	1921	2166	2421	2684	2958	3529	4133
	16	L	11'1"	11'8"	12'3"	12'10"	13'4"	13'10"	14'4"	14'10"	15'8"	16'7"
		E	1084	1270	1465	1669	1883	2103	2333	2570	3067	3592
	24	L	9'0"	9'6"	10'0"	10'5"	10'10"	11'3"	11'8"	12'0"	12'9"	13'5"
		E	873	1022	1179	1344	1514	1693	1877	2068	2467	2900
3 x 10	12	L	16'4"	17'3"	18'1"	18'10"	19'7"	20'4"	21'1"	21'9"	23'1"	24'4"
		E	1248	1462	1686	1921	2166	2421	2684	2958	3529	4133
	16	L	14'2"	14'11"	15'8"	16'4"	17'0"	17'8"	18'3"	18'11"	20'1"	21'1"
		E	1084	1270	1465	1669	1883	2103	2333	2570	3067	3592
	24	L	11'6"	12'2"	12'9"	13'3"	13'10"	14'4"	14'10"	15'4"	16'3"	17'2"
		E	873	1022	1179	1344	1514	1693	1877	2068	2467	2900
3 x 12	12	L	19'11"	20'11"	21'11"	22'11"	23'10"	24'9"	25'8"	26'6"	28'1"	29'7"
		E	1248	1462	1686	1921	2166	2421	2684	2958	3529	4133
	16	L	17'3"	18'2"	19'1"	19'11"	20'9"	21'6"	22'3"	23'0"	24'4"	25'8"
		E	1084	1270	1465	1669	1883	2103	2333	2570	3067	3592
	24	L	14'0"	14'9"	15'6"	16'2"	16'10"	17'6"	18'1"	18'7"	19'9"	20'10"
		E	873	1022	1179	1344	1514	1693	1877	2068	2467	2900
3 x 14	12	L	23'4"	24'7"	25'10"	27'0"	28'1"	29'2"	30'2"	31'2"	33'1"	34'11"
		E	1248	1462	1686	1921	2166	2421	2684	2958	3529	4133
	16	L	20'3"	21'4"	22'5"	23'5"	24'5"	25'4"	26'2"	27'0"	28'8"	30'3"
		E	1084	1270	1465	1669	1883	2103	2333	2570	3067	3592
	24	L	16'6"	17'4"	18'7"	19'0"	19'9"	20'6"	21'3"	22'0"	23'4"	24'7"
		E	873	1022	1179	1344	1514	1693	1877	2068	2467	2900
4 x 8	12	L	15'2"	16'0"	16'10"	17'7"	18'3"	18'11"	19'7"	20'3"	21'6"	22'7"
		E	1490	1745	2015	2295	2588	2891	3207	3533	4217	4939
	16	L	13'2"	13'11"	14'7"	15'3"	15'11"	16'6"	17'1"	17'7"	18'7"	19'7"
		E	1300	1533	1757	2002	2257	2522	2799	3082	3676	4306
	24	L	10'9"	11'4"	11'11"	12'5"	12'11"	13'5"	13'11"	14'4"	15'2"	16'0"
		E	1054	1234	1425	1625	1831	2046	2268	2500	2922	3492
4 x 10	12	L	19'5"	20'5"	21'5"	22'5"	22'4"	24'2"	25'0"	25'10"	27'5"	28'9"
		E	1490	1745	2015	2295	2588	2891	3207	3533	4217	4939
	16	L	16'10"	17'9"	18'7"	19'5"	20'3"	21'0"	21'9"	22'5"	23'10"	25'1"
		E	1300	1533	1757	2002	2257	2522	2799	3082	3676	4306
	24	L	13'8"	14'5"	15'2"	15'10"	16'6"	17'1"	17'8"	18'3"	19'3"	20'5"
		E	1054	1234	1425	1625	1831	2046	2268	2500	2922	3492
4 x 12	12	L	23'7"	24'10"	26'1"	27'3"	28'4"	29'5"	30'5"	31'5"	33'4"	35'2"
		E	1490	1745	2015	2295	2588	2891	3207	2533	4217	4939
	16	L	20'6"	21'7"	22'7"	23'7"	24'7"	25'6"	26'5"	27'4"	28'5"	30'6"
		E	1300	1533	1757	2002	2257	2522	2799	3082	3676	4306
	24	L	16'8"	17'7"	18'5"	19'3"	20'1"	20'10"	21'6"	22'2"	23'6"	24'10"
		E	1054	1234	1425	1625	1831	2046	2268	2500	2922	3492

(Courtesy of National Forest Products Association.)

TABLE 3.5 Maximum Spans for Floor Joists (Continued)

Symbols used in the tables are as follows:

F_b = Unit design stress at extreme fiber in bending, psi.
L = Clear distance between supports, feet.
E = Minimum required modulus of elasticity, 1000 psi, if deflection under the live load is limited to ℓ/360.

Joist sizes are expressed as nominal but calculations are based on dressed sizes

Size of Joist	Spacing C. to C.		F_b									
			900	1000	1100	1200	1300	1400	1500	1600	1800	2000
2 x 12	12	L	13'1"	13'10"	14'6"	15'2"	15'9"	16'4"	16'11"	17'6"	18'6"	19'6"
		E	855	1002	1156	1317	1384	1660	1841	2028	2419	2833
	16	L	11'4"	12'0"	12'7"	13'2"	13'8"	14'2"	14'8"	15'2"	16'1"	17'0"
		E	743	871	1005	1145	1291	1443	1600	1762	2103	2463
	24	L	9'3"	9'9"	10'3"	10'8"	11'1"	11'6"	11'11"	12'4"	13'1"	13'11"
		E	599	701	808	920	1039	1161	1287	1417	1692	1981
2 x 14	12	L	15'5"	16'3"	17'1"	17'10"	18'7"	19'3"	19'11"	20'7"	21'10"	23'0"
		E	855	1002	1156	1317	1384	1660	1841	2028	2419	2833
	16	L	13'4"	14'1"	14'10"	15'6"	16'1"	16'8"	17'3"	17'10"	18'11"	20'0"
		E	743	871	1005	1145	1291	1443	1600	1762	2103	2463
	24	L	10'11"	11'6"	12'1"	12'7"	13'1"	13'7"	14'1"	14'6"	15'4"	16'2"
		E	599	701	808	920	1039	1161	1287	1417	1692	1981
3 x 8	12	L	10'11"	11'6"	12'1"	12'7"	13'1"	13'7"	14'1"	14'7"	15'5"	16'3"
		E	1104	1294	1492	1700	1917	2142	2375	2618	3123	3658
	16	L	9'6"	10'0"	10'6"	10'11"	11'4"	11'9"	12'2"	12'7"	13'4"	14'1"
		E	960	1124	1297	1477	1666	1861	2065	2274	2714	3179
	24	L	7'7"	8'1"	8'6"	8'11"	9'3"	9'7"	9'11"	10'3"	10'10"	11'5"
		E	772	904	1043	1189	1340	1498	1661	1830	2183	2557
3 x 10	12	L	13'10"	14'7"	15'4"	16'1"	16'9"	17'5"	18'0"	18'7"	19'8"	20'9"
		E	1104	1294	1492	1700	1917	2142	2375	2618	3123	3658
	16	L	12'1"	12'9"	13'4"	13'11"	14'6"	15'1"	15'7"	16'1"	17'1"	18'0"
		E	960	1124	1297	1477	1666	1861	2065	2274	2714	3179
	24	L	9'10"	10'4"	10'10"	11'4"	11'10"	12'3"	12'8"	13'1"	13'10"	14'7"
		E	772	904	1043	1189	1340	1498	1661	1830	2183	2557
3 x 12	12	L	16'11"	17'10"	18'9"	19'7"	20'4"	21'1"	21'10"	22'6"	23'11"	25'3"
		E	1104	1294	1492	1700	1917	2142	2375	2618	3123	3658
	16	L	14'8"	15'6"	16'3"	16'11"	17'7"	18'3"	18'11"	19'6"	20 9"	21'11"
		E	960	1124	1297	1477	1666	1861	2065	2274	2714	3179
	24	L	11'11"	12'7"	13'2"	13'9"	14'4"	14'10"	15'4"	15'10"	16'10"	17'9"
		E	772	904	1043	1189	1340	1498	1661	1830	2183	2557
3 x 14	12	L	19'11"	21'0"	22'0"	23'0"	23'11"	24'10"	25'9"	26'7"	28'2"	29'9"
		E	1104	1294	1492	1700	1917	2142	2375	2618	3123	3658
	16	L	17'3"	18'2"	19'1"	19'11"	20'9"	21'7"	22'4"	23'1"	24'5"	25'9"
		E	960	1124	1297	1477	1666	1861	2065	2274	2714	3179
	24	L	14'1"	14'10"	15'7"	16'3"	16'11"	17'7"	18'2"	18'9"	19'10"	20'11"
		E	772	904	1043	1189	1340	1498	1661	1830	2183	2557
4 x 8	12	L	13'0"	13'8"	14'4"	15'0"	15'7"	16'2"	16'9"	17'3"	18'4"	19'4"
		E	1320	1546	1785	2033	2292	2561	2841	3129	3735	4375
	16	L	11'3"	11'10"	12'5"	13'0"	13'6"	14'0"	14'6"	15'0"	15'11"	16'9"
		E	1151	1349	1556	1773	1999	2234	2479	2730	3256	3816
	24	L	9'1"	9'7"	10'1"	10'7"	11'0"	11'5"	11'10"	12'3"	12'11"	13'7"
		E	934	1093	1262	1439	1622	1812	2009	2214	2588	3093
4 x 10	12	L	16'6"	17'5"	18'3"	19'1"	19'10"	20'7"	21'4"	22'1"	23'4"	24'7"
		E	1320	1546	1785	2033	2296	2561	2841	3129	3735	4375
	16	L	14'4"	15'1"	15'10"	16'7"	17'3"	17'11"	18'7"	19'2"	20'5"	21'5"
		E	1151	1349	1556	1773	1999	2234	2479	2730	3256	3814
	24	L	11'8"	12'4"	12'11"	13'6"	14'1"	14'7"	15'1"	15'7"	16'5"	17'5"
		E	934	1093	1262	1439	1622	1812	2009	2214	2588	3093
4 x 12	12	L	20'1"	21'2"	22'3"	23'3"	24'2"	25'0"	25'10"	26'8"	28'5"	30'0"
		E	1320	1546	1785	2033	2292	2561	2841	3121	3735	4375
	16	L	17'5"	18'5"	19'4"	20'2"	21'0"	21'9"	22'6"	23'3"	24'8"	26'0"
		E	1151	1349	1556	1773	1999	2234	2479	2730	3256	3814
	24	L	14'3"	15'0"	15'9"	16'5"	17'1"	17'9"	18'4"	18'11"	20'0"	21'2"
		E	934	1093	1262	1439	1622	1812	2009	2214	2588	3093
4 x 14	12	L	24'2"	25'5"	26'8"	27'10"	29'0"	30'1"	31'2"	32'2"	34'2"	36'0"
		E	1320	1546	1785	2033	2292	2561	2841	3129	3735	4375
	16	L	21'0"	22'1"	23'2"	24'2"	25'2"	26'1"	27'0"	24'11"	29'7"	31'3"
		E	1151	1349	1556	1773	1999	2234	2479	2730	3256	3814
	24	L	17'1"	18'0"	18'10"	19'8"	20'6"	21'3"	22'0"	22'9"	24'2"	25'5"
		E	934	1093	1262	1439	1622	1812	2009	2214	2588	3093

(Courtesy of National Forest Products Association.)

stresses. The tabulated spans are based on the bending strength using the live load indicated in each heading, plus a dead load of 10 lb/ft^2.

Planks

For scaffolding, wood planks 2 in. thick or thicker are laid on the flat and span from beam to beam (much as a floor would be laid). The beams are spaced 5 to 12 ft on center, depending on the load to be supported and the kind of wood used.

To compute the correct thickness of a plank, assume a width of 12 in. and apply beam design methods. Plank thicknesses together with other wood section properties are given in Table 3.6. The table gives the allowable spans for various live loads for plank thicknesses of 2½, 3, 4, and 5 in. Scaffold planks should be unsurfaced, and actual dry dimensions should not be less than specified:

2 × 9	1⅞ × 8¾
2 × 10	1⅞ × 9
2 × 12	1⅞ × 11¾

Planks should have strength ratios of at least 80% of the strength of a theoretically flawless plank.

When finally accepted, the plank should be immediately branded on its ends, or permanently marked suited for scaffolding. The use for scaffolding of other planks than those specifically accepted for scaffolding should be forbidden. Inspected planks should be carefully handled and not dropped in bulk from the delivery truck.

Timber Columns

Most timber columns, or posts, have square or rectangular cross sections and are single lengths of timber. Although columns built up from pieces of timber and joined by nails, bolts, or other mechanical fasteners are often used, they do not have the same strength as a one-piece member of comparable material and dimensions. The strength of built-up columns must be reduced by the percentages given in Table 3.7 to provide a section equal to the strength of a one-piece column of the same dimension and quality.

The slenderness ratio of a column is a measure of its stiffness and has an important bearing on the load a column will support. It is computed by dividing the column's laterally unsupported length by the appropriate cross-sectional dimension, both in inches (see Fig. 3.5).

To determine the laterally unsupported length of a column, measure the column's distance parallel to its longitudinal axis, between the supports that restrain the column against any lateral movement. A short column, one having a slenderness ratio less than 11, will support a load equal to the area of the cross section multiplied by the full allowable compressive strength of the particular type and grade of wood. However, if the length is increased, then the cross-sectional area must also be increased.

There are numerous timber column formulas used in finding the permissible unit stress. The one recommended by the National Forest Products Association is

$$F = C\left(1 - \frac{l}{80d}\right)$$

where F = maximum permissible unit stress for column cross section, lb/in.^{2}
C = maximum allowable unit stress parallel to grain for short blocks, lb/in.^{2}
l = unsupported length of column, inches
d = least width or diameter of cross section, inches

Timber columns are designed by trial and error, first assuming a cross section and then testing to see whether or not it meets the specified requirements. For example:

TABLE 3.6 Plank and Laminated Floors and Roofs
Safe Uniform Loads—Type III

Use of Tabular Data

The tabular data are presented in simplified form for those spans and loads most frequently encountered. The design load per square foot, w, is calculated on the basis of $M = wL^2/8$ which, as previously indicated, is appropriate for any of the four types of plank or lamination arrangement. For values of F_b intermediate to those listed, the design load per square foot, w, may be determined by direct interpolation.

The minimum required modulus of elasticity design value, E, if deflection under load w is limited to $\ell/240$, is calculated for Types III and IV arrangements of planks. For other arrangements of planks or laminations, the required E value for $\ell/240$ may be determined by direct interpolation using the appropriate constants listed for Types I or II arrangements. For a deflection limit of $\ell/180$, the tabulated E value may be multiplied by 0.75.

The design load per square foot, w, includes the weight of the deck. This weight should be subtracted from the tabulated load w to determine the load which may be applied to the floor or roof.

(*Courtesy of National Forest Products Association.*)

TABLE 3.6 Plank and Laminated Floors and Roofs
Safe Uniform Loads—Type III (Continued)

Symbols used in the table are as follows:

F_b = Unit design stress at extreme fiber in bending, psi
w = Load per square foot of deck, pounds
E = Minimum required modulus of elasticity, 1000 psi, if deflection under load w is limited to $\ell/240$, where $E = 314.2\ wL^3/I$ when $\Delta = \ell/240$

Deck thicknesses are expressed as nominal but calculations are based on dressed sizes.

Span	Thickness		F_b									
			900	1000	1100	1200	1300	1400	1500	1600	1800	2000
6'- 0"	2"	w	75	83	92	100	108	117	125	133	150	167
		E	1506	1674	1848	2009	2170	2350	2511	2672	3013	3355
	3"	w	208	231	256	278	300	325	347	369	417	464
		E	904	1003	1112	1208	1303	1412	1507	1603	1811	2016
7'- 0"	2"	w	55	61	68	73	79	86	92	98	110	123
		E	1755	1946	2169	2329	2520	2743	2935	3126	3509	3924
	3"	w	153	170	188	204	220	239	255	271	306	341
		E	1055	1173	1297	1407	1518	1649	1759	1869	2111	2352
8'- 0"	2"	w	42	47	52	56	61	66	70	75	84	94
		E	2000	2238	2476	2667	2905	3143	3333	3572	4000	4476
	3"	w	117	130	144	156	169	183	195	207	234	261
		E	1205	1339	1483	1606	1740	1884	2008	2131	2409	2687
	4"	w	230	254	282	306	330	358	383	407	459	511
		E	863	953	1058	1148	1238	1343	1437	1527	1722	1917
9'- 0"	3"	w	92	103	114	123	133	144	154	164	185	206
		E	1349	1510	1671	1803	1950	2111	2258	2404	2712	3020
	4"	w	182	201	223	242	261	283	303	322	363	404
		E	972	1074	1191	1293	1394	1512	1619	1720	1939	2158
10'- 0"	3"	w	75	83	92	100	108	117	125	133	150	167
		E	1508	1669	1850	2011	2172	2353	2514	2675	3017	3358
	4"	w	147	163	181	196	211	229	245	261	294	327
		E	1077	1195	1326	1436	1546	1678	1795	1913	2154	2396
11'- 0"	3"	w	62	69	76	83	90	97	103	110	124	138
		E	1660	1847	2034	2222	2382	2596	2757	2944	3319	3694
	4"	w	122	135	149	163	175	189	203	215	243	270
		E	1190	1307	1453	1580	1707	1844	1980	2097	2370	2634
12'- 0"	3"	w	52	58	64	70	75	81	87	92	104	116
		E	1807	2016	2224	2433	2606	2815	3023	3197	3614	4031
	4"	w	102	113	125	136	147	159	170	181	204	227
		E	1292	1431	1583	1722	1862	2013	2153	2292	2583	2875
	6"	w	252	279	309	336	363	393	420	447	504	561
		E	822	910	1008	1096	1185	1282	1371	1459	1645	1831
13'- 0"	3"	w	44	50	55	59	64	69	74	79	89	99
		E	1944	2165	2430	2607	2828	3048	3269	3490	3932	4374
	4"	w	87	96	107	116	125	136	145	154	174	194
		E	1401	1546	1723	1868	2013	2190	2335	2479	2801	3123
	6"	w	215	238	263	286	309	335	358	381	429	478
		E	892	987	1091	1186	1282	1390	1485	1580	1730	1983
14'- 0"	4"	w	75	83	92	100	108	117	126	133	150	167
		E	1508	1669	1850	2011	2172	2353	2534	2674	3016	3358
	6"	w	185	205	227	247	266	289	308	328	370	412
		E	959	1062	1176	1280	1378	1497	1596	1699	1917	2135
15'- 0"	4"	w	65	72	80	87	94	102	109	116	130	145
		E	1608	1781	1979	2152	2325	2523	2696	2869	3215	3586
	6"	w	161	179	198	215	232	252	269	286	323	359
		E	1026	1141	1262	1370	1478	1605	1714	1822	2058	2288
16'- 0"	4"	w	58	64	71	77	83	90	96	102	115	128
		E	1741	1921	2131	2311	2491	2701	2887	3062	3452	3842
	6"	w	142	157	174	189	204	221	236	251	283	315
		E	1098	1214	1346	1462	1578	1709	1825	1941	2188	2436
	8"	w	246	273	302	329	355	384	411	437	493	549
		E	830	921	1019	1110	1198	1296	1387	1474	1663	1852
17'- 0"	6"	w	126	139	154	167	181	196	209	223	251	279
		E	1169	1289	1429	1549	1679	1818	1939	2069	2328	2588
	8"	w	218	242	267	291	314	340	364	387	436	486
		E	882	980	1081	1178	1271	1376	1474	1567	1765	1967
18'- 0"	6"	w	112	124	137	149	161	175	187	199	224	249
		E	1233	1366	1508	1641	1773	1927	2059	2191	2467	2741
	8"	w	194	216	239	260	281	303	325	345	390	434
		E	932	1038	1149	1250	1351	1456	1562	1658	1874	2086

(Courtesy of National Forest Products Association.)

TABLE 3.6 Plank and Laminated Floors and Roofs

Safe Uniform Loads—Type III (Continued)

Symbols used in the table are as follows:

F_b = Unit design stress at extreme fiber in bending, psi

w = Load per square foot of deck, pounds

E = Minimum required modulus of elasticity, 1000 psi, if deflection under load w is limited to $\ell/240$, where $E = 345.6\ wL^3/I$ when $\Delta = \ell/240$

Deck thicknesses are expressed as nominal but calculations are based on dressed sizes.

Span	Thickness		F_b									
			900	1000	1100	1200	1300	1400	1500	1600	1800	2000
6'- 0"	2"	w	75	83	92	100	108	117	125	133	150	167
		E	1657	1841	2033	2210	2387	2585	2762	2939	3314	3691
	3"	w	208	231	256	278	300	325	347	369	417	464
		E	994	1103	1223	1329	1433	1553	1658	1763	1992	2218
7'- 0"	2"	w	55	61	68	73	79	86	92	98	110	123
		E	1931	2141	2386	2562	2772	3017	3229	3339	3860	4316
	3"	w	153	170	188	204	220	239	255	271	306	341
		E	1161	1290	1427	1548	1670	1814	2056	2056	2322	2587
8'- 0"	2"	w	42	47	52	56	61	66	70	75	84	94
		E	2200	2462	2724	2934	3196	3457	3666	3929	4400	4924
	3"	w	117	130	144	156	169	183	195	207	234	261
		E	1326	1473	1631	1767	1914	2072	2209	2344	2650	2956
	4"	w	230	254	282	306	330	358	383	407	459	511
		E	949	1048	1164	1263	1362	1477	1581	1680	1894	2109
9'- 0"	3"	w	92	103	114	123	133	144	154	164	185	206
		E	1484	1661	1838	1983	2145	2323	2484	2644	2983	3322
	4"	w	182	201	223	242	261	283	303	322	363	404
		E	1069	1181	1310	1422	1533	1663	1781	189[illegible]	2133	2374
10'- 0"	3"	w	75	83	92	100	108	117	125	133	150	16[illegible]
		E	1659	1836	2035	2212	2389	2588	2765	2943	3319	3694
	4"	w	147	163	181	196	211	229	245	261	294	327
		E	1185	1315	1459	1580	1701	1846	1975	2104	2369	2636
11'- 0"	3"	w	62	69	76	83	90	97	103	110	124	138
		E	1826	2032	2237	2444	2620	2856	3033	3238	3651	4063
	4"	w	122	134	149	162	175	189	203	215	243	270
		E	1309	1438	1598	1738	1878	2028	2178	2307	2607	2897
12'- 0"	3"	w	52	58	64	70	75	81	87	92	104	116
		E	1988	2218	2446	2676	2867	3095	3225	3517	3975	4434
	4"	w	102	113	125	136	147	159	170	181	204	227
		E	1421	1574	1741	1894	2048	2214	2368	2521	2841	3163
	6"	w	252	279	309	336	363	393	420	447	504	561
		E	904	1001	1109	1206	1304	1410	1508	1605	1810	2014
13'- 0"	3"	w	44	49	55	59	64	69	74	79	89	99
		E	2138	2382	2673	2868	3111	3353	3596	3839	4325	4811
	4"	w	87	96	107	116	125	136	145	154	174	194
		E	1541	1701	1895	2055	[illegible]214	2409	2569	2727	3081	3435
	6"	w	215	238	263	286	309	335	358	381	429	478
		E	981	1086	1200	1305	1410	1529	1634	1738	1958	2181
14'- 0"	4"	w	75	83	92	100	108	117	126	133	150	167
		E	1659	1836	2035	2212	2389	2588	2787	2941	3318	3694
	6"	w	185	205	227	247	266	289	308	328	370	412
		E	1055	1168	1294	1408	1516	1647	1756	1869	2109	2349
15'- 0"	4"	w	65	73	80	87	95	102	109	116	131	145
		E	1780	1977	2175	2373	2571	2768	2966	3164	3559	3955
	6"	w	162	180	197	215	233	251	269	287	323	359
		E	1133	1258	1384	1510	1636	1762	1887	2013	2265	2517
16'- 0"	4"	w	58	64	71	77	83	90	96	102	115	128
		E	1915	2113	2344	2542	2740	2971	3[illegible]76	3369	3797	4226
	6"	w	142	157	174	189	204	221	236	251	283	315
		E	1208	1335	1481	1608	1736	1880	2008	2135	2407	2680
	8"	w	246	273	302	329	355	384	411	437	493	549
		E	913	1013	1121	1221	1318	1426	1526	1621	1829	2037
17'- 0"	6"	w	126	139	154	167	181	196	209	223	251	279
		E	1286	1418	1572	1704	1847	2000	2133	2276	2561	2847
	8"	w	218	242	267	291	314	340	364	387	436	486
		E	970	1078	1189	1296	1398	1514	1621	1724	1942	2164
18'- 0"	6"	w	112	124	137	149	161	175	187	199	224	249
		E	1356	1503	1659	1805	1950	2120	2265	2410	2714	3015
	8"	w	194	216	239	260	281	303	325	345	390	434
		E	1025	1142	1264	1375	1486	1602	1718	1824	2061	2295

(Courtesy of National Forest Products Association.)

TABLE 3.7 Reduction-Factors for Calculating the Strength of Built-up Wood Columns

Slenderness	Percent reduction
6	18
10	23
14	29
18	35
22	26
26	18

1. Assume a cross-sectional area of approximate proper dimensions.
2. Solve for F in the timber column formula, substituting for l the length of the column in inches and for d the least width or diameter in inches.
3. The F value thus computed is in units of pounds per square inch, and is multiplied by the number of square inches in the assumed column's cross-sectional area to obtain the allowable safe load for the column.
4. Compare the allowable load found in step 3 with the actual load to be supported.
5. If the actual load is greater, then the assumed cross section is too small. Assume a larger cross section and test again.

While timber columns may be designed by use of a formula, the rigger will find it more convenient to use precomputed tables, such as Table 3.8, to make the proper selection of column sizes.

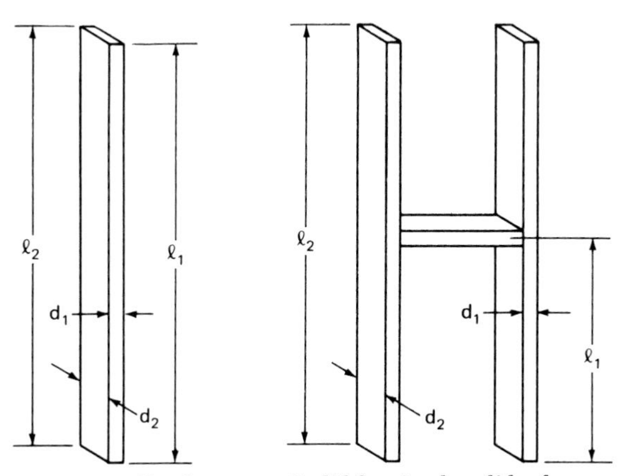

Figure 3.5 Slenderness ratio l/d for simple solid columns.

TABLE 3.8 Wood Columns

Use of Tabular Column Data

The tabular data included herein for unit axial stresses provides a simplified and accurate method for calculating design loads on columns of any size and length. The load is determined by multiplying the appropriate tabular unit stress by the cross-sectional area of the member, based on net dimensions. Where the degree of refinement so indicates, the weight of the column should be deducted to determine the design load which may be applied.

Unit axial stresses are provided for simple solid columns, spaced columns with end condition "a" and spaced columns with end condition "b".

Ratio of ℓ/d

The ℓ/d ratio is calculated in the manner previously described in the text on wood columns. Values of F_c' for ℓ/d ratios intermediate to those given may be determined by straight line interpolation. For example, for a simple, solid column having an F_c of 1200 psi and an E of 1,600,000 psi, the F_c' for an ℓ/d of 28 is 612 psi and the F_c' for an ℓ/d of 29 is 571 psi. For an ℓ/d of 28.4, the F_c' is 571 + 0.6 (612-571) = 595.6 psi.

Design Values of E and F_c

Modulus of elasticity, E, and compression parallel to grain, F_c, design values for the species and grade of wood to be used may be obtained from the National Design Specification for Wood Construction. If appropriate, E and F_c should be adjusted as previously described for the conditions under which the column will be used.

Tabular values of F_c' are provided for a range of E values from 2,100,000 to 900,000 psi, for F_c values between 300 and 2,800 psi as appropriate for each E. Values of F_c' for F_c values intermediate to those tabulated may be determined by straight line interpolation. For example, for an ℓ/d of 25 and E of 1,400,000 psi, the F_c' for an F_c' of 900 psi is 661 psi and the F_c' for an F_c' of 800 psi is 571 psi. For an F_c of 875 psi, the interpolated F_c' is $571 + \frac{75}{100}(661-571) = 638.5$ psi.

Use of Tabular Data for Round Columns

Unit axial loads for simple solid columns of square cross section may be converted to unit loads for round columns. First, multiply the column diameter by 0.886 to determine the dimension, d, and then calculate the ℓ/d ratio. From the tabular data obtain the applicable F_c' value for that ℓ/d ratio and multiply this by the cross-sectional area of the round column to determine the design load for the column.

Conversely, to determine the diameter of a round column required to carry the same total load as a square column, multiply the dimension d of the square column by 1.128.

(Courtesy of National Forest Products Association.)

TABLE 3.8 Wood Columns (Continued)

Obtain design values for E and F_c from the National Design Specification for Wood Construction.
Modify F_c for different load duration, if applicable.
Calculate ℓ/d where ℓ = unsupported length of column in inches and d = applicable least actual dimension of column cross section.
Determine value of F_c' from table.

Total design load on column = cross-sectional area in square inches times F_c' value.

E	F_c	ℓ/d																			
		11+	12	13	14	15	16	17	18	19	20	21	22	23	24	25	26	27	28	29	30
2,100,000	2800	2680	2630	2566	2485	2385	2263	2116	1940	1745	1575	1429	1302	1191	1094	1008	932	864	804	749	700
	2700	2592	2548	2490	2418	2328	2219	2086	1929	"	"	"	"	"	"	"	"	"	"	"	"
	2600	2504	2464	2413	2348	2268	2170	2052	1911	"	"	"	"	"	"	"	"	"	"	"	"
	2500	2415	2379	2333	2276	2205	2118	2013	1888	1740	"	"	"	"	"	"	"	"	"	"	"
	2400	2324	2293	2253	2202	2139	2062	1969	1858	1728	"	"	"	"	"	"	"	"	"	"	"
	2300	2234	2206	2170	2126	2070	2002	1921	1823	1708	1573	"	"	"	"	"	"	"	"	"	"
	2200	2142	2118	2086	2047	1999	1940	1868	1783	1682	1564	"	"	"	"	"	"	"	"	"	"
	2100	2049	2028	2001	1967	1925	1873	1811	1737	1650	1547	1428	"	"	"	"	"	"	"	"	"
	2000	1956	1938	1915	1885	1849	1804	1751	1687	1611	1522	1419	"	"	"	"	"	"	"	"	"
	1900	1863	1847	1827	1802	1770	1732	1686	1631	1566	1490	1402	1300	"	"	"	"	"	"	"	"
	1800	1768	1755	1738	1716	1690	1657	1618	1571	1516	1452	1377	1290	"	"	"	"	"	"	"	"
	1700	1673	1662	1648	1630	1607	1580	1547	1507	1461	1407	1343	1270	1187	"	"	"	"	"	"	"
	1600	1578	1568	1556	1541	1523	1500	1472	1440	1401	1355	1303	1242	1172	1093	"	"	"	"	"	"
	1500	1482	1474	1464	1452	1436	1417	1395	1368	1336	1298	1255	1205	1147	1082	"	"	"	"	"	"
	1400	1385	1379	1371	1361	1348	1333	1314	1292	1267	1236	1201	1160	1113	1060	1000	"	"	"	"	"
2,000,000	2800	2668	2613	2542	2453	2343	2208	2045	1852	1662	1500	1361	1240	1134	1042	960	888	823	765	713	667
	2700	2581	2532	2469	2389	2290	2169	2023	1850	"	"	"	"	"	"	"	"	"	"	"	"
	2600	2494	2450	2393	2322	2234	2126	1996	1841	"	"	"	"	"	"	"	"	"	"	"	"
	2500	2406	2367	2316	2253	2174	2079	1963	1825	"	"	"	"	"	"	"	"	"	"	"	"
	2400	2317	2282	2238	2181	2112	2027	1925	1803	1659	"	"	"	"	"	"	"	"	"	"	"
	2300	2227	2196	2157	2108	2047	1972	1882	1774	1647	"	"	"	"	"	"	"	"	"	"	"
	2200	2136	2109	2075	2032	1978	1913	1834	1740	1629	1499	"	"	"	"	"	"	"	"	"	"
	2100	2044	2021	1991	1954	1907	1850	1782	1700	1603	1490	"	"	"	"	"	"	"	"	"	"
	2000	1952	1932	1906	1874	1833	1784	1725	1654	1571	1473	1360	"	"	"	"	"	"	"	"	"
	1900	1859	1841	1819	1792	1757	1715	1664	1604	1532	1448	1351	"	"	"	"	"	"	"	"	"
	1800	1765	1750	1731	1708	1679	1643	1600	1548	1487	1416	1333	1238	"	"	"	"	"	"	"	"
	1700	1670	1658	1642	1622	1598	1567	1531	1488	1437	1377	1307	1226	"	"	"	"	"	"	"	"
	1600	1575	1565	1552	1535	1515	1490	1459	1423	1380	1330	1272	1205	1128	"	"	"	"	"	"	"
	1500	1480	1471	1460	1447	1430	1409	1384	1354	1319	1278	1230	1175	1111	1039	"	"	"	"	"	"
	1400	1383	1377	1368	1357	1343	1326	1306	1281	1253	1219	1180	1135	1084	1025	959	"	"	"	"	"
1,900,000	2800	2653	2592	2514	2415	2293	2144	1964	1759	1579	1425	1293	1178	1078	990	912	843	782	727	678	633
	2700	2569	2514	2444	2355	2246	2112	1950	"	"	"	"	"	"	"	"	"	"	"	"	"
	2600	2483	2434	2371	2292	2194	2075	1931	"	"	"	"	"	"	"	"	"	"	"	"	"
	2500	2396	2352	2297	2226	2139	2033	1905	1752	"	"	"	"	"	"	"	"	"	"	"	"
	2400	2308	2269	2220	2158	2081	1987	1874	1738	"	"	"	"	"	"	"	"	"	"	"	"
	2300	2219	2185	2142	2087	2019	1936	1837	1718	1577	"	"	"	"	"	"	"	"	"	"	"
	2200	2129	2099	2061	2013	1954	1882	1794	1690	1567	"	"	"	"	"	"	"	"	"	"	"
	2100	2038	2012	1979	1938	1886	1823	1747	1657	1550	1424	"	"	"	"	"	"	"	"	"	"
	2000	1947	1924	1896	1860	1815	1761	1695	1617	1525	1416	"	"	"	"	"	"	"	"	"	"
	1900	1854	1835	1811	1780	1742	1695	1639	1572	1492	1400	1292	"	"	"	"	"	"	"	"	"
	1800	1761	1745	1724	1698	1665	1626	1578	1521	1453	1375	1283	"	"	"	"	"	"	"	"	"
	1700	1667	1654	1636	1614	1587	1553	1513	1465	1408	1342	1264	1175	"	"	"	"	"	"	"	"
	1600	1573	1561	1547	1528	1505	1478	1444	1404	1357	1301	1237	1162	1077	"	"	"	"	"	"	"
	1500	1477	1468	1456	1441	1422	1399	1371	1338	1299	1254	1201	1139	1069	"	"	"	"	"	"	"
	1400	1382	1374	1364	1352	1337	1318	1295	1269	1237	1200	1157	1107	1050	985	"	"	"	"	"	"
1,800,000	2800	2637	2569	2481	2372	2235	2069	1869	1667	1496	1350	1224	1116	1021	938	864	799	741	689	642	600
	2700	2554	2493	2414	2316	2194	2045	1865	"	"	"	"	"	"	"	"	"	"	"	"	"
	2600	2469	2415	2345	2257	2148	2015	1854	"	"	"	"	"	"	"	"	"	"	"	"	"
	2500	2384	2335	2273	2195	2098	1980	1837	"	"	"	"	"	"	"	"	"	"	"	"	"
	2400	2297	2254	2199	2130	2044	1940	1813	1663	"	"	"	"	"	"	"	"	"	"	"	"
	2300	220	2172	2123	2063	1987	1895	1784	1651	"	"	"	"	"	"	"	"	"	"	"	"
	2200	2121	2088	2045	1992	1926	1845	1748	1632	1495	"	"	"	"	"	"	"	"	"	"	"
	2100	2031	2002	1966	1919	1862	1792	1707	1606	1487	"	"	"	"	"	"	"	"	"	"	"
	2000	1940	1916	1884	1844	1794	1734	1661	1573	1470	"	"	"	"	"	"	"	"	"	"	"
	1900	1849	1828	1800	1766	1724	1672	1609	1534	1446	1342	"	"	"	"	"	"	"	"	"	"
	1800	1757	1739	1715	1686	1650	1606	1553	1489	1414	1326	"	"	"	"	"	"	"	"	"	"
	1700	1663	1648	1629	1604	1574	1536	1492	1438	1375	1301	1215	"	"	"	"	"	"	"	"	"
	1600	1570	1557	1541	1520	1495	1464	1426	1382	1329	1267	1195	1113	"	"	"	"	"	"	"	"
	1500	1475	1464	1451	1434	1413	1388	1357	1320	1277	1226	1167	1098	1020	"	"	"	"	"	"	"
	1400	1380	1371	1360	1346	1329	1309	1284	1254	1218	1177	1129	1073	1010	937	"	"	"	"	"	"
	1300	1284	1277	1268	1257	1243	1227	1207	1181	1155	1121	1083	1039	988	930	"	"	"	"	"	"
	1200	1187	1182	1175	1166	1156	1142	1127	1108	1086	1060	1029	994	954	909	857	"	"	"	"	"
	1100	1090	1086	1081	1074	1066	1056	1044	1029	1012	992	968	942	911	876	836	791	"	"	"	"
	1000	993	989	985	980	974	967	958	947	934	919	901	881	858	831	802	768	730	688	"	"

(Courtesy of National Forest Products Association.)

TABLE 3.8 Wood Columns (Continued)

Obtain design values for E and F_c from the National Design Specification for Wood Construction.
Modify F_c for different load duration, if applicable.
Calculate ℓ/d where ℓ = unsupported length of column in inches and d = applicable least actual dimension of column cross section.
Determine value of F_c' from table.

Total design load on column = cross-sectional area in square inches times F_c' value.

E	F_c	ℓ/d																			
		11+	12	13	14	15	16	17	18	19	20	21	22	23	24	25	26	27	28	29	30
1,700,000	2400	2285	2237	2175	2098	2001	1884	1742	1574	1413	1275	1156	1054	964	885	816	754	700	651	606	567
	2300	2199	2156	2102	2034	1949	1846	1721	1573	"	"	"	"	"	"	"	"	"	"	"	"
	2200	2111	2074	2027	1967	1893	1803	1693	1563	"	"	"	"	"	"	"	"	"	"	"	"
	2100	2023	1991	1949	1897	1833	1754	1659	1546	"	"	"	"	"	"	"	"	"	"	"	"
	2000	1933	1906	1870	1825	1769	1701	1619	1522	1406	"	"	"	"	"	"	"	"	"	"	"
	1900	1843	1819	1788	1750	1702	1644	1574	1490	1391	"	"	"	"	"	"	"	"	"	"	"
	1800	1751	1731	1705	1672	1632	1582	1523	1451	1367	1269	"	"	"	"	"	"	"	"	"	"
	1700	1659	1642	1620	1592	1558	1517	1466	1406	1335	1252	"	"	"	"	"	"	"	"	"	"
	1600	1566	1552	1533	1510	1482	1447	1405	1355	1296	1227	1146	"	"	"	"	"	"	"	"	"
	1500	1472	1460	1445	1426	1403	1374	1339	1298	1249	1192	1126	1050	"	"	"	"	"	"	"	"
	1400	1377	1368	1355	1340	1321	1298	1269	1236	1196	1150	1096	1034	963	"	"	"	"	"	"	"
	1300	1282	1274	1264	1252	1237	1218	1195	1169	1137	1100	1057	1007	950	"	"	"	"	"	"	"
	1200	1186	1180	1172	1162	1150	1135	1118	1097	1072	1043	1009	969	925	873	"	"	"	"	"	"
	1100	1089	1084	1078	1071	1062	1050	1037	1020	1001	979	953	922	888	848	804	"	"	"	"	"
	1000	992	988	984	978	971	963	952	940	926	909	889	867	841	811	778	740	697	"	"	"
	900	894	891	888	884	879	873	865	856	846	834	819	803	784	762	738	710	679	645	"	"
	800	796	794	792	789	785	781	776	769	762	753	743	732	718	703	686	667	645	621	594	564
1,600,000	2000	1925	1893	1853	1802	1740	1663	1570	1460	1330	1200	1088	992	907	833	768	710	658	612	571	533
	1900	1835	1809	1774	1731	1667	1611	1532	1437	1325	"	"	"	"	"	"	"	"	"	"	"
	1800	1745	1722	1693	1656	1610	1554	1487	1406	1311	"	"	"	"	"	"	"	"	"	"	"
	1700	1654	1634	1610	1579	1540	1493	1436	1368	1288	1195	"	"	"	"	"	"	"	"	"	"
	1600	1561	1545	1525	1499	1467	1427	1380	1324	1257	1179	"	"	"	"	"	"	"	"	"	"
	1500	1468	1455	1438	1417	1390	1358	1319	1272	1217	1153	1078	"	"	"	"	"	"	"	"	"
	1400	1374	1363	1350	1332	1311	1284	1253	1215	1170	1118	1057	987	"	"	"	"	"	"	"	"
	1300	1279	1271	1260	1246	1228	1207	1182	1152	1116	1074	1025	969	905	"	"	"	"	"	"	"
	1200	1184	1177	1168	1157	1144	1127	1107	1083	1055	1022	984	940	889	831	"	"	"	"	"	"
	1100	1087	1082	1076	1067	1057	1044	1029	1010	988	963	934	900	861	816	766	"	"	"	"	"
	1000	991	987	982	975	967	958	946	933	916	897	875	849	820	787	749	706	"	"	"	"
	900	893	890	887	882	876	869	861	851	839	825	809	790	769	744	717	686	651	"	"	"
	800	795	793	791	787	783	778	773	765	757	747	736	723	708	691	671	650	625	598	567	"
1,500,000	2000	1914	1879	1833	1775	1704	1616	1511	1386	1247	1125	1020	930	851	781	720	666	617	574	535	500
	1900	1827	1796	1757	1707	1646	1571	1481	1373	"	"	"	"	"	"	"	"	"	"	"	"
	1800	1738	1712	1678	1636	1584	1520	1444	1352	1244	"	"	"	"	"	"	"	"	"	"	"
	1700	1647	1625	1597	1562	1518	1464	1400	1323	1232	"	"	"	"	"	"	"	"	"	"	"
	1600	1556	1538	1514	1485	1448	1404	1350	1285	1209	1121	"	"	"	"	"	"	"	"	"	"
	1500	1464	1449	1429	1405	1375	1318	1294	1241	1178	1105	"	"	"	"	"	"	"	"	"	"
	1400	1371	1358	1343	1323	1298	1268	1232	1189	1138	1079	1010	"	"	"	"	"	"	"	"	"
	1300	1276	1267	1254	1238	1219	1195	1166	1131	1091	1043	987	923	"	"	"	"	"	"	"	"
	1200	1181	1174	1164	1151	1136	1117	1094	1067	1035	998	954	904	846	"	"	"	"	"	"	"
	1100	1086	1080	1072	1063	1051	1036	1019	998	973	944	911	872	828	777	"	"	"	"	"	"
	1000	989	985	979	972	963	952	939	923	905	883	858	829	795	757	714	"	"	"	"	"
	900	892	889	885	880	873	865	855	844	830	815	796	775	751	723	692	656	"	"	"	"
	800	795	792	789	786	781	775	769	761	751	740	727	712	695	676	654	629	601	570	"	"
	700	696	695	693	690	687	684	679	674	667	660	651	641	630	617	602	585	567	546	523	497
	600	598	597	595	594	592	590	587	583	579	575	569	563	556	548	538	528	516	503	488	472
1,400,000	2800	2530	2418	2273	2092	1867	1641	1453	1296	1163	1050	952	868	794	729	672	621	576	536	499	467
	2700	2458	2357	2228	2065	1863	1641	"	"	"	"	"	"	"	"	"	"	"	"	"	"
	2600	2384	2294	2178	2033	1853	1641	"	"	"	"	"	"	"	"	"	"	"	"	"	"
	2500	2308	2228	2125	1996	1836	1641	"	"	"	"	"	"	"	"	"	"	"	"	"	"
	2400	2230	2159	2068	1954	1812	1639	"	"	"	"	"	"	"	"	"	"	"	"	"	"
	2300	2150	2088	2008	1907	1783	1630	"	"	"	"	"	"	"	"	"	"	"	"	"	"
	2200	2069	2015	1945	1856	1747	1614	"	"	"	"	"	"	"	"	"	"	"	"	"	"
	2100	1986	1939	1878	1801	1706	1590	1450	"	"	"	"	"	"	"	"	"	"	"	"	"
	2000	1902	1861	1808	1742	1660	1560	1439	"	"	"	"	"	"	"	"	"	"	"	"	"
	1900	1816	1781	1735	1679	1608	1522	1419	1295	"	"	"	"	"	"	"	"	"	"	"	"
	1800	1728	1698	1660	1612	1552	1479	1391	1286	"	"	"	"	"	"	"	"	"	"	"	"
	1700	1640	1614	1582	1541	1491	1430	1355	1267	1162	"	"	"	"	"	"	"	"	"	"	"
	1600	1550	1529	1502	1468	1426	1375	1313	1239	1152	"	"	"	"	"	"	"	"	"	"	"
	1500	1459	1441	1419	1391	1357	1314	1263	1202	1131	1046	"	"	"	"	"	"	"	"	"	"
	1400	1366	1352	1334	1311	1283	1249	1208	1158	1100	1031	"	"	"	"	"	"	"	"	"	"
	1300	1273	1262	1247	1229	1207	1179	1146	1106	1060	1005	941	"	"	"	"	"	"	"	"	"
	1200	1179	1170	1159	1144	1127	1105	1079	1048	1011	968	918	860	"	"	"	"	"	"	"	"
	1100	1084	1077	1068	1057	1043	1027	1007	983	954	921	883	838	787	"	"	"	"	"	"	"
	1000	988	983	976	968	957	945	930	912	891	866	837	803	765	721	"	"	"	"	"	"
	900	891	887	883	876	869	860	849	836	820	802	781	757	729	697	661	620	"	"	"	"
	800	794	791	788	783	778	772	764	755	744	731	716	699	680	657	632	604	571	"	"	"
	700	696	694	692	689	685	681	676	670	662	654	644	633	619	604	587	568	547	523	496	"
	600	597	596	595	593	591	588	585	581	576	571	565	558	549	540	529	517	504	488	472	453
	500	498	498	497	496	495	493	491	489	486	483	480	475	471	465	459	452	444	435	426	415

(Courtesy of National Forest Products Association.)

TABLE 3.8 Wood Columns (Continued)

Obtain design values for E and F_c from the National Design Specification for Wood Construction.
Modify F_c for different load duration, if applicable.
Calculate ℓ/d where ℓ = unsupported length of column in inches and d = applicable least actual dimension of column cross section.
Determine value of F_c' from table.

Total design load on column = cross-sectional area in square inches times F_c' value.

E	F_c	ℓ/d																			
		11-	12	13	14	15	16	17	18	19	20	21	22	23	24	25	26	27	28	29	30
1,300,000	2300	2126	2054	1962	1845	1700	1523	1349	1204	1080	975	884	806	737	677	624	577	535	497	464	433
	2200	2048	1985	1904	1802	1675	1520	..	..	..	..	..	..	..	..	..	..	..	..	..	..
	2100	1968	1913	1842	1753	1643	1509	..	..	..	..	..	..	..	..	..	..	..	..	..	..
	2000	1886	1838	1777	1701	1606	1489	..	..	..	..	..	..	..	..	..	..	..	..	..	..
	1900	1802	1761	1709	1643	1562	1462	1342	..	..	..	..	..	..	..	..	..	..	..	..	..
	1800	1717	1682	1638	1582	1512	1428	1326	..	..	..	..	..	..	..	..	..	..	..	..	..
	1700	1633	1601	1563	1516	1458	1386	1300	1198	..	..	..	..	..	..	..	..	..	..	..	..
	1600	1542	1517	1486	1447	1398	1339	1267	1181	..	..	..	..	..	..	..	..	..	..	..	..
	1500	1452	1432	1406	1374	1334	1285	1225	1155	1072	..	..	..	..	..	..	..	..	..	..	..
	1400	1361	1345	1324	1297	1265	1225	1177	1119	1052	972	..	..	..	..	..	..	..	..	..	..
	1300	1269	1256	1239	1218	1192	1160	1121	1075	1021	958	..	..	..	..	..	..	..	..	..	..
	1200	1175	1165	1152	1135	1115	1090	1059	1023	981	931	873	..	..	..	..	..	..	..	..	..
	1100	1081	1073	1063	1050	1034	1015	992	964	931	893	848	796	..	..	..	..	..	..	..	..
	1000	[illegible]	980	972	963	951	936	919	898	873	844	811	772	727	..	..	..	..	..	..	..
	900	[illegible]	885	880	873	864	853	841	825	807	786	762	734	701	664	623	..	..	..	..	..
	800	793	790	786	781	775	767	758	748	735	720	703	683	660	635	605	572	..	..	..	..
	700	695	693	690	687	683	678	672	665	656	647	635	622	607	589	569	547	522	495	..	..
	600	597	596	594	592	589	586	582	578	573	566	559	551	541	530	518	504	488	471	451	430
	500	498	497	497	495	494	492	490	487	484	481	476	471	466	460	452	444	435	425	414	401
	400	399	399	398	398	397	396	395	393	392	390	388	385	383	379	376	372	367	362	356	350
1,200,000	1700	1618	1584	1540	1484	1416	1332	1231	1111	997	900	816	744	681	625	576	533	494	459	428	400
	1600	1531	1503	1466	1420	1363	1293	1209	1108	..	..	..	..	..	..	..	..	..	..	..	..
	1500	1444	1420	1390	1352	1305	1247	1178	1095	..	..	..	..	..	..	..	..	..	..	..	..
	1400	1354	1335	1310	1279	1241	1194	1138	1071	991	..	..	..	..	..	..	..	..	..	..	..
	1300	1263	1248	1228	1204	1173	1135	1090	1036	973	898	..	..	..	..	..	..	..	..	..	..
	1200	1171	1159	1144	1124	1100	1071	1035	993	943	884	..	..	..	..	..	..	..	..	..	..
	1100	1078	1068	1057	1042	1023	1000	973	940	902	857	804	..	..	..	..	..	..	..	..	..
	1000	983	976	967	956	942	925	905	880	851	817	778	732	680	..	..	..	..	..	..	..
	900	888	883	876	868	858	845	830	813	791	767	738	705	667	624	..	..	..	..	..	..
	800	791	788	783	778	770	762	751	739	724	706	686	663	636	606	571	..	..	..	..	..
	700	694	692	689	685	680	674	667	659	649	637	624	608	590	570	547	521	492	..	..	..
	600	596	595	593	591	588	584	579	574	568	560	552	542	531	518	504	487	469	448	425	..
	500	498	497	496	495	493	491	488	485	481	477	472	467	460	453	444	435	424	412	399	384
	400	399	398	398	397	396	395	394	392	390	388	386	383	380	376	371	367	361	355	348	341
1,100,000	1700	1602	1561	1509	1443	1362	1262	1142	1019	914	825	748	682	624	573	528	488	453	421	392	367
	1600	1518	1484	1441	1386	1318	1235	1135	..	..	..	..	..	..	..	..	..	..	..	..	..
	1500	1433	1405	1369	1324	1268	1199	1117	1018	..	..	..	..	..	..	..	..	..	..	..	..
	1400	1345	1323	1293	1257	1211	1155	1088	1008	..	..	..	..	..	..	..	..	..	..	..	..
	1300	1256	1238	1215	1185	1149	1104	1050	986	910	..	..	..	..	..	..	..	..	..	..	..
	1200	1166	1151	1133	1110	1081	1046	1004	953	894	824	..	..	..	..	..	..	..	..	..	..
	1100	1073	1062	1048	1030	1008	981	949	910	864	810	..	..	..	..	..	..	..	..	..	..
	1000	980	972	961	948	931	911	886	857	823	782	735	681	..	..	..	..	..	..	..	..
	900	885	879	872	862	850	835	817	796	771	741	707	668	622	..	..	..	..	..	..	..
	800	790	786	780	773	765	754	742	727	709	689	665	637	605	569	..	..	..	..	..	..
	700	693	690	687	682	676	669	661	651	639	625	609	591	569	545	518	487	..	..	..	..
	600	596	594	592	589	585	581	575	569	562	553	543	531	518	503	485	466	444	419	..	..
	500	498	496	495	493	491	489	486	482	478	473	467	460	452	444	[illegible]	422	410	395	380	362
	400	399	398	398	397	396	394	393	391	389	386	383	380	376	371	366	360	354	346	338	329
	300	299	299	299	299	298	298	297	296	295	294	293	291	290	288	286	283	280	277	274	270
1,000,000	1300	1247	1225	1197	1161	1117	1063	998	920	831	750	680	620	567	521	480	444	412	383	357	333
	1200	1158	1141	1119	1091	1056	1014	962	901	829	..	..	..	..	..	..	..	..	..	..	..
	1100	1068	1055	1037	1016	989	956	917	870	814	749	..	..	..	..	..	..	..	..	..	..
	1000	976	966	953	937	917	892	863	827	785	737	..	..	..	..	..	..	..	..	..	..
	900	882	875	866	854	839	821	800	774	744	708	667	619	..	..	..	..	..	..	..	..
	800	788	783	776	768	757	745	730	712	690	665	636	603	564	..	..	..	..	..	..	..
	700	692	688	684	678	671	663	653	641	626	610	590	568	542	513	479	..	..	..	..	..
	600	595	593	590	586	582	577	570	563	554	543	531	517	501	482	461	438	411	..	..	..
	500	497	496	494	492	490	487	483	478	473	467	460	452	442	432	420	406	391	374	354	..
	400	398	398	397	396	395	393	391	389	386	383	380	375	371	365	359	352	344	335	325	315
	300	299	299	299	298	298	297	296	295	294	293	291	290	288	285	283	280	276	273	269	264
900,000	1200	1149	1127	1100	1065	1022	970	907	831	748	675	612	558	510	469	432	399	370	344	321	300
	1100	1060	1044	1023	996	963	923	874	816	747	..	..	..	..	..	..	..	..	..	..	..
	1000	970	958	942	922	897	867	830	787	735	..	..	..	..	..	..	..	..	..	..	..
	900	878	869	858	843	825	803	776	744	707	663	..	..	..	..	..	..	..	..	..	..
	800	785	778	770	760	747	732	713	691	664	634	598	556	..	..	..	..	..	..	..	..
	700	690	686	680	673	665	654	642	627	609	588	564	537	505	..	..	..	..	..	..	..
	600	594	591	587	583	578	571	563	554	543	530	515	497	477	454	429	..	..	..	..	..
	500	496	495	493	490	487	483	479	473	467	459	451	440	429	416	401	384	365	..	..	..
	400	398	397	396	395	393	391	389	386	383	379	375	370	364	357	349	341	331	320	308	295
	300	299	299	298	298	297	296	295	294	293	291	289	287	285	282	279	275	271	266	261	256

(Courtesy of National Forest Products Association.)

Rules of Thumb for Design of Wood Structural Members

1. The safe load on a beam varies directly with the beam's width. (Doubling the width of the beam doubles the safe load.)
2. The safe load on a beam varies as the square of the beam's depth. (Doubling the depth gives a safe load four times as great.)
3. Placing two planks on top of each other gives twice the strength of one plank.
4. Securely nailing or doweling plank ends together, to prevent slipping of one plank on the other, gives four times the strength of one plank; or twice the strength of the two planks placed on top of each other (unsecured).
5. The safe load on a beam varies indirectly with the beam's span. (Doubling the span cuts the safe load by one-half.)
6. A beam of a given size, material, and span will carry a uniformly distributed load twice as great as a concentrated load applied at the center of the span (provided the unbraced span is relatively short).
7. The deflection of a beam of a given size, material, and span varies directly as the load placed on the beam. (Doubling the load doubles the deflection.)
8. The deflection of a beam of given size, material, and loading varies as the cube of the beam's span. (Doubling the span gives $2 \times 2 \times 2$, or eight, times the deflection.)

Use of Tables

To use the tables, locate the appropriate bending design value, F_b, and refer to the span length involved. If the total load W is known read down the column under the appropriate F_b heading to find a matching design load W and then read across the page to see the required beam size. If the beam size is known read across the page to the column under the appropriate F_b heading to find the design load W.

Before selecting a size of beam it is advisable to check the board measure, *bm*, in several sizes which qualify in order to find the one which has the least amount of lumber and thus is the most efficient.

After determining the required beam size, or design load, W, in the manner just described, it is necessary to check the horizontal shear, F_v, and the modulus of elasticity, E, to make sure that the induced or required values do not exceed the respective values allowed for the species and grade of lumber to be used.

Chapter

4

Metal Structural Members

The selection of an adequately sized steel or aluminum section for a rigging or scaffolding operation is a relatively simple task using the tables in this chapter. However, it is to the rigger's advantage to understand the basic terminology and design fundamentals used in selecting the required beam or column.

Steel Structural Shapes

Among the many changes that have occurred in structural steel technology since the late 1960s, the most important has been the withdrawal of the common grade ASTM A7 structural steel, a material specification that had been fundamental to steel construction since the beginning of this century. In its place, ASTM A36 structural steel, with greater yield strength and improved qualities of modern fabrication, has become the most commonly used grade of steel.

Another significant change was the redesignation of two of the more common structural shapes: the WF, or wide-flange, section, now designated W shape; and the American standard I beam, now designated S shape. Other structural shapes used in rigging operations, such as channels, angles, and pipe sections (now used extensively for scaffolding), also fabricated of higher-strength steels, are still designated by their common names.

Aluminum Structural Shapes

When aluminum structural shapes were first introduced, they were patterned after the familiar steel shapes, which were formed by the rolled manufacturing process. However, it soon became apparent that the extrusion process used to manufacture aluminum sections permitted the placement of metal where it is needed, which resulted in improved section properties and reduced weight of the member. Although aluminum standard structural shapes are made from stock dies, extrusions can be made in almost any shape.

The characteristics of aluminum structural shapes are outlined in the following.

Flanges are thicker than the web, and the section modulus (resistance to bending) in the *X* axis is approximately 11% greater for the same weight per foot.

The shapes with section properties similar to the American standard shapes weigh 11% less.

The increases in *Y*-axis section properties are even more dramatic, ranging from 40 to 150%.

The radius of gyration shows similar improved properties.

Wider flanges make an extruded aluminum beam more stable.

Straight flanges make joining easier.

General Beam Characteristics

A standard W beam has flanges that are wider than the beam's depth and uniform in thickness. The S beam instead has relatively narrow flanges, which taper in thickness toward the flange's toe. Channel sections resemble half an S beam, with flanges on only one side of the web. Angles are L-shaped sections, having either equal or unequal leg lengths (see Figs. 4.1 to 4.3).

The distance between W or S beams is measured from the center of the section's web; for channels, it is measured from the back of the web or the heel of the flanges. The span of a simple beam supported at or near its ends is the distance between supports and is always measured in inches. The length of span depends to a degree on the nature of the material upon which the beam rests. If the supports are very firm, such as heavy steel or concrete, the effective span is measured from edge to edge of the supports (Fig. 4.4*a*). If the beam rests upon timbers that compress slightly under the bearing load (Fig. 4.4*b*), or if the steel beams upon which it rests can rotate slightly under the beam (Fig. 4.4*c*), then the span is measured from the center of one bearing to the center of the other bearing.

When selecting a metal beam, keep in mind the following rules.

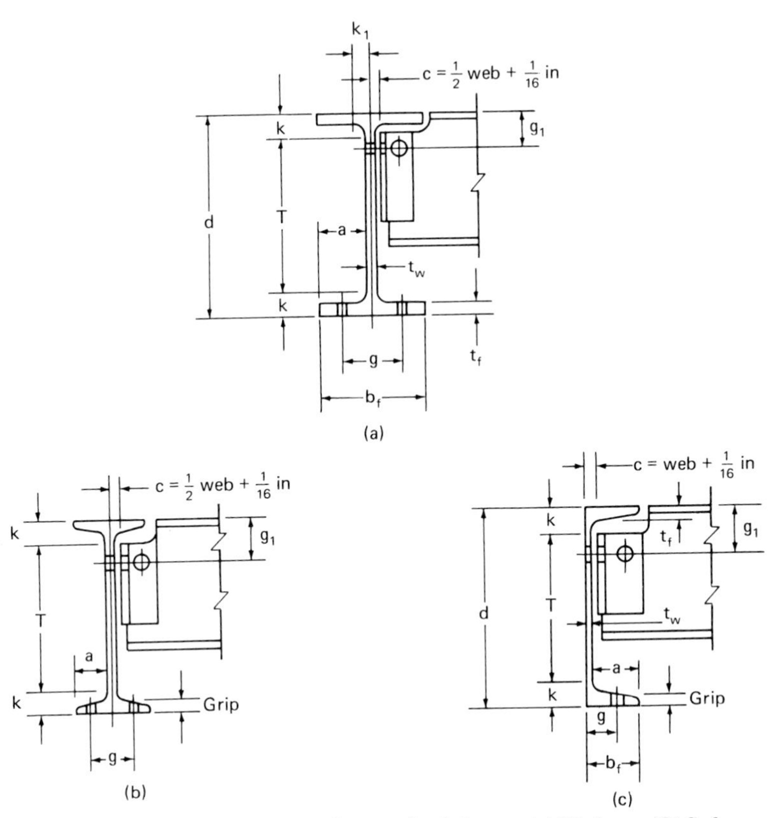

Figure 4.1 Detailing dimensions for standard shapes. (*a*) W shape. (*b*) S shape. (*c*) Channel.

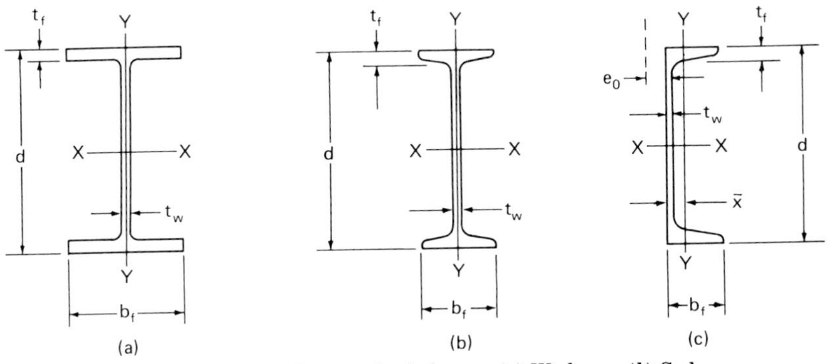

Figure 4.2 Design properties for standard shapes. (*a*) W shape. (*b*) S shape. (*c*) Channel.

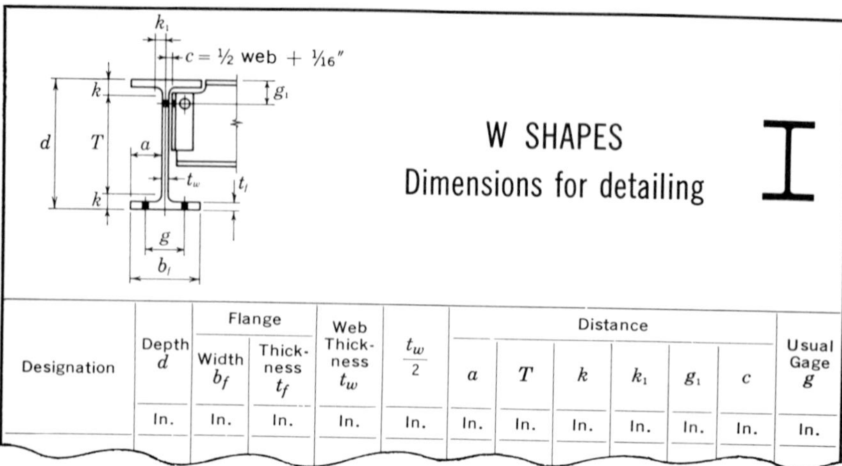

Designation	Depth d	Flange		Web Thickness t_w	$\frac{t_w}{2}$	Distance						Usual Gage g
		Width b_f	Thickness t_f			a	T	k	k_1	g_1	c	
	In.	In.	In.	In.	In.	In.	In.	In.	In.	In.	In.	In.

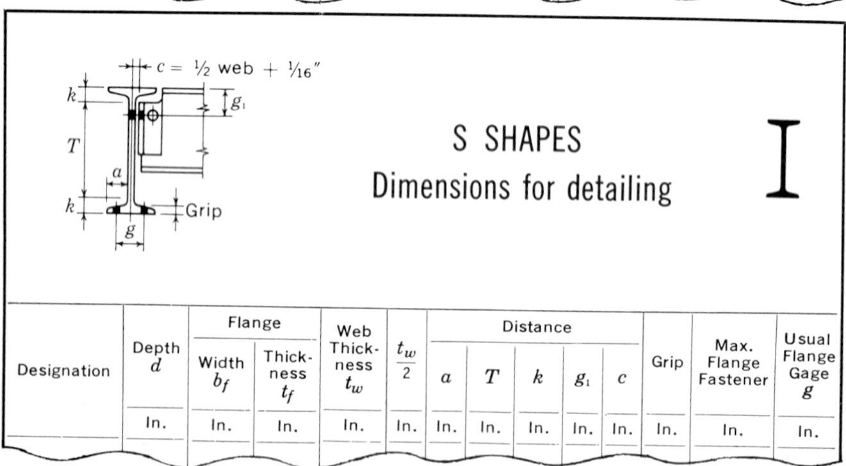

Designation	Depth d	Flange		Web Thickness t_w	$\frac{t_w}{2}$	Distance					Grip	Max. Flange Fastener	Usual Flange Gage g
		Width b_f	Thickness t_f			a	T	k	g_1	c			
	In.	In.	In.	In.	In.	In.	In.	In.	In.	In.	In.	In.	In.

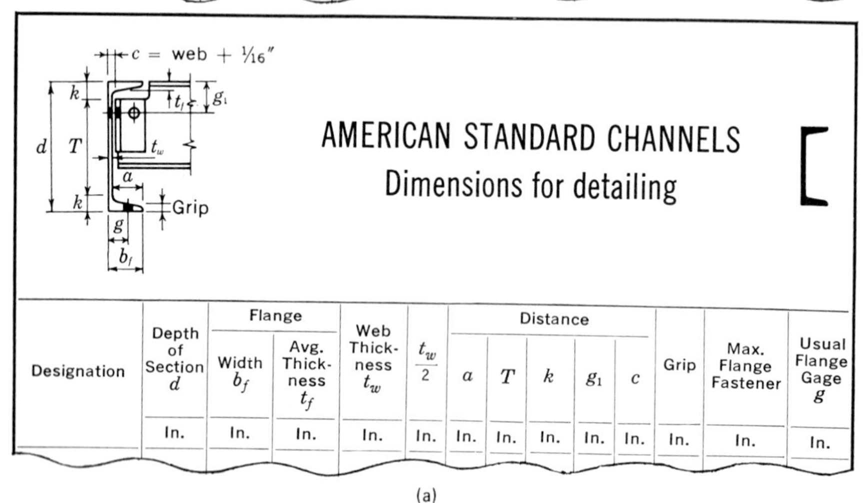

Designation	Depth of Section d	Flange		Web Thickness t_w	$\frac{t_w}{2}$	Distance					Grip	Max. Flange Fastener	Usual Flange Gage g
		Width b_f	Avg. Thickness t_f			a	T	k	g_1	c			
	In.	In.	In.	In.	In.	In.	In.	In.	In.	In.	In.	In.	In.

(a)

Figure 4.3 (*a*) Typical available detailing data. (*b*) Typical available designing data. (From *Manual of Steel Construction,* American Institute of Steel Construction.)

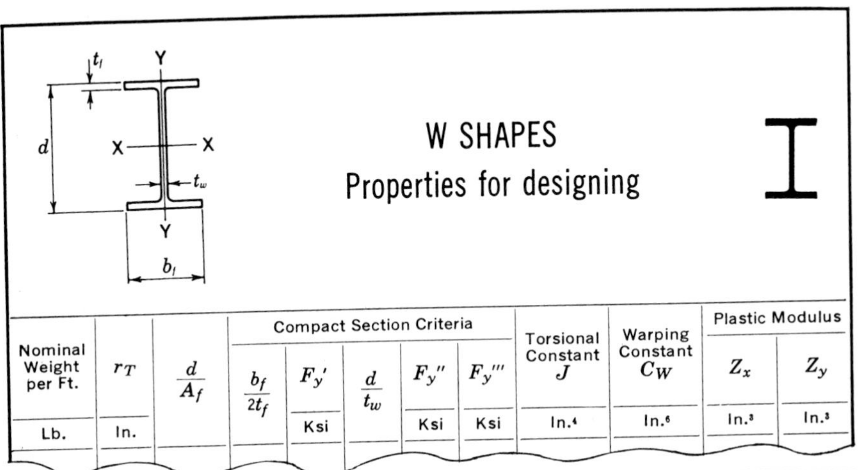

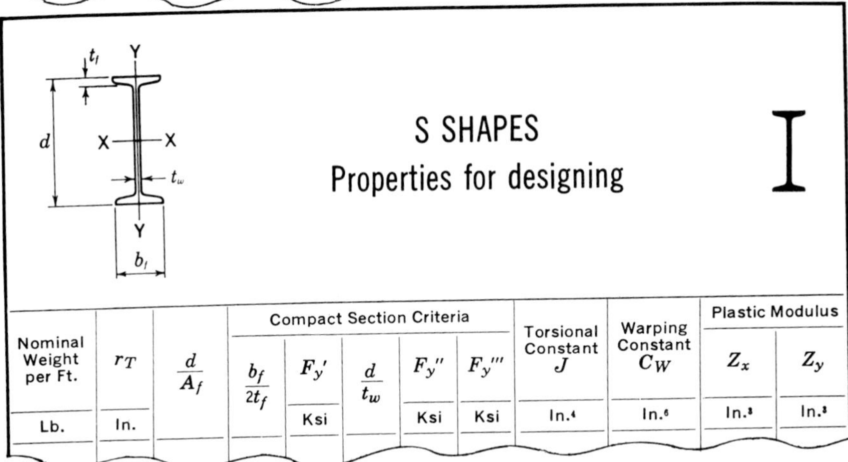

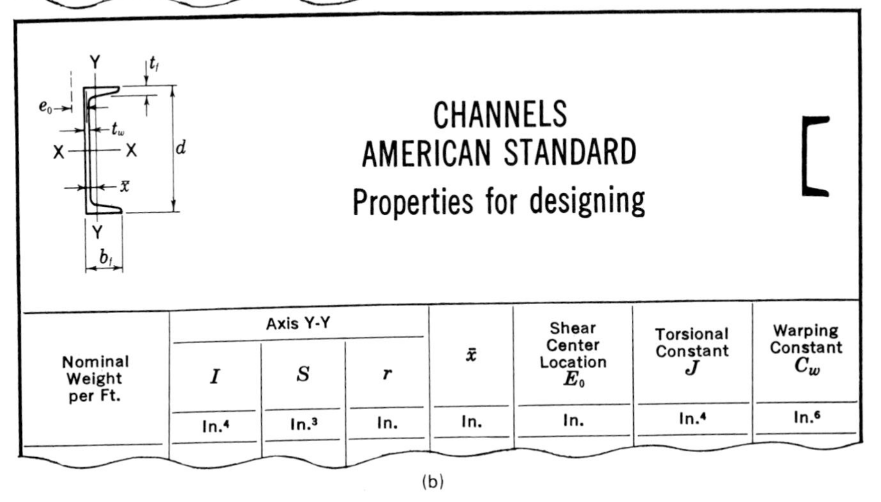

(b)

Figure 4.3 (Continued)

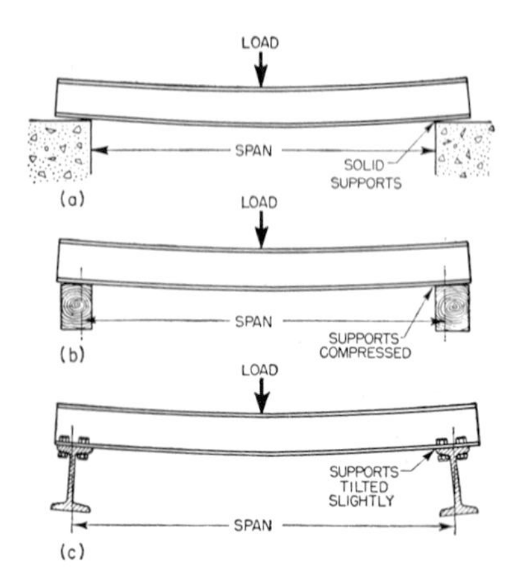

Figure 4.4 Measuring the length of the span of a beam.

1. A beam of given size and span will carry a uniformly distributed load twice as great as a load concentrated at the center of the span. In other words, if the table indicates that a certain beam on a given span will safely carry a concentrated load of 7,500 lb, then if the load is spread out over the entire length of the beam, 15,000 lb may be safely applied. Likewise, on a cantilever beam the allowable distributed load is twice the allowable concentrated load.

2. The safe load on a simple beam varies indirectly as the span. Therefore, if the span of a beam of given size is doubled, the safe load is reduced to one-half. On a cantilevered beam, if the lever arm is reduced to one-half, the load may be doubled.

3. The deflection or bending of a simple or a cantilever beam under load varies directly as the load. Thus, doubling the load will ordinarily double the deflection.

4. The deflection of a simple beam of a given size and load varies as the cube of the span. For example, if the span is multiplied by 2, the deflection is multiplied by 2^3, or 8 ($2 \times 2 \times 2$); if the span is multiplied by 2½, then the deflection is multiplied by $(2½)^3$, or 15⅝ ($2½ \times 2½ \times 2½$).

5. The load applied to a beam must include not only the useful load to be lifted, but also the weight of the slings, the hoist tackle or chain hoist, the pull on the hauling line or on the hand chain, and (if the beam is very heavy) the weight of the beam itself. For scaffolding, it would be necessary to consider not only the weight of the workers, but also that of scaffold planking and that of any material stored on the planks or equipment being used. The most commonly used

structural steel today is the ASTM A36 type, which has a high yield point (36,000 lb/in.^{2}). Because of this strength, smaller lighter sections can be used for many applications. All structural members made of A36 steel may be riveted, bolted, or welded for erection, as may the aluminum sections most frequently used today.

Reference Manuals

At this point in the chapter it is essential to point out the importance of employing supplementary information. It exists in the form of well-established manuals—for steel and for aluminum—which offer vital data for the selection of structural members that have sufficient strength to ensure safe, adequate loading.

The *Steel Construction Manual* (American Institute of Steel Construction, Chicago, Ill.) furnishes complete tabular data and methodology for choosing and dimensioning every variety of standard structural steel shape, covering even pipe and tubing, bars, and plating. Comparable facts and figures for structural aluminum appear in the manual *Specifications for Aluminum Structures* (The Aluminum Association, Washington, D.C.).

The use in rigging of steel or aluminum beams is today almost universal where new equipment is to be selected. The highly practical and useful wooden beam is, of course, well suited to many applications. The steel and aluminum manuals referenced above are, however, essential to any installation where either of these metals is to be part of the rigging activity.

Beam Calculations

Although the cited manuals will provide most of the required data, it is necessary that certain procedures be followed. To calculate beam loading, certain data must be known.

To calculate:	You must know:
Safe load	Span of beam, size, material
Beam size	Span of beam, load, material
Maximum beam span	Size of beam, load, material

The same unit stress of 36,000 lb/in.^{2} is used for all steel beams, and of 14,000 lb/in.^{2} for all aluminum beams. Tables in the manuals referred to list the product of unit stress times section modulus under the heading of maximum bending moment, thus simplifying the calculations.

Bending moment

A key factor in proper beam selection is the bending moment. The formulas for determining this bending moment when the size of the beam is to be calculated or, if the beam size is known, to find the maximum span or load are shown in Fig. 4.5. Unless the beam is very narrow relative to the span, all calculations are quite simple. On beams with a high slenderness ratio, additional investigations must be made. The slenderness ratio is the length of the span between lateral bracing of the compression flange, divided by the width of the beam flanges.

In a W or S beam used as a simple beam, the upper flange resists compressive stresses and the beam has the same tendency to buckle as does a loaded column. In selecting a beam, make sure the member has a cross-sectional area large enough to resist adequately all bending, shear, and deflection.

It is essential to take the slenderness ratio into consideration, for if the beam is very slender, there is a tendency for it to deflect sideways, then twist and roll over. The beam being much weaker on its side, it will fail and allow the load to drop. Bracing the compression flange against deflection sideways so that the compression flange of the longest unbraced part of the span is 15 in. or less, will prevent this from happening, and the full calculated load can be safely carried.

Bending

To design a steel beam for bending, simply use the flexure formula

$$\frac{M}{f} = S$$

First determine the type of steel being used and its unit stress, plus the laterally unsupported length of span. Then compute the maximum bending moment (as previously explained) of the beam. Next divide this moment by the allowable extreme fiber stress. (Building codes permit 22,000 or 24,000 lb/in.^{2} for A36 steel.) This gives the required section modulus S of the beam. Referring to the tables of properties of steel shapes, simply select a beam with a section modulus equal to or greater than that which is required. Generally, the section with the lightest weight per linear foot is the most economical.

In using the flexure formula, be careful not to mix units of measure: if f is in pounds per square inch, M must be in pound-inches, not in pound-feet.

This method of calculating the beam size is applicable to any type of loading, any material. However, in most applications, beams have uniformly distributed loads. Thus when designing for bending, it is more convenient to use the tables covering this type of loading to select the proper size of beam directly. All that is needed is the load and the span.

Shear

Since most beams large enough to resist bending are also large enough to resist shear, this step is usually omitted.

Deflection

To avoid floor and ceiling cracking in building construction, beam deflection should not exceed 1/360 of the span. However, for rigging and scaffolding applications the only serious problem with beam deflection might be one of having to provide sufficient clearance for equipment. Usually, because beams used for this type of operation have short lengths, this step can be disregarded. However, a long-span beam designed to carry a relatively light load should be investigated to make sure that deflection does not exceed 1/360 span.

Because simple beams with uniformly distributed loads occur so frequently in practice, tables of maximum loads for specific spans are of great convenience to the designer. The reader is referred to the appropriate manuals. The loads in these tables are given in 1,000 lb (kips), and the extreme fiber stress used in computing the load is 24,000 $lb/in.^2$

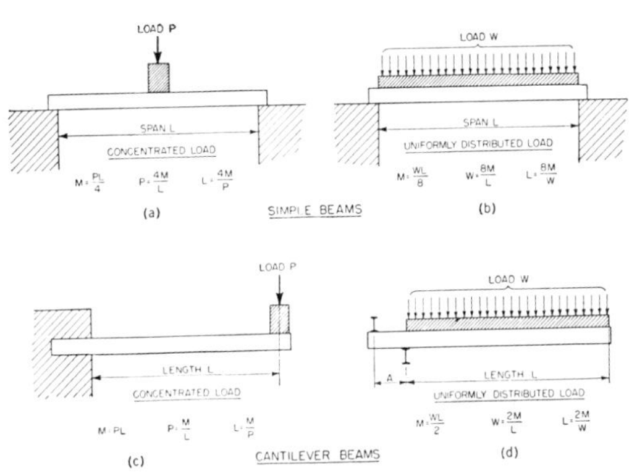

Figure 4.5 Calculating the bending moment of a beam under various types of loading. M—bending moment, pound-inches; P—concentrated load, pounds; W—total uniformly distributed load, pounds; L—length of span, inches.

Equivalent distributed loads

Although common practice generally dictates the use of tables covering uniform loading, such tables can be abridged for beams subjected to direct loading. Table 4.1 shows how this adjustment can be made for both load and deflection.

Since equivalent distributed loads found by this method do not include the weight of the beam, beam sections determined this way should also be checked for shear and deflection when used for structural purposes.

Bearing

A steel bearing plate is usually required to distribute the beam load over the bearing area of a masonry wall or concrete pier support to prevent the beam from crushing the masonry or concrete. The plate uniformly distributes the beam load on the masonry and also helps seat the beam at its proper elevation (see Fig. 4.6).

The size of the plate required is found by dividing the beam load on the masonry support by the allowable bearing unit stress,

$$A = \frac{P}{F_p}$$

TABLE 4.1 Equivalent Uniform Loads

Type of loading: equal loads, equal spaces	Equivalent uniform load	Deflection coefficient
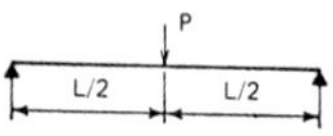 P; L/2, L/2	2.00P	0.80
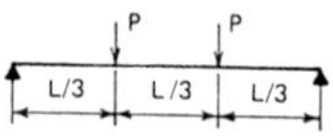 P, P; L/3, L/3, L/3	2.67P	1.02
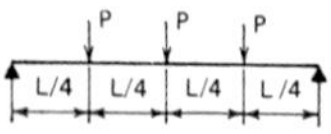 P, P, P; L/4, L/4, L/4, L/4	4.00P	0./95
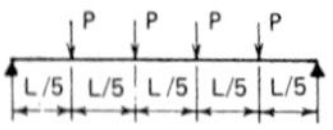 P, P, P, P; L/5, L/5, L/5, L/5, L/5	4.90P	1.01

SOURCE: *Manual of Steel Construction*, American Institute of Steel Construction.

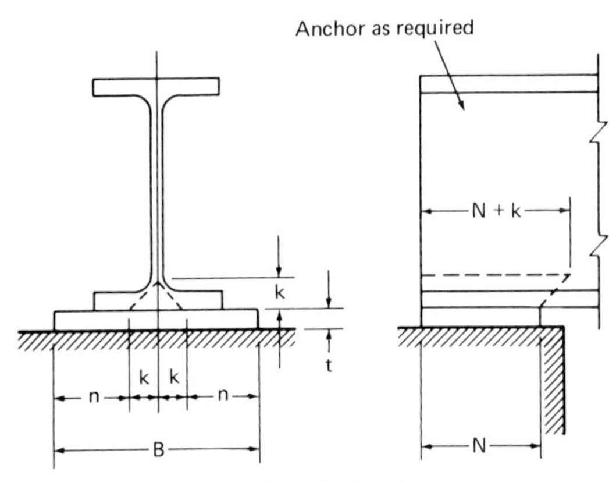

Figure 4.6 Bearing-plate design data.

where A = area of bearing plate = $B \times C$, square inches
P = load from beam, pounds
F_p = allowable bearing stress of masonry, lb/in.^{2}

Table 4.2 provides the allowable bearing capacity for standard masonry walls.

The wall thickness will determine the plate dimension parallel to the beam length. The plate dimension parallel to the length of the wall is B. Both dimensions are usually given in even inches. To determine the thickness of the plate, take the maximum bending moment at a distance n from the edge of the plate. Then use the formula

$$t = \sqrt{\frac{3pn^2}{F_b}}$$

where t = thickness of bearing plate, inches
p = actual bearing pressure of plate on masonry, lb/in.^{2}

TABLE 4.2 Allowable Bearing Capacity of Masonry Walls, lb/in.^{2}

Stone concrete, depending on quality	600–800
Common brick, lime mortar	100
Common brick, lime-cement mortar	200
Common brick, cement mortar	250
Hard brick, cement mortar	300
Rubble, cement mortar	150
Rubble, lime-cement mortar	100
Hollow T.C. blocks, cement mortar	80
Hollow cinder blocks, cement mortar	80

F_b = allowable bending stress in bearing plate, lb/in.^{2} AISC gives $F_b = 0.75F_y$, F_y being the yield-point stress of the steel plate.

$n = (B - k)/2$ inches

k = distance from bottom of flange to web toe of fillet, inches. Values of k may be found in the AISC manual under dimensions for detailing.

Since A36 steel is commonly used for bearing plates, the formula reduces to

$$t = \sqrt{\frac{3pn^2}{27,000}} = \sqrt{\frac{pn^2}{9,000}}$$

Open-Web Steel Joists

Open-web steel joists are shop-fabricated lightweight steel trusses used principally to support floors and roof panels between main supporting beams, girders, trusses, or walls. They may also be used to support wood decking for scaffolding.

Four series of joists are manufactured and used in construction today: J, H, LA, and LH. The most commonly used are the J joists, which are manufactured in standard depths of 8 to 24 in., in 2-in. increments and up to 48-in. lengths.

Steel Columns

The load a steel column can carry safely depends not only on the number of square inches in the cross section, but also on the shape of the column section. (This is not true for a wood column, which is always a solid shape.) However, the cross section of a steel column is seldom symmetrical with respect to both major axes, and therefore when axially loaded, such a column tends to bend in a plane perpendicular to the axis of the cross section about which the moment of inertia is the least.

In an I beam the moment of inertia about the axis parallel to the web is considerably smaller than that about the axis parallel to the flanges. Therefore, despite the amount of material in such a cross section, an I beam is not an economical column section.

Pipe columns and structural tubing with their symmetrical shapes have equal moments of inertia about each major axis.

Failure of short columns occurs by crushing, that of long slender columns, by stresses resulting from bending. In short members, the average unit stress is $f = P/A$. But for small cross-section steel columns, long enough to have a tendency to bend, the stresses are not equally distributed over the cross section and therefore the average unit stress must be less than 20,000 lb/in.[2] This average stress is dependent on the slenderness ratio of the column, its end conditions, and its cross-sectional area. While the slenderness ratio for timber columns is l/d (the unbraced column length divided by the dimension of the least side, both in inches), for steel columns it is l/r (the unbraced column length divided by the least radius of gyration, both in inches).

The least radius of gyration describes a steel column's measure of effectiveness in resisting bending. Given the least moment of inertia and the area of the cross section, the radius of gyration r may be readily computed. However, tables of properties of steel sections provide the r value, thus eliminating the need to compute it.

Steel Pipe Columns

Steel pipe sections are frequently used as columns, especially for scaffolding. Properties and allowable axial loads for various sizes of standard steel pipe columns are given in the AISC handbook.

Column Base Plates

All steel columns must be supported at their bases on a rolled steel slab or billet to distribute the column load over an adequate area and prevent crushing of the supporting concrete or masonry foundation. The base of the column and the steel slab must be in absolute contact, attached by welding or by angle sections secured to the masonry footing by anchor bolts. The area of the slab or billet is found by dividing the column load by the allowable unit compressive stress for the concrete footing. Determine the thickness of the billet by assuming the plate to be an inverted cantilever with the maximum bending moment at the edge of the column.

Chapter

5

Useful Formulas, Tables, and Conversion Factors

TABLE 5.1 Trigonometric Functions

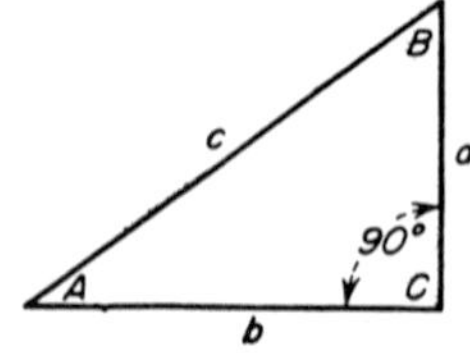

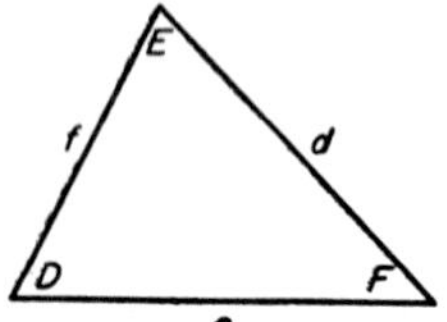

Given	Required	Formulas		
a, c	A, B, b	$\sin A = \frac{a}{c}$;	$\cos B = \frac{a}{c}$;	$b = \sqrt{c^2 - a^2}$
a, c	Area	$\frac{a}{2}\sqrt{c^2 - a^2}$		
a, b	A, B, c	$\tan A = \frac{a}{b}$;	$\tan B = \frac{b}{a}$;	$c = \sqrt{a^2 + b^2}$
a, b	Area	$\frac{ab}{2}$		
A, a	B, b, c	$B = 90° - A$;	$b = a \cot A$;	$c = \frac{a}{\sin A}$
A, a	Area	$\frac{a^2 \cot A}{2}$		
A, b	B, a, c	$B = 90° - A$;	$a = b \tan A$;	$c = \frac{b}{\cos A}$
A, b	Area	$\frac{b^2 \tan A}{2}$		
A, c	B, a, b	$B = 90° - A$;	$a = c \sin A$;	$b = c \cos A$
A, c	Area	$\frac{c^2 \sin 2A}{4}$		
d, e, f	D	$\sin \frac{D}{2} = \sqrt{\frac{(s-e)(s-f)}{ef}}$;	$s = \frac{d+e+f}{2}$	
d, e, f	E	$\sin \frac{E}{2} = \sqrt{\frac{(s-d)(s-f)}{df}}$		
d, e, f	F	$\sin \frac{F}{2} = \sqrt{\frac{(s-d)(s-e)}{de}}$		

TABLE 5.1 Trigonometric Functions (Continued)

Given	Required	Formulas
d, e, f	Area	$\sqrt{s(s-d)(s-e)(s-f)}$
d, D, E	e, f	$e = \dfrac{d \sin E}{\sin D}$; $\quad f = \dfrac{d \sin F}{\sin D}$
d, D, E	Area	$\dfrac{de \sin F}{2}$ (e from above formula)
d, e, D	E	$\sin E = \dfrac{e \sin D}{d}$
d, e, D	f	$f = \dfrac{e \sin F}{\sin E}$
d, e, F	D	$\tan D = \dfrac{d \sin F}{e - d \cos F}$
d, e, F	f	$f = \dfrac{d \sin F}{\sin D}$ (D from above formula)
d, e, F	Area	$\dfrac{de \sin F}{2}$

TABLE 5.2 Trigonometric Solutions of Right Triangles

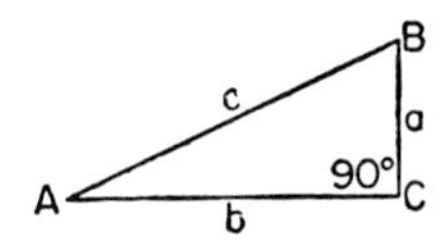

Known	Required				
	A	B	a	b	c
a, b	$\tan A = \dfrac{a}{b}$	$\cot B = \dfrac{a}{b}$			$\sqrt{a^2 + b^2}$
a, c	$\sin A = \dfrac{a}{c}$	$\cos B = \dfrac{a}{c}$		$\sqrt{c^2 - a^2}$	
b, c	$\cos A = \dfrac{b}{c}$	$\sin B = \dfrac{b}{c}$	$\sqrt{c^2 - b^2}$		
A, a		$90° - A$		$a \cot A$	$\dfrac{a}{\sin A}$
A, b		$90° - A$	$b \tan A$		$\dfrac{b}{\cos A}$
A, c		$90° - A$	$c \sin A$	$c \cos A$	

TABLE 5.3 Areas of Plane Figures

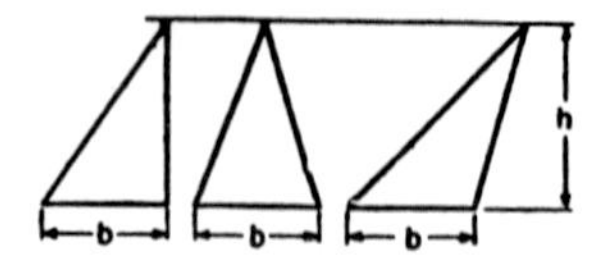

Triangle: Area $= \dfrac{bh}{2}$

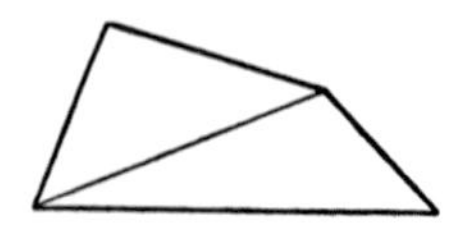

Trapezium: Area = Sum of areas of component triangles

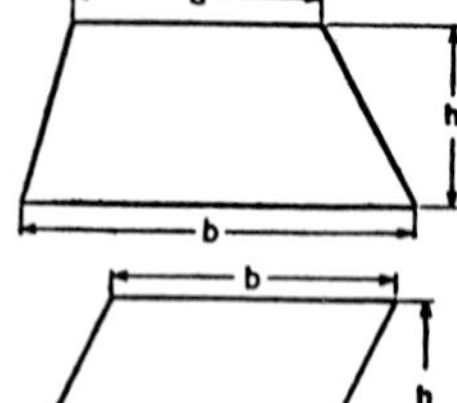

Trapezoid: Area $= \dfrac{(a+b)h}{2}$

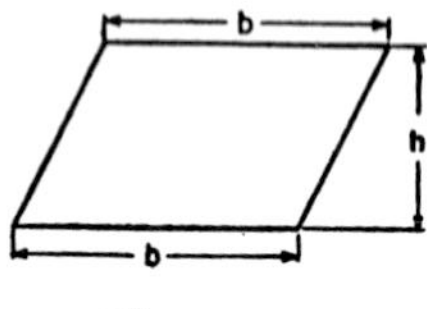

Parallelogram: Area $= bh$

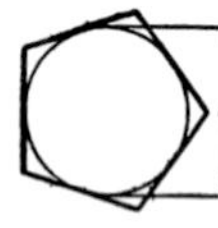

Regular polygon: Area $= \dfrac{d \times \text{perimeter}}{4}$

Circle: Area $= 0.7854d^2$

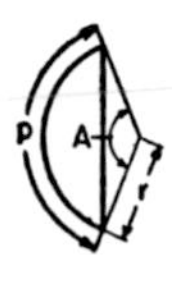

Segment: Area $= \dfrac{r^2}{2}(0.0175A - \sin A)$

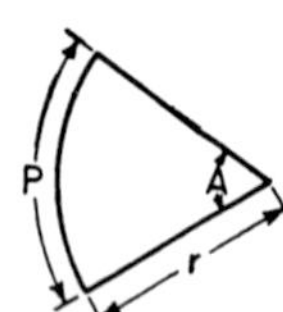

Sector: Area $= 0.00873r^2A = \dfrac{Pr}{2}$

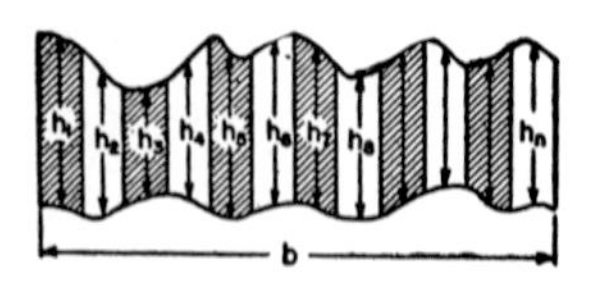

Irregular shape: divide length into parallel strips of equal width.

Area $= b\dfrac{h_1 + h_2 + h_3 + \cdots + h_n}{n}$ (approx.)

TABLE 5.4 Volumes of Solid Figures

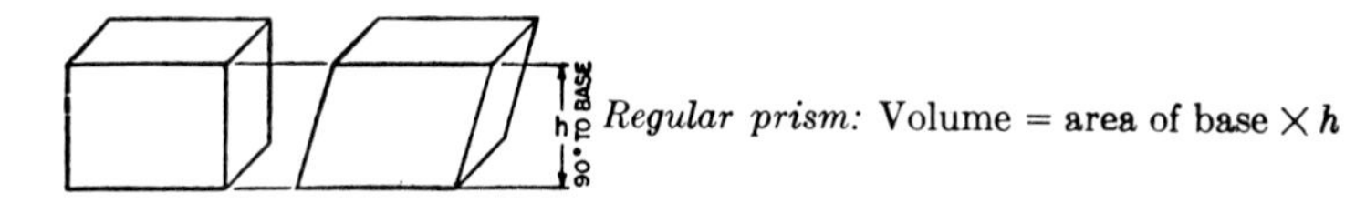

Regular prism: Volume = area of base × h

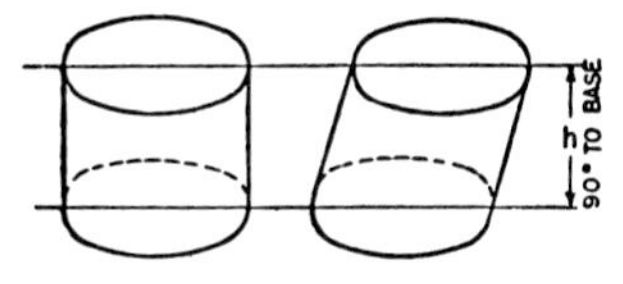

Cylinder: Volume = area of base × h

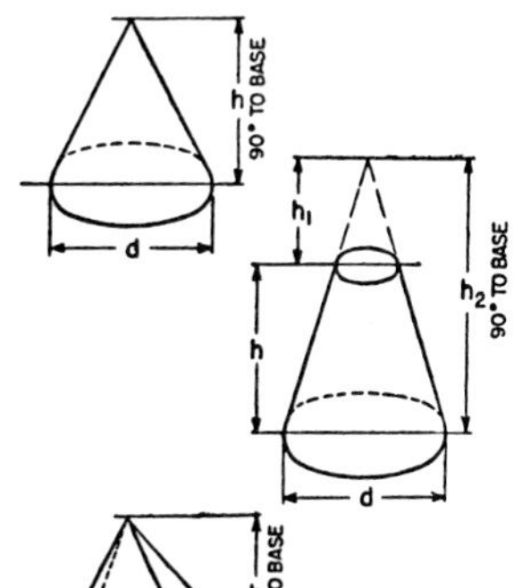

Cone: Volume $= \dfrac{\text{area of base} \times h}{3}$

Frustum of cone: Volume = volume of cone of height h_2 − volume of cone of height h_1

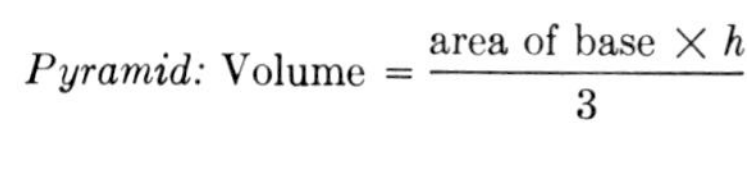

Pyramid: Volume $= \dfrac{\text{area of base} \times h}{3}$

Frustum of pyramid: Volume = volume of pyramid of height h_2 − volume of con of height h_1

Sphere: Volume = $0.524d^3$

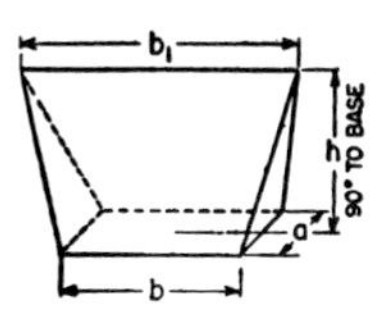

Wedge: (Rectangular base, b_1 parallel to b)

Volume $= \dfrac{ha(2b + b_1)}{6}$

Ring: Volume = $2.46Dd^2$

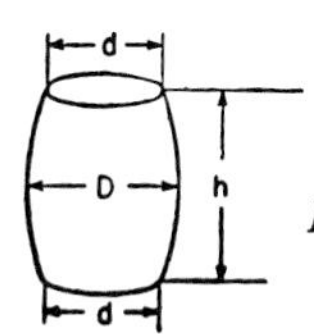

Barrel: Volume = $0.262h(2D^2 + d^2)$

TABLE 5.5 Circumferences and Areas of Circles*

Diameter	Circum-ference	Area	Diameter	Circum-ference	Area	Diameter	Circum-ference	Area	Diameter	Circum-ference	Area
			7/8	2.749	0.6013	4	12.57	12.57	9	28.27	63.62
1/64	0.04909	.00019	57/64	2.798	0.6230	1/16	12.76	12.96	1/8	28.67	65.40
1/32	0.09817	.00077	29/32	2.847	0.6450	1/8	12.96	13.36	1/4	29.06	67.20
3/64	0.1473	.00173	59/64	2.896	0.6675	3/16	13.16	13.77	3/8	29.45	69.03
1/16	0.1963	.00307	15/16	2.945	0.6903	1/4	13.35	14.19	1/2	29.85	70.88
5/64	0.2454	.00479	61/64	2.994	0.7135	5/16	13.55	14.61	5/8	30.24	72.76
3/32	0.2945	.00690	31/32	3.043	0.7371	3/8	13.74	15.03	3/4	30.63	74.66
7/64	0.3436	.00940	63/64	3.093	0.7610	7/16	13.94	15.47	7/8	31.02	76.59
1/8	0.3927	.01227	1	3.142	0.7854	1/2	14.14	15.90	10	31.42	78.54
9/64	0.4418	.01553	1/16	3.338	0.8866	9/16	14.33	16.35	1/8	31.81	80.52
5/32	0.4909	.01917	1/8	3.534	0.9940	5/8	14.53	16.80	1/4	32.20	82.52
11/64	0.5400	.02320	3/16	3.731	1.108	11/16	14.73	17.26	3/8	32.59	84.54
3/16	0.5890	.02761	1/4	3.927	1.227	3/4	14.92	17.72	1/2	32.99	86.59
13/64	0.6381	.03241	5/16	4.123	1.353	13/16	15.12	18.19	5/8	33.38	88.66
7/32	0.6872	.03758	3/8	4.320	1.485	7/8	15.32	18.67	3/4	33.77	90.76
15/64	0.7363	.04314	7/16	4.516	1.623	15/16	15.51	19.15	7/8	34.16	92.89
1/4	0.7854	.04909	1/2	4.712	1.767	5	15.71	19.63	11	34.56	95.03
17/64	0.8345	.05542	9/16	4.909	1.917	1/16	15.90	20.13	1/8	34.95	97.21
9/32	0.8836	.06213	5/8	5.105	2.074	1/8	16.10	20.63	1/4	35.34	99.40
19/64	0.9327	.06922	11/16	5.301	2.237	3/16	16.30	21.14	3/8	35.74	101.6
5/16	0.9817	.07670	3/4	5.498	2.405	1/4	16.49	21.65	1/2	36.13	103.9
21/64	1.031	.08456	13/16	5.694	2.580	5/16	16.69	22.17	5/8	36.52	106.1
11/32	1.080	.09281	7/8	5.890	2.761	3/8	16.89	22.69	3/4	36.91	108.4
23/64	1.129	.1014	15/16	6.087	2.948	7/16	17.08	23.22	7/8	37.31	110.8
3/8	1.178	.1104	2	6.283	3.142	1/2	17.28	23.76	12	37.70	113.1
25/64	1.227	.1198	1/16	6.480	3.341	9/16	17.48	24.30	1/8	38.09	115.5
13/32	1.276	.1296	1/8	6.676	3.547	5/8	17.67	24.85	1/4	38.48	117.9
27/64	1.325	.1398	3/16	6.872	3.758	11/16	17.87	25.41	3/8	38.88	120.3
7/16	1.374	.1503	1/4	7.069	3.976	3/4	18.06	25.97	1/2	39.27	122.7
29/64	1.424	.1613	5/16	7.265	4.200	13/16	18.26	26.53	5/8	39.66	125.2
15/32	1.473	.1726	3/8	7.461	4.430	7/8	18.46	27.11	3/4	40.06	127.7
31/64	1.522	.1843	7/16	7.658	4.666	15/16	18.65	27.69	7/8	40.45	130.2
1/2	1.571	.1963	1/2	7.854	4.909	6	18.85	28.27	13	40.84	132.7
33/64	1.620	.2088	9/16	8.050	5.157	1/8	19.24	29.46	1/8	41.23	135.3
17/32	1.669	.2217	5/8	8.247	5.412	1/4	19.63	30.68	1/4	41.63	137.9
35/64	1.718	.2349	11/16	8.443	5.673	3/8	20.03	31.92	3/8	42.02	140.5
9/16	1.767	.2485	3/4	8.639	5.940	1/2	20.42	33.18	1/2	42.41	143.1
37/64	1.816	.2625	13/16	8.836	6.213	5/8	20.81	34.47	5/8	42.80	145.8
19/32	1.865	.2769	7/8	9.032	6.492	3/4	21.21	35.78	3/4	43.20	148.5
39/64	1.914	.2916	15/16	9.228	6.777	7/8	21.60	37.12	7/8	43.59	151.2
5/8	1.963	.3068	3	9.425	7.069	7	21.99	38.48	14	43.98	153.9
41/64	2.013	.3223	1/16	9.621	7.366	1/8	22.38	39.87	1/8	44.37	156.7
21/32	2.062	.3382	1/8	9.817	7.670	1/4	22.78	41.28	1/4	44.77	159.5
43/64	2.111	.3545	3/16	10.01	7.980	3/8	23.17	42.72	3/8	45.16	162.3
11/16	2.160	.3712	1/4	10.21	8.296	1/2	23.56	44.18	1/2	45.55	165.1
45/64	2.209	.3883	5/16	10.41	8.618	5/8	23.95	45.66	5/8	45.95	168.0
23/32	2.258	.4057	3/8	10.60	8.946	3/4	24.35	47.17	3/4	46.34	170.9
47/64	2.307	.4236	7/16	10.80	9.281	7/8	24.74	48.71	7/8	46.73	173.8
3/4	2.356	.4418	1/2	11.00	9.621	8	25.13	50.27	15	47.12	176.7
49/64	2.405	.4604	9/16	11.19	9.968	1/8	25.53	51.85	1/8	47.52	179.7
25/32	2.454	.4794	5/8	11.39	10.32	1/4	25.92	53.46	1/4	47.91	182.7
51/64	2.503	.4987	11/16	11.58	10.68	3/8	26.31	55.09	3/8	48.30	185.7
13/16	2.553	.5185	3/4	11.78	11.04	1/2	26.70	56.75	1/2	48.69	188.7
53/64	2.602	.5386	13/16	11.98	11.42	5/8	27.10	58.43	5/8	49.09	191.7
27/32	2.651	.5591	7/8	12.17	11.79	3/4	27.49	60.13	3/4	49.48	194.8
55/64	2.700	.5800	15/16	12.37	12.18	7/8	27.88	61.86	7/8	49.87	197.9

* From Marks, *Mechanical Engineers' Handbook*, 4th ed., McGraw-Hill, 1941.

TABLE 5.5 Circumferences and Areas of Circles (Continued)

Diameter	Circum-ference	Area	Diameter	Circum-ference	Area	Diameter	Circum-ference	Area	Diameter	Circum-ference	Area
16	50.27	201.1	19 ½	61.26	298.6	23	72.26	415.5	29	91.11	660.5
⅛	50.66	204.2	⅝	61.65	302.5	⅛	72.65	420.0	¼	91.89	672.0
¼	51.05	207.4	¾	62.05	306.4	¼	73.04	424.6	½	92.68	683.5
⅜	51.44	210.6	⅞	62.44	310.2	⅜	73.43	429.1	¾	93.46	695.1
½	51.84	213.8	20	62.83	314.2	½	73.83	433.7	30	94.25	706.9
⅝	52.23	217.1	⅛	63.22	318.1	⅝	74.22	438.4	¼	95.03	718.7
¾	52.62	220.4	¼	63.62	322.1	¾	74.61	443.0	½	95.82	730.6
⅞	53.01	223.7	⅜	64.01	326.1	⅞	75.01	447.7	¾	96.60	742.6
17	53.41	227.0	½	64.40	330.1	24	75.40	452.4	31	97.39	754.8
⅛	53.80	230.3	⅝	64.80	334.1	¼	76.18	461.9	¼	98.17	767.0
¼	54.19	233.7	¾	65.19	338.2	½	76.97	471.4	½	98.96	779.3
⅜	54.59	237.1	⅞	65.58	342.2	¾	77.75	481.1	¾	99.75	791.7
½	54.98	240.5	21	65.97	346.4	25	78.54	490.9	32	100.5	804.2
⅝	55.37	244.0	⅛	66.37	350.5	¼	79.33	500.7	¼	101.3	816.9
¾	55.76	247.4	¼	66.76	354.7	½	80.11	510.7	½	102.1	829.6
⅞	56.16	250.9	⅜	67.15	358.8	¾	80.90	520.8	¾	102.9	842.4
18	56.55	254.5	½	67.54	363.1	26	81.68	530.9	33	103.7	855.3
⅛	56.94	258.0	⅝	67.94	367.3	¼	82.47	541.2	¼	104.5	868.3
¼	57.33	261.6	¾	68.33	371.5	½	83.25	551.5	½	105.2	881.4
⅜	57.73	265.2	⅞	68.72	375.8	¾	84.04	562.0	¾	106.0	894.6
½	58.12	268.8	22	69.12	380.1	27	84.82	572.6	34	106.8	907.9
⅝	58.51	272.4	⅛	69.51	384.5	¼	85.61	583.2	¼	107.6	921.3
¾	58.90	276.1	¼	69.90	388.8	½	86.39	594.0	½	108.4	934.8
⅞	59.30	279.8	⅜	70.29	393.2	¾	87.18	604.8	¾	109.2	948.4
19	59.69	283.5	½	70.69	397.6	28	87.96	615.8	35	110.0	962.1
⅛	60.08	287.3	⅝	71.08	402.0	¼	88.75	626.8	¼	110.7	975.9
¼	60.48	291.0	¾	71.47	406.5	½	89.54	637.9	½	111.5	989.8
⅜	60.87	294.8	⅞	71.86	411.0	¾	90.32	649.2	¾	112.3	1003.8

TABLE 5.6 Areas of Circles*
Diameters in Feet and Inches, Areas in Square Feet

Feet	Inches											
	0	1	2	3	4	5	6	7	8	9	10	11
0	.0000	.0055	.0218	.0491	.0873	.1364	.1963	.2673	.3491	.4418	.5454	.6600
1	.7854	.9218	1.069	1.227	1.396	1.576	1.767	1.969	2.182	2.405	2.640	2.885
2	3.142	3.409	3.687	3.976	4.276	4.587	4.909	5.241	5.585	5.940	6.305	6.681
3	7.069	7.467	7.876	8.296	8.727	9.168	9.621	10.08	10.56	11.04	11.54	12.05
4	12.57	13.10	13.64	14.19	14.75	15.32	15.90	16.50	17.10	17.72	18.35	18.99
5	19.63	20.29	20.97	21.65	22.34	23.04	23.76	24.48	25.22	25.97	26.73	27.49
6	28.27	29.07	29.87	30.68	31.50	32.34	33.18	34.04	34.91	35.78	36.67	37.57
7	38.48	39.41	40.34	41.28	42.24	43.20	44.18	45.17	46.16	47.17	48.19	49.22
8	50.27	51.32	52.38	53.46	54.54	55.64	56.75	57.86	58.99	60.13	61.28	62.44
9	63.62	64.80	66.00	67.20	68.42	69.64	70.88	72.13	73.39	74.66	75.94	77.24
10	78.54	79.85	81.18	82.52	83.86	85.22	86.59	87.97	89.36	90.76	92.18	93.60
11	95.03	96.48	97.93	99.40	100.9	102.4	103.9	105.4	106.9	108.4	110.0	111.5
12	113.1	114.7	116.3	117.9	119.5	121.1	122.7	124.4	126.0	127.7	129.4	131.0
13	132.7	134.4	136.2	137.9	139.6	141.4	143.1	144.9	146.7	148.5	150.3	152.1
14	153.9	155.8	157.6	159.5	161.4	163.2	165.1	167.0	168.9	170.9	172.8	174.8

TABLE 5.7 From Inches and Fractions of an Inch to Decimals of a Foot*

Inches	1	2	3	4	5	6	7	8	9	10	11
Feet	0.0833	0.1667	0.2500	0.3333	0.4167	0.5000	0.5833	0 6667	0.7500	0.8333	0.9167
Inches	⅛	¼	⅜	½	⅝	¾	⅞				
Feet	0.0104	0.0208	0.0313	0.0417	0.0521	0.0625	0.0729				

EXAMPLE: 5 ft 7⅜ in. = 5.0 + 0.5833 + 0.0313 = 5.6146 ft

* From Marks, *Mechanical Engineers' Handbook*, 4th ed., McGraw-Hill, 1941.

TABLE 5.8 Natural Sines and Cosines *

Natural Sines at intervals of 0°.1, or 6′.

Degrees	°.0 = (0′)	°.1 (6′)	°.2 (12′)	°.3 (18′)	°.4 (24′)	°.5 (30′)	°.6 (36′)	°.7 (42′)	°.8 (48′)	°.9 (54′)			Average difference
											0.0000	90°	
0°	0.0000	0017	0035	0052	0070	0087	0105	0122	0140	0157	0175	89	17
1	0175	0192	0209	0227	0244	0262	0279	0297	0314	0332	0349	88	17
2	0349	0366	0684	0401	0419	0436	0454	0471	0488	0506	0523	87	17
3	0523	0541	0558	0576	0593	0610	0628	0645	0663	0680	0698	86	17
4	0698	0715	0732	0750	0767	0785	0802	0819	0837	0854	0.0872	85	17
5	0.0872	0889	0906	0924	0941	0958	0976	0993	1011	1028	1045	84	17
6	1045	1063	1080	1097	1115	1132	1149	1167	1184	1201	1219	83	17
7	1219	1236	1253	1271	1288	1305	1323	1340	1357	1374	1392	82	17
8	1392	1409	1426	1444	1461	1478	1495	1513	1530	1547	1564	81	17
9	1564	1582	1599	1616	1633	1650	1668	1685	1702	1719	0.1736	80°	17
10°	0.1736	1754	1771	1788	1805	1822	1840	1857	1874	1891	1908	79	17
11	1908	1925	1942	1959	1977	1994	2011	2028	2045	2062	2079	78	17
12	2079	2096	2113	2130	2147	2164	2181	2198	2215	2233	2250	77	17
13	2250	2267	2284	2300	2317	2334	2351	2368	2385	2402	2419	76	17
14	2419	2436	2453	2470	2487	2504	2521	2538	2554	2571	0.2588	75	17
15	0.2588	2605	2622	2639	2656	2672	2689	2706	2723	2740	2756	74	17
16	2756	2773	2790	2807	2823	2840	2857	2874	2890	2907	2924	73	17
17	2924	2940	2957	2974	2990	3007	3024	3040	3057	3074	3090	72	17
18	3090	3107	3123	3140	3156	3173	3190	3206	3223	3239	3256	71	17
19	3256	3272	3289	3305	3322	3338	3355	3371	3387	3404	0.3420	70°	16
20°	0.3420	3437	3453	3469	3486	3502	3518	3535	3551	3567	3584	69	16
21	3584	3600	3616	3633	3649	3665	3681	3697	3714	3730	3746	68	16
22	3746	3762	3778	3795	3811	3827	3843	3859	3875	3891	3907	67	16
23	3907	3923	3939	3955	3971	3987	4003	4019	4035	4051	4067	66	16
24	4067	4083	4099	4115	4131	4147	4163	4179	4195	4210	0.4226	65	16
25	0.4226	4242	4258	4274	4289	4305	4321	4337	4352	4368	4384	64	16
26	4384	4399	4415	4431	4446	4462	4478	4493	4509	4524	4540	63	16
27	4540	4555	4571	4586	4602	4617	4633	4648	4664	4679	4695	62	16
28	4695	4710	4726	4741	4756	4772	4787	4802	4818	4833	4848	61	15
29	4848	4863	4879	4894	4909	4924	4939	4955	4970	4985	0.5000	60°	15
30°	0.5000	5015	5030	5045	5060	5075	5090	5105	5120	5135	5150	59	15
31	5150	5165	5180	5195	5210	5225	5240	5255	5270	5284	5299	58	15
32	5299	5314	5329	5344	5358	5373	5388	5402	5417	5432	5446	57	15
33	5446	5461	5476	5490	5505	5519	5534	5548	5563	5577	5592	56	15
34	5592	5606	5621	5635	5650	5664	5678	5693	5707	5721	0.5736	55	14
35	0.5736	5750	5764	5779	5793	5807	5821	5835	5850	5864	5878	54	14
36	5878	5892	5906	5920	5934	5948	5962	5976	5990	6004	6018	53	14
37	6018	6032	6046	6060	6074	6088	6101	6115	6129	6143	6157	52	14
38	6157	6170	6184	6198	6211	6225	6239	6252	6266	6280	6293	51	14
39	6293	6307	6320	6334	6347	6361	6374	6388	6401	6414	0.6428	50°	13
40°	0.6428	6441	6455	6468	6481	6494	6508	6521	6534	6547	6561	49	13
41	6561	6574	6587	6600	6613	6626	6639	6652	6665	6678	6691	48	13
42	6691	6704	6717	6730	6743	6756	6769	6782	6794	6807	6820	47	13
43	6820	6833	6845	6858	6871	6884	6896	6909	6921	6934	6947	46	13
44	6947	6959	6972	6984	6997	7009	7022	7034	7046	7059	0.7071	45°	12
45°	0.7071												
		°.9 = (54′)	°.8 (48′)	°.7 (42′)	°.6 (36′)	°.5 (30′)	°.4 (24′)	°.3 (18′)	°.2 (12′)	°.1 (6′)	°.0 (0′)	Degrees	

Natural Cosines

* From Marks, *Mechanical Engineers' Handbook*, 4th ed., McGraw-Hill, 1941.

TABLE 5.8 Natural Sines and Cosines (Continued)

Natural Sines at intervals of 0°.1, or 6′.

Degrees	°.0 =(0′)	°.1 (6′)	°.2 (12′)	°.3 (18′)	°.4 (24′)	°.5 (30′)	°.6 (36′)	°.7 (42′)	°.8 (48′)	°.9 (54′)			Average difference
											0.7071	45°	
45°	0.7071	7083	7096	7108	7120	7133	7145	7157	7169	7181	7193	44	12
46	7193	7206	7218	7230	7242	7254	7266	7278	7290	7302	7314	43	12
47	7314	7325	7337	7349	7361	7373	7385	7396	7408	7420	7431	42	12
48	7431	7443	7455	7466	7478	7490	7501	7513	7524	7536	7547	41	12
49	7547	7559	7570	7581	7593	7604	7615	7627	7638	7649	0.7660	40°	1[illegible]
50°	0.7660	7672	7683	7694	7705	7716	7727	7738	7749	7760	7771	39	11
51	7771	7782	7793	7804	7815	7826	7837	7848	7859	7869	7880	38	11
52	7880	7891	7902	7912	7923	7934	7944	7955	7965	7976	7986	37	11
53	7986	7997	8007	8018	8028	8039	8049	8059	8070	8080	8090	36	10
54	8090	8100	8111	8121	8131	8141	8151	8161	8171	8181	0.8192	35	10
55	0.8192	8202	8211	8221	8231	8241	8251	8261	8271	8281	8290	34	10
56	8290	8300	8310	8320	8329	8339	8348	8358	8368	8377	8387	33	10
57	8387	8396	8406	8415	8425	8434	8443	8453	8462	8471	8480	32	9
58	8480	8490	8499	8508	8517	8526	8536	8545	8554	8563	8572	31	9
59	8572	8581	8590	8599	8607	8616	8625	8634	8643	8652	0.8660	30°	9
60°	0.8660	8669	8678	8686	8695	8704	8712	8721	8729	8738	8746	29	9
61	8746	8755	8763	8771	8780	8788	8796	8805	8813	8821	8829	28	8
62	8829	8838	8846	8854	8862	8870	8878	8886	8894	8902	8910	27	8
63	8910	8918	8926	8934	8942	8949	8957	8965	8973	8980	8988	26	8
64	8988	8996	9003	9011	9018	9026	9033	9041	9048	9056	0.9063	25	7
65	0.9063	9070	9078	9085	9092	9100	9107	9114	9121	9128	9135	24	7
66	9135	9143	9150	9157	9164	9171	9178	9184	9191	9198	9205	23	7
67	9205	9212	9219	9225	9232	9239	9245	9252	9259	9265	9272	22	7
68	9272	9278	9285	9291	9298	9304	9311	9317	9323	9330	9336	21	6
69	9336	9342	9348	9354	9361	9367	9373	9379	9385	9391	0.9397	20°	6
70°	0.9397	9403	9409	9415	9421	9426	9432	9438	9444	9449	9455	19	6
71	9455	9461	9466	9472	9478	9483	9489	9494	9500	9505	9511	18	6
72	9511	9516	9521	9527	9532	9537	9542	9548	9553	9558	9563	17	5
73	8563	9568	9573	9578	9583	9588	9593	9598	9603	9608	9613	16	5
74	9613	9617	9622	9627	9632	9636	9641	9646	9650	9655	0.9659	15	5
75	0.9659	9664	9668	9673	9677	9681	9686	9690	9694	9699	9703	14	4
76	9703	9707	9711	9715	9720	9724	9728	9732	9736	9740	9744	13	4
77	9744	9748	9751	9755	9759	9763	9767	9770	9774	9778	9781	12	4
78	9781	9785	9789	9792	9796	9799	9803	9806	9810	9813	9816	11	3
79	9816	9820	9823	9826	9829	9833	9836	9839	9842	9845	0.9848	10°	3
80°	0.9848	9851	9854	9857	9860	9863	9866	9869	9871	9874	9877	9	3
81	9877	9880	9882	9885	9888	9890	9893	9895	9898	9900	9903	8	3
82	9903	9905	9907	9910	9912	9914	9917	9919	9921	9923	9925	7	2
83	9925	9928	9930	9932	9934	9936	9938	9940	9942	9943	9945	6	2
84	9945	9947	9949	9951	9952	9954	9956	9957	9959	9960	0.9962	5	2
85	0.9962	9963	9965	9966	9968	9969	9971	9972	9973	9974	9976	4	1
86	9976	9977	9978	9979	9980	9981	9982	9983	9984	9985	9986	3	1
87	9986	9987	9988	9989	9990	9990	9991	9992	9993	9993	9994	2	1
88	9994	9995	9995	9996	9996	9997	9997	9997	9998	9998	0.9998	1	0
89	0.9998	9999	9999	9999	9999	0000	0000	0000	0000	0000	1.0000	0°	0
90°	1.0000												
		°.9 = (54′)	°.8 (48′)	°.7 (42′)	°.6 (36′)	°.5 (30′)	°.4 (24′)	°.3 (18′)	°.2 (12′)	°.1 (6′)	°.0 (0′)	Degrees	

Natural Cosines

TABLE 5.9 Natural Tangents and Cotangents*

Natural Tangents at intervals of 0°.1, or 6′.

Degrees	°.0 =(0′)	°.1 (6′)	°.2 (12′)	°.3 (18′)	°.4 (24′)	°.5 (30′)	°.6 (36′)	°.7 (42′)	°.8 (48′)	°.9 (54′)			Average difference
											0.0000	90°	
0°	0.0000	0017	0035	0052	0070	0087	0105	0122	0140	0157	0175	89	17
1	0175	0192	0209	0227	0244	0262	0279	0297	0314	0332	0349	88	17
2	0349	0367	0384	0402	0419	0437	0454	0472	0489	0507	0524	87	17
3	0524	0542	0559	0577	0594	0612	0629	0647	0664	0682	0699	86	18
4	0699	0717	0734	0752	0769	0787	0805	0822	0840	0857	0.0875	85	18
5	0.0875	0892	0910	0928	0945	0963	0981	0998	1016	1033	1051	84	18
6	1051	1069	1086	1104	1122	1139	1157	1175	1192	1210	1228	83	18
7	1228	1246	1263	1281	1299	1317	1334	1352	1370	1388	1405	82	18
8	1405	1423	1441	1459	1477	1495	1512	1530	1548	1566	1584	81	18
9	1584	1602	1620	1638	1655	1673	1691	1709	1727	1745	0.1763	80°	18
10°	0.1763	1781	1799	1817	1835	1853	1871	1890	1908	1926	1944	79	18
11	1944	1962	1980	1998	2016	2035	2053	2071	2089	2107	2126	78	18
12	2126	2144	2162	2180	2199	2217	2235	2254	2272	2290	2309	77	18
13	2309	2327	2345	2364	2382	2401	2419	2438	2456	2475	2493	76	18
14	2493	2512	2530	2549	2568	2586	2605	2623	2642	2661	0.2679	75	19
15	0.2679	2698	2717	2736	2754	2773	2792	2811	2830	2849	2867	74	19
16	2867	2886	2905	2924	2943	2962	2981	3000	3019	3038	3057	73	19
17	3057	3076	3096	3115	3134	3153	3172	3191	3211	3230	3249	72	19
18	3249	3269	3288	3307	3327	3346	3365	3385	3404	3424	3443	71	19
19	3443	3463	3482	3502	3522	3541	3561	3581	3600	3620	0.3640	70°	20
20°	0.3640	3659	3679	3699	3719	3739	3759	3779	3799	3819	3839	69	20
21	3839	3859	3879	3899	3919	3939	3959	3979	4000	4020	4040	68	20
22	4040	4061	4081	4101	4122	4142	4163	4183	4204	4224	4245	67	21
23	4245	4265	4286	4307	4327	4348	4369	4390	4411	4431	4452	66	21
24	4452	4473	4494	4515	4536	4557	4578	4599	4621	4642	0.4663	65	21
25	0.4663	4684	4706	4727	4748	4770	4791	4813	4834	4856	4877	64	21
26	4877	4899	4921	4942	4964	4986	5008	5029	5051	5073	5095	63	22
27	5095	5117	5139	5161	5184	5206	5228	5250	5272	5295	5317	62	22
28	5317	5340	5362	5384	5407	5430	5452	5475	5498	5520	5543	61	23
29	5543	5566	5589	5612	5635	5658	5681	5704	5727	5750	0.5774	60°	23
30°	0.5774	5797	5820	5844	5867	5890	5914	5938	5961	5985	6009	59	24
31	6009	6032	6056	6080	6104	6128	6152	6176	6200	6224	6249	58	24
32	6249	6273	6297	6322	6346	6371	6395	6420	6445	6469	6494	57	25
33	6494	6519	6544	6569	6594	6619	6644	6669	6694	6720	6745	56	25
34	6745	6771	6796	6822	6847	6873	6899	6924	6950	6976	0.7002	55	26
35	0.7002	7028	7054	7080	7107	7133	7159	7186	7212	7239	7265	54	26
36	7265	7292	7319	7346	7373	7400	7427	7454	7481	7508	7536	53	27
37	7536	7563	7590	7618	7646	7673	7701	7729	7757	7785	7813	52	28
38	7813	7841	7869	7898	7926	7954	7983	8012	8040	8069	8098	51	28
39	8098	8127	8156	8185	8214	8243	8273	8302	8332	8361	0.8391	50°	29
40°	0.8391	8421	8451	8481	8511	8541	8571	8601	8632	8662	8693	49	30
41	8693	8724	8754	8785	8816	8847	8878	8910	8941	8972	9004	48	31
42	9004	9036	9067	9099	9131	9163	9195	9228	9260	9293	9325	47	32
43	9325	9358	9391	9424	9457	9490	9523	9556	9590	9623	0.9657	46	33
44	0.9657	9691	9725	9759	9793	9827	9861	9896	9930	9965	1.0000	45°	34
45°	1.0000												
		°.9 = (54′)	°.8 (48′)	°.7 (42′)	°.6 (36′)	°.5 (30′)	°.4 (24′)	°.3 (18′)	°.2 (12′)	°.1 (6′)	°.0 (0′)	Degrees	

Natural Cotangents

* From Marks, *Mechanical Engineers' Handbook*, 4th ed., McGraw-Hill, 1941.

TABLE 5.9 Natural Tangents and Cotangents (Continued)

Natural Tangents at intervals of 0°.1, or 6′.

Degrees	°.0 =(0′)	°.1 (6′)	°.2 (12′)	°.3 (18′)	°.4 (24′)	°.5 (30′)	°.6 (36′)	°.7 (42′)	°.8 (48′)	°.9 (54′)			Average difference
											1.0000	45°	
45°	1.0000	0035	0070	0105	0141	0176	0212	0247	0283	0319	0355	44	35
46	0355	0392	0428	0464	0501	0538	0575	0612	0649	0686	0724	43	37
47	0724	0761	0799	0837	0875	0913	0951	0990	1028	1067	1106	42	38
48	1106	1145	1184	1224	1263	1303	1343	1383	1423	1463	1504	41	40
49	1504	1544	1585	1626	1667	1708	1750	1792	1833	1875	1.1918	40°	41
50°	1.1918	1960	2002	2045	2088	2131	2174	2218	2261	2305	2349	39	43
51	2349	2393	2437	2482	2527	2572	2617	2662	2708	2753	2799	38	45
52	2799	2846	2892	2938	2985	3032	3079	3127	3175	3222	3270	37	47
53	3270	3319	3367	3416	3465	3514	3564	3613	3663	3713	3764	36	49
54	3764	3814	3865	3916	3968	4019	4071	4124	4176	4229	1.4281	35	52
55	1.4281	4335	4388	4442	4496	4550	4605	4659	4715	4770	4826	34	55
56	4826	4882	4938	4994	5051	5108	5166	5224	5282	5340	5399	33	57
57	5399	5458	5517	5577	5637	5697	5757	5818	5880	5941	6003	32	60
58	6003	6066	6128	6191	6255	6319	6383	6447	6512	6577	6643	31	64
59	1.6643	6709	6775	6842	6909	6977	7045	7113	7182	7251	1.7321	30°	67
60°	1.732	1.739	1.746	1.753	1.760	1.767	1.775	1.782	1.789	1.797	1.804	29	7
61	1.804	1.811	1.819	1.827	1.834	1.842	1.849	1.857	1.865	1.873	1.881	28	8
62	1.881	1.889	1.897	1.905	1.913	1.921	1.929	1.937	1.946	1.954	1.963	27	8
63	1.963	1.971	1.980	1.988	1.997	2.006	2.014	2.023	2.032	2.041	2.050	26	9
64	2.050	2.059	2.069	2.078	2.087	2.097	2.106	2.116	2.125	2.135	2.145	25	9
65	2.145	2.154	2.164	2.174	2.184	2.194	2.204	2.215	2.225	2.236	2.246	24	10
66	2.246	2.257	2.267	2.278	2.289	2.300	2.311	2.322	2.333	2.344	2.356	23	11
67	2.356	2.367	2.379	2.391	2.402	2.414	2.426	2.438	2.450	2.463	2.475	22	12
68	2.475	2.488	2.500	2.513	2.526	2.539	2.552	2.565	2.578	2.592	2.605	21	13
69	2.605	2.619	2.633	2.646	2.660	2.675	2.689	2.703	2.718	2.733	2.747	20°	14
70°	2.747	2.762	2.778	2.793	2.888	2.824	2.840	2.856	2.872	2.888	2.904	19	16
71	2.904	2.921	2.937	2.954	2.971	2.989	3.006	3.024	3.042	3.860	2.078	18	17
72	3.078	3.096	3.115	3.133	3.152	3.172	3.191	3.211	3.230	3.251	3.271	17	19
73	3.271	3.291	3.312	3.333	3.354	3.376	3.398	3.420	3.442	3.465	3.487	16	22
74	3.487	3.511	3.534	3.558	3.582	3.606	3.630	3.655	3.681	3.706	3.732	15	24
75	3.732	3.758	3.785	3.812	3.839	3.867	3.895	3.923	3.952	3.981	4.011	14	28
76	4.011	4.041	4.071	4.102	4.134	4.165	4.198	4.230	4.264	4.297	4.331	13	32
77	4.331	4.366	4.402	4.437	4.474	4.511	4.548	4.586	4.625	4.665	4.705	12	37
78	4.705	4.745	4.787	4.829	4.872	4.915	4.959	5.005	5.050	5.097	5.145	11	44
79	5.145	5.193	5.242	5.292	5.343	5.396	5.449	5.503	5.558	5.614	5.671	10°	53
80°	5.671	5.730	5.789	5.850	5.912	5.976	6.041	6.107	6.174	6.243	6.314	9	
81	6.314	6.386	6.460	6.535	6.612	6.691	6.772	6.855	6.940	7.026	7.115	8	
82	7.115	7.207	7.300	7.396	7.495	7.596	7.700	7.806	7.916	8.028	8.144	7	
83	8.144	8.264	8.386	8.513	8.643	8.777	8.915	9.058	9.205	9.357	9.514	6	
84	9.514	9.677	9.845	10.02	10.20	10.39	10.58	10.78	10.99	11.20	11.43	5	
85	11.43	11.66	11.91	12.16	12.43	12.71	13.00	13.30	13.62	13.95	14.30	4	
86	14.30	14.67	15.06	15.46	15.90	16.35	16.83	17.34	17.89	18.46	19.08	3	
87	19.08	19.74	20.45	21.20	22.02	22.90	23.86	24.90	26.03	27.27	28.64	2	
88	28.64	30.14	31.82	33.69	35.80	38.19	40.92	44.07	47.74	52.08	57.29	1	
89	57.29	63.66	71.62	81.85	95.49	114.6	143.2	191.0	286.5	573.0	∞	0°	
90°	∞												
		°.9 = (54′)	°.8 (48′)	°.7 (42′)	°.6 (36′)	°.5 (30′)	°.4 (24′)	°.3 (18′)	°.2 (12′)	°.1 (6′)	°.0 (0′)	Degrees	

Natural Cotangents

TABLE 5.10 Decimal Equivalents of Common Fractions*

8ths	16ths	32nds	64ths	Exact decimal values	8ths	16ths	32nds	64ths	Exact decimal values
					4	8	16	32	0.50
			1	0.01 5625				33	.51 5625
		1	2	.03 125			17	34	.53 125
			3	.04 6875				35	.54 6875
	1	2	4	.06 25		9	18	36	.56 25
			5	.07 8125				37	.57 8125
		3	6	.09 375			19	38	.59 375
			7	.10 9375				39	.60 9375
1	2	4	8	.12 5	5	10	20	40	.62 5
			9	.14 0625				41	.64 0625
		5	10	.15 625			21	42	.65 625
			11	.17 1875				43	.67 1875
	3	6	12	.18 75		11	22	44	.68 75
			13	.20 3125				45	.70 3125
		7	14	.21 875			23	46	.71 875
			15	.23 4375				47	.73 4375
2	4	8	16	.25	6	12	24	48	.75
			17	.26 5625				49	.76 5625
		9	18	.28 125			25	50	.78 125
			19	.29 6875				51	.79 6875
	5	10	20	.31 25		13	26	52	.81 25
			21	.32 8125				53	.82 8125
		11	22	.34 375			27	54	.84 375
			23	.35 9375				55	.85 9375
3	6	12	24	.37 5	7	14	28	56	.87 5
			25	.39 0625				57	.89 0625
		13	26	.40 625			29	58	.90 625
			27	.42 1875				59	.92 1875
	7	14	28	.43 75		15	30	60	.93 75
			29	.45 3125				61	.95 3125
		15	30	.46 875			31	62	.96 875
			31	.48 4375				63	.98 4375

TABLE 5.11 Strength of U.S. Standard Bolts from ¼ to 3 in. Diameter*

Bolt		Areas		Tensile strength, lb			Shearing strength, lb			
							Full bolt		Bottom of thread	
Diameter of bolt, in.	Number of threads per in.	Full bolt, sq in.	Bottom of thread, sq in.	At 10,000 lb per sq in.	At 12,500 lb per sq in.	At 17,500 lb per sq in.	At 7,500 lb per sq in.	At 10,000 lb per sq in.	At 7,500 lb per sq in.	At 10,000 lb per sq in.
¼	20	0.049	0.027	270	340	470	380	490	200	270
5/16	18	0.077	0.045	450	570	790	580	770	340	450
⅜	16	0.110	0.068	680	850	1,190	830	1,100	510	680
7/16	14	0.150	0.093	930	1,170	1,630	1,130	1,500	700	930
½	13	0.196	0.126	1,260	1,570	2,200	1,470	1,960	940	1,260
9/16	12	0.248	0.162	1,620	2,030	2,840	1,860	2,480	1,220	1,620
⅝	11	0.307	0.202	2,020	2,520	3,530	2,300	3,070	1,510	2,020
¾	10	0.442	0.302	3,020	3,770	5,290	3,310	4,420	2,270	3,020
⅞	9	0.601	0.419	4,190	5,240	7,340	4,510	6,010	3,150	4,190
1	8	0.785	0.551	5,510	6,890	9,640	5,890	7,850	4,130	5,510
1⅛	7	0.994	0.693	6,990	8,660	12,130	7,450	9,940	5,200	6,930
1¼	7	1.227	0.890	8,890	11,120	15,570	9,200	12,270	6,670	8,900
1⅜	6	1,435	1.054	10,540	13,180	18,450	11,140	14,850	7,910	10,540
1½	6	1.767	1.294	12,940	16,170	22,640	13,250	17,670	9,700	12,940
1⅝	5½	2.074	1.515	15,150	18,940	26,510	15,550	20,740	11,360	15,150
1¾	5	2.405	1.745	17,450	21,800	30,520	18,040	24,050	13,080	17,440
1⅞	5	2.761	2.049	20,490	25,610	35,860	20,710	27,610	15,370	20,490
2	4½	3.142	2.300	23,000	28,750	40,500	23,560	31,420	17,250	23,000
2¼	4½	3.976	3.021	30,210	37,770	52,870	29,820	39,760	22,660	30,210
2½	4	4.909	3.716	37,160	46,450	65,040	36,820	49,090	27,870	37,160
2¾	4	5.940	4.620	46,200	57,750	80,840	44,580	59,400	34,650	46,200
3	3½	7.069	5.428	54,280	67,850	94,990	53,020	70,690	40,710	54,280

* From Marks, *Mechanical Engineers' Handbook*, 4th ed., McGraw-Hill, 1941.

TABLE 5.12 Square and Hexagonal Regular Bolt Heads*

(All dimensions in inches.)

Bolt diameter	Rough and semifinished					Finished				
	Width across flats		Min width across corners		Height	Width across flats		Min width across corners		Height
	Max	Min	Hex	Square		Max	Min	Hex	Square	
1/4	3/8	0.363	0.414	0.498	11/64	7/16	0.428	0.488	0.588	3/16
5/16	1/2	0.484	0.552	0.665	13/64	9/16	0.552	0.629	0.758	15/64
3/8	9/16	0.544	0.620	0.747	1/4	5/8	0.613	0.699	0.842	9/32
7/16	5/8	0.603	0.687	0.828	19/64	3/4	0.737	0.840	1.012	21/64
1/2	3/4	0.725	0.827	0.995	21/64	13/16	0.799	0.911	1.097	3/8
9/16	7/8	0.847	0.966	1.163	3/8	7/8	0.861	0.982	1.182	27/64
5/8	15/16	0.906	1.033	1.244	27/64	15/16	0.922	1.051	1.266	15/32
3/4	1 1/8	1.088	1.240	1.494	1/2	1 1/8	1.108	1.263	1.521	9/16
7/8	1 5/16	1.269	1.447	1.742	19/32	1 5/16	1.293	1.474	1.775	21/32
1	1 1/2	1.450	1.653	1.991	21/32	1 1/2	1.479	1.686	2.031	3/4
1 1/8	1 11/16	1.631	1.859	2.239	3/4	1 11/16	1.665	1.898	2.286	27/32
1 1/4	1 7/8	1.813	2.067	2.489	27/32	1 7/8	1.850	2.109	2.540	15/16
1 1/2	2 1/4	2.175	2.480	2.986	1	2 1/4	2.222	2.533	3.051	1 1/8
1 3/4	2 5/8	2.538	2.893	3.485	1 5/32	2 5/8	2.593	2.956	3.560	1 5/16
2	3	2.900	3.306	3.982	1 11/32	3	2.964	3.379	4.070	1 1/2
2 1/4	3 3/8	3.263	3.720	4.480	1 1/2	3 3/8	3.335	3.802	4.579	1 11/16
2 1/2	3 3/4	3.625	4.133	4.977	1 21/32	3 3/4	3.707	4.226	5.090	1 7/8
2 3/4	4 1/8	3.988	4.546	5.476	1 53/64	4 1/8	4.078	4.649	5.599	2 1/16
3	4 1/2	4.350	4.959	5.973	2	4 1/2	4.449	5.072	6.108	2 1/4

Regular nuts (rough, semifinished, and finished) have a maximum width across flats of $1\frac{1}{2}D$ except for $D = \frac{1}{4}$ to $\frac{9}{16}$ when the width $= 1\frac{1}{2}D + \frac{1}{16}$. D is bolt diameter. Tolerance for width is $-0.050D$. Thickness is $\frac{7}{8}D$.

* From Marks, *Mechanical Engineers' Handbook*, 4th ed., McGraw-Hill, 1941.

TABLE 5.13 Wire and Sheet-Metal Gauges*
(Diameters and thicknesses in decimel parts of an inch)

Gauge No.	American wire gauge, or Brown & Sharpe (for nonferrous sheet and wire)	Steel wire gauge or Washburn & Moen or Roebling (for steel wire)	Birmingham wire gauge (B.W.G.) or Stubs iron wire (for steel rods or sheets)	Stubs steel wire gauge	British Imperial standard wire gauge (S.W.G.)	U.S. standard gauge for wrought iron sheet (480 lb per cu ft)	U.S. standard gauge for steel and open-hearth iron sheet (489.6 lb per cu ft)	British standard gauge for iron and steel sheets and hoops, 1914 (B.G.)
0000000	......	0.4900	......	.....	0.500	0.500	0.4902	0.6666
000000	......	0.4615	......	.....	0.464	0.469	0.4596	0.6250
00000	......	0.4305	......	.....	0.432	0.438	0.4289	0.5883
0000	0.460	0.3938	0.454	.....	0.400	0.406	0.3983	0.5416
000	0.410	0.3625	0.425	.....	0.372	0.375	0.3676	0.5000
00	0.365	0.3310	0.380	.....	0.348	0.344	0.3370	0.4452
0	0.325	0.3065	0.340	.....	0.324	0.312	0.3064	0.3964
1	0.289	0.2830	0.300	0.227	0.300	0.281	0.2757	0.3532
2	0.258	0.2625	0.284	0.219	0.276	0.266	0.2604	0.3147
3	0.229	0.2437	0.259	0.212	0.252	0.250	0.2451	0.2804
4	0.204	0.2253	0.238	0.207	0.232	0.234	0.2298	0.2500
5	0.182	0.2070	0.220	0.204	0.212	0.219	0.2145	0.2225
6	0.162	0.1920	0.203	0.201	0.192	0.203	0.1991	0.1981
7	0.144	0.1770	0.180	0.199	0.176	0.188	0.1838	0.1764
8	0.128	0.1620	0.165	0.197	0.160	0.172	0.1685	0.1570
9	0.114	0.1483	0.148	0.194	0.144	0.156	0.1532	0.1398
10	0.102	0.1350	0.134	0.191	0.128	0.141	0.1379	0.1250
11	0.091	0.1205	0.120	0.188	0.116	0.125	0.1225	0.1113
12	0.081	0.1055	0.109	0.185	0.104	0.109	0.1072	0.0991
13	0.072	0.0915	0.095	0.182	0.092	0.094	0.0919	0.0882
14	0.064	0.0800	0.083	0.180	0.080	0.078	0.0766	0.0785
15	0.057	0.0720	0.072	0.178	0.072	0.070	0.0689	0.0699
16	0.051	0.0625	0.065	0.175	0.064	0.062	0.0613	0.0625
17	0.045	0.0540	0.058	0.172	0.056	0.056	0.0551	0.0556
18	0.040	0.0475	0.049	0.168	0.048	0.050	0.0490	0.0495
19	0.036	0.0410	0.042	0.164	0.040	0.0438	0.0429	0.0440
20	0.032	0.0348	0.035	0.161	0.036	0.0375	0.0368	0.0392
21	0.0285	0.0317	0.032	0.157	0.032	0.0344	0.0337	0.0349
22	0.0253	0.0286	0.028	0.155	0.028	0.0312	0.0306	0.0313
23	0.0226	0.0258	0.025	0.153	0.024	0.0281	0.0276	0.0278
24	0.0201	0.0230	0.022	0.151	0.022	0.0250	0.0245	0.0248
25	0.0179	0.0204	0.020	0.148	0.020	0.0219	0.0214	0.0220
26	0.0159	0.0181	0.018	0.146	0.018	0.0188	0.0184	0.0196
27	0.0142	0.0173	0.016	0.143	0.0164	0.0172	0.0169	0.0175
28	0.0126	0.0162	0.014	0.139	0.0148	0.0156	0.0153	0.0156
29	0.0113	0.0150	0.013	0.134	0.0136	0.0141	0.0138	0.0139
30	0.0100	0.0140	0.012	0.127	0.0124	0.0125	0.0123	0.0123
31	0.0089	0.0132	0.010	0.120	0.0116	0.0109	0.0107	0.0110
32	0.0080	0.0128	0.009	0.115	0.0108	0.0102	0.0100	0.0098
33	0.0071	0.0118	0.008	0.112	0.0100	0.0094	0.0092	0.0087
34	0.0063	0.0104	0.007	0.110	0.0092	0.0086	0.0084	0.0077
35	0.0056	0.0095	0.005	0.108	0.0084	0.0078	0.0077	0.0069
36	0.0050	0.0090	0.004	0.106	0.0076	0.0070	0.0069	0.0061
37	0.0045	0.0085	......	0.103	0.0068	0.0066	0.0065	0.0054
38	0.0040	0.0080	......	0.101	0.0060	0.0062	0.0061	0.0048
39	0.0035	0.0075	......	0.099	0.0052	0.0059	0.0057	0.0043
40	0.0031	0.0070	......	0.097	0.0048	0.0055	0.0054	0.0039
41	......	0.0066	......	0.095	0.0044	0.0053	0.0052	0.0034
42	......	0.0062	......	0.092	0.0040	0.0051	0.0050	0.0031
43	......	0.0060	......	0.088	0.0036	0.0049	0.0048	0.0027
44	......	0.0058	......	0.085	0.0032	0.0047	0.0046	0.0024
45	......	0.0055	......	0.081	0.0028	......	......	0.0022
46	......	0.0052	......	0.079	0.0024	......	......	0.0019
47	......	0.0050	......	0.077	0.0020	......	......	0.0017
48	......	0.0048	......	0.075	0.0016	......	......	0.0015
49	......	0.0046	......	0.072	0.0012	......	......	0.0014
50	......	0.0044	......	0.069	0.0010	......	......	0.0012

* From Marks, *Mechanical Engineers' Handbook*, 4th ed., McGraw-Hill, 1941.

TABLE 5.14 Weights of Square and Bound Steel Bars*
(For iron, subtract 2%)

Size, in.	Weight, lb per lin ft		Size, in.	Weight, lb per lin ft		Size, in.	Weight, lb per lin ft		Size, in.	Weight, lb per lin ft	
	Square	Round		Square	Round		Square	Round		Square	Round
0	.....	.....	3	30.60	24.03	6	122.4	96.1	9	275.4	216.3
1/16	0.013	0.010	1/16	31.89	25.05	1/16	125.0	98.2	1/16	279.2	219.3
1/8	0.053	0.042	1/8	33.20	26.08	1/8	127.6	100.2	1/8	283.1	222.4
3/16	0.120	0.094	3/16	34.54	27.13	3/16	130.2	102.2	3/16	287.0	225.4
1/4	0.213	0.167	1/4	35.91	28.21	1/4	132.8	104.3	1/4	290.9	228.5
5/16	0.332	0.261	5/16	37.31	29.30	5/16	135.5	106.4	5/16	294.9	231.6
3/8	0.478	0.376	3/8	38.73	30.42	3/8	138.2	108.5	3/8	298.8	234.7
7/16	0.651	0.511	7/16	40.18	31.55	7/16	140.9	110.7	7/16	302.8	237.8
1/2	0.850	0.668	1/2	41.65	32.71	1/2	143.7	112.8	1/2	306.9	241.0
9/16	1.076	0.845	9/16	43.15	33.89	9/16	146.4	115.0	9/16	310.9	244.2
5/8	1.328	1.043	5/8	44.68	35.09	5/8	149.2	117.2	5/8	315.0	247.4
11/16	1.607	1.262	11/16	46.23	36.31	11/16	152.1	119.4	11/16	319.1	250.6
3/4	1.913	1.502	3/4	47.81	37.55	3/4	154.9	121.7	3/4	323.2	253.9
13/16	2.245	1.763	13/16	49.42	38.81	13/16	157.8	123.9	13/16	327.4	257.1
7/8	2.603	2.044	7/8	51.05	40.10	7/8	160.7	126.2	7/8	331.6	260.4
15/16	2.988	2.347	15/16	52.71	41.40	15/16	163.6	128.5	15/16	335.8	263.7
1	3.400	2.670	4	54.40	42.73	7	166.6	130.9	10	340.0	267.0
1/16	3.838	3.015	1/16	56.11	44.07	1/16	169.6	133.2	1/16	344.3	270.4
1/8	4.303	3.380	1/8	57.85	45.44	1/8	172.6	135.6	1/8	348.6	273.8
3/16	4.795	3.766	3/16	59.62	46.83	3/16	175.6	137.9	3/16	352.9	277.1
1/4	5.313	4.172	1/4	61.41	48.23	1/4	178.7	140.4	1/4	357.2	280.6
5/16	5.857	4.600	5/16	63.23	49.66	5/16	181.8	142.8	5/16	361.6	284.0
3/8	6.428	5.049	3/8	65.08	51.11	3/8	184.9	145.2	3/8	366.0	287.4
7/16	7.026	5.518	7/16	66.95	52.58	7/16	188.1	147.7	7/16	370.4	290.9
1/2	7.650	6.008	1/2	68.85	54.07	1/2	191.3	150.2	1/2	374.9	294.4
9/16	8.301	6.519	9/16	70.78	55.59	9/16	194.5	152.7	9/16	379.3	297.9
5/8	8.978	7.051	5/8	72.73	57.12	5/8	197.7	155.3	5/8	383.8	301.5
11/16	9.682	7.604	11/16	74.71	58.67	11/16	200.9	157.8	11/16	388.4	305.0
3/4	10.413	8.178	3/4	76.71	60.25	3/4	204.2	160.4	3/4	392.9	308.6
13/16	11.170	8.773	13/16	78.74	61.85	13/16	207.5	163.0	13/16	397.5	312.2
7/8	11.953	9.388	7/8	80.80	63.46	7/8	210.9	165.6	7/8	402.1	315.8
15/16	12.763	10.024	15/16	82.89	65.10	15/16	214.2	168.2	15/16	406.7	319.5
2	13.600	10.681	5	85.00	66.76	8	217.6	170.9	11	411.4	323.1
1/16	14.463	11.359	1/16	87.14	68.44	1/16	221.0	173.6	1/16	416.1	326.8
1/8	15.353	12.058	1/8	89.30	70.14	1/8	224.5	176.3	1/8	420.8	330.5
3/16	16.270	12.778	3/16	91.49	71.86	3/16	227.9	179.0	3/16	425.5	334.2
1/4	17.213	13.519	1/4	93.71	73.60	1/4	231.4	181.8	1/4	430.3	338.0
5/16	18.182	14.280	5/16	95.96	75.36	5/16	234.9	184.5	5/16	435.1	341.7
3/8	19.178	15.062	3/8	98.23	77.15	3/8	238.5	187.3	3/8	439.9	345.5
7/16	20.201	15.866	7/16	100.53	78.95	7/16	242.1	190.1	7/16	444.8	349.3
1/2	21.250	16.690	1/2	102.85	80.78	1/2	245.7	192.9	1/2	449.7	353.2
9/16	22.326	17.534	9/16	105.20	82.62	9/16	249.3	195.8	9/16	454.6	357.0
5/8	23.428	18.400	5/8	107.58	84.49	5/8	252.9	198.7	5/8	459.5	360.9
11/16	24.557	19.287	11/16	109.98	86.38	11/16	256.6	201.5	11/16	464.4	364.8
3/4	25.713	20.195	3/4	112.41	88.29	3/4	260.3	204.5	3/4	469.4	368.7
13/16	26.895	21.123	13/16	114.87	90.22	13/16	264.0	207.4	13/16	474.4	372.6
7/8	28.103	22.072	7/8	117.35	92.17	3/8	267.8	210.3	3/8	479.5	376.6
15/16	29.338	23.042	15/16	119.86	94.14	15/16	271.6	213.3	15/16	484.5	380.5

* From Marks, *Mechanical Engineers' Handbook*, 4th ed., McGraw-Hill, 1941.

TABLE 5.15 Standard Pipe and Line Pipe*

Diameter			Nominal thickness, in.	Circumference		Transverse areas			Length of pipe per sq ft of		Length of pipe containing 1 cu ft, ft	Nominal weight per ft, lb	No. of threads per in. of screw
Nominal internal, in.	Actual external, in.	Approx internal diam, in.		External, in.	Internal, in.	External, sq in.	Internal, sq in.	Metal, sq in.	External surface, ft	Internal surface, ft			
⅛	0.405	0.27	0.068	1.27	0.85	0.13	0.06	0.07	9.44	14.15	2513.00	0.24	27
¼	0.540	0.36	0.088	1.70	1.14	0.23	0.10	0.12	7.08	10.49	1383.30	0.42	18
⅜	0.675	0.49	0.091	2.12	1.55	0.36	0.19	0.17	5.66	7.76	751.20	0.57	18
½	0.840	0.62	0.109	2.63	1.95	0.55	0.30	0.25	4.55	6.15	472.40	0.85	14
¾	1.050	0.82	0.113	3.30	2.59	0.87	0.53	0.33	3.64	4.64	270.00	1.13	14
1	1.315	1.05	0.134	4.13	3.29	1.36	0.86	0.50	2.90	3.65	166.90	1.68	11½
1¼	1.660	1.38	0.140	5.22	4.34	2.16	1.50	0.67	2.30	2.77	96.25	2.27	11½
1½	1.900	1.61	0.145	5.97	5.06	2.84	2.04	0.80	2.01	2.37	70.66	2.72	11½
2	2.375	2.07	0.154	7.46	6.49	4.43	3.36	1.07	1.61	1.85	42.91	3.65	11½
2½	2.875	2.47	0.204	9.03	7.75	6.49	4.78	1.71	1.33	1.55	30.10	5.79	8
3	3.500	3.07	0.217	11.00	9.63	9.62	7.39	2.24	1.09	1.25	19.50	7.57	8
3½	4.000	3.55	0.226	12.57	11.15	12.57	9.89	2.68	0.96	1.08	14.57	9.11	8
4	4.500	4.03	0.237	14.14	12.65	15.90	12.73	3.18	0.85	0.95	11.31	10.79	8
5	5.563	5.05	0.259	17.48	15.85	24.31	19.99	4.32	0.69	0.76	7.20	14.62	8
6	6.625	6.07	0.280	20.81	19.05	34.47	28.89	5.59	0.58	0.63	4.98	18.97	8
8	8.625	8.07	0.276	27.10	25.35	58.43	51.15	7.28	0.44	0.47	2.82	24.69	8
8	8.625	7.98	0.322	27.10	25.07	58.43	50.02	8.41	0.44	0.48	2.88	28.55	8
9	9.625	8.94	0.344	30.24	28.08	72.76	62.72	10.04	0.40	0.43	2.29	33.91	8
10	10.750	10.19	0.278	33.77	32.01	90.76	81.55	9.21	0.36	0.37	1.76	31.20	8
10	10.750	10.14	0.306	33.77	31.86	90.76	80.75	10.01	0.36	0.38	1.78	34.24	8
10	10.750	10.02	0.366	33.77	31.47	90.76	78.82	11.94	0.36	0.38	1.82	40.48	8
12	12.750	12.09	0.328	40.06	37.98	127.68	114.80	12.88	0.30	0.32	1.25	43.77	8
12	12.750	12.00	0.375	40.06	37.70	127.68	113.10	14.59	0.30	0.32	1.27	49.56	8

TABLE 5.16 Extra-Strong Pipe*

Diameter			Nominal thickness, in.	Circumference		Transverse areas			Length of pipe per sq ft of		Nominal weight per ft, lb
Nominal internal, in.	Actual external, in.	Approx internal diam, in.		External, in.	Internal, in.	External, sq in.	Internal, sq in.	Metal, sq in.	External surface, ft	Internal surface, ft	
⅛	0.405	0.21	0.100	1.27	0.64	0.13	0.03	0.10	9.43	18.63	0.31
¼	0.540	0.29	0.123	1.70	0.92	0.23	0.07	0.16	7.08	12.99	0.54
⅜	0.675	0.42	0.127	2.12	1.32	0.36	0.14	0.22	5.66	9.07	0.74
½	0.840	0.54	0.149	2.64	1.70	0.55	0.23	0.32	4.55	7.05	1.09
¾	1.050	0.74	0.157	3.30	2.31	0.87	0.43	0.44	3.64	5.11	1.47
1	1.315	0.95	0.182	4.13	2.99	1.36	0.71	0.65	2.90	4.02	2.17
1¼	1.660	1.27	0.194	5.22	4.00	2.16	1.27	0.89	2.30	3.00	2.99
1½	1.900	1.49	0.203	5.97	4.69	2.84	1.75	1.08	2.01	2.56	3.63
2	2.375	1.93	0.221	7.46	6.07	4.43	2.94	1.50	1.61	1.98	5.02
2½	2.875	2.32	0.280	9.03	7.27	6.49	4.21	2.28	1.33	1.65	7.66
3	3.500	2.89	0.304	11.00	9.09	9.62	6.57	3.05	1.09	1.33	10.25
3½	4.000	3.36	0.321	12.57	10.55	12.57	8.86	3.71	0.96	1.14	12.50
4	4.500	3.82	0.341	14.14	12.00	15.90	11.45	4.46	0.85	1.00	14.98
5	5.563	4.81	0.375	17.48	15.12	24.31	18.19	6.11	0.69	0.79	20.78
6	6.625	5.75	0.437	20.81	18.07	34.47	25.98	8.50	0.58	0.66	28.57
8	8.625	7.63	0.500	27.10	23.96	58.43	45.66	12.76	0.44	0.50	43.34
10	10.750	9.75	0.500	33.77	30.63	90.76	74.66	16.10	0.36	0.40	54.73
12	12.750	11.75	0.500	40.06	36.91	127.68	108.43	19.25	0.30	0.33	65.41

* From Marks, *Mechanical Engineers' Handbook*, 4th ed., McGraw-Hill, 1941.

TABLE 5.17 Double-Extra-Strong Pipe*

Nominal pipe size, in.	Outside diameter, in.	Nominal wall thickness, in.		Wt per ft, lb, plain ends	Nominal pipe size, in.	Outside diameter, in.	Nominal wall thickness, in.		Wt per ft, lb, plain ends
		Wrought iron	Steel				Wrought iron	Steel	
½	0.840	0.307	0.294	1.714	2½	2.875	0.565	0.552	13.695
¾	1.050	0.318	0.308	2.440	3	3.500	0.615	0.600	18.583
1	1.315	0.369	0.358	3.659	4	4.500	0.690	0.674	27.541
1¼	1.660	0.393	0.382	5.214	5	5.563	0.768	0.750	38.552
1½	1.900	0.411	0.400	6.408	6	6.625	0.884	0.864	53.160
2	2.375	0.447	0.436	9.029	8	8.625	0.895	0.875	72.424

* From Marks, *Mechanical Engineers' Handbook*, 4th ed., McGraw-Hill, 1941.

TABLE 5.18 Metric System of Measures and Weights

Linear Measure		
1 millimeter	=	0.03937 inch
1 centimeter	=	0.3937 inch
1 decimeter	=	3.937 inches
1 meter	=	39.37 inches 3.28083 feet 1.09361 yards
1 kilometer	=	3280.83 feet 1093.61 yards 0.62137 mile
1 inch	=	25.4 millimeters 2.54 centimeters 0.254 decimeter 0.0254 meter
1 foot	=	0.3048 meter
1 yard	=	0.9144 meter
1 mile	=	1.60935 kilometers

Square Measure		
1 square millimeter	=	0.00155 square inch 1973.5 circular mils
1 square centimeter	=	00.155 square inch
1 square decimeter	=	15.5 square inches
1 square meter	=	1550 square inches 10.7639 square feet 1.196 square yards
1 square kilometer	=	0.386109 square mile 247.11 acres
1 square myriameter	=	38.6109 square miles
1 square inch	=	645.2 square millimeters 6.452 square centimeters 0.06452 square decimeter
1 square foot	=	0.0929 square meter
1 square yard	=	0.836 square meter

Cubic Measure		
1 cubic centimeter	=	0.061 cubic inch
1 cubic decimeter	=	61.0234 cubic inches 0.035314 cubic foot
1 cubic meter	=	35.314 cubic feet 1.308 cubic yards

TABLE 5.18 Metric System of Measures and Weights (Continued)

Weight		
1 gram	=	15.432 grains
1 kilogram	=	2.204622 pounds
1 metric ton, 1000 kilograms	=	2204.6 pounds 0.9842 gross ton 1.1023 net tons
1 grain	=	0.0648 gram
1 pound	=	0.4536 kilogram
Capacity		
1 liter	=	61.0234 cubic inches 0.03531 cubic foot 0.2642 gallon
1 cubic foot	=	28.317 liters
1 gallon	=	3.785 liters
Compound Units		
1 gram per square millimeter	=	1.422 pounds per square inch
1 kilogram per square millimeter	=	1422.32 pounds per square inch
1 kilogram per square centimeter	=	14.2232 pounds per square inch
1 kilogram per square meter	=	0.2048 pound per square foot 1.8433 pounds per square yard
1 kilogram-meter	=	7.2330 foot-pounds
1 kilogram per meter	=	0.6720 pound per foot
1 pound per square inch	=	0.07031 kilogram per square centimeter
1 pound per square foot	=	0.0004882 kilogram per square centimeter 0.006944 pound per square inch
1 pound per cubic inch	=	27679.7 kilograms per cubic meter
1 pound per cubic foot	=	16.0184 kilograms per cubic meter
1 kilogram per cubic meter	=	0.06243 pound per cubic foot
1 foot per second	=	0.30480 meter per second
1 meter per second	=	3.28083 feet per second 2.23693 miles per hour

TABLE 5.19 Beaufort Scale of Wind Force (Compiled by U.S. Weather Bureau, 1955)

Beaufort number	Miles per hour	Knots	Wind effects observed on land	Terms used in USWB forecasts
0	Less than 1	Less than 1	Calm; smoke rises vertically	
1	1–3	1–3	Direction of wind shown by smoke drift; but not by wind vanes	Light
2	4–7	4–6	Wind felt on face; leaves rustle; ordinary vane moved by wind	
3	8–12	7–10	Leaves and small twigs in constant motion; wind extends light flag	Gentle
4	13–18	11–16	Raises dust, loose paper; small branches are moved	Moderate
5	19–24	17–21	Small trees in leaf begin to sway; created wavelets form on inland waters	Fresh
6	25–31	22–27	Large branches in motion; whistling heard in telegraph wires; umbrellas used with difficulty	Strong
7	32–38	28–33	Whole trees in motion; inconvenience felt walking against wind	
8	39–46	34–40	Breaks twigs off trees; generally impedes progress	Gale
9	47–54	41–47	Slight structural damage occurs; chimney pots, slates removed	
10	55–63	48–55	Seldom experienced inland; trees uprooted; considerable structural damage occurs	Whole gale
11	64–72	56–63	Very rarely experienced; accompanied by widespread damage	Hurricane
12 or more	73 or more	64 or more	Very rarely experienced; accompanied by widespread damage	

TABLE 5.20 Table of Temperature Equivalents, Fahrenheit and Celsius Scales

Farenheit	Celsius	Farenheit	Celsius	Farenheit	Celsius	Farenheit	Celsius
−35	−35	32	0	95	35	158	70
−24	−30	41	5	104	40	167	75
−13	−25	50	10	113	45	176	80
−4	−20	59	15	122	50	185	85
5	−15	68	20	131	55	194	90
14	−10	77	25	140	60	203	95
23	−5	86	30	149	65	212	100

TABLE 5.21 Weights of Various Materials

Substance	Weight, lb/ft^3
Metals, Alloys, Ores	
Aluminum, cast-hammered	165
Aluminum, bronze	481
Brass, cast-rolled	534
Bronze, 7.9 to 14%Sn	509
Bronze, phosphor	554
Copper, cast-rolled	556
Copper ore, pyrites	262
Gold coin (U.S.)	1073
Iron, gray cast	442
Iron, cast, pig	450
Iron, wrought	485
Iron ore, hematite	325
Iron ore, limonite	237
Iron ore, magnetite	315
Iron slag	172
Lead	710
Lead ore, galena	465
Manganese	475
Manganese ore, pyrolusite	259
Mercury	849
Nickel	537
Platinum, cast-hammered	1330
Silver, cast-hammered	656
Steel, cold-drawn	489
Tin, cast-hammered	459
Tin ore, cassiterite	418
Tungsten	1200
Zinc, cast-rolled	440
Zinc ore, blende	253
Timber, U.S. Seasoned (Moisture 15 to 20%)	
Ash, white-red	40
Birch	32
Cedar, white-red	22
Cypress	30
Fir, Douglas spruce	32
Fir, eastern	25
Elm, white	45
Hemlock	29
Hickory	48
Mahogany	44
Maple, hard	43
Maple, white	33
Oak, chestnut	54
Oak, live	59
Oak, red, black	41
Oak, white	46
Pine, Oregon	32
Pine, red	30
Pine, white	26
Pine, yellow, long-leaf	44
Pine, yellow, short-leaf	38

TABLE 5.21 Weights of Various Materials (Continued)

Substance	Weight, lb/ft^3
Timber (Continued)	
Poplar	27
Redwood, California	26
Spruce, white, black	27
Walnut, black	42
Walnut, white	26
Water, 4°C, maximum density	62.428
Water, 100C	59.830
Water, ice	56
Water, snow, fresh fallen	8
Water, seawater	64
Various Building Materials	
Ashes, cinders	40–45
Cement, portland, loose	90
Cement, portland, set	183
Lime, gypsum, loose	53–64
Mortar, set	103
Slags, bank slag	67–72
Slags, bank screenings	98–117
Slags, machine slag	96
Slags, slag sand	49–55
Earth, etc. Excavated	
Clay, dry	63
Clay, damp, plastic	110
Clay and gravel, dry	100
Earth, dry, loose	76
Earth, dry, packed	95
Earth, mud, packed	115
Riprap, limestone	80–85
Riprap, sandstone	90
Riprap, shale	105
Sand, gravel, dry, loose	90–105
Sand, gravel, dry, packed	100–120
Sand, gravel, wet	118–120
Excavations in Water	
Sand or gravel	60
Sand or gravel and clay	65
Clay	80
River mud	90
Soil	70
Stone riprap	65
Stone, Quarried, Piled	
Basalt, granite, gneiss	96
Limestone, marble, quartz	95
Sandstone	82
Shale	92

TABLE 5.21 Weights of Various Materials (*Continued*)

Coal and Coke, Piled	
Coal, anthracite	47–58
Coal, bituminous, lignite	40–54
Coal, peat, turf	20–26
Coal, charcoal	10–14
Coal, coke	23–32

TABLE 5.22 Metric Equivalents of Standard Rope Sizes*

Diameter		Circumference		Diameter		Circumference	
Inches	Millimeters	Inches	Millimeters	Inches	Millimeters	Inches	Millimeter
1/8	3.2	3/8	10.0	15/16	23.8	3	76.2
5/32	4.0	15/32	12	1	25.4	3-1/8	79.4
3/16	4.8	9/16	15	1-1/16	27.0	3-3/8	85.7
7/32	5.6	11/16	17	1-1/8	28.6	3-1/2	88.9
1/4	6.3	3/4	20	1-3/16	30.2	3-3/4	95.2
5/16	7.9	1	25.4	1-1/4	31.7	3-7/8	98.4
3/8	9.5	1-1/8	28.6	1-3/8	34.9	4-3/8	111
7/16	11.1	1-3/8	34.9	1-7/16	36.5	4-1/2	114
1/2	12.7	1-5/8	41.3	1-1/2	38.1	4-3/4	121
9/16	14.3	1-3/4	44.4	1-5/8	41.3	5-1/8	130
5/8	15.9	2	50.8	1-11/16	42.9	5-1/4	133
11/16	17.5	2-1/4	57.1	1-3/4	44.4	5-1/2	140
3/4	19.0	2-3/8	60.3	1-13/16	46.0	5-3/4	146
13/16	20.6	2-1/2	63.5	1-7/8	47.6	5-7/8	149
7/8	22.2	2-3/4	69.8	1-15/16	49.2	6-1/8	156
				2	50.8	6-1/4	159

*U.S. sizes designed by diameters. Circumferences shown are for comparison only.

TABLE 5.23 Basic Formulas (For Computing Loads on Block and Tackle)

Adopted symbol	Description	Formula
M	Mechanical advantage	$M = N$
N	Number of parts of line supporting load	Known quantity
W	Load being hoisted, pounds	$W = P \times N \times E$
P	Lead-line pull, pounds	$P = W / (N \times E)$
E	Efficiency of tackle	Refer to chart below
F	Load on standing block, pounds	$F = W + P$
L	Length of lead line (hauled in), feet	$L = D \times N$
D	Distance load is hoisted, feet	Known distance
T	Time required to hoist load, minutes	Known time
A	Accomplishment (work put in), foot pounds	$A = P \times L$
H	Horsepower (work expressed mathematically) (1 hp =33,000 lb. of work done in 1 min)	$H = (P \times L) / (33{,}000 \times T)$

Number of parts of line supporting load N	Efficiency of tackle* E
1	0.96
2	0.922
3	0.885
4	0.849
5	0.815
6	0.783
7	0.751
8	0.722
9	0.693
10	0.664
11	0.638
12	0.612

*The actual efficiency is determined by many variables, such as type and condition of bearings, type and construction of wire rope, size and lubrication of sheave pin, and diameter of sheave. Figures listed are very conservative averages and in most cases can be considered applicable to both drums and sheaves.

Example

Problem: Find the horsepower required to lift 40,000 lb to a height of 40 ft within average time intervals of 2 min, using a 3-sheave head block and a 2-sheave fall block (5 parts of line). Blocks, hooks, and shackles to be of predetermined safe working load for a 40,000-lb load.

Solution: The formula for horsepower being $H = (P \times L)/(33{,}000 \times T)$, it is necessary to determine the values of P and L. These are obtained by applying the above formulas to the known factors, which are W = 40,000 lb, D = 40 ft, T = 2 min, N = 5 parts of line, E = 0.815 efficiency of tackle.

$P = W/(N \times E)$ or, in this instance, P= 40,000/ 5 × 0.815 = 9,800 lb. In addition, the lead line passes over one lead sheave. An additional 4% friction loss must be considered. Therefore total P= 9,800/.96 = 10,230 lb. L= $D \times N$ or, in this instance, L= 40 × 5 200 ft. Since P and L have been computed, it is now possible to determine the horsepower required:

$$H = \frac{P \times L}{33{,}000 \times T} = \frac{10{,}230 \times 200}{33{,}000 \times 2}$$
$$= 31 \text{ horsepower}$$

Section
2

Rigging Tools

Chapter

6

Fiber Rope (Natural and Synthetic)

Rope probably is one of humans' oldest tools. For thousands of years, vegetable materials were used for hauling and lifting loads. At first primitive people used vine stems and woven rushes twisted together to form crude strands that had sufficient strength to meet job requirements, yet were flexible and could be tied into knots. In time, other materials were used, such as fibrous bark of the palm tree, hair from the coconut, camel hair, horse hair, thongs cut from hides, and fibers of cotton, jute, sisal, flax, wild and cultivated hemp, and manila (from the abaca, a wild banana plant that grows in the Philippines).

The introduction of nylon filaments in the 1940s expanded rope-making capabilities significantly. Today synthetic-fiber ropes of nylon, polyester, polyethylene, and polypropylene, as well as various combinations of these materials, have replaced natural-fiber ropes in many applications. Table 6.1 compares the characteristics of various fibers used in cordage.

Fiber Rope

The function of any rope basically is to transmit a tensile force from its point of origin to a point of application. But for rope to provide this service effectively, it must have a closely packed structural form that will remain compact and maintain its cross-sectional dimension throughout its serviceable life. The degree and effectiveness of this compactness will depend on the amount of twist applied, the tension in

TABLE 6.1 Comparative Characteristics of Various Fibers Used in Cordage

	3- and 8-strand, plaited		Double braid	
	Manila		Core and cover nylon	
Rope diameter	1 in.	2 in.	1 in.	2 in.
Strength characteristics				
Tensile strength, dry (approximate average), lb	9,000	31,000	31,300	117,000
Recommended factor of safety	5	5	5 to 8*	5 to 8*
Working strength, lb	1,800	6,200	To 6,260*	To 23,400*
Wet strength compared to dry strength, %	To 120	To 120	88	88
Strength per unit of weight or "breaking length" (breaking strength), lb/ft	33,000	29,000	118,000	110,000
Cyclic loading characteristics	Poor	Poor	Excellent	Excellent
Individual filament or fiber strength, grams per denier	6.0 to 7.5	6.0 to 7.5	7.5 to 8.3	7.5 to 8.3
Weight and density characteristics				
Weight per 100 ft, lb	27.0	108.0	26.0	106.0
Specific gravity of fiber	1.38	1.38	1.14	1.14
Ability to float	No	No	No	No
Elasticity, stretch (approximate)				
Permanent elongation at working load (20% of breaking strength), %	4.8	4.8	6.5 (11.0 wet)	
Working elasticity (temporary stretch under load) at working load (20% of breaking strength), %	5.0	5.0	6.5 (11.0 wet)	
Elongation at 100% load (at break) for broken-in ropes, %	13	13	23.5	
Individual filament or fiber elongation, %	2 to 3	2 to 3	16 to 20	
Surface characteristics				
Rendering qualities, ability to ease out smoothly under load over bitts	Poor		Good, requires fewer wraps	
Hand (feeling of rope to the touch)	Some harshness due to hairs; after use considerable harshness due to broken fiber ends		Smooth; after use becomes fuzzy with a softer feel	

Water absorbed into fiber (Some water will be held between fibers of all ropes)	Up to 100% of weight of rope	5 to 6%
Resistance to rot, mildew, and attack by marine organisms (Some marine organisms will attach themselves to any submerged object, including synthetic ropes)	Poor	100% resistant
Deterioration		
Due to aging (stored ropes, ideal conditions)	About 1% per year	Zero
Due to exposure to sunlight	Some slight	Some slight
Resistance to chemicals		
Acids	Very poor	Fair, except to concentrated sulfuric and hydrochloric acids
Alkalis	Very poor	Excellent
Solvents	Good	Good
Wear		
Resistance to surface abrasion	Good	Excellent
Resistance to internal wear from flexing	Good	Excellent
Resistance to cutting (toughness)	Good	Excellent
High- and low-temperature properties		
Melting point	Loses strength rapidly over 180°F	480°F; progressive strength loss above 350°F
Low-temperature properties	No change	No change
Flammability	Burns like wood	Burns with difficulty

*Depending on application.
SOURCE: *Sampson Rope Manual*, Sampson Ocean Systems, Inc.

TABLE 6.1 Comparative Characteristics of Various Fibers Used in Cordage (Continued)

	3- and 8-strand, plaited			
	Nylon		Polyester	
Rope diameter	1 in.	2 in.	1 in.	2 in.
Strength characteristics				
Tensile strength, dry (approximate average), lb	25,000	92,000	20,000 to 22,000	65,000 to 80,000
Recommended factor of safety	9	9	9	9
Working strength, lb	2,890	10,000	2,220 to 2,450†	7,200 to 8,900†
Wet strength compared to dry strength, %	90 to 95	90 to 95	100	100
Strength per unit of weight or "breaking length" (breaking strength), lb/ft	96,000	97,000	65,000 to 72,000	55,000 to 64,000
Cyclic loading characteristics	Good	Good	Good to excel-lent	Good to excel-lent
Individual filament or fiber strength, grams per denier	7.5 to 8.3	7.5 to 8.3	6.0 to 7.0	6.0 to 7.0
Weight and density characteristics				
Weight per 100 ft, lb	26.0	95.0	30.5	118.0
Specific gravity of fiber	1.14	1.14	1.38	1.38
Ability to float	No	No	No	No
Elasticity, stretch (approximate)				
Permanent elongation at working load (20% of breaking strength), %	8.0	8.0	6.0 to 6.2	6.0 to 6.2
Working elasticity (temporary stretch under load) at working load (20% of breaking strength), %	16.0	16.0	5.9 to 6.5	5.9 to 6.5
Elongation at 100% load (at break) for broken-in ropes, %	35	35	20 to 22	20 to 22
Individual filament or fiber elongation, %	16 to 20		11 to 13	11 to 13
Surface characteristics				
Rendering qualities, ability to ease out smoothly under load over bitts	Poor		Good	
Hand (feeling of rope to the touch)	Smooth; after use becomes fuzzy with a softer feel		Smooth and hard, not slippery; after use becomes fuzzy with a softer feel	

Water absorbed into fiber (Some water will be held between fibers of all ropes)	5 to 6%	Less than 1%
Resistance to rot, mildew, and attack by marine organisms (Some marine organisms will attach themselves to any submerged object, including synthetic ropes)	100% resistant	100% resistant
Deterioration		
Due to aging (stored ropes, ideal conditions)	Zero	Zero
Due to exposure to sunlight	Some slight	Almost none
Resistance to chemicals		
Acids	Fair, except to concentrated sulfuric and hydrochloric acids	Very good to excellent
Alkalis	Excellent	Very good, except to concentrated sodium hydroxide at high temperatures
Solvents	Good	Very good to excellent
Wear		
Resistance to surface abrasion	Very good	Excellent
Resistance to internal wear from flexing	Excellent	Very good to excellent
Resistance to cutting (toughness)	Excellent	Very good to excellent
High- and low-temperature properties		
Melting point	480°F; progressive strength loss above 350°F	480°F; progressive strength loss above 350°F
Low-temperature properties	No change	No change
Flammability	Burns with difficulty	Burns with difficulty

*Depending on application.
†Depending on grade of polyester used.
SOURCE: *Sampson Rope Manual*, Sampson Ocean Systems, Inc.

TABLE 6.1 Comparative Characteristics of Various Fibers Used in Cordage (Continued)

	3-strand		12-strand single braid	
	Polyethlene		Polyester and polypropylene	
Rope diameter	1 in.	2 in.	1 in.	2 in.
Strength characteristics				
Tensile strength dry, (approximate average), lb	12,600	47,700	23,500	88,000
Recommended factor of safety	6	6	5 to 8*	5 to 8*
Working strength, lb	2,100	7,900	To 4,700*	To 17,600*
Wet strength compared to dry strength, %	100	100	100	100
Strength per unit of weight or "breaking length" (breaking strength), lb/ft	66,000	75,500	94,000	89,000
Cyclic loading characteristics	Fair	Fair	Excellent	
Individual filament or fiber strength, grams per denier	3.0 to 7.0	3.0 to 7.0	Polyester 7.5 to 9.0; monopoly 3.0 to 7.0	
Weight and density characteristics				
Weight per 100 ft, lb	18.5	72.5	25.0	99.0
Specific gravity of fiber	0.95	0.95	1.14	1.14
Ability to float	Yes	Yes	No	No
Elasticity, stretch (approximate)				
Permanent elongation at working load (20% of breaking strength), %	5.8	5.8	3.9	3.9
Working elasticity (temporary stretch under load) at working load (20% of breaking strength), %	5.9	5.9	3.0	3.0
Elongation at 100% load (at break) for broken-in ropes, %	22	22	16	16
Individual filament or fiber elonation, %	10 to 12	10 to 12	11 to 20	
Surface characteristics				
Rendering qualities, ability to ease out smoothly under load over bitts	Good but requires extra wraps		Good	
Hand (feeling of rope to the touch)	Smooth and very slippery; after use becomes slightly harsh due to broken fiber ends		Smooth, not slippery; after use, fuzz and a few broken fiber ends but little harshness	

Water absorbed into fiber (Some water will be held between fibers of all ropes)	Zero	Less than 1%
Resistance to rot, mildew, and attack by marine organisms (Some marine organisms will attach themselves to any submerged object, including synthetic ropes)	100% resistant	100% resistant
Deterioration		
Due to aging (stored ropes, ideal conditions)	Zero	Zero
Due to exposure to sunlight	Some; black resists best	Some slight
Resistance to chemicals		
Acids	Excellent except to concentrated sulfuric acid	Very good to excellent
Alkalis	Good	Good, except to concentrated sodium hydroxide at high temperatures
Solvents	Good	Good to excellent
Wear		
Resistance to surface abrasion	Good	Very good
Resistance to internal wear from flexing	Very good	Very good
Resistance to cutting (toughness)	Good	Very good
High- and low-temperature properties		
Melting point	280°F; softens above 250°F	330°F; softens above 300°F
Low-temperature properties	Brittle below −150°F	No change
Flammability	Burns with difficulty	Burns with difficulty

*Depending on application.
SOURCE: *Sampson Rope Manual*, Sampson Ocean Systems, Inc.

TABLE 6.1 Comparative Characteristics of Various Fibers Used in Cordage (Continued)

	3- and 8-strand, plaited			
	Polypropylene monofilament		Polypropylene multifilament	
Rope diameter	1 in.	2 in.	1 in.	2 in.
Strength characteristics				
Tensile strength, dry (approximate average), lb	14,000	52,000	15,700	55,500
Recommended factor of safety	6	6	6	6
Working strength, lb	2,330	8,700	2,620*	9,250*
Wet strength compared to dry strength, %	102 to 105	102 to 105	102 to 105	102 to 105
Strength per unit of weight or "breaking length" (breaking strength), lb/ft	77,800	75,500	82,000	75,000
Cyclic loading characteristics	Excellent	Excellent	Excellent	Excellent
Individual filament or fiber strength, grams per denier	3.0 to 7.0	3.0 to 7.0	3.0 to 7.0	3.0 to 7.0
Weight and density characteristics				
Weight per 100 ft, lb	18.0	69.0	19.2	74.0
Specific gravity of fiber	0.91	0.91	0.91	0.91
Ability to float	Yes	Yes	Yes	Yes
Elasticity, stretch (approximate)				
Permanent elongation at working load (20% of breaking strength), %	3.8	3.8	7.5	7.5
Working elasticity (temporary stretch under load) at working load (20% of breaking strength), %	8.9	8.9	10.5	10.5
Elongation at 100% load (at break) for broken-in ropes, %	24	24	36	36
Individual filament or fiber elongation, %	16 to 20	10 to 12	22 to 28	22 to 28
Surface characteristics				
Rendering qualities, ability to ease out smoothly under load over bitts	Poor		Fair	
Hand (feeling of rope to the touch)	Smooth but not slippery; after use becomes harsh due to broken fiber ends		Smooth and soft with some natural fuzziness; remains same after use	

Water absorbed into fiber (Some water will be held between fibers of all ropes)	Zero	Zero
Resistance to rot, mildew, and attack by marine organisms (Some marine organisms will attach themselves to any submerged object, including synthetic ropes)	100% resistant	100% resistant
Deterioration		
Due to aging (stored ropes, ideal conditions)	Zero	Zero
Due to exposure to sunlight	White some; black resists best	White some; black resists best
Resistance to chemicals		
Acids	Excellent	Excellent
Alkalis	Good	Good
Solvents	Good	Good
Wear		
Resistance to surface abrasion	Good	Good
Resistance to internal wear from flexing	Very good	Very good
Resistance to cutting (toughness)	Good	Very good
High- and low-temperature properties		
Melting point	330°F, softens above 300°F	330°F, softens above 300°F
Low-temperature properties	No change	No change
Flammability	Burns with difficulty	Burns with difficulty

*Depending on application.
SOURCE: *Sampson Rope Manual*, Sampson Ocean Systems, Inc.

TABLE 6.1 Comparative Characteristics of Various Fibers Used in Cordage (Continued)

	3- and 8-strand, plaited			
	Core polypropylene; cover polyester and polyethlene		Core polyethlene; cover veneer of polyethylene and dacron	
Rope diameter	1 in.	2 in.	1 in.	2 in.
Strength characteristics				
Tensile strength, dry (approximate average), lb	15,000	56,500	14,000	60,000
Recommended factor of safety	6¾	6¾	6¾	6¾
Working strength, lb	2,500	9,400	2,300	10,000
Wet strength compared to dry strength, %	100	100	100	100
Strength per unit of weight or "breaking length" (breaking strength), lb/ft	69,800	68,000	53,000	63,000
Cyclic loading characteristics	Very good	Very good	Good	Good
Individual filament or fiber strength, grams per denier	—	—	—	—
Weight and density characteristics				
Weight per 100 ft, lb	21.5	83.0	26.5	95.0
Specific gravity of fiber	Varies with rope size	Varies with rope size	Varies with rope size	Varies with rope size
Ability to float	Varies with rope size	Varies with rope size	No	No
Elasticity, stretch (approximate)				
Permanent elongation at working load (20% of breaking strength), %	5.9	5.9	4.7	4.7
Working elasticity (temporary stretch under load) at working load (20% of breaking strength), %	7.1	7.1	5.1	5.1
Elongation at 100% load (at break) for broken-in ropes, %	27	27	21	21
Individual filament or fiber elongation, %	—	—	—	—

Surface characteristics		
Rendering qualities, ability to ease out smoothly under load over bitts	Good	Good
Hand (feeling of rope to the touch)	Smooth and hard; after use bristles and fuzz appear but little harshness	Smooth and hard, not slippery; after use becomes fuzzy but not harsh
Water absorbed into fiber (Some water will be held between fibers of all ropes)	Almost zero	Almost zero
Resistance to rot, mildew, and attack by marine organisms (Some marine organisms will attach themselves to any submerged object, including synthetic ropes)	100% resistant	100% resistant
Deterioration		
Due to aging (stored ropes, ideal conditions)	Zero	Zero
Due to exposure to sunlight	Slight	Some slight
Resistance to chemicals		
Acids	Good to excellent	Excellent, except to concentrated nitric acid
Alkalis	Excellent	Excellent
Solvents	Good to excellent	Good to excellent
Wear		
Resistance to surface abrasion	Very good	Very good
Resistance to internal wear from flexing	Very good	Very good
Resistance to cutting (toughness)	Very good	Very good
High- and low-temperature properties		
Melting point	Progressive strength loss above 250°F	Progressive strength loss above 250°F
Low-temperature properties	Some brittleness below −150°F	Some brittleness below −150°F
Flammability	Burns with difficulty	Burns with difficulty

SOURCE: *Sampson Rope Manual*, Sampson Ocean Systems, Inc.

TABLE 6.1 Comparative Characteristics of Various Fibers Used in Cordage (Continued)

	Double braid			
	Core polypropylene; cover nylon		Core and cover polyester	
Rope diameter	1 in.	2 in.	1 in.	2 in.
Strength characteristics				
Tensile strength, dry (approximate average), lb	28,400	106,000	28,400	106,000
Recommended factor of safety	5 to 8*	5 to 8*	5 to 8*	5 to 8*
Working strength, lb	To 5,700*	To 21,200*	To 5,700*	To 21,200*
Wet strength compared to dry strength, %	95	95	100	100
Strength per unit of weight or "breaking length" (breaking strength), lb/ft	120,000	112,000	88,700	82,800
Cyclic loading characteristics	Excellent		Excellent	
Individual filament or fiber strength, grams per denier	Cover 7.5 to 8.3; core 3.0 to 7.0		Cover 6.0 to 7.0; core 3.0 to 7.0	
Weight and density characteristics				
Weight per 100 ft, lb	24.0	95.0	32.0	126.0
Specific gravity of fiber	1.02	1.02	1.38	1.38
Ability to float	Yes	Yes	No	No
Elasticity, stretch (approximate)				
Permanent elongation at working load (20% of breaking strength), %	6 (11.0 wet)		3 (5 wet)	
Working elasticity (temporary stretch under load) at working load (20% of breaking strength), %	6 (11.0 wet)		3 (5 wet)	
Elongation at 100% load (at break) for broken-in ropes, %	30		17	
Individual filament or fiber elonation, %	16 to 30		11 to 13	
Surface characteristics				
Rendering qualities, ability to ease out smoothly under load over bitts	Good		Good	
Hand (feeling of rope to the touch)	Smooth; after use becomes fuzzy with a softer feel		Smooth, not slippery; after use becomes fuzzy with a softer feel	

Water absorbed into fiber (Some water will be held between fibers of all ropes)	3.0%	Less than 1%
Resistance to rot, mildew, and attack by marine organisms (Some marine organisms will attach themselves to any submerged object, including synthetic ropes)	100% resistant	100% resistant
Deterioration		
Due to aging (stored ropes, ideal conditions)	Zero	Zero
Due to exposure to sunlight	Some slight	Almost none
Resistance to chemicals		
Acids	Fair, except to concentrated sulfuric and hydrochloric acids	Very good to excellent
Alkalis	Excellent	Very good, except to concentrated sodium hydroxide at high temperatures
Solvents	Good	Very good to excellent
Wear		
Resistance to surface abrasion	Very good to excellent	Excellent
Resistance to internal wear from flexing	Good	Very good to excellent
Resistance to cutting (toughness)	Very good to excellent	Very good to excellent
High- and low-temperature properties		
Melting point	Good	480°F; progressive strength loss above 350°F
Low-temperature properties	Very good	No change
Flammability	Burns with difficulty	Burns with difficulty

*Depending on application.
SOURCE: *Sampson Rope Manual*, Sampson Ocean Systems, Inc.

various components during twisting operations, and the compression that the tubes and dies impart to a rope's components while being twisted (see Fig. 6.1).

The construction most commonly encountered in fiber ropes is one in which the fibers are first compacted by a right-hand twist into yarns; the yarns are then compacted by a left-hand twist into strands; and finally the strands are compacted by a right-hand twist into ropes. (Cables are constructed by twisting these ropes in a left-hand direction.)

With its components twisted in opposing directions, a rope will maintain its compactness during use, but only as long as the opposing twists remain in equilibrium. Any use of a rope that disrupts this equilibrium will alter the rope's properties.

Improper coiling and handling of rope will remove the turns, eventually inducing strand kinking, and thus will reduce a rope's serviceability.

Repeated stretching may alter the rope and the strand-twist relationship.

Swelling, as a result of wetting, has a strong effect on the equilibrium of twists.

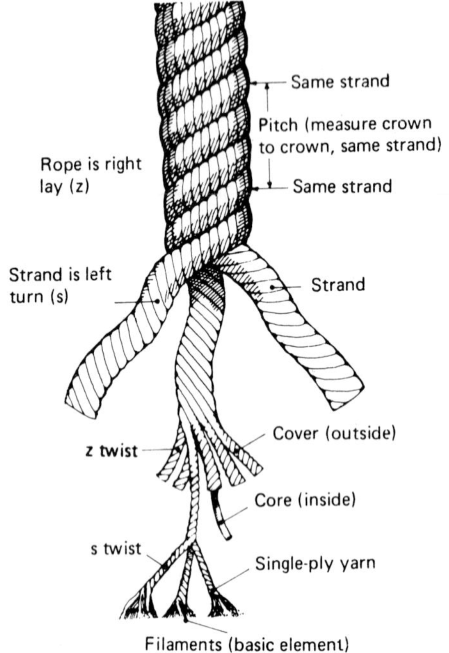

Figure 6.1 Fiber rope construction. (*a*) Three-strand rope. (*b*) Eight-strand plaited rope. (*c*) Double-braided rope. (*Courtesy of The Cordage Group, Auburn, N.Y.*)

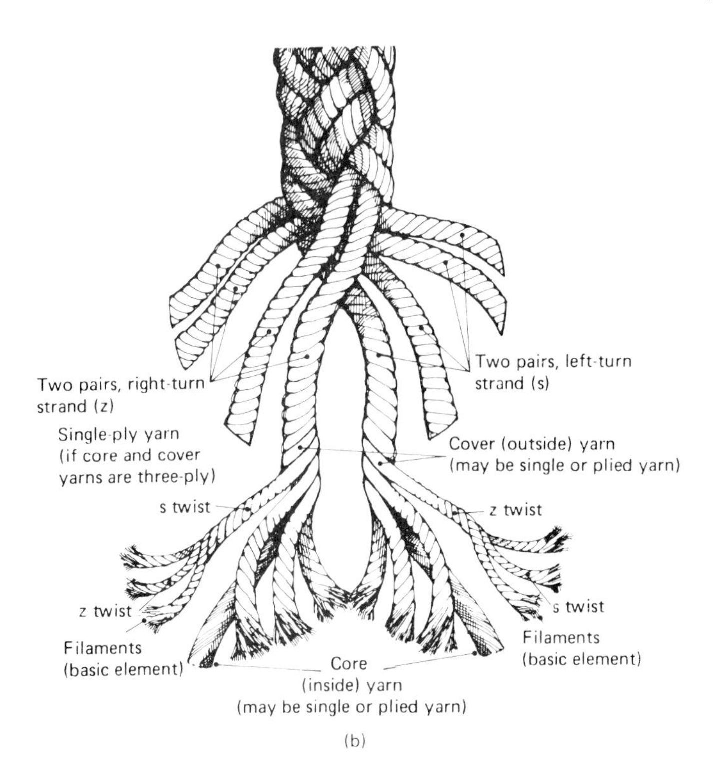

(b)

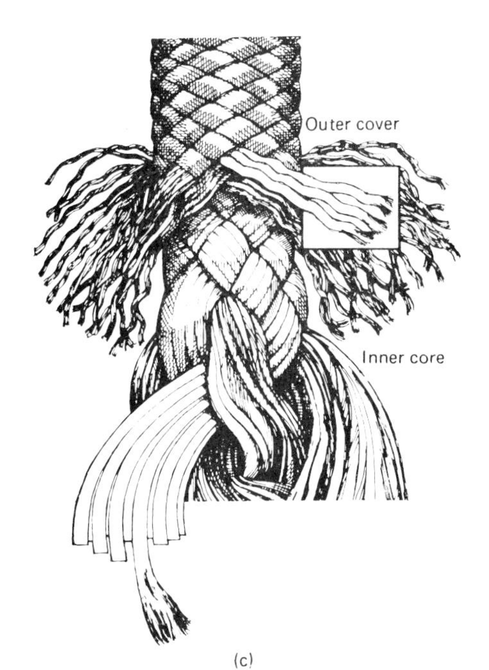

(c)

Figure 6.1 *(Continued)*

Stresses

A rope in use is subject to three basic mechanical stresses: tensile pull, structural friction associated with bending and flexing, and surface friction that contributes to abrasion. These forces act on a rope under dynamic operating conditions and repeated loadings. Secondary stresses resulting from acceleration and heat are usually also involved.

Tensile pull or load. Acceleration stresses, resulting from reversing a rope's motion momentarily, increase the tensile load on a rope by as much as 100%. When a rope bends on sheaves, the tensile load will increase by as much as 10% for each sheave. Tensile stress applied and maintained during a pull will stretch a rope continually, reducing its diameter until the rope breaks. Of course, the amount of stretch will depend on the fiber, amount of twist, and size of rope, with the shortest components—the rope's core or inner fibers—stretching to their limit and breaking first.

Bending and flexing. When a rope is bent, its individual strands roll over one another, contracting for the inner part of the bend and pulling away from each other to form its outer part. The strands rub against each other and produce powdering and chafing on the surface. If the rope is excessively overloaded, then internal chafing will result where strands are in contact. Since strand movement depends on the bend radius, a larger sheave will minimize the wear. Sheave diameters should always be five times a rope's diameter or more, depending on the speed and frequency of strand-bending adjustments. No frictional damage occurs to a rope under tensile stress when subjected to static bending, such as in a knot. Load failure instead results from a shearing force applied to the rope by the knot's loop.

Surface friction. A rope in use normally is subjected to surface friction as it rubs against itself and against sharp edges, both in straight tension and in bending moments. It is also subject to internal and external wear from imbedded grit, resulting in the major cause of most rope failure—surface abrasion.

Strength

Breaking strength is a measure of a rope's serviceability. However, strength as a property of the rope is significant only in relation to its size. Rope strengths are not measured directly in terms of stress per unit area, as with most structural materials, but rather on the basis of strength and length-to-weight ratio. Table 6.2 gives the rope strength for various rope types.

TABLE 6.2 Rope Strength for Various Types of Rope

(Three-Strand Laid and Eight-Strand Plaited, Standard Construction)

Nominal size		Manila			
Diameter	Circumference	Linear density,* lb/100 ft	New rope tensile strength,† lb	Safety factor	Working load,‡ lb
3/16	5/8	1.50	406	10	41
1/4	3/4	2.00	540	10	54
5/16	1	2.90	900	10	90
3/8	1 1/8	4.10	1,220	10	122
7/16	1 1/4	5.25	1,580	9	176
1/2	1 1/2	7.50	2,380	9	264
9/16	1 3/4	10.4	3,100	8	388
5/8	2	13.3	3,960	8	496
3/4	2 1/4	16.7	4,860	7	695
13/16	2 1/2	19.5	5,850	7	835
7/8	2 3/4	22.4	6,950	7	995
1	3	27.0	8,100	7	1,160
1 1/16	3 1/4	31.2	9,450	7	1,350
1 1/8	3 1/2	36.0	10,800	7	1,540
1 1/4	3 3/4	41.6	12,200	7	1,740
1 5/16	4	47.8	13,500	7	1,930
1 1/2	4 1/2	60.0	16,700	7	2,380
1 5/8	5	74.5	20,200	7	2,880
1 3/4	5 1/2	89.5	23,800	7	3,400
2	6	108.	28,000	7	4,000
2 1/8	6 1/2	125.	32,400	7	4,620
2 1/4	7	146.	37,000	7	5,300
2 1/2	7 1/2	167.	41,800	7	5,950
2 5/8	8	191.	46,800	7	6,700
2 7/8	8 1/2	215.	52,000	7	7,450
3	9	242.	57,500	7	8,200
3 1/4	10	298.	69,500	7	9,950
3 1/2	11	366.	82,000	7	11,700
4	12	434.	94,500	7	13,500

*Linear density (lb/100 ft) shown is average. Maximum is 5% higher.

†New rope tensile strengths are based on tests of new and unused rope of standard construction in accordance with Cordage Institute standard test methods.

‡Working loads are for rope in good condition with appropriate splices, in noncritical applications, and under normal service conditions. Working loads should be exceeded only with expert knowledge of conditions and professional estimates of risk. Working load should be reduced where life, limb, or valuable property is involved, or for exceptional service conditions such as shock loads or sustained loads. Working loads are not necessarily intended to apply in those applications where a thorough engineering analysis of all conditions of use has been made and these conditions will not be exceeded in service. In such cases tensile strength, elongation, energy absorption, behavior under long-term or cyclic loading, and other pertinent properties of a rope may be evaluated to allow the selection of the rope best suited to the requirements of the application and an analysis of any risk involved in its use.

SOURCE: The Cordage Institute.

TABLE 6.2 Rope Strength for Various Types of Rope (Continued)

(Three-Strand Laid and Eight-Strand Plaited, Standard Construction)

Nominal size		Propylene			
Diameter	Circumference	Linear density,* lb/100 ft	New rope tensile strength,† lb	Safety factor	Working load,‡ lb
3/16	5/8	.70	720	10	72
1/4	3/4	1.20	1,130	10	113
5/16	1	1.80	1,710	10	171
3/8	1 1/8	2.80	2,440	10	244
7/16	1 1/4	3.80	3,160	9	352
1/2	1 1/2	4.70	3,780	9	420
9/16	1 3/4	6.10	4,600	8	575
5/8	2	7.50	5,600	8	700
3/4	2 1/4	10.7	7,650	7	1,090
13/16	2 1/2	12.7	8,900	7	1,270
7/8	2 3/4	15.0	10,400	7	1,490
1	3	18.0	12,600	7	1,800
1 1/16	3 1/4	20.4	14,400	7	2,060
1 1/8	3 1/2	23.8	16,500	7	2,360
1 1/4	3 3/4	27.0	18,900	7	2,700
1 5/16	4	30.4	21,200	7	3,020
1 1/2	4 1/2	38.4	26,800	7	3,820
1 5/8	5	47.6	32,400	7	4,620
1 3/4	5 1/2	59.0	38,800	7	5,550
2	6	69.0	46,800	7	6,700
2 1/8	6 1/2	80.0	55,000	7	7,850
2 1/4	7	92.0	62,000	7	8,850
2 1/2	7 1/2	107.	72,000	7	10,300
2 5/8	8	120.	81,000	7	11,600
2 7/8	8 1/2	137.	91,000	7	13,000
3	9	153.	103,000	7	14,700
3 1/4	10	190.	123,000	7	17,600
3 1/2	11	232.	146,000	7	20,800
4	12	276.	171,000	7	24,400

*Linear density (lb/100 ft) shown is average. Maximum is 5% higher.

†New rope tensile strengths are based on tests of new and unused rope of standard construction in accordance with Cordage Institute standard test methods.

‡Working loads are for rope in good condition with appropriate splices, in noncritical applications, and under normal service conditions. Working loads should be exceeded only with expert knowledge of conditions and professional estimates of risk. Working load should be reduced where life, limb, or valuable property is involved, or for exceptional service conditions such as shock loads or sustained loads. Working loads are not necessarily intended to apply in those applications where a thorough engineering analysis of all conditions of use has been made and these conditions will not be exceeded in service. In such cases tensile strength, elongation, energy absorption, behavior under long-term or cyclic loading, and other pertinent properties of a rope may be evaluated to allow the selection of the rope best suited to the requirements of the application and an analysis of any risk involved in its use.

SOURCE: The Cordage Institute.

TABLE 6.2 Rope Strength for Various Types of Rope (Continued)

(Three-Strand Laid and Eight-Strand Plaited, Standard Construction)

Nominal size		Polyester			
Diameter	Circumference	Linear density,* lb/100 ft	New rope tensile strength,† lb	Safety factor	Working load,‡ lb
3/16	5/8	1.20	900	10	90
1/4	3/4	2.00	1,490	10	149
5/16	1	3.10	2,300	10	230
3/8	1 1/8	4.50	3,340	10	334
7/16	1 1/4	6.20	4,500	9	500
1/2	1 1/2	8.00	5,750	9	640
9/16	1 3/4	10.2	7,200	8	900
5/8	2	13.0	9,000	8	1,130
3/4	2 1/4	17.5	11,300	7	1,610
13/16	2 1/2	21.0	14,000	7	2,000
7/8	2 3/4	25.0	16,200	7	2,320
1	3	30.4	19,800	7	2,820
1 1/16	3 1/4	34.4	23,000	7	3,280
1 1/8	3 1/2	40.0	26,600	7	3,800
1 1/4	3 3/4	46.2	29,800	7	4,260
1 5/16	4	52.5	33,800	7	4,820
1 1/2	4 1/2	67.0	42,200	7	6,050
1 5/8	5	82.0	51,500	7	7,350
1 3/4	5 1/2	98.0	61,000	7	8,700
2	6	118.	72,000	7	10,300
2 1/8	6 1/2	135.	83,000	7	11,900
2 1/4	7	157.	96,500	7	13,800
2 1/2	7 1/2	181.	110,000	7	15,700
2 5/8	8	204.	123,000	7	17,600
2 7/8	8 1/2	230.	139,000	7	19,900
3	9	258.	157,000	7	22,400
3 1/4	10	318.	189,000	7	27,000
3 1/2	11	384.	228,000	7	32,600
4	12	454.	270,000	7	38,600

*Linear density (lb/100 ft) shown is average. Maximum is 5% higher.

†New rope tensile strengths are based on tests of new and unused rope of standard construction in accordance with Cordage Institute standard test methods.

‡Working loads are for rope in good condition with appropriate splices, in noncritical applications, and under normal service conditions. Working loads should be exceeded only with expert knowledge of conditions and professional estimates of risk. Working load should be reduced where life, limb, or valuable property is involved, or for exceptional service conditions such as shock loads or sustained loads. Working loads are not necessarily intended to apply in those applications where a thorough engineering analysis of all conditions of use has been made and these conditions will not be exceeded in service. In such cases tensile strength, elongation, energy absorption, behavior under long-term or cyclic loading, and other pertinent properties of a rope may be evaluated to allow the selection of the rope best suited to the requirements of the application and an analysis of any risk involved in its use.

SOURCE: The Cordage Institute.

TABLE 6.2 Rope Strength for Various Types of Rope (Continued)
(Three-Strand Laid and Eight-Strand Plaited, Standard Construction)

Nominal size		Composite‖			
Diameter	Circumference	Linear density,* lb/100 ft	New rope tensile strength,† lb	Safety factor	Working load,‡ lb
3/16	5/8	.94	720	10	72
1/4	3/4	1.61	1,130	10	113
5/16	1	2.48	1,710	10	171
3/8	1 1/8	3.60	2,430	10	243
7/16	1 1/4	5.00	3,150	9	350
1/2	1 1/2	6.50	3,960	9	440
9/16	1 3/4	8.00	4,860	8	610
5/8	2	9.50	5,760	8	720
3/4	2 1/4	12.5	7,560	7	1,080
13/16	2 1/2	15.2	9,180	7	1,310
7/8	2 3/4	18.0	10,800	7	1,540
1	3	21.8	13,100	7	1,870
1 1/16	3 1/4	25.6	15,200	7	2,170
1 1/8	3 1/2	29.0	17,400	7	2,490
1 1/4	3 3/4	33.4	19,800	7	2,830
1 5/16	4	35.6	21,200	7	3,020
1 1/2	4 1/2	45.0	26,800	7	3,820
1 5/8	5	55.5	32,400	7	4,620
1 3/4	5 1/2	66.5	38,800	7	5,550
2	6	78.0	46,800	7	6,700
2 1/8	6 1/2	92.0	55,000	7	7,850
2 1/4	7	105.	62,000	7	8,850
2 1/2	7 1/2	122.	72,000	7	10,300
2 5/8	8	138.	81,000	7	11,600
2 7/8	8 1/2	155.	91,000	7	13,000
3	9	174.	103,000	7	14,700
3 1/4	10	210.	123,000	7	17,600
3 1/2	11	256.	146,000	7	20,800
4	12	300.	171,000	7	24,400

*Linear density (lb/100 ft) shown is average. Maximum is 5% higher.

†New rope tensile strengths are based on tests of new and unused rope of standard construction in accordance with Cordage Institute standard test methods.

‡Working loads are for rope in good condition with appropriate splices, in noncritical applications, and under normal service conditions. Working loads should be exceeded only with expert knowledge of conditions and professional estimates of risk. Working load should be reduced where life, limb, or valuable property is involved, or for exceptional service conditions such as shock loads or sustained loads. Working loads are not necessarily intended to apply in those applications where a thorough engineering analysis of all conditions of use has been made and these conditions will not be exceeded in service. In such cases tensile strength, elongation, energy absorption, behavior under long-term or cyclic loading, and other pertinent properties of a rope may be evaluated to allow the selection of the rope best suited to the requirements of the application and an analysis of any risk involved in its use.

‖Materials and construction of this polyester/polypropylene composite rope conform to MIL-R-43942 and MIL-R-43952. For other composite ropes, consult the manufacturer.

SOURCE: The Cordage Institute.

TABLE 6.2 Rope Strength for Various Types of Rope (Continued)

(Three-Strand Laid and Eight-Strand Plaited, Standard Construction)

Nominal size		Nylon			
Diameter	Circumference	Linear density,* lb/100 ft	New rope tensile strength,† lb	Safety factor	Working load,‡ lb
3/16	5/8	1.00	900	12	75
1/4	3/4	1.50	1,490	12	124
5/16	1	2.50	2,300	12	192
3/8	1 1/8	3.50	3,340	12	278
7/16	1 1/4	5.00	4,500	11	410
1/2	1 1/2	6.50	5,750	11	525
9/16	1 3/4	8.15	7,200	10	720
5/8	2	10.5	9,350	10	935
3/4	2 1/4	14.5	12,800	9	1,420
13/16	2 1/2	17.0	15,300	9	1,700
7/8	2 3/4	20.0	18,000	9	2,000
1	3	26.4	22,600	9	2,520
1 1/16	3 1/4	29.0	26,000	9	2,880
1 1/8	3 1/2	34.0	29,800	9	3,320
1 1/4	3 3/4	40.0	33,800	9	3,760
1 5/16	4	45.0	38,800	9	4,320
1 1/2	4 1/2	55.0	47,800	9	5,320
1 5/8	5	66.5	58,500	9	6,500
1 3/4	5 1/2	83.0	70,000	9	7,800
2	6	95.0	83,000	9	9,200
2 1/8	6 1/2	109.	95,500	9	10,600
2 1/4	7	129.	113,000	9	12,600
2 1/2	7 1/2	149.	126,000	9	14,000
2 5/8	8	168.	146,000	9	16,200
2 7/8	8 1/2	189.	162,000	9	18,000
3	9	210.	180,000	9	20,000
3 1/4	10	264.	226,000	9	25,200
3 1/2	11	312.	270,000	9	30,000
4	12	380.	324,000	9	36,000

*Linear density (lb/100 ft) shown is average. Maximum is 5% higher.

†New rope tensile strengths are based on tests of new and unused rope of standard construction in accordance with Cordage Institute standard test methods.

‡Working loads are for rope in good condition with appropriate splices, in noncritical applications, and under normal service conditions. Working loads should be exceeded only with expert knowledge of conditions and professional estimates of risk. Working load should be reduced where life, limb, or valuable property is involved, or for exceptional service conditions such as shock loads or sustained loads. Working loads are not necessarily intended to apply in those applications where a thorough engineering analysis of all conditions of use has been made and these conditions will not be exceeded in service. In such cases tensile strength, elongation, energy absorption, behavior under long-term or cyclic loading, and other pertinent properties of a rope may be evaluated to allow the selection of the rope best suited to the requirements of the application and an analysis of any risk involved in its use.

SOURCE: The Cordage Institute.

TABLE 6.2 Rope Strength for Various Types of Rope (Continued)

(Three-Strand Laid and Eight-Strand Plaited, Standard Construction)

Nominal size		Sisal			
Diameter	Circumference	Linear density,* lb/100 ft	New rope tensile strength,† lb	Safety factor	Working load,‡ lb
3/16	5/8	1.50	360	10	36
1/4	3/4	2.00	480	10	48
5/16	1	2.90	800	10	80
3/8	1 1/8	4.10	1,080	10	108
7/16	1 1/4	5.26	1,400	9	156
1/2	1 1/2	7.52	2,120	9	236
9/16	1 3/4	10.4	2,760	8	354
5/8	2	13.3	3,520	8	440
3/4	2 1/4	16.7	4,320	7	617
13/16	2 1/2	19.5	5,200	7	743
7/8	2 3/4	22.5	6,160	7	880
1	3	27.0	7,200	7	1,030
1 1/16	3 1/4	31.3	8,400	7	1,200
1 1/8	3 1/2	36.0	9,600	7	1,370
1 1/4	3 3/4	41.7	10,800	7	1,540
1 5/16	4	47.8	12,000	7	1,710
1 1/2	4 1/2	59.9	14,800	7	2,110
1 5/8	5	74.6	18,000	7	2,570
1 3/4	5 1/2	89.3	21,200	7	3,030
2	6	108.	24,800	7	3,540
2 1/8	6 1/2	—	—	7	—
2 1/4	7	146.	32,800	7	4,690
2 1/2	7 1/2	—	—	7	—
2 5/8	8	191.	41,600	7	5,940
2 7/8	8 1/2	—	—	7	—
3	9	242.	51,200	7	7,300
3 1/4	10	299.	61,600	7	8,800
3 1/2	11	—	—	7	—
4	12	435.	84,000	7	13,000

*Linear density (lb/100 ft) shown is average. Maximum is 5% higher.

†New rope tensile strengths are based on tests of new and unused rope of standard construction in accordance with Cordage Institute standard test methods.

‡Working loads are for rope in good condition with appropriate splices, in noncritical applications, and under normal service conditions. Working loads should be exceeded only with expert knowledge of conditions and professional estimates of risk. Working load should be reduced where life, limb, or valuable property is involved, or for exceptional service conditions such as shock loads or sustained loads. Working loads are not necessarily intended to apply in those applications where a thorough engineering analysis of all conditions of use has been made and these conditions will not be exceeded in service. In such cases tensile strength, elongation, energy absorption, behavior under long-term or cyclic loading, and other pertinent properties of a rope may be evaluated to allow the selection of the rope best suited to the requirements of the application and an analysis of any risk involved in its use.

SOURCE: The Cordage Institute.

The degree of twist in a rope and its strands is the primary factor influencing a rope's strength. The greater the twist, the lower a rope's strength—a tightly twisted hard-lay rope will be as much as 20% lower in strength than the corresponding standard-lay rope.

Wetting a rope has a pronounced effect on its strength, with wet natural fiber ropes tending to swell in size and shrink in length. Swelling increases the angle of twist, thus making the rope harder and reducing its strength; synthetic fibers in particular lose considerable strength when wet. However, manila and cotton ropes tend to increase in strength.

Repeated tensile loading will not adversely affect rope strength. In fact, if a rope's internal yarns are not broken, repeated loading can actually increase breaking strengths by as much as 20%, because yarn and strand tensions become better adjusted as the rope is stretched repeatedly.

A rope's strength is only about two-thirds the potential strength of its combined components. Since rope efficiency depends on the degree of twist, a slacker twist results in higher strength. But as the number of strands increases for a given size, a rope's strength will decrease.

Weather also affects a rope's serviceable life by degrading the fibers and weakening them. Alternate wetting, drying, and stretching structurally loosens a rope, and this in turn increases elongation even during minor loading.

Elongation, elasticity, and creep

The degree to which a rope elongates depends on fiber type and rope size. Elongation also depends on a rope's construction. Hard-fiber ropes do not stretch as much as some made of soft fibers, small-diameter ropes stretch less than larger-sized ones when loaded to the breaking point, and the more its twist (the harder the lay), the more a rope will stretch.

After initial loading, each time a rope is used within a given loading range it will behave more and more like a rigid material. But unlike rigid materials, ropes do not display true elasticity. Once stretched by an appreciable load, a rope remains more or less permanently elongated.

Some contraction of the rope begins to take place immediately upon release of a load. More than half the total recovery that will ever be attained occurs in a matter of minutes. The extent of a rope's recovery from stretch will depend on the magnitude of the applied load, rope size, nature of fiber, and structure of rope.

A rope loaded lightly in comparison to its breaking strength will recover faster than when loaded heavily.

A larger-sized rope, when stretched, tends to readjust internally more or less permanently and thereby has less of a chance to recover from stretch.

A four-strand rope shows less recovery than a three-strand rope.

A hard-lay rope, one with more twist, will recover less than a soft-lay rope.

A rope loaded repeatedly, thereby compacted to its maximum tensile load, will ultimately show complete recovery from stretch, provided the applied load does not exceed the load used in prestretching the rope.

A wet rope will not show as marked a recovery as a dry rope, because the wet one has a higher initial stretch, and wetting internally adjusts the rope's yarn and strand tension relationship.

Rope loaded for any sustained time will continue to stretch beyond its initial elongation when the load was first applied. This slow, continual stretch under load is called creep. A rope pulled taut and secured to maintain tension will stretch until it breaks, even though the tension is well below its normal breaking strength.

Impact loading

Although a rope's ultimate elongation remains about the same, regardless of the rate of loading, its breaking strength will increase with a faster loading rate. However, with impact loading a rope can absorb considerably more energy than it can during a slow steadily applied load. Tests indicate that under extreme impact loading, a new hard-fiber rope will absorb as much as twice the amount of energy it would absorb from a slow steady loading.

Flexibility

A rope's flexibility depends on the degree with which strands that are normally compressed against each other can overcome the frictional forces binding them together, and thus adjust to a new position.

As the rope flexes (bends), lubricant is forced to the outer surface of the strands, which tends to restrict yarn and strand movement and thus results in poor rope performance. Improving both the type and amount of lubrication will improve a rope's flexing endurance. Hard-fiber rope requires a minimum lubricant content of about 10 to 15%. Sticky ingredients in lubricants will have an adverse effect on a rope's flexing endurance.

A twisted-fiber rope will have better flexing endurance for a comparable loading than a braided one of the same size and material. And increasing the twist will help improve a rope's flexing endurance. However, an increase in the number of strands generally will have no effect.

Remember, all the forces acting on a rope when it is pulled in tension are also present when the rope is bent. As this tension increases, the compacting action of the rope's strand surface tends to reduce rope flexibility. In addition, the outer surface compression that develops in a bent rope, in relation to its internal compressive forces, tends to create shearing stresses that further damage the rope.

Strand surface condition and fiber rigidity will also influence knot strength. Where strand surfaces are smooth, the frictional forces involved in bending are reduced and a knot will break more readily. When a rope is wet, knot strength efficiency ordinarily will increase; this is particularly true of manila rope.

Working load

Because of the wide range of rope use, rope condition, exposure to the different factors that affect rope behavior, and the degree of risk to life and property involved, it is impossible to make blanket recommendations for safe working loads. To provide the rigger with some guidelines for using rope under normal service conditions in noncritical applications, the Cordage Institute has tabulated safe working loads for rope in good condition with appropriate splices. However, where uses of rope involve a serious risk of injury to personnel or damage to valuable property, or if there is any question about the loads involved or the conditions of use, the working load should be reduced substantially and the rope properly inspected. Approximate safe working loads are given in Table 6.3.

Caution

1. Working loads are only guidelines for rope in good condition with appropriate splices, in noncritical applications, and under normal service conditions.

2. Working loads should be reduced where life, limb, or valuable property are involved, or for exceptional service conditions such as shock loads or sustained loads.

3. Working loads are not applicable when rope is subject to significant dynamic loading resulting from picking up, stopping, moving, or swinging a load rapidly or suddenly, such as when rope is used as towing lines, lifelines, safety lines, and climbing ropes.

TABLE 6.3 Approximate Safe Working Loads, Pounds

New Three-Strand Fiber Ropes					
Nominal rope diameter, in.	Manila	Nylon	Poly-propy-lene	Polyes-ter	Poly-ethyl-ene
3/16	100	200	150	200	150
1/4	120	300	250	300	250
5/16	200	500	400	500	350
3/8	270	700	500	700	500
1/2	530	1,250	830	1,200	800
5/8	880	2,000	1,300	1,900	1,050
3/4	1,080	2,800	1,700	2,400	1,500
7/8	1,540	3,800	2,200	3,400	2,100
1	1,800	4,800	2,900	4,200	2,500
1 1/8	2,400	6,300	3,750	5,600	3,300
1 1/4	2,700	7,200	4,200	6,300	3,700
1 1/2	3,700	10,200	6,000	8,900	5,300
1 5/8	4,500	12,400	7,300	10,800	6,500
1 3/4	5,300	15,000	8,700	12,900	7,900
2	6,200	17,900	10,400	15,200	9,500

Braided Synthetic Fiber Ropes			
Nominal rope diameter (inches)	Nylon cover, nylon core	Nylon cover, polypropy-lene core	Polyester cover, poly-propylene core
1/4	420	—	380
5/16	640	—	540
3/8	880	680	740
7/16	1,200	1,000	1,060
1/2	1,500	1,480	1,380
9/16	2,100	1,720	—
5/8	2,400	2,100	2,400
3/4	3,500	3,200	2,860
7/8	4,800	4,150	3,800
1	5,700	4,800	5,600
1 1/8	8,000	7,000	—
1 1/4	8,800	8,000	—
1 1/2	12,800	12,400	—
1 5/8	16,000	14,000	—
1 3/4	19,400	18,000	—
2	23,600	20,000	—

*Safety factor = 5.

Natural-Fiber Ropes

Natural-fiber ropes are made with short, overlapping fibers that come from plants and include manila, hemp, sisal, jute, and cotton. Each type of rope has its own advantages and disadvantages, and varying grades of a particular type of rope are used for different operating conditions.

Strength among natural-fiber ropes varies considerably and depends on such diverse factors as climate, curing, manufacturing, and maintenance.

The two most commonly used natural-fiber ropes are manila and sisal. Manila, a light yellow rope with a smooth waxy surface, is strong and durable. It should be used for operations requiring a dependable rope that stands up under severe use and resists weathering. Sisal, a whitish rather coarse rope, is less strong and durable. It should be used where rope requirements are less demanding and cost is a major consideration.

Note: Only number 1 grade manila fiber rope should be specified for rigging operations.

Manila rope is available in a number of grades.

Yacht. Extremely light-color smooth manila rope of the highest quality. It is costly and thus is used on jobs where appearance is most important.

Bolt. Very light-color high-grade manila rope that is about 10 to 15% stronger than number 1 grade manila.

Number 1. Very light-color standard high-grade manila hoisting rope specified for most rigging jobs.

Number 2. Slightly darker manila rope with about the same initial strength as number 1 grade, but that loses strength very rapidly through use.

Hardware store. Darkest of all manila ropes, a very poor grade, with many fiber ends protruding from the strands and having low strength and short life.

When purchasing hoisting rope, be sure to specify that new manila rope be made of first-grade manila fibers. Most rope manufacturers identify their first-grade manila rope by twisting one or more colored yarns, or a colored string, in a strand. Rope without such a fiber marking should not be used for hoisting.

Remember: It is up to the purchaser to specify the type of rope construction best suited to the specific job requirements.

Most manila rope used in rigging operations is of three-strand construction. However, a four-strand rope with a rope core is recommended for work where abrasion is the key factor. Such a rope is rounder in cross section and only slightly less strong than the more standard three-strand rope. It is used almost exclusively for power hoisting.

Manila ropes are chemically treated to resist mildew and dry rot, thus permitting their use for prolonged periods under wet and dirty conditions. Nevertheless, all manila ropes are subject to mildew and dry rot and therefore must be cleaned and dried at regular intervals.

Proper maintenance of manila rope will extend serviceability. To slow down rotting, wash rope in cold water after use and hang in loosely formed coils over wooden pegs to dry. Allow air to circulate around stored ropes. Carefully check slings and safety lines frequently for signs of mildew or rot.

Keep manila rope protected from prolonged exposure to sunlight, which also causes deterioration. If a rope stiffens with storage, it can be made pliable again with a thin coating of warm lubricating oil.

Synthetic-Fiber Ropes

Ropes of synthetic fibers, unlike those constructed of natural fibers, consist of individual fibers running the entire length of a rope, instead of short overlapping fibers. These longer fibers impart a greater strength to the rope, and as a result, ropes of synthetic fibers such as nylon, polyester, polyethylene, and polypropylene have replaced manila rope for many uses.

Although synthetic-fiber ropes cost considerably more than do their natural-fiber counterparts, numerous advantages justify the added expense. Synthetic-fiber ropes are stable under the most adverse conditions; are impervious to rot, mildew, and fungus; and have a high resistance to chemicals. These ropes are also lighter than natural-fiber ropes, are more flexible and easier to handle, and have excellent impact and fatigue characteristics. Nylon absorbs little moisture; polypropylene none. Thus, these two types of rope do not stiffen when wet, do not freeze, and have good dielectric properties when clean and dry.

However, synthetic-fiber ropes tend to soften at high temperatures. Hence, they should never be used under conditions of extreme heat, such as near welding or other open fires, or where high and concentrated friction might cause the fibers to melt.

Nylon

Nylon-fiber rope has a chalky white texture with a smooth surface. It is soft, pliant, and feels elastic to the touch. Continuous filaments running the full length provide nylon rope with high strength and high resistance to creep under sustained loading.

Nylon rope stretches permanently upon initial loading. However, once this permanent stretch occurs, nylon rope recovers completely from subsequent stretching under load. To compensate for this, all new nylon ropes should be broken in and allowed to stretch before being put into use.

Because nylon fibers are highly elastic, ropes are capable of absorbing high-energy shock loading. This can be disadvantageous where lifting headroom is restricted.

Nylon rope's high resistance to abrasion results from a protective fuzz coating that develops after its outer fibers are initially abraded. However, although highly impervious to alkalis, nylon fibers are easily weakened by chemicals, most acids, paints, and linseed oil. If contact with chemicals is suspected, wash rope thoroughly in cold water and carefully examine fibers to detect any possible damage.

Because nylon fibers absorb moisture readily, nylon rope when wet becomes slippery and loses about 10% of its strength. When dry again, nylon rope regains its full strength. Among its other characteristics, nylon rope is not susceptible to mildew or dry rot and thus can be stored wet or dry. It retains its physical properties almost to its melting point, permitting use at temperatures up to 300°F without loss of strength.

Braided nylon

A braided nylon cover over braided nylon construction provides a soft flexible rope that will not twist or kink and displays the highest possible strength. With this type of construction, stress from loading is divided equally between the rope's cover and core. Thus, regardless of cover damage, 50% of the rope's initial strength still remains in the core.

Polyester (dacron and terylene)

Although almost the same as nylon in appearance, polyester rope does not have the elastic feeling of nylon. In comparable sizes, polyester rope is much heavier than nylon, but not as strong.

However, like nylon, polyester rope stretches permanently under initial loading and recovers fully from subsequent loading stretch. Hence it must be broken in to allow initial stretching before use. Under normal loading, polyester-fiber rope stretches approximately the same as manila rope. While polyester rope's ability to absorb shock loading is only about two-thirds that of nylon, it is considerably more than that of manila rope.

Polyester has an abrasion resistance similar to that of nylon, and can be used under similar hot working conditions (up to 330°F) for long periods without loss of strength. However, unlike nylon, polyester rope absorbs almost no water and thus does not lose its strength when wet. It is highly resistant to damage from sun and weather, mildew and dry rot, acids and alkalies, but its fibers can be damaged by contact with chemicals. Polyester-fiber slings should be washed frequently in cold water.

Polypropylene and polyethylene

These two synthetic-fiber ropes, nearly identical in appearance, are pliant and have a smooth, almost slippery feel. Both ropes have considerably lower strength than nylon or polyester.

Polypropylene ropes stretch considerably more than do polyester ropes. Polyethylene stretches about the same as polyester. Shock loading properties are about half that of nylon.

Although not as tough as nylon or polyester, polypropylene rope does not lose strength when wet. However, polyethylene rope strength is much lower compared with that of other synthetic-fiber ropes. Both ropes soften progressively with rising temperature and thus are unsuitable for use in high-temperature operations. Both are highly resistant to attack by acids and alkalies, mildew and rot, but are weakened by prolonged exposure to sunlight.

Handling and Care

When opening a coil of new rope, follow carefully the instructions printed on the tag. Fiber rope, like wire rope, can be damaged if it is removed from the shipping coil incorrectly.

After the burlap wrapping has been opened or removed, lay the coil of rope flat on the floor with the inside rope end at the bottom of the core. Cut the lashings that bind the coil together, then reach down inside the core and pull the end of the rope up through the core (see Fig. 6.2). As the rope comes out of the coil, it will unwind in a counterclockwise direction. Even though the rope is unwound properly, loops and

Figure 6.2 Uncoiling rope, unwinding it counterclockwise from coil. (*Courtesy of Columbian Rope Company.*)

kinks may form in it. These should be removed carefully to prevent their being pulled tight, thereby damaging the rope.

If for any reason a rope must be uncoiled from the outside, place the coil so that as the rope pays out, it will unwind in a counterclockwise direction. If a rope has a large number of kinks, coil it counterclockwise on the floor, then pass the end through the coil and proceed to uncoil.

When not in use, rope should always be coiled up. Make sure to recoil rope in a clockwise direction, removing any kinks as they form during recoiling. Start by looping the rope over the left arm a number of times until about 15 ft of rope remains. Then, starting about 1 ft from the top of the coil, wrap the rope about six times around the loops by rolling them in the left hand. Next extend the left hand through the coil and pull the bight back through the loops. Finally, tie two half-hitches around the bight, leaving a short end for carrying or for tying to a peg or supporting bar.

When rope is cut to a required length, the rope ends must be bound or whipped with yarn to prevent the strands from untwisting and the ends from fraying. If the ends are not properly whipped, the strands will slip in relation to each other, causing one of them to assume more or less than its share of the load, thus shortening the rope's useful life.

To whip the ends of a rope, loop the end of the yarn and place the loop at the end of the rope. Wind the standing part of the yarn around the rope until it covers all but a small part of the loop. Pass the remainder of the standing end up through the small loop, then pull the dead end of the yarn, thus pulling the standing end and the small loop (through which it is threaded) back underneath the whipping until they reach a point midway underneath the whipping. Trim both ends of the yarn close up against the loops to finish off the whipping.

To keep ropes in good condition and extend their working life, protect them from undue exposure to weather, dampness, and sunlight. Keep them away from all sources of heat and exhaust gases that will cause rapid deterioration of fibers.

Always thaw out a frozen rope before using it to prevent breaking the frozen fibers. Wet and frozen ropes should be allowed to dry naturally, since too much heat will cause the fibers to become brittle.

Wet ropes should be dried and cleaned before being stored. Make sure ropes are stored in a dry cool room with good air circulation, at a temperature of 50 to 70°F and at a humidity of 40 to 60%. Do not store ropes on the floor, in boxes, or on shelves where air circulation is restricted. Instead, hang them up in loose coils on large-diameter wooden pegs, well above the floor. Whenever a natural-fiber rope smells sour, make sure to air it until the odor disappears.

Inspection

Regularly inspect every foot of a rope's length to determine its safety, life expectancy, and load-carrying ability. Look for external wear and cutting, internal wear between the strands, and deterioration of the fibers. The section of the rope showing the most deterioration should be the determining factor when estimating a rope's condition. Be careful not to distort a rope's lay during inspection.

First check the surface of the rope for signs of abrasion, cuts, broken fibers and yarns, burns, any unlaying of the twist, or a reduction in the rope's diameter. Remember, broken fibers or yarns or any other failures represent a loss of rope strength.

Do not be misled by dirt on the rope's surface. Any rope that has been used will be dirty on the outside. Carefully open up the rope by untwisting the strands to observe the condition of the inside of the rope. Try not to kink the strands. The rope's interior should be as bright and clean as when the rope was new. Check for broken yarns inside, excessively loose strands and yarns, or any accumulation of powder like dust that results from excessive strand wear.

If the rope is large enough to permit it, open up a strand and with a pencil or other blunt instrument try to pull out one of the inside yarns, keeping in mind that, if a rope has been overloaded, it is the interior yarns that will have failed first. Excessive oil on the outside of a new rope also indicates that the rope has been overloaded.

If it is a four-strand rope with a heart, try gently to pull out the heart. If the heart readily comes out in short pieces, the rope has been overloaded and should not be used for hoisting.

If possible, pull out a couple of long fibers from the end of the rope and try to break them. The finer fibers are relatively stronger than the coarser ones, and all should be broken only with difficulty. Some fibers have a tensile strength as high as 30,000 $lb/in.^2$

If the inside of the rope is dirty, if the strands have begun to unlay, or if the rope has lost its life and elasticity, it should not be used for hoisting purposes. If the rope is high-stranded and presents a spiral appearance, or if the heart protrudes, the load will not be equally distributed on the strands and a very short life may be expected.

Any rope that feels dry and brittle, has sections that are glazed or fused, or is discolored from exposure to acid fumes should always be discarded.

If a rope is weak in just one spot, cut out that portion and splice the rope. Make sure that all splices are properly served or taped. Do not allow any tuck to become undone, since every tuck is necessary for optimum splice efficiency.

If thimbles are loose in eyes because of rope stretch, they can be retightened in the eye by seizing the eye. Never allow a thimble to become so loose that it rocks in the eye. This leads to chafing of the rope inside the eye, with ultimate breakdown.

If there is any doubt as to whether the rope is fit for use, replace it at once. Never risk danger to life or damage to property by taking a chance.

Note: *When a rope has been condemned, it should be destroyed at once or cut up into short hand lines so that it cannot be used again for hoisting purposes.*

Splices, Knots, Bends, and Hitches

Splices

Splicing is a method of joining the ends of two ropes or the ends of one rope with its standing part by interweaving their strands together. Although splicing a rope may be necessary for a particular rope application, any splice will reduce the load-carrying ability of that rope by 10 to 15%.

Three basic splices are used for rigging operations: the short, the long, and the eye splice.

Short splice. Although it is the strongest of all splices, the short splice is not suitable for a rope that must run over sheaves or pass through blocks, since the diameter of the rope is almost doubled at the point of joining. The short splice is most suitable for making endless slings.

For the short splice (see Fig. 6.3), the ends of the two ropes are unlaid

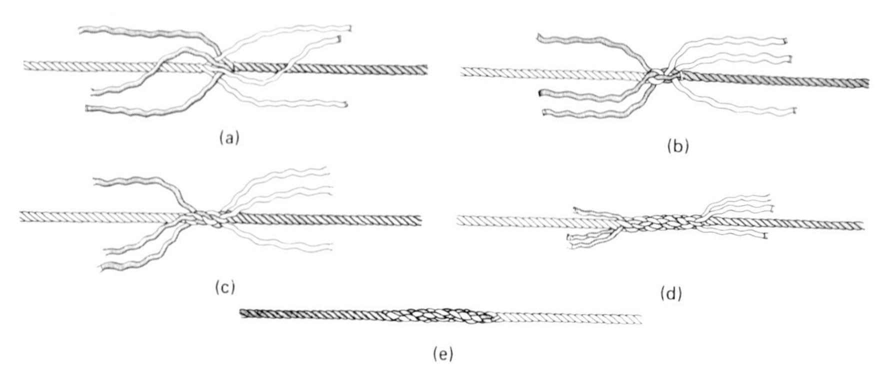

Figure 6.3 Making a short splice. (*Courtesy of The Cordage Group, Auburn, N.Y.*)

for a sufficient distance and placed together with the strands of one rope passing alternately between the strands of the other. Then the two ropes are jammed closely together. Holding the end of one rope and the strands of the other rope firmly in the left hand, pass the middle strand over the strand of the second rope, which goes down to the left of it; then tuck it under and haul taut.

Do the same to each of the other strands in succession, putting them over the strand next to them and under the next one beyond. Turn the rope around and do the same to the other set of strands. Repeat on both ropes. Take care not to bring two strands up through the same interval in the rope. Each strand should come up separately between the two strands of the rope into which they are passed.

To finish the splice, taper it by dividing the yarns after the ends have been interwoven, passing one-half as before and then cutting off the other half. To bring the rope to shape again after splicing, roll it under the foot. For small cord use a piece of flat wood.

Long splice. Although the long splice is slightly weaker than the short one, it has definite advantages for specific applications. Since it increases the rope diameter only slightly, it allows the rope to run through a sheave or block easily. A long splice also lessens wear and chafing of the rope's fibers at the point of splicing. A long splice should be made only with ropes of the same size. It is not suitable for endless slings because of their lower strength.

To make a long splice (see Fig. 6.4), unlay the end of two ropes for a much greater distance than with the short splice, and put the ends together. Unlay one strand for some length and fill up the space left by its removal with the opposite strand from the other rope. Do the same with two more strands. Make an overhand knot with the two remaining strands, taking care that the ends follow the lay of the rope and do not cross the strands. Divide both strands into halves and pass one half over

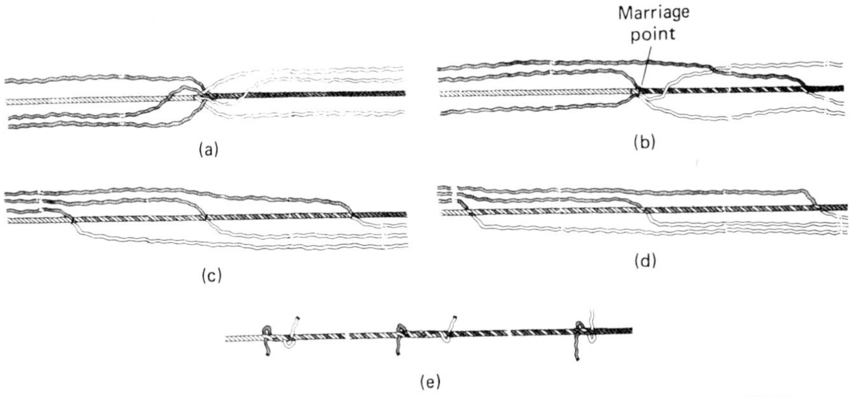

Figure 6.4 Making a long splice. (*Courtesy of The Cordage Group, Auburn, N.Y.*)

the next strand and under the following one; do this two or three times and cut all the ends off close. Work the remaining two pairs of strands the same way, and the splice is finished. The rope should be well stretched before the ends of the strands are cut off.

Eye splice. The eye of a side splice is normally used to form an eye in the end of a fiber rope by splicing the end back into its own side. This splice is made like the short splice, except that only one rope is used. Metal or nylon thimbles should be fitted to all eye splices for lifting.

To form an eye at the end of a rope (see Fig. 6.5), unlay the strands and place them on the standing part so as to form the eye. Then put one strand under the strand next to it, and pass the next one over this strand and under the second. The last strand must go through the third strand on the other side of the rope. Taper an eye splice similarly to a long splice by halving the strands and tucking them securely.

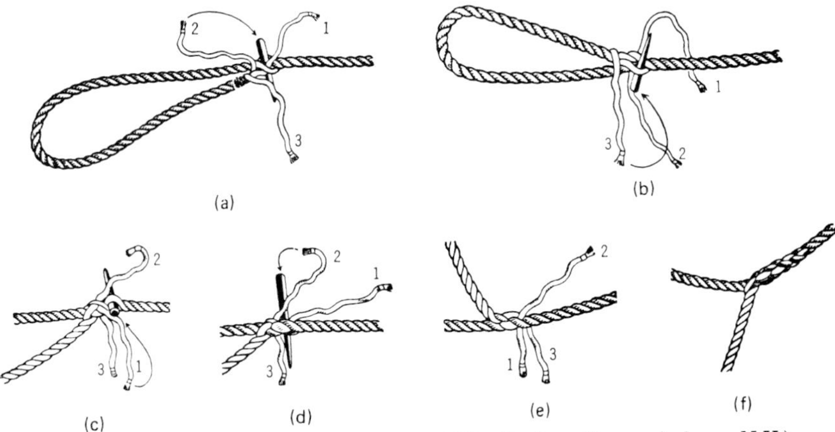

Figure 6.5 Making an eye splice. (*Courtesy of The Cordage Group, Auburn, N.Y.*)

With short and eye splices, four full tucks tapered to finish with one-half and one-quarter tucks will give full efficiency. With medium-sized and large ropes always taper the splice to obtain maximum strength. Where very heavy and rapidly fluctuating loads are involved, increase the number of tucks as a safety measure. The normal splices can be made on synthetic ropes, but because of the smooth surface of the yarns it is essential to insert extra tucks to maintain full efficiency and prevent slippage. Care must be taken when splicing synthetic-fiber rope to avoid losing the twist and to maintain the form and lay of yarns and strands, which will separate easily because of their smooth nature.

When tapering a splice, make sure to leave the yarn ends of the strands with long tails, preferably not less than 1 or 2 in., and simply seized to the finished rope.

Knots, bends, and hitches

There are hundreds, if not thousands, of different rope knots, bends, and hitches, but the average rigger can get along with the knowledge of a comparatively few. The distinction between knots, bends, and hitches is generally accepted as follows:

A knot is the intertwining of the end of a rope within a portion of the rope.

A bend is the intertwining of the ends of two ropes or of the same rope to make one continuous rope or endless rope.

A hitch is the attachment of a rope to a post, pole, ring, hook, or other object.

A good knot, bend, or hitch is one that can be tied with speed and ease and which, when tied, will hold. Their prime requirements are suitability, strength, and security against slippage. Remember, a rope fastening is never as strong as the original rope; knots, bends, and hitches reduce the strength of the rope. The proper fastening must be selected for the job to be done and tied correctly to obtain the maximum strength.

Chapter

7

Wire Strand Rope

Wire rope consists of a number of single wires that are laid (not twisted) into a number of strands. These are often laid, or helically bent, around a core, which may be either of fiber (natural or synthetic) or of wire. The core, which can be either an independent wire rope core (IWRC) or a wire strand core (WSC), depending on its construction, serves to support the outer strands and prevents them from crushing when the wire is wrapped around a drum. In addition, the core helps keep outer strands in proper position during use (see Fig. 7.1).

The durability and performance of wire rope is determined primarily by the type and size of wire used, the number of wires and the strand arrangement, the lay of wires and strands, the type of core, and any special treatment applied to the rope, such as preforming or prestressing. Proper handling and care are essential to maintain wire rope in a durable, safe, and serviceable condition.

Type and Size

Wire ropes are manufactured in several tensile strength grades to meet the varied requirements of many applications. Each grade provides a different combination of tensile strength, toughness, ability to withstand abrasion, and resistance to fatigue from bending and flexing.

Type

Wire rope classifications—in increasing degree of strength—are iron, traction steel, mild plow steel, plow steel, improved plow steel, and extra improved plow steel.

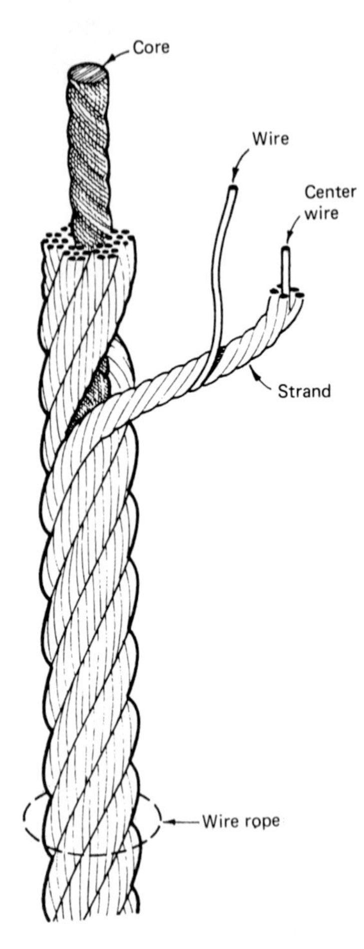

Figure 7.1 Basic components of a typical wire rope. (*Courtesy of American Iron and Steel Institute.*)

Although today all these grades are made from steel, the iron rope classification takes its name from the rope originally made from wrought or puddled iron. High-strength steel rope, which was later developed from the iron grade, was used extensively on steam plows in England in the nineteenth century, thus giving rise to the term plow steel. Iron and traction steel ropes are of lower strength than plow steel, and are used primarily for elevators. Each grade of plow steel is about 15% stronger than the preceding grade.

Any of these grades are furnished in specially treated or coated versions, where corrosion resistance is of utmost importance, including iron-chromium-nickel composition, zinc-coated, and plastic-jacketed steel ropes.

Size

The size of a wire rope is determined by its diameter. Make sure, when measuring the diameter, to use the gage properly to get the correct measurement (see Fig. 7.2). The actual diameter of a wire rope, when new, is usually slightly larger than its nominal size. Any variation in diameter is always over, never under, and it can range from 1⁄64 in. for smaller-sized ropes to 1⁄8 in. for ropes over 2 in. in diameter.

Wire ropes come in various sizes depending on their use:

General hoisting ropes	3⁄16 to 4 in. diameter
Rigging and guy ropes	1⁄4 to 1 1⁄4 in. diameter
Running ropes	3⁄4 to 2 1⁄16 in. diameter
Mooring lines	3⁄8 to 3 1⁄2 in. diameter
Shovel and dragline hoist ropes	3⁄4 to 5 in. diameter
Elevator ropes	3⁄16 to 1 1⁄16 in. diameter

Number and Arrangement

The smaller (and more numerous) the number of wires used in a rope, the more flexible the rope will be. However, such a rope has considerably less abrasion resistance than does a larger-sized rope.

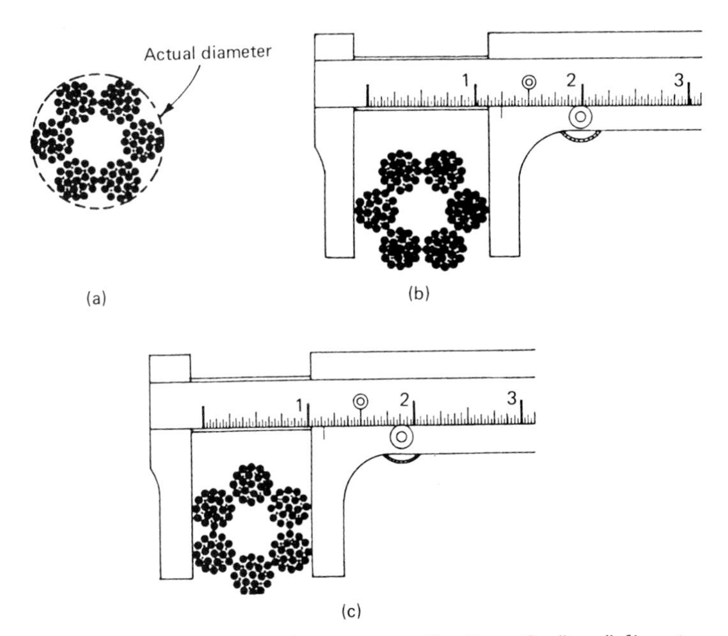

Figure 7.2 Measuring a wire rope correctly. Since the "true" diameter (*a*) lies within the circumscribed circle, always measure the larger dimension (*b*). (*c*) Incorrect. (*Courtesy of American Iron and Steel Institute.*)

Wire ropes are designated primarily by their type of construction, that is, the number of wires in each strand and the number and arrangement of strands in the rope. Constructions having similar properties and commonly listed strengths and weights are grouped in Table 7.1 into a series of basic classifications.

Wire ropes are also identified by their strand shape—round, flattened, locked coil, and concentric. Special constructions are listed in Table 7.2.

Most rigging and hoisting ropes are of the round-strand classification, and these are further classified according to the number of wires per strand and the arrangement (or geometric construction) of wires within the strands. The wires in a round strand may be all the same size or a mixture of sizes. The four common arrangements of wires and strands are illustrated in Fig. 7.3:

1. *Ordinary.* One size of wire used throughout; each strand has 4 to 18 outer wires
2. *Warrington.* Two sizes of outer wires used, alternately laid, large and small; each strand has 10 to 18 outer wires
3. *Seale.* Two sizes of wire used, smaller wires on the inside to provide flexibility and larger wires on the outside to resist abrasion; each strand has 8 to 18 outer wires

TABLE 7.1 Wire Rope Classifications Based on Nominal Number of Wires in Each Strand

Classification	Description
6 × 7	Contains 6 strands that are made up of 3 through 14 wires, of which no more than 9 are outside wires.
6 × 19	Contains 6 strands that are made up of 15 through 26 wires, of which no more than 12 are outside wires.
6 × 37	Contains 6 strands that are made up of 27 through 49 wires, of which no more than 18 are outside wires.
6 × 61	Contains 6 strands that are made up of 50 through 74 wires, of which no more than 24 are outside wires.
6 × 91	Contains 6 strands that are made up of 75 through 109 wires, of which no more than 30 are outside wires.
6 × 127	Contains 6 strands that are made up of 110 or more wires, of which no more than 36 are outside wires.
8 × 19	Contains 8 strands that are made up of 15 through 26 wires, of which no more than 12 are outside wires.
19 × 7, 18 × 7	Contains 19 strands, each strand made up of 7 wires. It is manufactured by covering an inner rope of 7 × 7 left-lang-lay construction with 12 strands in right regular lay. (The rotation-resistant property that characterizes this highly specialized construction is a result of the counter torques developed by the two layers.) When the steel wire core strand is replaced by a fiber core, the designation becomes 18 × 7.

SOURCE: American Iron and Steel Institute.

TABLE 7.2 Special Wire Rope Constructions

3 × 7	Guardrail rope
3 × 19	Slusher
6 × 12	Running rope
6 × 24	Hawsers
6 × 30	Hawsers
6 × 42	Tiller rope (6 × 6 × 7)
6 × 3 × 19	Spring lay
5 × 19	Marlin clad
6 × 19	Marlin clad
6 × 25B	Flattened strand
6 × 27H	Flattened strand
6 × 30G	Flattened strand

SOURCE: American Iron and Steel Institute.

4. *Filler.* Two sizes of wire used, very small wires filling valleys between outer and inner rows of the same-size wires to provide good abrasion and fatigue resistance; each strand has 8 to 18 wires of the same size

As the number of wires per strand increases, combinations of Filler, Seale, and Warrington arrangements are used to provide the rope with specific characteristics and produce varying degrees of resistance to bending, fatigue, and abrasion.

Ordinary and Seale-type constructions may have an even or an odd number of outer wires. Warrington and Filler types always have an even number in their outer layer. The number of outer wires determines in a general way the relative flexibility of the wire rope. By increasing the number of outside wires the reserve strength of a rope may be increased as shown in Table 7.3.

Rope Lay

In addition to wire and strand size and pattern, the lay of a rope can vary, that is, the direction in which wires and strands are twisted within the

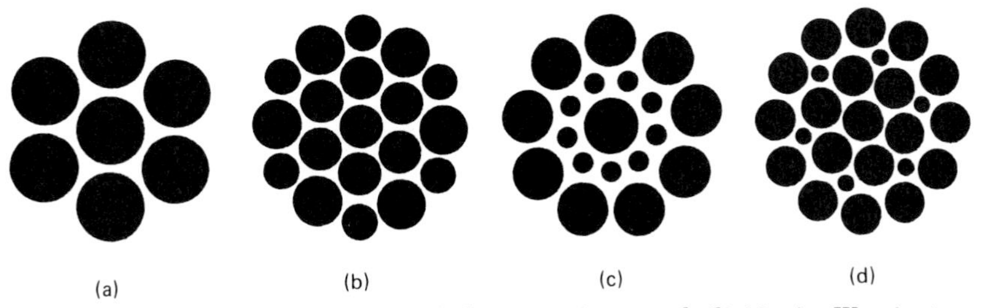

Figure 7.3 Basic strand patterns. (*a*) Ordinary 7-wire strand. (*b*) 19-wire Warrington. (*c*) 19-wire Seale. (*d*) 25-wire filler. (*Courtesy of American Iron and Steel Institute.*)

TABLE 7.3 Reserve Strength through Increasing Number of Outside Wires

Number of outside wires	Percent of reserve strength
3	0
4	5
5	13
6	18
7	22
8	27
9	32
10	36
12	43
14	49
16	54
18	58

Source: American Iron and Steel Institute.

rope. Strand lay refers to the twist of individual wires composing a strand; rope lay refers to the twist of strands around the core (see Fig. 7.4).

1. *Right lay (or left lay).* Manner in which the strands of rope rotate, that is, to the right (similar to a right-hand screw thread) or to the left while receding from the observer and when viewed from above. Right lay is the standard lay of wire rope, with a very few types of installations requiring use of left-lay rope.
2. *Regular lay.* Rope lay of either direction, with wires in the individual strands laid in the opposite direction from that of the rope itself. Because of this, regular-lay ropes are less likely to kink and untwist in use. In addition, the outer wires in regular-lay ropes are approximately parallel to the longitudinal axis of the rope, thus exposing a shorter length of outer wires to damage from crushing and distortion.
3. *Lang lay.* Rope in which the wires composing the strands are laid in the same direction as the rope (either right or left lay). The outer wires run diagonally across the longitudinal axis of the rope and are exposed for longer lengths than in regular lay, thus presenting a greater wearing surface. Lang-lay ropes have increased resistance to abrasion, are more flexible, and have more resistance to fatigue than do regular-lay ropes. However, greater care must be exercised when handling lang-lay ropes since they are more likely to kink and untwist than regular-lay ropes. Before using lang-lay ropes, fasten both ends permanently to prevent untwisting.

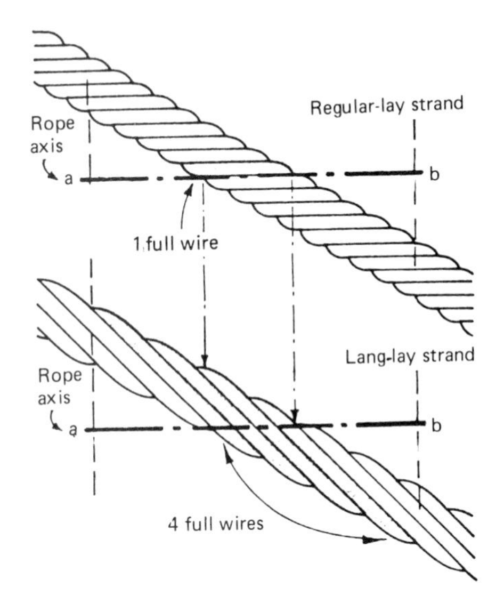

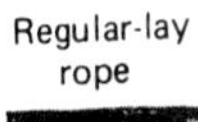

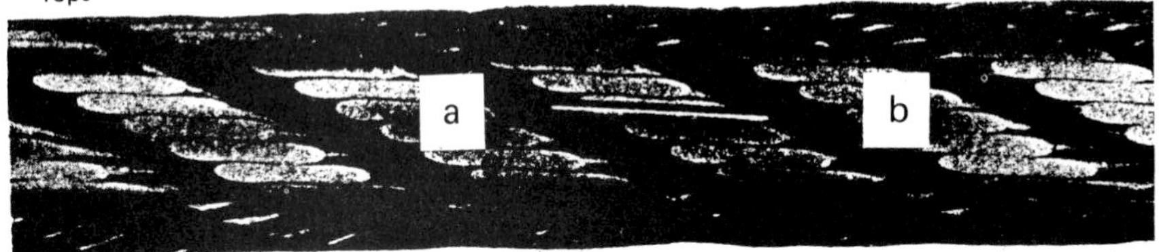

Lang-lay
rope

Figure 7.4 Comparison of wear characteristics between regular-lay and lang-lay ropes. Lines *a–b* indicate the rope axis. (*Courtesy of American Iron and Steel Institute.*)

4. *Alternate lay.* Rope in which the wires in alternate strands are laid right lay and left lay.
5. *Herringbone.* Rope in which two right-hand lang-lay strands alternate with two left-hand regular-lay strands.

Rope Core

At the center of a wire rope is a core, made of fiber, wire, plastic, paper, or asbestos, around which the main strands are laid. The core's principal function is to support the strands and maintain their proper position when loads are applied. Three types of core are commonly used in wire ropes (see Fig. 7.5):

1. *Fiber cores (FC).* Usually made of sisal and, occasionally, of manila fibers. However, cores are also manufactured from synthetic fibers such as polypropylene or nylon, which are less susceptible to compacting than natural fibers, especially when the rope is wet. NEVER use fiber core ropes where the temperatures may damage the core, nor when multilayer winding is used, since the ropes are susceptible to crushing.

2. *Independent wire rope core (IWRC).* Consists of a separate wire rope that increases the main rope's resistance to crushing by supporting its circular cross section as it bends around sheaves and drums under heavy loading.

3. *Wire strand core (WSC).* Consists of a single strand, usually but not always of the same construction as the main rope strands. (When a six-strand rope has a wire strand core of a different construction, it is designated as six-strand rope with WSC; if the wire strand core is of the same construction, the rope is usually simply called a seven-strand rope.) Strand cores are used only with standing ropes and small-diameter running ropes.

Wire cores stretch less than fiber ones, and actually increase a rope's overall strength by about 7½%. However, wire core ropes tend to be less resilient and less capable of absorbing shock loads than do fiber core ropes.

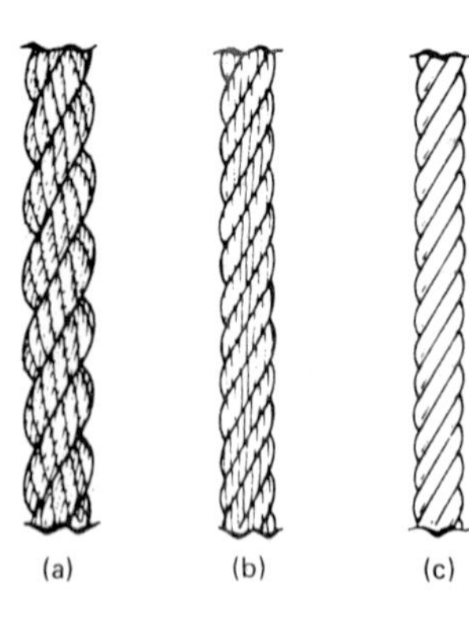

Figure 7.5 Basic wire rope cores. (*a*) Fiber core (FC). (*b*) Independent wire rope core (IWRC). (*c*) Wire strand core (WSC). In selecting the most appropriate core for a given application, wire rope manufacturer should be called on for guidance. (*Courtesy of American Iron and Steel Institute.*)

Preforming and Prestressing

In preforming, wires and strands are shaped in the manufacturing process to fit their position in the finished rope. This removes the tendency of wires and strands to straighten themselves out, and leaves them relaxed in their normal position in the rope. This means that preformed wire rope is free from twisting and less likely to kink or foul in use. Cut or broken ends do not untwist and therefore need not be seized. However, because broken wires tend to lie flat, preformed wire rope requires careful inspection to detect damages. Other advantages of preforming are that each strand in a preformed wire rope carries an equal share of the load, and preformed rope causes less wear on sheaves and drums.

In prestressing, a predetermined load is applied to the rope for a length of time sufficient to permit the component parts of the rope to adjust to the load. This removes the structural stretch in a rope before applying a load. Prestressing is particularly valuable for wires used as guys for towers, and for other applications where only a limited amount of elongation under loading is permissible. However, for most rigging operations, prestressing is not necessary.

Selection Criteria

Selecting the proper lay of rope is possible only after having compared the advantages and disadvantages of each. The most important characteristics to consider are flexibility, fatigue resistance, and abrasion resistance.

Flexibility

Individual wires in a rope bend more sharply in regular-lay than in lang-lay rope because the outside wires in regular-lay rope lie parallel with the rope axis and can only take a curvature equal to the radius of the sheave plus one rope diameter. With lang-lay rope the outside wires lie at an angle with the axis of the rope, and therefore the rope's radius of curvature is greater—thus its bending stress is less. Where small-diameter sheaves are involved, the difference in the radius of curvature of individual wires can amount to as much as 15% in favor of lang-lay ropes.

However, when sheave diameters fall below the economical minimum, other conditions make the use of lang-lay rope impractical. For instance, the length of wire exposed on the outside of the rope is about twice as long in lang lay as in regular lay. Thus, when a lang-lay rope is bent over a very small sheave, there is a tendency for the wires on the underside next to the sheave to loosen and spring away from the rope. This action is cumulative, so that with repeated use these wires

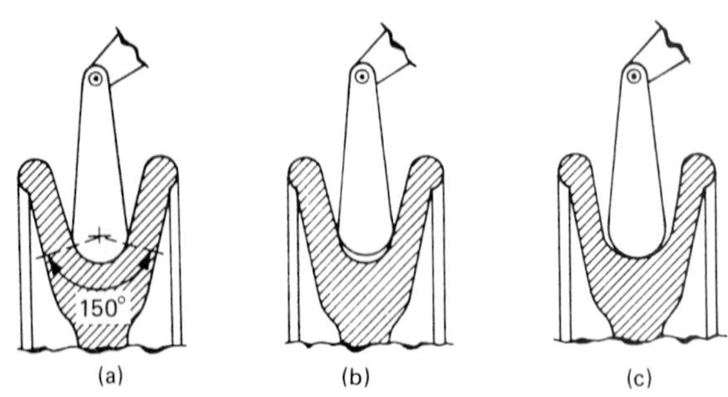

Figure 7.6 Cross sections of three sheave-groove conditions. (*a*) Correct. (*b*) Too tight. (*c*) Too loose. (*Courtesy of American Iron and Steel Institute.*)

will not return to their proper position in the strand. They become flattened and cut by adjacent wires, which deteriorates the rope rapidly.Therefore, with lang-lay rope, sheaves should be used that are at least as large as the economical minimum for the type of service involved (see Figs. 7.6 to 7.8).

Fatigue resistance

As a load is applied to a rope, each wire in the rope is stretched and drawn tightly together. Then when the rope bends around a sheave or drum, the tightly compressed wires move, inducing high stresses that continually reverse as the rope moves through a system of sheaves, with the result that individual wires fail due to fatigue.

Repeated bending and straightening of individual wires passing over sheaves and drums produces small cracks in the wires, which increase in number and grow larger with continued bending. This process occurs in all ropes, but is accelerated if drums and sheaves are too small or if the rope is relatively inflexible.

Lower bending stresses found in lang-lay ropes enable such ropes to bend over a sheave for a longer period before wires break from fatigue.

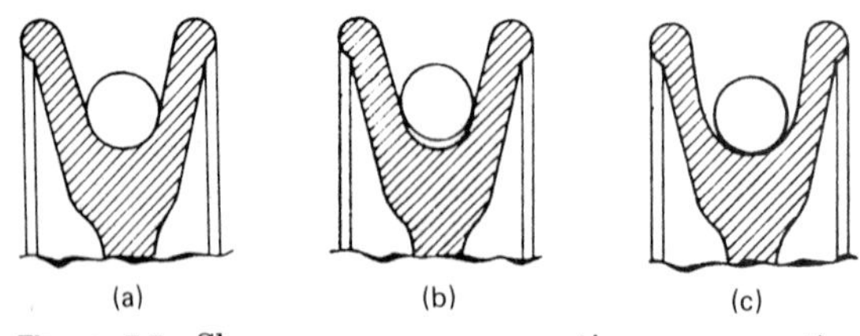

Figure 7.7 Sheave-groove cross sections representing three wire rope seating conditions. (*a*) New rope in a new groove. (*b*) New rope in a worn groove. (*c*) Worn rope in a worn groove. (*Courtesy of American Iron and Steel Institute.*)

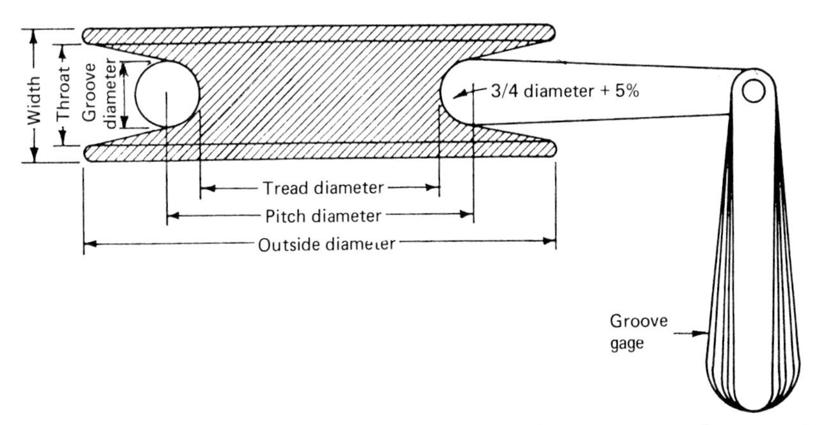

Figure 7.8 Various dimensions of a sheave and use of a groove gage. (*Courtesy of American Iron and Steel Institute.*)

Abrasion resistance

Lang-lay rope has more abrasion resistance than regular-lay rope, because wear is distributed differently in the two types of rope. In lang-lay ropes each wire is worn for a greater length, but there are fewer worn places in a given length of wire; in regular-lay ropes the worn places are shorter and there are more of them.

Since the bending stresses are greatest where the cross section is least, and the reduced cross section in lang-lay rope is about three times that in regular-lay rope, these bending stresses are distributed over a greater length of wire and thus are not as great at any one place. Therefore, in lang-lay rope more of the metal will be worn away before the outer wires break due to bending stresses.

Although wire rope of Seale construction resists abrasion better than do other types, because of its large outer wires which take longer to wear through, the larger stiffer wires also tend to crack and break. This type of rope should not be used unless the manufacturer specifically recommends using it for a particular application.

Safety factors

To guard against possible rope failure, the actual load on a wire rope should be only a fraction of the manufacturer's rated breaking strength. Applying a factor of safety to this value provides the margin of strength necessary to handle loads safely and prevent accidents.

The maximum safe working load of a rope can be determined by dividing the manufacturer's rated breaking strength of the particular rope by the appropriate factor of safety. For rigging ropes, the minimum acceptable safety factor is 5; for ropes used on equipment that hoists people, the factor is 10. Table 7.4 lists safety factors for different applications.

Note: A factor of safety is NOT a reserve strength. It CANNOT be used to provide additional rope capacity; and it must NEVER be lowered.

The actual working load of a wire rope includes not only the static or dead load, but also the loading that results from acceleration, retardation, and shock load. This total load in turn is influenced by the size of the sheaves and drums, the method of reeving, and the rope's end attachments. Further, the load can vary, within limits, with individual installation conditions. Thus, no fixed or arbitrary values for working loads can be properly applied to all of the numerous classifications of service to which wire rope is subjected.

When the rigger is confronted with a load problem and references for wire rope are not available, either of these two approximate methods will give safe serviceable results:

1. Change the rope diameter into eighths of an inch; square the number of eighths; and multiply this result by 250 to determine the working strength in *pounds*.
2. Square the diameter of rope in inches; then multiply this result by 8 to determine the working strength in *tons*.

Care and Maintenance

Proper care and maintenance of wire rope is essential to its satisfactory performance, long life, and adequate safety. Use this checklist to assure that rigging rope meets these criteria:

Inspect all rope regularly and thoroughly.

Check for local wear in rope.

Never use wire rope that is cut, badly kinked, or crushed.

TABLE 7.4 Minimum Factors of Safety

Wire rope	Minimum factor of safety
Track cables	3.2
Guys	3.5
Mine shafts	4.0–8.0
Hoisting equipment	5.0
Haulage ropes	6.0
Cranes and derricks	6.0
Electric and air hoists	7.0
Hot-ladle cranes	8.0
Personnel hoists	10.0

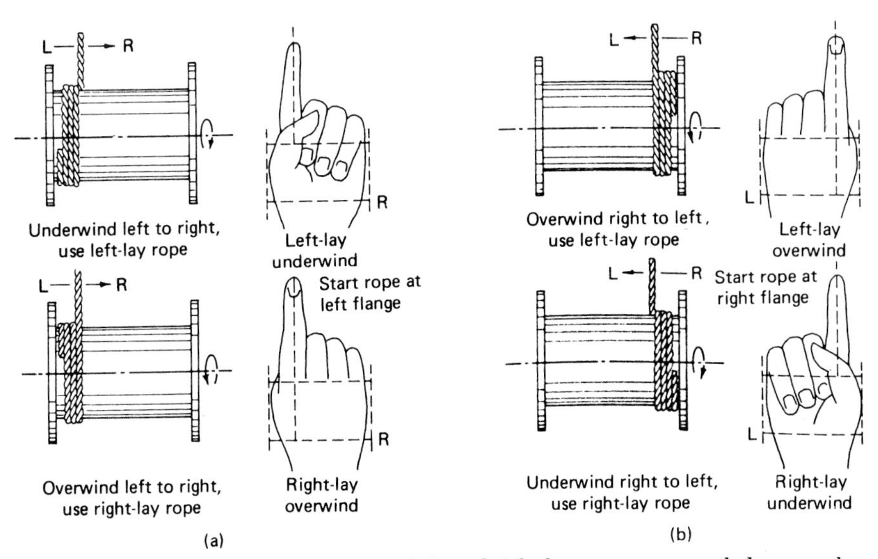

Figure 7.9 Proper procedure for applying left- and right-lay rope on a smooth drum can be easily determined by holding the right or left hand with index finger extended, palm up or palm down. (*a*) Winding left to right. (*b*) Winding right to left. (*Courtesy of American Iron and Steel Institute.*)

Always use the correct rope for every job.

Use a larger-sized rope when the size and severity of loading are unknown, or when lifting operations are severe or hazardous.

Do not overload a rope.

Avoid shock loading, especially in cold weather.

Never use frozen ropes.

Make sure that drums and sheaves have large enough diameters for the rope being used.Check to see that ropes do not bind in sheaves.

Always spool rope on a drum correctly, and make sure the rope does not cross-wind (see Figs. 7.9 and 7.10).

Never wind more rope on a drum than the drum can properly take.

Avoid bending ropes in reverse when reeving through blocks or sheaves.

Check and correct lines for whipping or vibration.

Always use protective padding where a rope passes over sharp edges or around sharp corners.

Do not pull rope out from under loads or drag it over obstacles.

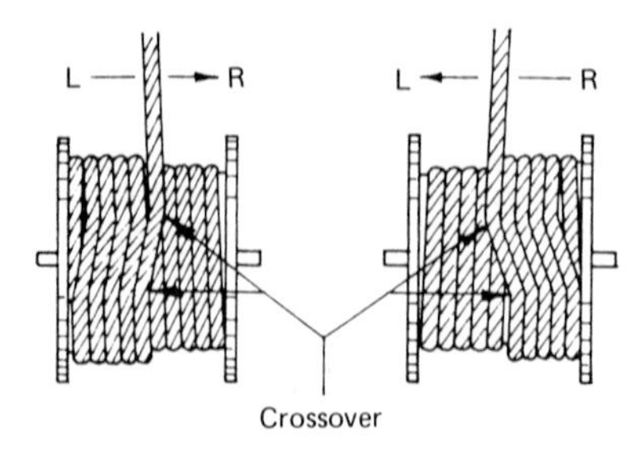

Figure 7.10 After the first layer is wound on a drum, the point at which the rope winds back for each wrap is called crossover. (*Courtesy of American Iron and Steel Institute.*)

Make sure loops in slack lines are not pulled taut and formed into kinks.

Never roll a load with ropes passing under the loads.

Avoid dropping rope from heights.

Always store unused rope in a clean, dry place.

Lubricate wire rope regularly.

Make sure all rope ends are properly seized.

Use thimbles in all eye fittings.

Repair or replace faulty guides and rollers.

Replace sheaves if grooves are deeply worn or scored, if rims are cracked or broken, or if bearings are worn or damaged.

Repair faulty clutches.

Inspection

Do not wait for a rope to fail before inspecting ropes in rigging spread. Check all wire rope each time it is put into service and periodically during normal operations. Visually inspect such rope each week and thoroughly examine it at least once a month. Any rope that has been idle for more than a month should be thoroughly inspected before it is put back into service. Figs. 7.11 to 7.31 illustrate various forms of damage that has occurred to wire rope.

Close examination of a rope will indicate not only when it is time to remove the rope from service, but also whether the rope is suited to the job it is doing.

Rope that has broken wire, but shows no wear, indicates that the rope has been subject to excessive bending. This usually results from using sheaves and drums that are too small for the rope diameter, or from using a rope that is too coarse.

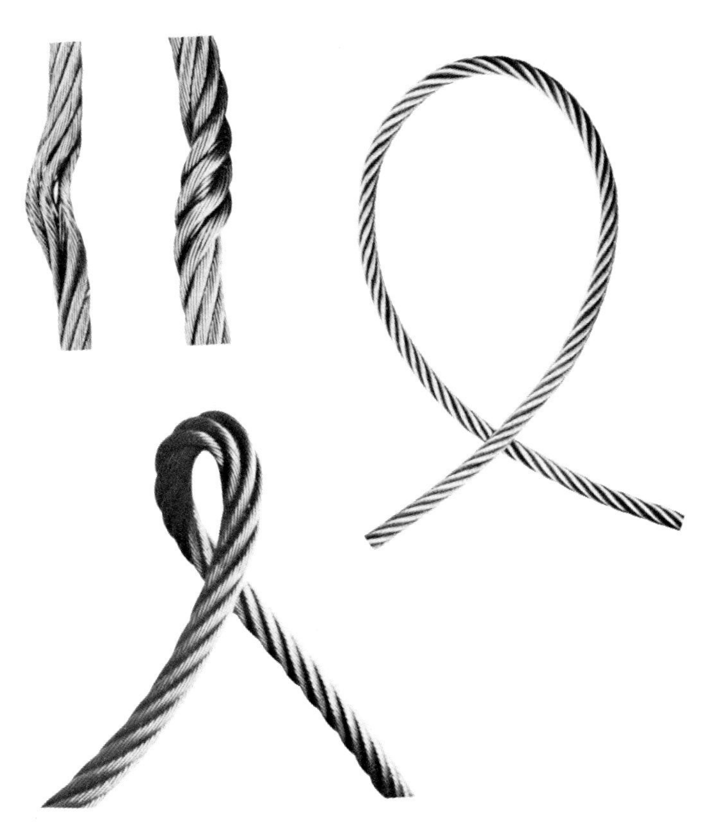

Figure 7.11 Improper handling helps create (*a*) open or (*b*) closed kinks. The open kink will open the rope lay, the closed kink will close it. (*c*) Starting loop. Do not allow rope to form a small loop. If, however, a loop does form and is removed at the stage shown, a kink can be avoided. (*d*) Kink. The looped rope was put under tension, the kink was formed, and the rope is permanently damaged and must be removed. (*Courtesy of American Iron and Steel Institute.*)

Rope that is crushed or flattened indicates that it has been under too much pressure, which usually is the result of improper winding. It is also possible that the rope has not been designed to withstand the conditions to which it has been exposed.

When whole strands or the entire rope break soon after installation, either the rope is overloaded, or a stronger rope or a larger-diameter rope is required.

Safe rope operation depends on the rope's strength as it nears the end of its useful life. Where rope is consistently in use, it should be replaced automatically, regardless of condition, after a predetermined amount of time such as several hundred hours, several weeks, or several months.

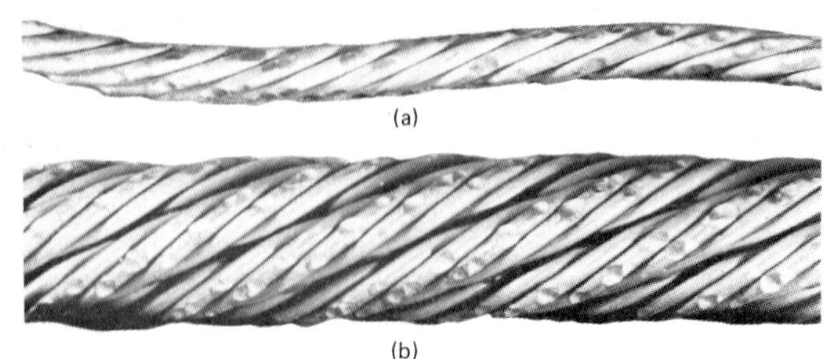

Figure 7.12 Outer strand (*a*) from a 19 × 7 rope shows nicking between adjacent strands as well as between strands and inner rope (*b*). (*Courtesy of American Iron and Steel Institute.*)

Figure 7.13 Single strand removed from wire rope subjected to strand nicking. Resulting from adjacent strands rubbing against one another, this condition is usually caused by core failure due to continued operation under high tensile load. Ultimately there will be individual wire breaks in the valleys of the strands. (*Courtesy of American Iron and Steel Institute.*)

Figure 7.14 Wire rope abuses during shipment, such as improper fastening of rope end to reel as by nailing through rope end, create serious problems. Acceptable methods are illustrated. (*a*) One end of a wire "noose" holds the rope while the other end is secured to the reel. (*b*) The rope end is held in place by a J bolt fixed to the reel. (*Courtesy of American Iron and Steel Institute.*)

Figure 7.15 When a reel has been damaged in transit, it is a safe assumption that irreparable damage may have been done to rope. (*Courtesy of American Iron and Steel Institute.*)

Figure 7.16 Rope damaged on the reel while being rolled over a sharp object. (*Courtesy of American Iron and Steel Institute.*)

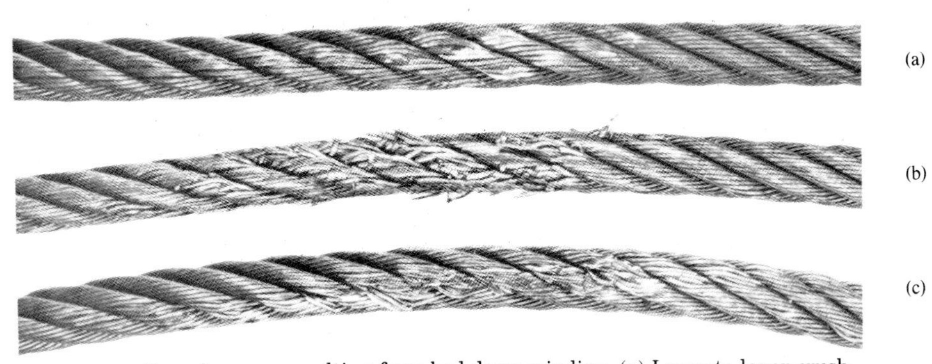

Figure 7.17 Rope damages resulting from bad drum winding. (*a*) Layer-to-layer crushing. (*b*) Scrubbing at crossover or flange turnback. (*c*) Layer-to-layer crushing. (*Courtesy of American Iron and Steel Institute.*)

Figure 7.18 Rope subjected to drum crushing. The individual wires are distorted and displaced from their normal position, which is usually caused by the rope scrubbing on itself. (*Courtesy of American Iron and Steel Institute.*)

Figure 7.19 Drum crushing and spiraling in a winch line, caused by small drums, high loads, and multiple-layer winding conditions. (*Courtesy of American Iron and Steel Institute.*)

Figure 7.20 Tightly spiraled pig-tail rope, resulting from rope being pulled around an object of small diameter. (*Courtesy of American Iron and Steel Institute.*)

Figure 7.21 Rope condition called a dog leg. (*Courtesy of American Iron and Steel Institute.*)

Figure 7.22 Rope condition called a popped core. (*Courtesy of American Iron and Steel Institute.*)

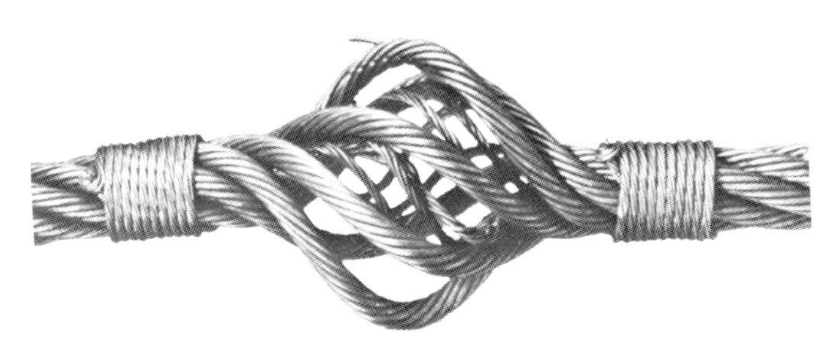

Figure 7.23 This condition, called a bird cage, is caused by a sudden release of tension and the resulting rebound of the rope from its overloaded condition. The strands and wires will not return to their original positions. (*Courtesy of American Iron and Steel Institute.*)

Figure 7.24 Strand wires were snagged and gouged. (*Courtesy of American Iron and Steel Institute.*)

Figure 7.25 Fatigue fractures in a wire rope subjected to heavy loads while on small sheaves. The usual crown breaks are accompanied by breaks in the valleys between strands. The latter breaks are caused by strand nicking, a result of heavy loads. (*Courtesy of American Iron and Steel Institute.*)

Figure 7.26 Typical tension break, a result of overloading. (*Courtesy of American Iron and Steel Institute.*)

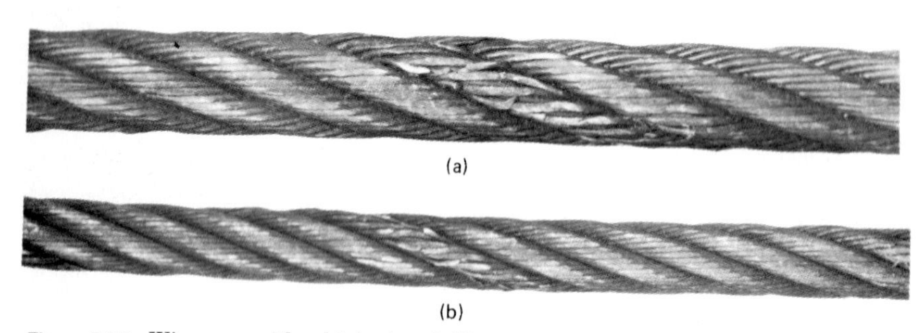

Figure 7.27 Wire rope with a high strand. Here one or two strands are worn before adjoining strands due to improper socketing or seizing, kinks, or dog legs. (*a*) Close view of concentration of wear. (*b*) This recurs in every sixth strand (in a six-strand rope). (*Courtesy of American Iron and Steel Institute.*)

Figure 7.28 Typical example of localized wear. Development of this condition might be reduced if suitable cut-off practice were employed. (*Courtesy of American Iron and Steel Institute.*)

Figure 7.29 Deeply corrugated sheave. (*Courtesy of American Iron and Steel Institute.*)

Figure 7.30 Wire rope has jumped a sheave. The deformation is in the shape of a curl, as if it had been bent around a circular shaft. The two types of breaks—normal tensile cup and cone, and shear break—appear as having been cut with a cold chisel on an angle. (*Courtesy of American Iron and Steel Institute.*)

Figure 7.31 Rope was subjected to repeated bending over sheaves while under normal load, resulting in fatigue breaks in individual wires. These breaks are square and usually on the crown of the strand. (*Courtesy of American Iron and Steel Institute.*)

This will eliminate the risk of rope failure from fatigue. However, only inspection will permit determining a rope's condition and whether it should be removed from service sooner. Table 7.5 is a diagnostic guide to common wire rope degradation.

Rope safety, and the need for its possible replacement in rigging operations, should be questioned when any of the following conditions exist: reduction in rope diameter, excessive abrasion of outside wires, broken wires, internal wire damage, severe corrosion, or rope stretch or elongation.

Reduced rope diameter

This indicates critical deterioration. It may be due to excessive abrasion of outside wires, collapse of core support, or external or internal wire corrosion. Replace any rope whose diameter is reduced by more than:

3⁄64 in. for rope up to and including 3⁄4-in. diameter

1⁄16 in. for rope between 7⁄8 -in. and 1 1⁄8-in. diameter

3⁄32 in. for rope of 1 1⁄4-in. to 1 1⁄2-in. diameter

Excessive wire abrasion

Wear due to friction on sheaves, rollers, or drums flattens outer wires and reduces the rope diameter. The smaller cross section reduces the degree of internal lubrication available to a wire rope, which in turn further accelerates deterioration. Replace any rope whose diameter has been reduced by more than one-third its nominal size.

Broken wires

Broken wires indicate possible fatigue. The effect of broken wires on the strength of a rope can be estimated by taking into account the number of broken wires and their distribution among various strands. Remove from service and replace ropes if one of the following conditions occurs:

1. *Running ropes.* If there are six or more randomly broken wires in one rope lay, or if there are three or more broken wires in one strand within one rope lay
2. *Standing ropes.* If there are three or more broken wires in one rope lay
3. *Any rigging rope.* If it has one or more broken wires near an attached fitting

TABLE 7.5 Diagnostic Guide to Common Wire Rope Degradation

Mode	Symptoms	Possible causes
Fatigue	Wire break is transverse, either straight across or Z shape. Broken ends will appear grainy.	Check for rope bent around too small a radius; vibration or whipping; wobbly sheaves; rollers too small; reverse bends; bent shafts; tight grooves; corrosion; small drums and sheaves; incorrect rope construction; improper installation; poor end terminations. In the absence of other modes of degradation, all rope will eventually fail in fatigue.
Tension	Wire break reveals a mixture of cup and cone fracture and shear breaks.	Check for overloads; sticky, grabby clutches; jerky conditions; loose bearing on drum; fast starts; fast stops; broken sheave flange; wrong rope size and grade; poor end terminations. Check for too great a strain on rope after factors of degradation have weakened it.
Abrasion	Wire break mainly displays outer wires worn smooth to knife-edge thinness. Wire broken by abrasion in combination with another factor will show a combination break.	Check for change in rope or sheave size; change in load; overburden change; frozen or stuck sheaves; soft rollers, sheaves, or drums; excessive fleet angle; misalignment of sheaves; kinks; improperly attached fittings; grit and sand; objects imbedded in rope; improper grooving.
Abrasion plus fatigue	Reduced cross section is broken off square, thereby producing a chisel shape.	A long-term condition normal to the operating process.
Abrasion plus tension	Reduced cross section is necked down as in a cup and cone configuration. Tensile break produces a chisel shape.	A long-term condition normal to the operating process.
Cut, gouged, or rough wire	Wire ends are pinched down, mashed, or cut in a rough diagonal shearlike manner.	Check on all the above conditions for mechanical abuse, or either abnormal or accidental forces during installation.
Torsion or twisting	Wire ends show evidence of twist or corkscrew effect.	Check on all the above conditions for mechanical abuse, or either abnormal or accidental forces during installation.
Mashing	Wires are flattened and spread at broken ends.	Check on all the above conditions for mechanical abuse, or either abnormal or accidental forces during installation. This is a common occurrence on the drum.
Corrosion	Wire surfaces are pitted, with break showing evidence of either fatigue tension or abrasion.	Indicates improper lubrication or storage, or a corrosive environment.

SOURCE: American Iron and Steel Institute.

Table 7.6 indicates when to replace a rope based on the number of broken wires.

Internal wire damage

This is a serious and dangerous condition, which is not visually apparent. Although lubrication may protect external wires in a rope, lack of penetration may seriously damage internal wires as well as the rope's core.

Severe corrosion

More dangerous than wear, corrosion can exist internally before it can be detected visually. The remaining strength cannot be calculated with safety. Corrosion can be controlled by proper lubrication. However, when corrosion is detected, the rope must be removed from service.

Rope stretch or elongation

Excessive stretch beyond a wire rope's initial elongation is a deterioration of the rope resulting either from overloading or from loss of strength, and is a reason for replacement. Permanent initial stretch,

TABLE 7.6 When to Replace Wire Rope, Based on Number of Broken Wires

		Number of broken wires in running ropes		Number of broken wires in standing ropes	
ANSI* standard	Equipment	In one rope lay	In one strand	In one rope lay	At end connection
B30.2	Overhead and gantry cranes	12	4	Not specified	
B30.4	Portal, tower and pillar cranes	6	3	3	2
B30.5	Crawler, locomotive and truck cranes	6	3	3	2
B30.6	Derricks	6	3	3	2
B30.7	Base-mounted drum hoists	6	3	3	2
B30.8	Floating cranes and derricks	6	3	3	2
B30.16	Overhead hoists	12	4	Not specified	
A10.4	Personnel hoists	6†	3	2†	2
A10.5	Material hoists	6†	Not specified	Not specified	

†Also remove for 1 valley break

SOURCE: American Iron and Steel Institute.

caused by wires and strands tightening into their respective cores, is about 6 in. for 100 ft of six-strand rope; about 9 to 10 in. for 100 ft of eight-strand rope.

Finally, check wire rope for the following conditions and remove the rope from service if found defective: crushed, flattened, or jammed strands, high stranding or unlaying, bird-caging, kinks, bulges, gaps or excessive clearance between strands, core protrusion, heat damage, torch burn, or electric arc strikes.

Where splices are found damaged or inadequate, replace that section of the rope and resplice.

If end connections are corroded, cracked, bent, worn, or improperly applied, replace the fittings.

Lubrication

Make sure wire rope is kept properly lubricated, like any other piece of machinery. Use only manufacturer-specified lubricants designed for specific wire rope operating conditions. Various methods of lubricant application are illustrated in Fig. 7.32.

Used crankcase oil should never be applied to wire rope. It is acidic and contains metal particles that are highly abrasive and damaging to wire rope.

Before lubricating wire rope, make sure it is absolutely clean and dry. Then use a jet of air, steam, or a wire brush to clean the rope. Allow the rope to dry thoroughly before lubricating, otherwise the lu-

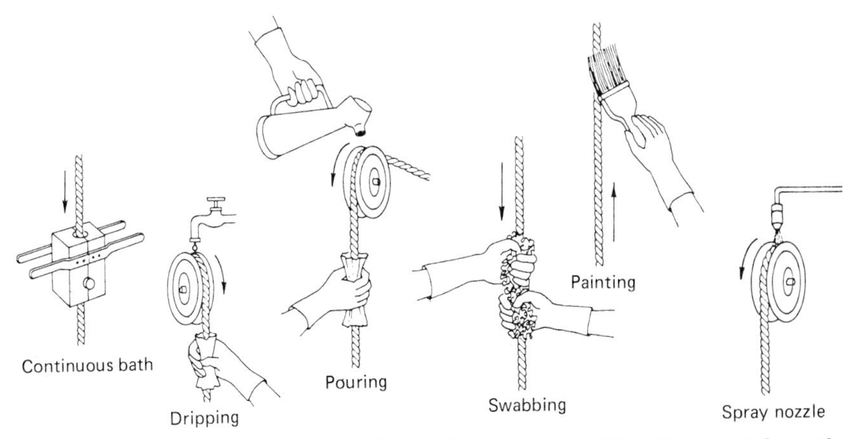

Figure 7.32 Methods of lubricant application in general use. When the rope is bent, the lubricant will penetrate much more easily. Arrows indicate the direction of the rope's movement. (*Courtesy of American Iron and Steel Institute.*)

bricant will not penetrate to the rope's core. Apply warm lubricant by brush, or run the rope through a trough filled with warm oil to obtain maximum penetration.

Glossary of Wire Rope Terms

Abrasion Frictional surface wear on the wires of a wire rope.

Acceleration Stress Additional stress imposed on a wire rope as a result of an increase in load velocity. *See* **Deceleration Stress.**

Aggregate Area *See* **Area, Metallic.**

Aggregate Strength Strength derived by totaling the individual breaking strengths of the elements of a strand or rope. This strength does not give recognition to the reduction in strength resulting from the angularity of the elements in the rope, or other factors that may affect efficiency.

Aircraft Cables Strands, cords, and wire ropes made of special-strength wire, designed primarily for use in various aircraft industry applications.

Alberts Lay *See* **Lay, Types.**

Alternate Lay *See* **Lay, Types.**

Area, Metallic Sum of the cross-sectional areas of all wires in either a wire rope or a strand.

Back-Stay Wire rope or strand guy used to support a boom or mast; or that section of a main cable, as on a suspension bridge, cableway, etc., leading from the tower to the anchorage.

Bail (1) U-shaped member of a bucket. (2) U-shaped portion of a socket or other fitting used on wire rope.

Bailing Line In well drilling, wire rope that operates the bailer that removes water and drill cuttings.

Barney Car Relatively small car permanently attached to a haulage rope that pushes cars along a haulage system.

Basket of Socket Conical portion of a socket into which a broomed rope end is inserted and then secured.

Becket End attachment to facilitate wire rope installation.

Becket Loop Loop of small rope or strand fastened to the end of a larger wire rope. Its function is to facilitate wire rope installation.

Bending Stress Stress imposed on the wires of a strand or rope by a bending or curving action.

Bicable Wire rope aerial tramway that has a fixed cable or strand to support the load, as well as a traction or haul rope that moves the load about the system.

Bird Cage Appearance of a wire rope forced into compression. The outer strands form a cage and, at times, displace the core.

Block Wire rope sheave (pulley) enclosed in side plates and fitted with some attachment such as a hook or shackle.

Boom Hoist Line Wire rope that operates the boom hoist system of derricks, cranes, draglines, shovels, etc.

Boom Pendant Nonoperating rope or strand with end terminations to support the boom.

Breaking Strength

Breaking strength: Ultimate load at which a tensile failure occurs in the sample of wire rope being tested. (*Note:* The term *breaking strength* is synonymous with actual strength.)

Minimum acceptance strength: Strength that is 2½% lower than the catalog or nominal strength. This tolerance is used to offset variables that occur during sample preparation and actual physical tests of a wire rope.

Nominal strength: Published (catalog) strength calculated by a standard procedure that is accepted by the wire rope industry. The wire rope manufacturer designs wire rope to this strength, and the user should consider this strength when making design calculations.

Bridge Cable (Structural Rope or Strand) All-metallic wire rope or strand used as catenary and suspenders on a suspension bridge.

Bridge Socket Wire rope or strand end termination made of forged or cast steel that is designed with baskets, having adjustable bolts, for securing rope ends. There are two styles: (1) The closed type has a U bolt with or without a bearing block in the U of the bolt. (2) The open type has two eyebolts and a pin.

Bridle Sling Multileg wire rope sling.

Bright Rope Wire rope fabricated from wires that are not coated.

Bronze Rope Wire rope fabricated from bronze wires.

Bull Wheel Large-diameter wire rope sheave, such as the sheaves at the end of a ski lift.

Button Conveyor Rope Wire rope to which buttons or disks are attached at regular intervals to move material as in a trough.

Cable Wire rope, wire strand, or electric conductor.

Cable-Laid Wire Rope Wire rope consisting of several wire ropes laid into a single wire rope, such as 6 × 42 (6 × 6 × 7) tiller rope.

Cable Tool Drilling Line Wire rope used to operate the cutting tools in the *cable tool* drilling method (i.e., rope drilling).

Cableway Aerial conveying system for transporting single loads along a suspended track cable.

Casing Line Wire rope used to install oil well casings.

Catenary Curve formed by a strand or wire rope when supported horizontally between two fixed points, such as the main spans on a suspension bridge.

Center Axial member of a strand about which the wires are laid.

Change of Layer Point Point in the traverse of a rope across the face of the drum where it reaches the flange, reverses direction, and begins forming the next layer; also referred to as the drum crossover or turn-back point.

Choker Rope Short wire rope sling that forms a slip noose around an object to be moved or lifted.

Circumference Measured perimeter of a circle that circumscribes either the wires of a strand or the strands of a wire rope.

Clamp, Strand Fitting for forming a loop at the end of a length of strand, consisting of two grooved plates and bolts.

Cleaning out Line Wire rope used in conjunction with tools that are used to clean an oil well.

Clevis *See* **Shackle**.

Clip Fitting for clamping two parts of wire rope to each other.

Closed Socket Wire rope end termination consisting of basket and bail made integral.

Closer Machine that lays strands around a core to form rope.

Closing Line Wire rope that performs two functions: (1) It closes a clamshell or orange peel bucket. (2) It operates as a hoisting rope.

Coarse-Laid Rope In oil fields, 6 × 7 wire rope.

Coil Circular bundle or package of wire rope not affixed to a reel.

Come Along Device for making temporary grip on a wire rope.

Conical Drum Grooved hoisting drum with varying diameter. *See* **Drum.**

Construction Geometric design description of the wire rope's cross section. This includes the number of strands, the number of wires per strand, and the pattern of wire arrangement in each strand.

Constructional Stretch Stretch that occurs when the rope is loaded. It is due to the helically laid wires and strands creating a constricting action that compresses the core and generally brings all of the rope's elements into close contact.

Continuous Bend Reeving of wire rope over sheaves and drums so that it bends in one direction, as opposed to reverse bend.

Conveyor Rope Endless wire rope used to carry material. *See* **Button Conveyor Rope.**

Cord Small-diameter specialty wire rope or strand.

Core Axial member of a wire rope about which the strands are laid.

Coring Line Wire rope used to operate the coring tool that is used to take core samples during oil well drilling.

Corrosion Chemical decomposition of wires in a rope through the action of moisture, acids, alkalines, or other destructive agents.

Corrosion-Resisting Steel Chrome-nickel steel alloys designed for increased resistance to corrosion.

Corrugated Condition of grooves of a sheave or drum after these have been worn down to a point where they show an impression of a wire rope.

Cotton Center *See* **Fiber Center.**

Cotton Core *See* **Fiber Core.**

Coupling Device for joining the ends of two lengths of track cable.

Cover Wires Outer layer of wires.

Cracker Manila rope spliced or otherwise attached to the end of a wire drilling line.

Creep Unique movement of a wire rope with respect to a drum surface or sheave surface resulting from the asymmetrical load between one side of the sheave (drum) and the other. It is not dissimilar to the action of a caterpillar moving over a flat surface. It should be distinguished from slip, which is yet another type of relative movement between rope and sheave or drum surface.

Critical Diameter For any given wire rope, diameter of the smallest bend that permits both wires and strands to adjust themselves by relative movement while retaining their normal cross-section position.

Cross Lay *See* **Lay, Types.**

Crowd Rope Wire rope used to drive or force a power shovel bucket into the material that is to be handled.

Cylindrical Drum Hoisting drum of uniform diameter. *See* **Drum.**

Dead Line In drilling, end of rotary drilling line fastened to anchor or dead-line clamp.

Deceleration Stress Additional stress imposed on a wire rope as a result of a decrease in load velocity. *See* **Acceleration Stress.**

Deflection (1) Sag of rope in a span; usually measured at midspan as the depth from the chord joining the tops of the two supports. (2) Any deviation from a straight line.

Design Factor In a wire rope, ratio of nominal strength to total working load.

Diameter Line segment which passes through the center of a circle and whose end points lie on the circle. As related to wire rope, it would be the diameter of a circle which circumscribes the wire rope.

Dog Leg Permanent bend or kink in a wire rope, caused by improper use or handling.

Dragline (1) Wire rope used for pulling excavating or drag buckets. (2) Specific type of excavator.

Drilling Line *See* **Cable Tool Drilling Line; Rotary Line.**

Drum Cylindrical flanged barrel, of either uniform or tapering diameter, on which rope is wound for either operation or storage. Its surface may be smooth or grooved.

Efficiency Ratio of a wire rope's actual breaking strength to the aggregate strength of all individual wires tested separately; usually expressed as a percentage.

Elastic Limit Stress limit above which permanent deformation will take place within material.

Elliptic Spool Endless-rope drive drum with a face in the shape of an elliptic arc.

Elongation *See* **Stretch.**

End Preparation Treatment of the end of a length of wire rope designed primarily as an aid for pulling the rope through a reeving system or tight drum opening. Unlike end terminations, these are not designed for use as a method for making a permanent connection.

End Termination Treatment at the end or ends of a length of wire rope, usually made by forming an eye or attaching a fitting and designed to be the permanent end termination on the wire rope that connects it to the load.

Endless Rope Rope with ends spliced together to form a single continuous loop.

Equalizing Sheave Sheave at the center of a rope system over which no rope movement occurs other than equalizing movement. It is frequently overlooked during crane inspections, with disastrous consequences. It can be a source of severe degradation.

Equalizing Slings Multiple-leg slings composed of wire rope and fittings that are designed to help distribute the load equally. *See* **Sling.**

Equalizing Thimbles Special type of load-distributing fitting used as a component of certain wire rope slings.

Extra Flexible Wire Rope Ambiguous and archaic term sometimes describing wire ropes in the 8 × 19 and 6 × 37 classes. The term is so indefinite as to be meaningless and is in disfavor today.

Extra High-Strength Strand Grade of galvanized strand.

Extra Improved Plow-Steel Rope Specific grade of wire rope.

Eye or Eye Splice Loop, with or without a thimble, formed at the end of a wire rope.

Factor of Safety Originally, ratio of nominal strength to total working load. The term is no longer used since it implies a permanent existence for this ratio when, in actuality, the rope strength begins to reduce the moment the rope is placed in service. *See* **Design Factor.**

Fatigue Usually, process of progressive fracture resulting from the bending of individual wires. These fractures may and usually do occur at bending stresses well below the ultimate strength of the material. It is not an abnormality, although it may be accelerated due to conditions in the rope such as rust or lack of lubrication.

Ferrule Metallic button, usually cylindrical in shape, normally fastened to a wire rope by swaging, but sometimes by spelter socketing.

Ferry Rope Wire rope suspended over water for the purpose of guiding a boat.

Fiber Center Cord or rope of vegetable or synthetic fiber used as axial member of a strand.

Fiber Core Cord or rope of vegetable or synthetic fiber used as axial member of a rope.

Filler Wire Small spacer wires within a strand which help position and support other wires. Also, type of strand pattern utilizing filler wires.

Fitting Functional accessory attached to a wire rope.

Flag Marker placed on a rope so as to locate the load position.

Flat Rope Wire rope made of a series of parallel, alternating right-lay and left-lay ropes, sewn together with relatively soft wires.

Flattened Strand Rope Wire rope made of either oval or triangular-shaped strands in order to form a flattened rope surface.

Fleet Angle Angle between the rope's position at the extreme end wrap on a drum and a line drawn perpendicular to the axis of the drum through the center of the nearest fixed sheave. *See* **Drum; Sheave.**

Flexible Wire Rope Archaic and imprecise term to differentiate one rope construction from another, such as 6 × 7 (least flexible) and 6 × 19 (somewhat more flexible) classification.

Galvanized Zinc coating for corrosion resistance.

Galvanized Rope Wire rope made of galvanized wire.

Galvanized Strand Strand made of galvanized wire.

Galvanized Wire Zinc-coated wire.

Grade Wire rope or strand classification by strength and/or type of material, such as improved plow steel, type 302 stainless, phosphor bronze. It does not imply a strength of the basic wire used to meet the rope's nominal strength.

Grades, Rope Classification of wire rope by the wire's metallic composition and the rope's nominal strength.

Grades, Strand Classification of strand by the wire's metallic composition and the strand's nominal strength. In the order of increasing nominal strengths, the grades are common, Siemens-Martin, high-strength, and extra-high-strength. A utilities grade is also made to meet special requirements and its strength is usually greater than high-strength.

Grain Shovel Rope 6 × 19 marline-clad rope used for handling grain in scoops.

Grommet Endless circle or ring fabricated from one continuous length of strand or rope.

Groove Depression, helical or parallel, in the surface of a sheave or drum, shaped to position and support the rope.

Grooved Drum Drum with a grooved surface that accommodates the rope and guides it for proper winding.

Guy Line Strand or rope, usually galvanized, for stabilizing or maintaining a structure in fixed position.

Haulage Rope Wire rope used for pulling movable devices such as cars that roll on a track.

Hawser Wire rope, usually galvanized, used for towing or mooring marine vessels.

High-Strength Strand Grade of galvanized strand.

Holding Line Wire rope on a clamshell or orange-peel bucket that suspends the bucket while the closing line is released to dump its load.

Idler Sheave or roller used to guide or support a rope. *See* **Sheave.**

Improved Plow-Steel Rope Specific grade of wire rope.

Incline Rope Rope used in the operation of cars on an inclined haulage.

Independent Wire Rope Core (IWRC) Wire rope used as axial member of a larger wire rope.

Inner Wires All wires of a strand except the outer or cover wires.

Internally Lubricated Wire rope or strand having all of its wire components coated with lubricants.

Ironing *See* **Milking.**

Iron Rope Specific grade of wire rope.

IWRC *See* **Independent Wire Rope Core.**

Kink Unique deformation of a wire rope caused by a loop of rope being pulled down tight. It represents irreparable damage to and an indeterminate loss of strength in the rope.

Lagging (1) External wood covering on a reel to protect wire rope or strand. (2) Grooved shell of a drum.

Lang-Lay Rope *See* **Lay, Types.**

Lay (1) Manner in which the wires in a strand or the strands in a rope are helically laid. (2) Distance measured parallel to the axis of the rope (or strand) in which a strand (or wire) makes one complete helical convolution about the core (or center). In this connection, lay is also referred to as lay length or pitch.

Lay, Types

Right lay: Direction of strand or wire helix corresponding to that of a right-hand screw thread.

Left lay: Direction of strand or wire helix corresponding to that of a left-hand screw thread.

Cross lay: Rope or strand in which one or more operations are performed in opposite directions. A multiple-operation product is described according to the direction of the outside layer.

Regular lay: Type of rope wherein the lay of the wires in the strand is in the opposite direction to the lay of the strand in the rope. The crowns of the wires appear to be parallel to the axis of the rope.

Lang lay: Type of rope in which the lay of the wires in the strand is in the same direction as the lay of the strand in the rope. The crowns of the wires appear to be at an angle to the axis of the rope.

Alternate lay: Lay of a wire rope in which the strands are alternately regular-lay and lang-lay.

Alberts lay: Old, rarely used term for lang lay.

Reverse lay: Alternate lay.

Spring lay: Specific wire rope construction; not definable as a unique lay.

Lay Length *See* **Lay.**

Lead Line Part of a rope tackle leading from the first, or fast, sheave to the drum. *See* **Drum; Sheave.**

Left Lay *See* **Lay, Types.**

Line Synonymous with wire rope.

Locked Coil Strand Smooth-surfaced strand ordinarily constructed of shaped, outer wires arranged in concentric layers around a center of round wires.

Loop 360° change of direction in the course of a wire rope which, when pulled down tight, will result in a kink. *See* **Eye; Eye Splice.**

Marline Prelubricated fiber material.

Marline-Clad Rope Rope with individual strands spirally wrapped with marline.

Marline Spike Tapered steel pin used as a tool for splicing wire rope.

Martensite Brittle microconstituent of steel formed when steel is heated above its critical temperature and rapidly quenched. This occurs in wire rope as a result of frictional heating and the mass cooling effect of the cold metal beneath. Martensite cracks very easily, and such cracks can propagate from the surface through the entire wire.

Messenger Strand Galvanized strand used as support for telephone and electric cables.

Metallic Core *See* **Wire Strand Core; Independent Wire Rope Core.**

Mild Plow-Steel Rope Specific grade of wire rope.

Milking Progressive movement of strands along rope axis, resulting from the rope's movement through a restricted passage such as a tight sheave; sometimes called ironing.

Modulus of Elasticity Mathematical quantity expressing the ratio, within the elastic limit, between a definite range of unit stress on a wire rope and the corresponding unit elongation.

Monocable Wire rope conveyance designed with a single wire rope that not only supports the load but conveys it as well.

Mooring Lines Galvanized wire rope, usually 6 × 12, 6 × 24, or 6 × 3 × 19 spring lay, for holding ships to dock.

Nonpreformed Rope or strand that is not preformed. *See* **Preformed Strands; Preformed Wire Rope.**

Nonrotating Wire Rope Term, now abandoned, referring to 19 × 7 or 18 × 7 rope. *See* **Rotation-Resistant Rope.**

Nonspinning Wire Rope *See* **Rotation-Resistant Rope.**

Open Socket Wire rope fitting that consists of a basket and two ears with a pin. *See* **Fitting.**

Outer Wires *See* **Cover Wires.**

Peening Permanent distortion resulting from cold plastic metal deformation of outer wires; usually caused by pounding against a sheave or machine member, or by heavy operating pressure between rope and sheave, rope and drum, or rope and adjacent wrap of rope.

Pitch *See* **Lay.**

Plow-Steel Rope Specific grade of wire rope.

Preformed Strands Strand in which the wires are permanently formed during fabrication into the helical shape they will assume in the strand.

Preformed Wire Rope Wire rope in which the strands are permanently formed during fabrication into the helical shape they will assume in the wire rope.

Pressed Fittings Fittings attached by means of cold forming on the wire rope.

Prestressing Incorrect reference to **Prestretching.**

Prestretching Subjecting a wire rope or strand to tension prior to its intended application, for an extent and over a period of time sufficient to remove most of the constructional stretch.

Proportional Limit Virtually synonymous with elastic limit. It is the end of the load versus elongation relationship at which an increase in load no longer produces a proportional increase in elongation and from which point recovery to the rope's original length is unlikely.

Rated Capacity Load that a new wire rope or wire rope sling may handle under given operating conditions and at an assumed design factor.

Reel Flanged spool on which wire rope or strand is wound for storage or shipment.

Reeve To pass a rope through a hole or around a system of sheaves.

Regular Lay *See* **Lay, Types.**

Reserve Strength Strength of a rope exclusive of the outer wires.

Reverse Bend Reeving a wire rope over sheaves and drums so that it bends in opposing directions. *See* **Reeve.**

Reverse Lay *See* **Lay, Types.**

Right Lay *See* **Lay, Types.**

Rollers Relatively small-diameter cylinders or wide-faced sheaves that serve as support for ropes.

Rotary Line On a rotary drilling rig, wire rope used for raising and lowering the drill pipe, as well as for controlling its position.

Rotation-Resistant Rope Wire rope consisting of an inner layer of strand laid in one direction covered by a layer of strand laid in the opposite direction. This has the effect of counteracting torque by reducing the tendency of finished rope to rotate.

Round-Wire Track Strand Strand composed of concentric layers of round wires, used as track cable, sometimes called smooth-coil track strand.

Running Rope 6 × 12 galvanized wire rope.

Safety Factor *See* **Design Factor.**

Safe Working Load Essentially that portion of the nominal rope strength that can be applied to either move or sustain a load. The term is misleading and in disfavor because it is only valid when the rope is new and equipment is in good condition. *See* **Rated Capacity.**

Sag *See* **Deflection.**

Sand Line *See* **Bailing Line.**

Sash Cord Small 6 × 7 wire rope, commonly made of iron wires.

Seale Type of strand pattern that has two adjacent layers laid in one operation with any number of uniform sized wires in the outer layer, and with the same number of uniform but smaller sized wires in the inner layer.

Seize To make a secure binding at the end of a wire rope or strand with seizing wire or seizing strand.

Seizing Strand Small-diameter strand usually made up of 7 wires.
Seizing Wire Wire for seizing. *See* **Seize.**
Serve To cover the surface of a wire rope or strand with a fiber cord or wire wrapping.
Sewing Wires *See* **Flat Rope.**
Shackle U- or anchor-shaped fitting with pin.
Sheave Grooved pulley for wire rope.
Siemens-Martin Strand Grade of galvanized strand.
Sling, Wire Rope Assembly fabricated from wire rope which connects the load to the lifting device.
Sling, Braided Flexible sling, the body of which is made up of two or more wire ropes braided together.
Smooth-Coil Track Strand Strand composed of concentric layers of round wires used as track cable; more commonly called round-wire track strand.
Smooth-Faced Drum Drum with a plain, ungrooved surface. *See* **Drum.**
Socket Type of wire rope fitting. *See* **Bridge Socket; Closed Socket; Open Socket; Wedge Socket.**
Special Flexible Wire Rope 6 × 37 classification wire rope.
Spin-Resistant Abandoned term referring to a rotation-resistant rope of the 8 × 19 classification.
Spiral Groove Continuous helical groove that follows a path on and around a drum face, similar to a screw thread. *See* **Drum.**
Splicing (1) Making a loop or eye in the end of a rope by tucking the ends of the strands back into the main body of the rope. (2) Formation of loops or eyes in a rope by means of mechanical attachments pressed onto the rope. (3) Joining of two rope ends so as to form a long or short splice in two pieces of rope.
Spring Lay *See* **Lay, Types.**
Stainless-Steel Rope Wire rope made of corrosion-resistant steel wires.
Standing Rope *See* **Guy Line.**
Stirrup Eyebolt attachment on a bridge socket. *See* **Socket.**
Stone Sawing Strand 2- or 3-wire strand used in stone and slate quarrying operations.
Stone Sawing Wire Shaped and twisted wire used in stone and slate quarrying operations.
Strand Plurality of round or shaped wires helically laid about an axis.
Strand Center *See* **Center.**
Strand Core *See* **Wire Strand Core.**
Strander Machine that lays wires together helically to form a strand.
Stress Force or resistance within any solid body against alteration of form. In the case of a solid wire it would be the load on the rope divided by the cross-sectional area of the wire.
Stretch Elongation of a wire rope under load.
Swab Line *See* **Cleaning out Line.**
Swaged Fitting Fitting into which wire rope can be inserted and then permanently attached by cold pressing (swaging) the shank that encloses the rope.
Tag Line Small wire rope used to prevent rotation of a load.
Tapered Drum *See* **Conical Drum.**
Tapering and Welding Reducing the diameter of a wire rope at its end, and

then welding the wires so as to facilitate reeving. *See* **End Preparation.**

Thimble Grooved metal fitting to protect the eye, or fastening loop, of a wire rope.

Tiller Rope Highly flexible rope constructed by cable-laying six 6 × 7 ropes around a fiber core.

Tinned Wire Wire coated with tin.

Track Cable On an aerial conveyor, suspended wire rope or strand along which the carriers move.

Traction Rope On an aerial conveyor or haulage system, wire rope that propels the carriages.

Traction Steel Rope Specific grade of wire rope.

Tramway Aerial conveying system for transporting multiple loads.

Turn Single wrap around a drum; synonymous with wrap.

Turn-Back Point *See* **Change of Layer Point.**

Warrington Type of strand pattern characterized by having one of its wire layers (usually the outer) made up of an arrangement of alternately large and small wires.

Wedge Socket Wire rope fitting wherein the rope end is secured by a wedge. *See* **Fitting.**

Whipping Synonymous with seizing.

Wire, Round Single continuous length of metal with circular cross section, cold-drawn from rod.

Wire, Shaped Single continuous length of metal with noncircular cross section, either cold-drawn or cold-rolled from rod.

Wire Rope Plurality of wire strands helically laid about an axis.

Wire Strand Core (WSC) Wire strand used as axial member of a wire rope.

Wrap *See* **Turn.**

Chapter

8

Slings and Hitches (Fiber, Wire, Chain, and Webbing)

Rigging in its truest sense is the handling of loads suspended from a crane, hoist, or derrick. But properly designing a piece of hoisting equipment to lift safely a given load on a hook is not the only problem that the rigger faces. The load must be secured to the hook, and the attachment must be of adequate strength to lift the load safely; thus the reason for using a sling and the need for proper sling design.

Various materials can be used for slings. For the lightest loads, an endless manila rope looped into a noose around the object to be lifted is satisfactory. Such a sling is inexpensive, lightweight, flexible, and easy to handle. It can bend around the comparatively sharp edges of boxes or crates, but should be padded when passing over the sharp machined edges of metal parts. Even on some heavier jobs, such as handling of steel shafts, which must not be scratched or burred, manila rope slings are useful.

Chains have a limited application as slings. They are used primarily in foundries, where they are exposed to high temperatures and where they must pick up rough castings that would quickly destroy other types of slings. Like manila rope, a chain should preferably be padded where bearing on sharp edges of metal parts. Otherwise some of the links may be subjected to severe bending stresses. The chain may also bruise or otherwise damage the edges of the load being lifted.

For special work, slings are sometimes made of other materials, such as synthetic-fiber webbing and steel-mesh webbing. But by far the greatest number of slings in use today are made of wire rope.

Although wire rope slings have become the workhorses of the sling field, with exceptionally sound safety features, they must be chosen carefully for the service in which they are to be placed since they also have certain limitations. Like manila rope and chain slings, wire rope slings should be padded where they bend over sharp edges of the load to be lifted. Although load edges made of soft material, such as wood crate or skid, will not cut the sling, individual wires are bent sharply and highly stressed, possibly beyond their elastic limit, even before the life load is applied.

Slings are manufactured in a wide variety of configurations. However, all can be identified in terms of just three elements:

1. Type of sling: eye and eye or endless
2. Form of hitch: vertical, bridle, choker, or basket
3. Angle of legs: vertical, included, or horizontal

Types of Slings

Eye and eye. Made by forming an eye in each end of a length of wire rope, using either a mechanical or a hand splice.

Endless (grommet). Made of either strand-laid or cable-laid construction. Strand-laid grommets are formed of a single length of wire strand that is laid or twisted helically around itself on each of six successive loops, with ends tucked into the space where a short length of the hemp center has been removed. Cable-laid grommets are formed of six wire ropes laid around a wire core, with ends joined together by hand or mechanically to form an endless body (see Fig. 8.1).

Forms of Hitches

Vertical. The three most common types of hitches are illustrated in Fig. 8.2. Permits the use of full rated capacity of a sling in a direct link between load and lifting device. However, no load rotation can be permitted with this type of sling, since there is a danger of the rope losing its twist or lay.

Bridle. A combination of two, three, or four single hitches, used together for hoisting an object that has the necessary lifting attachments. It provides excellent load stability when the load is distrib-

Figure 8.1 Typical endless sling.

uted equally among the legs, with the hook located directly over the center of gravity of the load. However, its use requires that sling angles be determined carefully to ensure that individual legs are not overloaded (see Fig. 8.3).

Choker. Used to hold irregular or unbalanced loads securely. However, rated capacities are reduced by 25% as compared to straight attachment since the noose configuration affects the ability of the component wires and strands in the rope to adjust during the lift. It can actually damage the body at the point of choke.

Basket. Used in pairs for better balance and greatest strength, distributing a load evenly between the two ends of a sling. However, since the two ends of such a sling usually are not vertical, the lifting capacity may not be double the capacity of a vertical hitch with the same sling.

Angle of Legs

Unless the sling angle is stated, the lifting capacities are misleading since the capacity of a sling leg is reduced as the rigging angle is re-

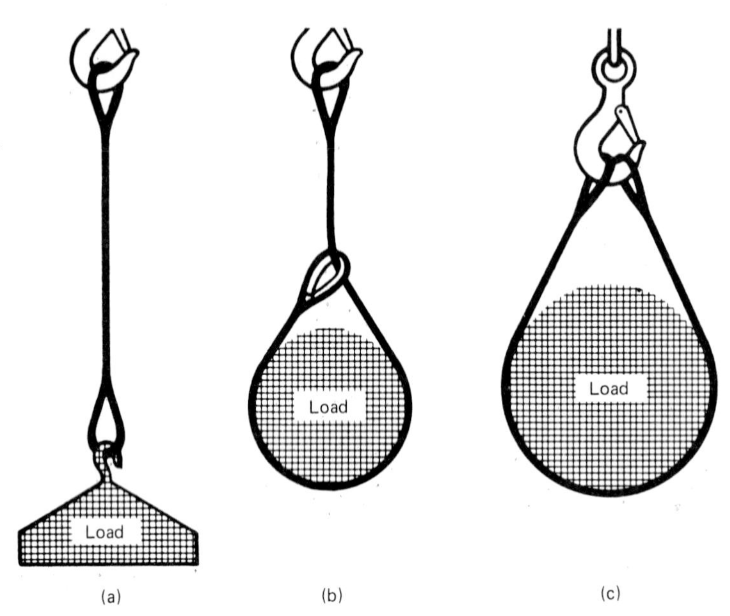

Figure 8.2 Forms of common hitches: vertical or straight, choker, basket. *(Courtesy of Paulsen Wire Rope Corporation.)*

duced. Comparing the rated capacities of a sling on manufacturers' charts readily indicates the loss of capacity between a straight sling with one vertical leg and a sling with angled legs.

Whenever possible, sling angles should be greater than 45° from the horizontal. The use of any sling with legs at angles less than 30° from

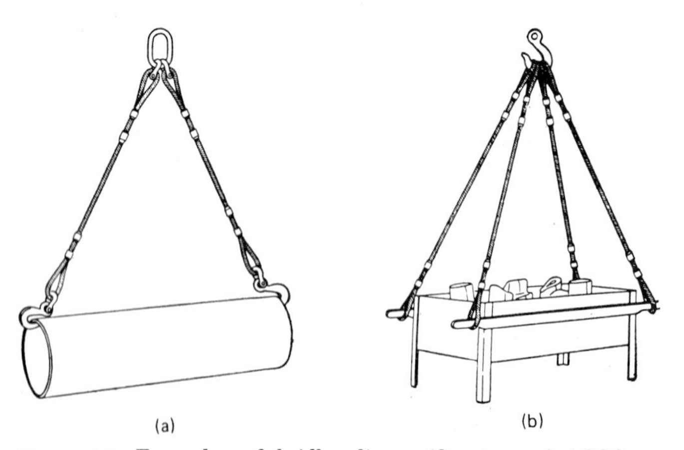

Figure 8.3 Examples of bridle slings. *(Courtesy of ACCO-American Chain Div.)*

the horizontal is extremely dangerous and is not recommended because of the effect on the load that would result from an error in sling angle measurement of as little as 5°.

Figure 8.4 shows that rigging the lifting device close to the load, thereby increasing the angle at which the sling legs are spread, that is, the included angle, sharply increases the tension placed on the legs of a sling. Although the length and width of the load, the sling length, and the available headroom will determine the sling angle that can be used to lift a particular load, it is imperative that the sling leg angle be kept as large as possible. As a rule of thumb, when the sling leg angle as measured from the horizontal, that is, the horizontal angle, is:

60°, the lifting capacity is reduced by about 15%

45°, the lifting capacity is reduced by about 30%

30°, the lifting capacity is reduced by about 50%

A sling leg angle less than 30° is neither economical nor good practice. Such angles not only build up the tension in the sling legs out of all proportion to the weight of the load, they also create a much greater "in pull" on the ends of the load, thus producing an eccentrically loaded

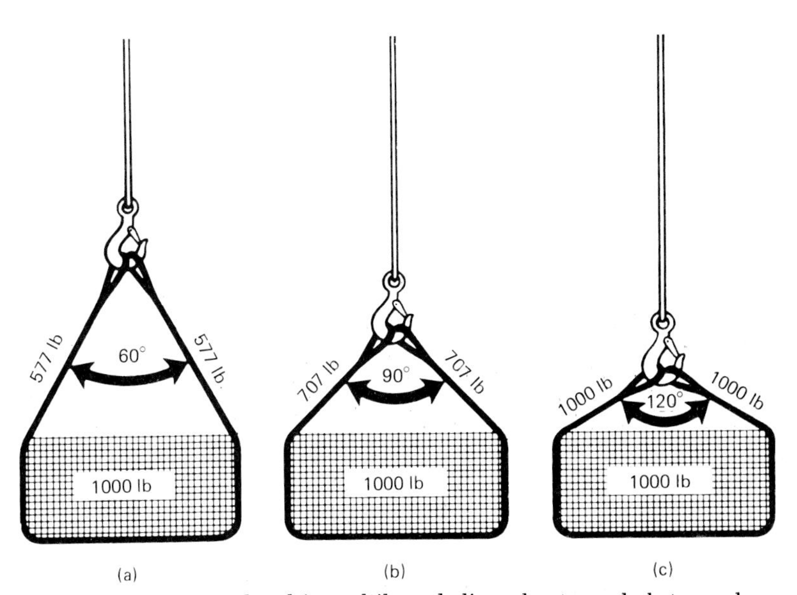

Figure 8.4 Stresses induced in multilegged slings due to angle between legs. *(Courtesy of Paulsen Wire Rope Corporation.)*

column effect (long, slender objects have a tendency to buckle). Sling capacity reduction factor for various vertical angles are given in Table 8.1.

The different ways in which manufacturers refer to sling leg angles are illustrated in Fig. 8.5.

Formulas for Determining Safe Working Loads

Always refer to the sling manufacturer's safe working load tables when selecting a sling. However, for estimating the loads in the most common sling configurations, the following formulas will provide approximates usable values. Each formula for a given sling configuration, material, and size is based on the safe working load of the single vertical hitch of that sling. Make sure the efficiencies of end fittings used are also taken into account when determining the capacity of the combination.

Bridle hitches

For two, three, and four-leg hitches,

$$\text{SWL} = \text{SWL}_v \times H/L \times 2$$

TABLE 8.1 Sling Capacity Reduction Factors

Largest angle between any leg and vertical*	Capacity reduction factor (cosine)
0°	1.000
5°	0.996
10°	0.985
15°	0.966
20°	0.940
25°	0.906
30°	0.866
35°	0.819
40°	0.766
45°	0.707
50°	0.643
55°	0.574
60°	0.500
65°	0.423
70°	0.342
75°	0.259
80°	0.174
85°	0.087
90°	0.000

*Normally a sling should not be used when the vertical angle exceeds 45°.

SOURCE: Paulsen Wire Rope Corporation.

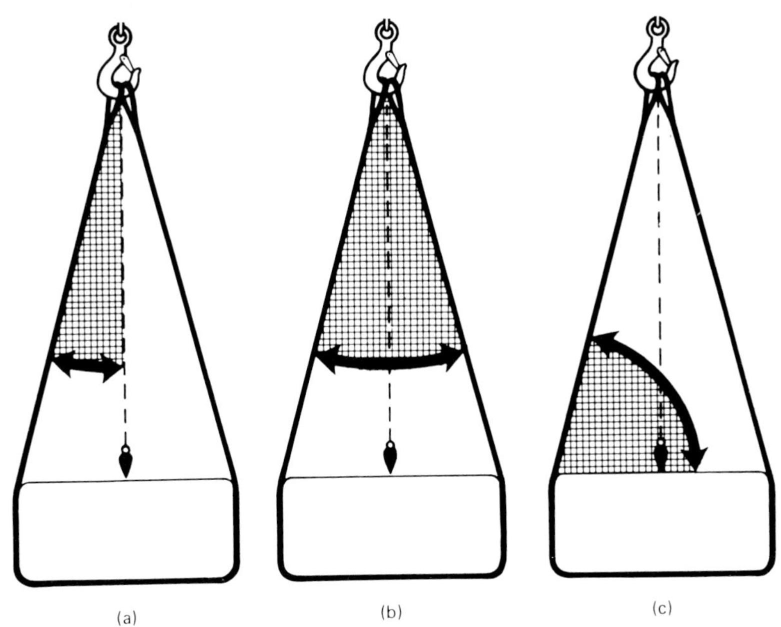

Figure 8.5 Typical designations of angle between legs. (*a*) Vertical or half-included angle. (*b*) Included angle. (*c*) Horizontal angle. *(Courtesy of Paulsen Wire Rope Corporation.)*

where SWL = safe working load
SWL_v = safe working load, single vertical hitch
L = length of sling legs
H = headroom between hook and load

Choker hitches

Single, angles greater than 45°,

$$SWL = SWL_v \times 3/4$$

Double, angles greater than 45°,

$$SWL = SWL_v \times 3/4 \times H/L \times 2$$

Basket hitches

Single, vertical legs,

$$SWL = SWL_v \times 2$$

Single, inclined legs,

$$\text{SWL} = \text{SWL}_v \times H/L \times 2$$

Double, vertical legs,

$$\text{SWL} = \text{SWL}_v \times 4$$

Double, inclined legs,

$$\text{SWL} = \text{SWL}_v \times H/L \times 4$$

Endless slings and grommets

$$\text{SWL} = 2 \times \text{values for previous configurations}$$

Never assume that a three- or four-leg hitch will safely lift a load equal to the safe load on one leg multiplied by the number of legs. There is no way of knowing that each leg is carrying its share of the load; two legs may take the full load, while the other legs serve only to balance the load.

How to Select a Sling

The following guidelines, developed by the Wire Rope Producers Committee of the American Iron and Steel Institute, are intended to help a rigger make the proper selection of a sling for a lift:

Determine weight. Make sure that the load to be lifted is known. If not, take proper steps to ensure the sling has more than adequate rated capacity.

Decide on hitch. Make sure that the hitch accommodates the load's shape and size as well as its weight. Selection must take into consideration any possible physical damage to the load as well as providing a positive attachment. The hitch chosen may affect the choice of sling construction.

Check lifting device. Make sure that the lifting device has sufficient capacity, is in proper working condition, and provides the maneuverability required once the load is hoisted—and that the sling fits on this device.

Consider room to lift. Make sure that the lifting device has sufficient headroom to pick up the load and handle it when the length of the sling is added to the hook.

Determine sling length. Make sure to use the longest sling possible for completing the lift, since the longest sling will provide the smallest angle of spread between the legs for minimum stress on the sling.

Check rated capacity. Ascertain the sling's safe working load capacity. Do not guess; use the rated capacity chart. Double-check that the sling length, type, and diameter chosen, when rigged at the selected angle, will accommodate the load to be lifted.

Determine leg styles. Make sure to use the correct leg style. Whether to use a single-part sling (one rope) or a multipart sling (several ropes) involves the handling characteristics of the sling more than any other factor. Considering the capacity alone, multipart slings will be more flexible, more easily handled, than single-part slings, and will often provide the only practical means for handling extremely heavy lifts.

Fiber Rope Slings

Although wire rope slings are the most commonly used type of slings in rigging today, fiber rope slings have their proper place on rigging jobs—for lifting comparatively light loads and on temporary work. Fiber rope slings should never be used for lifting objects having sharp edges that could cut the rope or in applications where a sling would be exposed to high temperatures, severe abrasion, or acids.

Choosing the correct fiber rope type and size depends on the application, the weight to be lifted, and the sling angle. However, before lifting any load with a fiber rope sling, be sure to check carefully for any deterioration of the rope.

Although it is very difficult to estimate the actual strength of fiber rope slings, the approximate load capacity can be determined by the formula

$$\text{Load capacity} = \frac{N \times A \times B \times (\text{minimum breaking strength of rope})}{\text{SF}}$$

where N = number of legs in rope sling
A = angle factor (sine of horizontal angle)
B = sling factor
SF = safety factor

Manufacturers' safe working loads for fiber rope slings are based on a 5:1 factor of safety and refer to new and unknotted ropes. Safe working load tables assume that the loads are symmetrical and balanced; the sling legs are of equal length; the rope, splices, and fittings used to couple the slings to the load are strong enough; and padding is used to protect the slings when lifting objects with sharp corners.

All tables are based on conventional three-strand ropes, ranging in diameter from ½ to 2 in. inclusive. Smaller sizes are not generally recommended for slings because of the serious effect of relatively minor surface damage and the susceptibility of small ropes to mishandling and overloading. Slings from ropes larger than 2⅝ in. in diameter are acceptable, but are seldom available.

Splicing

In manila ropes, eye splices should contain at least three full tucks; short splices, at least six full tucks (three on each side of the centerline of the splice). In synthetic-fiber ropes, eye splices should contain at least four full tucks; short splices, at least eight full tucks (four on each side of the centerline of the splice).

The strand end tails of splices should never be trimmed short (flush with the surface of the rope) immediately adjacent to the full tucks. For fiber ropes under 1 in. in diameter the tails should project at least six rope diameters beyond the last full tuck; for 1-in. diameters and larger, the tails should project at least 6 in. beyond the last full tuck.

The clear rope length (between eyes) in each leg of a fiber rope sling should be at least 10 times the rope diameter.

Synthetic Webbing Slings

Synthetic webbing slings are widely used for rigging loads that must not be marred or scratched, such as finely machined, highly polished, or painted surfaces (see Fig. 8.6). These webbing slings offer a number of advantages over slings made of other materials:

Softness and bearing width. They have less tendency to crush fragile loads.

Flexibility. They tend to mold to the shape of the load, thus gripping a load more securely.

Elasticity. They tend to stretch under load and are thus able to absorb heavy shocks and cushion loads.

Long life. They are not affected by moisture and certain chemicals.

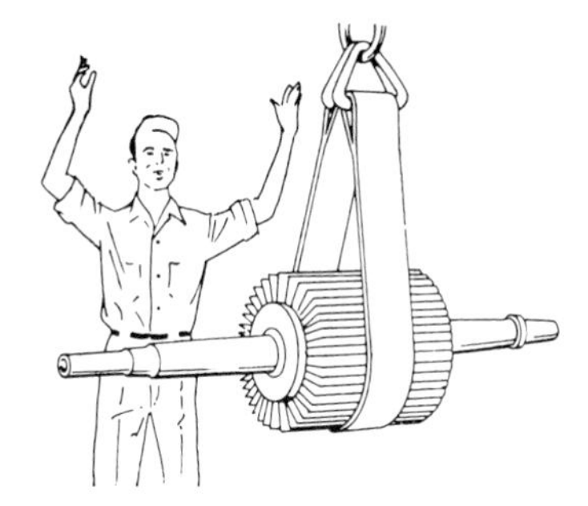

Figure 8.6 Typical fiber sling application.

Stability. They are less apt to twist and spin during lifts.

Nonsparking. They can be used safely in explosive atmosphere.

Nonrusting. They do not stain material being lifted.

Nylon

The most popular and best general-purpose synthetic webbing slings are made of nylon. They are unaffected by grease or oil and have good chemical resistance to aldehydes, ethers, and strong alkalies. Nylon slings should not be used with acids and bleaching agents. They are not suitable for use at temperatures that exceed 250°F. The stretch rated capacity is approximately 10%.

Polyester

Whenever acid conditions are present and a minimum stretch is desired, use polyester webbing slings. They are unaffected by common acids and hot bleaching solutions, but they should not be used with concentrated sulfuric acid and alkaline conditions. Polyester slings are not suitable for use at temperatures that exceed 250°F. The stretch rated capacity is approximately 3%.

Polypropylene

Use polypropylene webbing slings whenever acid or alkaline conditions exist, since the fibers are not affected by these conditions. These fibers are not suitable for use at temperatures exceeding 180°F. The stretch rated capacity is approximately 10%, as it is for nylon.

For specific applications, all synthetic webbing slings are available with protective coatings:

Latex. Seals out moisture and dirt and reduces the effects of abrasion

Neoprene. Increases wear resistance and the sling's coefficient of friction for a firmer grip on the load

Polyurethane. Increases sling life—by as much as five times that of untreated fabrics—and the coefficient of friction, providing greater gripping power

Synthetic webbing slings are available in one-, two-, three-, or four-ply webbing, with metal end fittings (triangle and choker, or triangles), standard eye and eye, twisted eye, and endless or grommet configurations. All slings have a 5:1 design factor.

Triangle and choker triangle

Used in chokers, vertical slings, or basket hitches, these slings are equipped with metal triangle and choker fittings at their ends. They are available in 2-to-12-in. web widths, 2,400-to-38,000-lb capacities.

Triangles

Used in vertical slings or basket hitches only, these slings are equipped with triangle fittings at both ends. They are available in 2-to-12-in. web widths, 3,200-to-38,400-lb capacities.

Sewn eyes

Used in vertical slings, chokers, or basket hitches, these slings are equipped with either flat tapered eyes or twisted eyes. They are available in 1-to-6-in. web widths, 4,600-to-28,600-lb capacities.

Endless or grommet

Used in vertical slings, chokers, or basket hitches, these slings are continuous web loops. They are available in 1-to-6-in. web widths, 2,600-to-57,200-lb capacities.

Wire Rope Slings

A wire rope sling is a complex machine with many moving parts, which takes the full brunt of the burden, plus additional stresses and forces every time a load is lifted. All wire rope slings should be made of improved plow steel and should have independent wire rope cores to reduce the possibility of the rope being crushed in service.

Only experienced splicers should make eye splices in wire rope slings. All eyes should be formed with the Flemish splice and should be properly served to conceal the sharp protruding ends of the wire, which can cause serious injury to the riggers handling them. Securing a splice by swaged or pressed mechanical sleeves will produce an eye as strong as the rope itself. While not absolutely necessary, it is good practice and strongly recommended that thimbles be placed in all spliced eyes.

Socketing should be done only by experienced persons. Care should be taken, when socketing a wire sling end, not to allow the twist or lay to come out of the rope as it enters the throat of the socket.

Two types of splices are used to form the eyes for single-part eye-and-eye slings, either a mechanical or a hand splice. In a mechanical splice, the eye is made by forming a loop and pressing or swaging one or more metallic sleeves over the rope junction. In hand splicing, the eye is made by forming a loop and tucking strands of the dead end into the live end of the rope. The efficiency loss in any type of splice is reflected in the rated capacity that the manufacturer applies to a sling.

Always rig a sling so that neither the load nor the lifting device can rotate, since the twisting motion may cause eye splices to pull out, or it may result in excessive stress in the splice.

Wire rope slings are most commonly made in diameters from ¼ to 2 in.—the greater the diameter, the greater the sling's strength—and in lengths from 5 to 20 ft. They are formed of either single-part (made from a single wire) or multipart (made from multiple-wire ropes) construction. A multipart sling can have from 3 to 32 parts or ropes. The influence of the angle of multipart slings on the load capacity can be seen from Table 8.2.

The three most common types of multipart wire rope slings are cable-laid, round-braided, and flat-braided, according to the Wire Rope Technical Board. They are defined as follows:

Cable-laid

Multipart cable-laid slings are similar in appearance to single-part slings. However, the body of a standard cable-laid sling consists of six individual wire ropes, each composed of six strands around a steel core. All six ropes are laid around another wire rope of the same construction, making a finished sling body of seven parts. Typical bodies might be described as $7 \times 7 \times 7$ or $7 \times 7 \times 9$.

Variations of the cable-laid type of sling use multiple strands or ropes,

TABLE 8.2 Safe Working Loads for Typical Slings, Pounds

Size of rope	Slip noose	Two end bridle sling, when angle is 75°	Two end bridle sling, 60°	Two end bridle sling, 45°	Three-end bride sling, when angle is 75°	Three-end bride sling, 60°	Three-end bride sling, 45°
3/8	3,200	3,000	2,800	2,200	4,600	4,200	3,400
1/2	5,400	5,200	4,800	4,000	7,800	7,200	6,000
5/8	8,400	8,000	7,200	6,000	12,000	10,800	9,000
3/4	12,000	11,400	10,200	8,400	17,000	15,400	12,600
3/8	16,000	16,000	14,000	11,400	24,000	21,000	17,200
1	22,000	20,000	18,000	15,000	30,000	26,000	22,000
1 1/8	26,000	26,000	23,000	19,000	40,000	36,000	28,000
1 1/4	32,000	32,000	28,000	23,000	48,000	42,000	36,000
1 3/8	40,000	38,000	34,000	28,000	58,000	52,000	42,000
1 1/2	46,000	44,000	40,000	32,000	66,000	60,000	48,000

SOURCE: AMHOIST—American Hoist & Derrick Co.

or both, in the same manner, but with varying parts in the sling body. They use both hand and mechanical splices to form the sling eyes, whereas in standard cable-laid slings, eyes are formed with mechanical splices only.

Since there are usually many more component wires in a cable-laid sling body than in a typical single-part wire rope sling with the same rated capacity, by comparison, the cable-laid body will exhibit phenomenal flexibility. For this reason, the D/d ratio can be less for cable-laid slings—they can be in a much tighter arc around a load. A minimum D/d ratio of 10 is recommended by most wire rope and sling manufacturers for the sling body in a cable-laid sling.

Round-braided

This type of sling is perhaps the more common braided construction, offering flexibility far beyond single-part and even some cable-laid slings. In addition, round-braided slings are easy to handle, resist slipping and kinking, and their multiplicity of parts provides early warning signals for inspectors by exhibiting damage among their many wires and parts before critical loss of capacity occurs. Close internal inspection is made easy by the braided design.

Because of their extreme flexibility, eight-part round-braided slings conform quickly to load contour, provide maximum gripping contact with the load, and securely hold slippery, irregular, and unbalanced loads.

Round-braided slings are usually fabricated by hand braiding component ropes. Properly balanced braids provide good control over residual torques in the rope parts. Therefore such a sling does not tend to rotate or unlay itself under load, as do single-part sling bodies. This reduces the requirements for tag lines or many lifts.

Flat-braided

Slings with six ropes or parts are the more popular of this type, although other constructions are in use. Flat braids provide a relatively wide bearing area for protection against slippage and marring of the load. These slings are hand fabricated, with special care being taken to balance individual ropes.

As with all multipart slings, flat braids consist of a large number of small wires. Since these component wires are comparatively small in diameter, they are highly susceptible to wear and breakage from abuse, abrasion, and nicking.

In both round- and flat-braided slings, as with cable-laid types, the number of component wires is greatly increased in comparison with single-part slings of the same diameter or rated capacity. The D/d ratio of braided slings is 20 times the component rope diameter. This allows lifts of heavy, compact loads that would not be possible with a comparable single-part sling.

Metal Mesh Slings

Wire or chain mesh slings are widely used in metalworking and other industries where loads are abrasive, hot, or will tend to cut fiber web slings. They are available in three mesh sizes—10-gage for heavy-duty general-purpose lifting, 12-gage for medium-duty, and 14-gage for light-duty lifting, in widths from 2 to 20 in., and in lengths from 5 to 20 ft.

The principal characteristics of metal mesh slings are their smooth, flat bearing surfaces and an ability to conform to irregular shapes, to grip loads tightly, not to kink, tangle, or whip, to resist corrosion, abrasion, and cutting, and to withstand temperatures up to 500°F.

For handling loads with finishes that could damage the metal mesh, slings can be coated with neoprene or clear PVC plastic that combines the softness of an elastic surface with the strength of steel. The capacity of metal mesh slings is determined by wire diameter, mesh width, and type of hitch. A 10-gage steel wire in a 20-in.-wide mesh can lift 12 tons in a direct connection, 24 tons in a basket hitch. However, because of improved load balance, 90% of all mesh slings are worked with a choker hitch (see Figs. 8.7 and 8.8).

Chain Slings

Only alloy steel chain is suitable for slings used in overhead lifting, and is identified by a letter (usually A) marked on each link. This type of sling is made of wire or bar steel links from ¼ to 1½ in. in diameter, in lengths of 5 to 20 ft.

The user of alloy steel chain slings now has available for lifting applications the new rating of the National Association of Chain Manufacturers (NACM)—grade 80 alloy steel chain. These grade 80 slings offer higher working load limits than shown in most chain capacity tables currently in use and are in compliance with OSHA standards when used in accordance with the chain manufacturer's recommendations.

Chain slings are used for lifting operations where the primary requirements are ruggedness, abrasion resistance, and high-temperature resistance. They are available in both single- and multiple-leg styles.

While a 1¼-in.-diameter single chain can lift 40 tons, a four-leg sling can easily handle more than 100 tons. However, when a chain sling is hooked back on itself (into the chain, rather than to the master link or

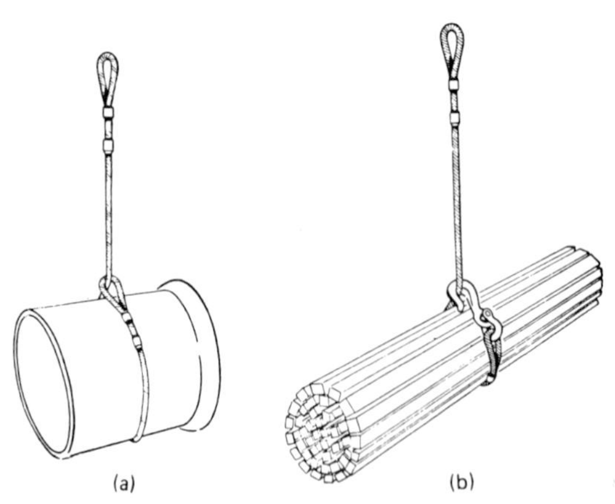

Figure 8.7 Typical choker sling applications. *(Courtesy of ACCO—American Chain Div.)*

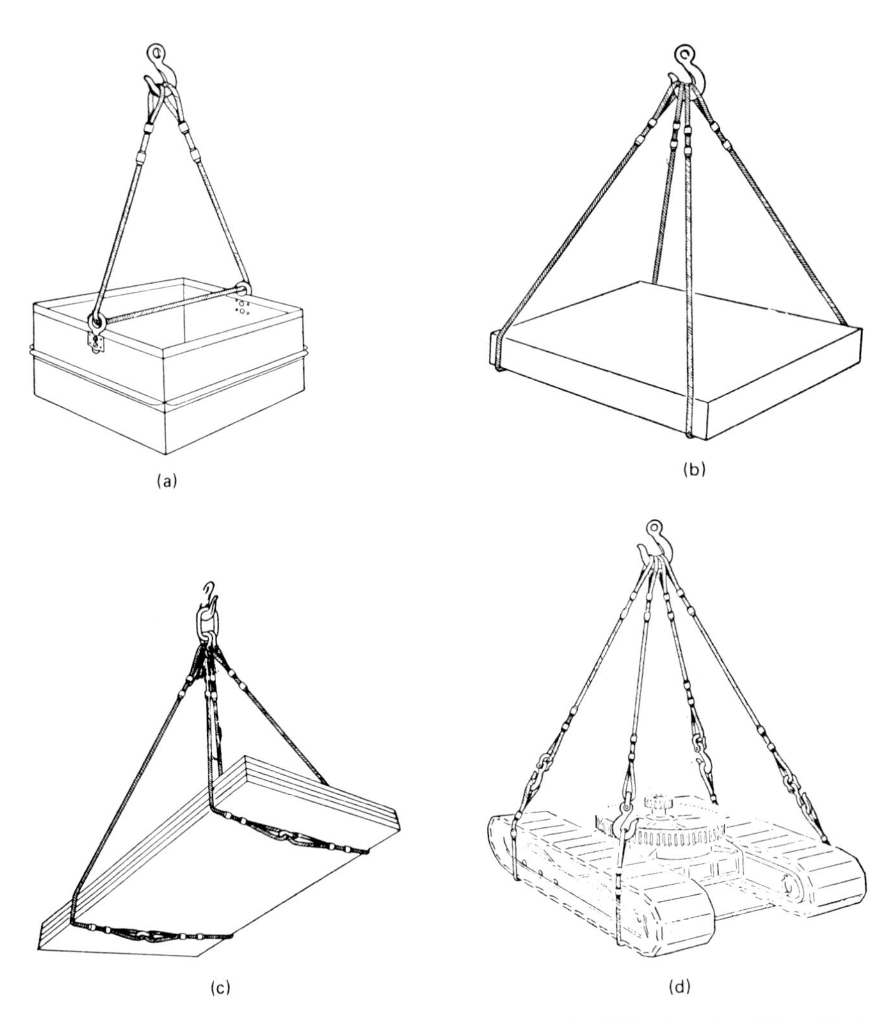

Figure 8.8 Typical basket sling applications. *(Courtesy of ACCO—American Chain Div.)*

master coupling link), the capacity of the sling should be reduced by one-fourth. Thus, choker hitch ratings in manufacturers' safe working load tables are less than ratings for vertical hitches.

Whenever using chain slings in a basket hitch configuration, make sure they are hooked back into the master link, not the chain or coupling link.

Chain slings should always be padded where they bear on sharp edges of metal parts to prevent links from being subjected to bending stresses for which they were not designed.

Proper Sling Rigging

Each lifting job varies in size, shape, weight, and the location of lift points for sling attachment. A good rigger must know and learn to recognize the difference between a good hitch and an unsafe one, where and how to attach a sling, and must be able to anticipate how a load will react when lifted in or on the sling. Improper rigging is not only hazardous, it can be costly in damage to both the load and the sling.

It is extremely important that everyone involved in the lifting and transporting of loads be familiar with the fundamentals of statics and the effect of the center of gravity of a body in relation to the lifting point, or the distribution of forces in the sling and body.

For example, in a uniform, rectangular load, as illustrated in Fig. 8.9*a*, the center of gravity is directly beneath the intersection of the diagnals. As a rule of thumb, when an irregular shaped object O as shown in Fig. 8.9*b* is to be lifted, it is helpful to visualize the load as "enclosed" by an imaginary rectangle, and the center of gravity can be perceived as a point where diagonals of this imaginary rectangle intersect. It must always be kept in mind, however, that such a general rule cannot apply to objects which have weight concentrated at one end or side.

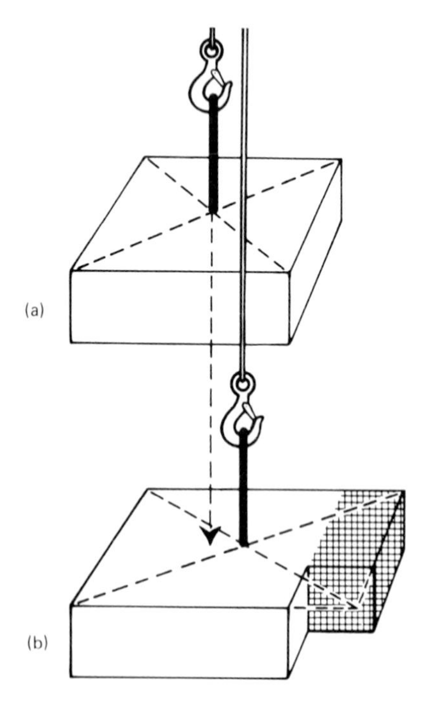

Figure 8.9 Compensating for center of gravity in irregularly shaped objects. *(Courtesy of Paulsen Wire Rope Corporation.)*

The following points illustrate the effects of such forces in equilibrium:

1. The center of gravity is under the crane hook. The angles of lift are equal and each sling leg is carrying the same load (see Fig. 8.10).
2. The center of gravity is not in line with the point of lift. This requires the use of unequal leg lengths to put the center of gravity under the point of lift and balance the load. The rated capacity of the sling must be based on the smallest horizontal angle (see Fig. 8.11).
3. The center of gravity of the load is above the points of attachment. When lifted, the load will rotate to bring the center of gravity to the lowest point. Whenever possible, have the points of attachment above the center of gravity of the load.
4. With a two-legged sling, the load will tilt and possibly slip from hooks unless precisely balanced.
5. With a three-legged sling, the third leg provides stability. However, one leg is carrying half the weight of the load and the capacity should be based on a two-legged sling at some horizontal lift angle *C*.
6. A four-legged sling provides good stability and the load is equally distributed on each leg.
7. When the load is lifted by a sling, the legs of the sling will exert

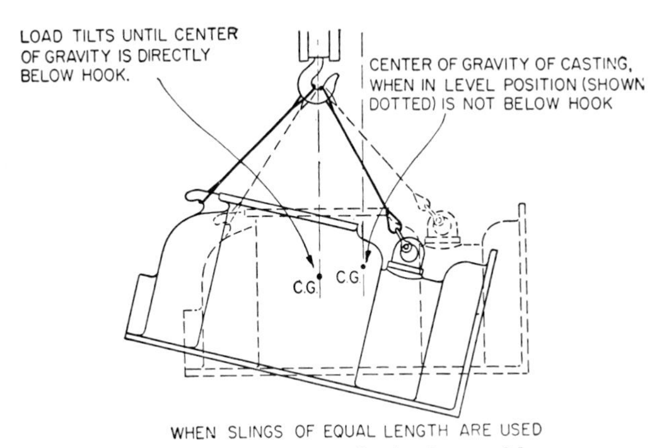

Figure 8.10 Center of gravity under hook; equal legs.

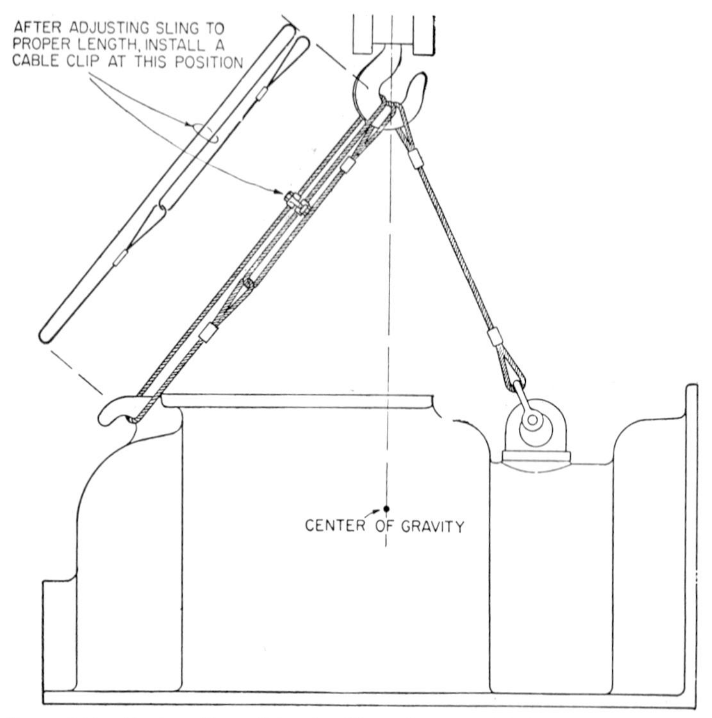

Figure 8.11 Center of gravity not midway between hitch points; unequal legs.

lifting forces on the load to raise it. The legs will also exert other forces, which are horizontal components of the loads on the sling legs. The forces can crush or deform the load if it cannot withstand these. When fragile loads are involved, a spreader beam is recommended.

The center of gravity is the point in a load at which all the weight can be said to be concentrated during a lift. It acts downward to bring the load to a position of equilibrium directly below the lifting hook, even though the load may not be level.

Before attempting to lift a load that may be much heavier on one end than on the other, or where eyebolts or lifting lugs may not have been located with respect to the center of gravity of the load, the rigger must estimate the location of the load's center of gravity and spot the crane hook directly over this theoretical point.

Next, it is important to use slings of proper strength and length to reach from the hook to the eyebolts or lifting lugs. This may require that one or more of the slings be of an odd length not readily available. If the load is lifted without regard to the position of the center of gravity, the load will tilt until the center of gravity is directly below the point of support that is the hook, and the load will be suspended at that angle.

Where an odd-length sling is required, but where it is not necessary to provide accurate leveling, a single sling may be rigged up and adjusted to the required length. Theoretically, this sling will slip under strain, but practice seems to indicate that friction is sufficient to hold its adjustment. However, just to be safe, it is recommended that a cable clip be installed.

When placing a sling on a load, be sure to pad all sharp corners, not merely by means of some burlap, but by protecting the edges by cable guards or bent plates of adequate thickness to hold their shape under load. A good rule to follow is to make sure that the length of arc of contact of the rope is at least equal to one rope lay (about seven times the rope diameter). When the bend is of this length, each of the strands has been on the inner and outer sides of the rope bend, and the slippage of the strands relative to each other minimizes the stress. On the other hand, if the bend is very short, the strands on the outer side of the rope bend will have to stretch, and this stretch will leave a sharp bend or kink in the sling. If a load is to be rotated so as to rest on its side instead of its end, the sling must be properly installed. Otherwise, when partially turned over, the load may flop over and slam down. A single sling used as a choker hitch should be arranged as shown in Fig. 8.12.

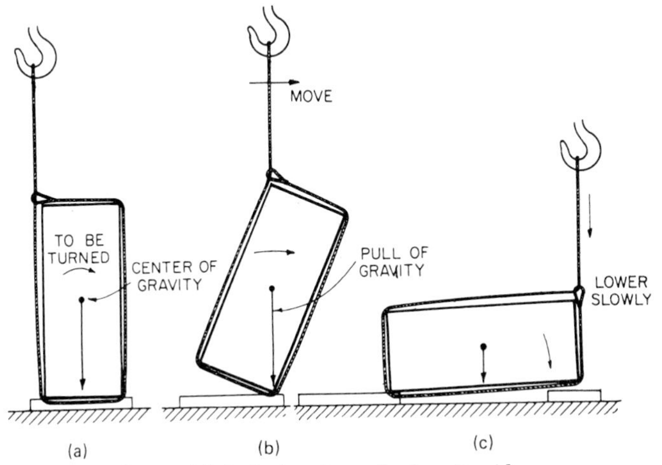

Figure 8.12 Proper hitch for turning a load on its side.

The part of the sling leading from the crane hook should go directly to the bottom of the load. A strain is taken on the sling, and the crane is moved as indicated in Fig. 8.12*b*. This causes rotation of the load. When a vertical line downward from the center of gravity moves past the corner of the load which rests on the ground, the rotation will continue by gravity as the crane operator lowers the hook (see Fig. 8.12*c*). If the hitch point is below the top corner of the load, an unsafe condition will be created.

When the gravity line passes the corner of the load resting on the ground, the load will turn rapidly, rocking back and forth, putting excessive stresses on the rope until the load finally comes to rest. If the hitch point is near the lower corner, the load will flop over and slam down.

Never use a basket sling in such a manner that the load can slip on it as it is being rotated, for the sharp edges will cause severe damage to the sling. Use a choker sling instead.

When using a spreader bar or rig for a two- or four-leg sling, the rigger must exercise extreme care to avoid spilling the load (see Fig. 8.13).

For example, when lifting a load such as an engine lathe mounted on wood skids, the load has a very high center of gravity.

In Fig. 8.13*a*, draw lines from the center of gravity (CG) to the points where the skids rest on the basket slings, angles *a* and *b*, respectively. Then compare these with angles *A* and *B* of the slings above the spreader. Angles *A* and *B* must be considerably greater than the respective angles *a* and *b* to ensure stability during lift. Figure 8.13*b* shows what may happen if this precaution is not taken, for the center of gravity will tend to move to the lowest point possible directly below the point of support, which is the hoist hook. The same precautions must be taken when using a rigid spreader, as shown in Fig. 8.13*c*.

Care and Inspection of Slings

When not in use, all slings should be hung up in an orderly manner on special hooks or brackets to keep them as straight as practicable.

All slings should be inspected periodically and condemned when found to be in unsafe condition. Inspect all slings in service daily, and remove a sling from service that shows any of the following defects:

Abnormal wear

Powdered fiber between strands (fiber rope)

Broken or cut fibers or strands

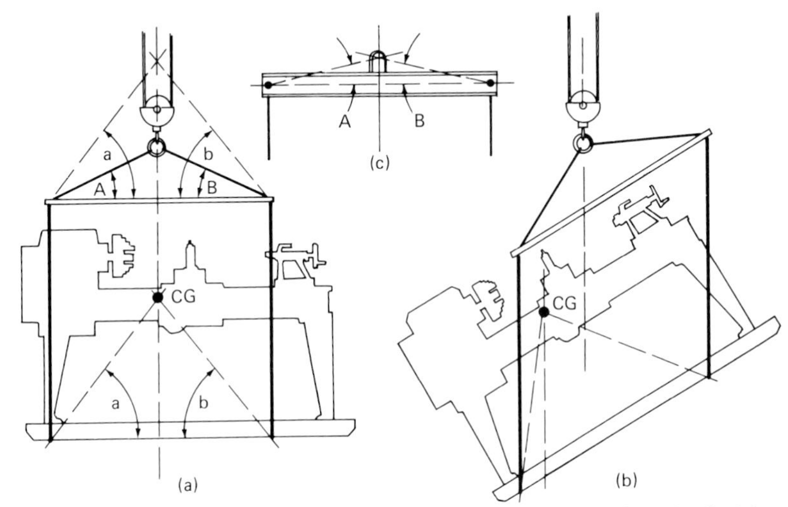

Figure 8.13 Proper sling angle must be maintained when handling load with a spreader beam.

Variations in size or roundness of strands

Discoloration or rotting (fiber rope)

Distortion of sling hardware

Slings, regardless of type, should be protected from corrosion (or decay) and from contact with injurious chemicals. With reasonable care, a wire rope sling should last indefinitely.

To protect load and sling from damage at sharp corners, always use wooden blocks and pads at lift contact points. A protective pad should be used any time a sling passes around a sharp corner.

Always examine every sling visually from end to end before every lift. Remember, manufacturer-rated capacity applies only to a new sling in unused condition. A sling should be carefully examined to determine that it is in as nearly new condition as practicable before each lift.

Wire rope slings must not be allowed to become kinked; this also causes severe bending stresses. In addition, the strands that have been displaced are subject to unequal distribution of the live load, some strands taking more than their normal load, in addition to the stress produced by the sharp bending. Even though an attempt is made to remove a kink, the damage done to the rope is usually permanent.

The very large sizes of slings are often made of 6×37 or $6 \times 6 \times 19$ construction to increase the flexibility. All slings that have at one time or another been bent sharply develop permanent deflections; when the strain is relieved, they will snarl in some unpredictable manner. This requires continual vigilance to keep them from developing kinks when taking a strain preparatory to lifting a load. Slings made of preformed rope behave somewhat better than those made in the conventional manner. They also have the advantage that broken wires will not readily wicker out and present a hand hazard to the rigger, but rather tend to lie dead in the original position.

Basic Sling Terms

D/d Ratio Ratio between the diameter of curvature *D*, where a load contacts a sling body, and the diameter of the sling rope *d*. For mechanically spliced slings, this ratio must be at least 20:1; for hand splices, at least 10:1. This means that the rated capacity of a given sling will be available only if the wire rope does not bend around an object or corner that is sharper than 20 (or 10) times the diameter of the rope.

Design Factor Factor by which a manufacturer adjusts the actual breaking strength of a component rope to arrive at the rated capacity for a sling. Because shock loads and the effects of wire rope deformation are unpredictable, a sling should always be used at or below its rated capacity.

Rated Capacity Maximum load that should be applied to a specific sling as determined by the sling manufacturer.

Chapter

9

End Attachments and Fittings

A rigger should know not only what types of end attachments and fittings to use for a particular hoisting or rigging application, but also what their safe working loads are in relationship to the rope or chain being used.

Although there is a wide variety of end attachments and fittings available, only forged alloy steel load-rated types should be used for overhead lifting to assure the highest degree of safety. Such attachments have their safe working load stamped directly on them.

Seizing

It is most important that all rope ends be tightly seized to prevent wires and strands from becoming slack with use. If the rope is to be cut, seizings must be placed on both sides of the point where the cut is to be made.

Proper seizing and cutting operations are not difficult to perform. However, only two methods of seizing are acceptable:

1. *Ropes over 1 in. in diameter.* Place one end of the seizing wire in the valley between two strands. Then turn its long end at a right angle to the rope and closely and tightly wind the wire back over itself and the rope until the proper length of seizing has been applied. Twist the two ends of wire together and by alternately pulling and twisting, draw the seizing ends tight (see Fig. 9.1*a*).

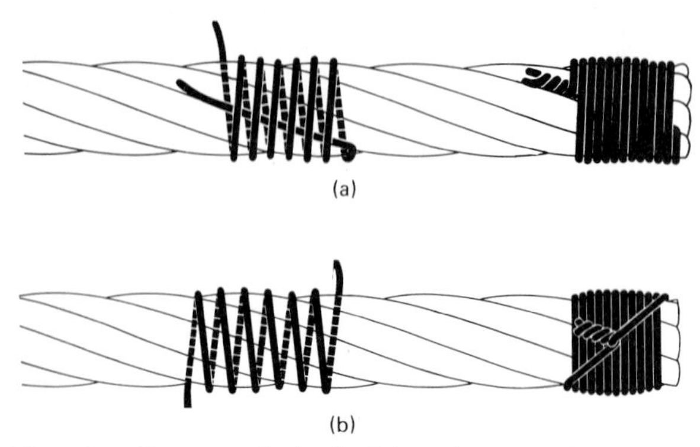

Figure 9.1 Proper methods of seizing wire rope. *(Courtesy of Bethlehem Steel Corp./Wire Rope Division.)*

2. *Ropes less than 1 in. in diameter.* Wind the seizing wire in a tight spiral around the rope end to the desired length of seizing. Twist the two ends of the seizing wire together at the approximate center of the seizing, alternately twisting and pulling until the proper tightness is achieved (see Fig. 9.1*b*).

Always use soft annealed wire for seizing. With galvanized rope, it should be zinc-coated steel wires.

Both the diameter of the seizing wire and the length of the seizing are determined by the rope diameter. The length of a seize should never be less than the rope diameter (see Table 9.1).

Sockets

The strongest wire rope end attachment is the zinc-poured socket which, if properly attached, will develop 100% of the wire rope's strength. However, *do not* think that the splice is stronger than the rope. Remember, the rope is always stronger than indicated in the manufacturer's

TABLE 9.1 Suggested Seizing Lengths and Wire Diameters

Rope diameter, in.	Seizing wire diameter, in.	Seizing length, in.
1/8 to 5/16	0.032	1/4
3/8 to 9/16/	0.048	1/2
5/8 to 15/16	0.063	3/4
1 to 1 5/16	0.080	1 1/4
1 3/8 to 1 11/16	0.104	1 3/4
1 3/4 to 2 1/2	0.124	2 1/2
2 9/16 to 3 1/2	0.124	2 1/2
3 5/8 to 4	0.124	4
4 1/8 to 4 1/2	0.124	4 1/2

SOURCE: Bethlehem Steel Corp./Wire Rope Div.

table. Therefore, it is possible for the splice to be as strong as the strength of the rope indicated in the catalog and still fail before the body of the rope fails.

In making a socket attachment, put a seizing on the rope at a distance from the end equal to the length of the socket basket, and two additional seizings spaced one and one-half rope diameters apart immediately in back of the first seizing. Unlay the strands of the rope and cut off the hemp center near the first seizing, then carefully unlay and broom out all the wires in the several strands. It is not necessary to straighten out the wires.

After separating the wires, clean them with a solvent (benzine or gasoline). Dry the wires with a clean cloth, then dip them for about three-quarters of their length into a 50% solution of commercial muriatic acid for about half a minute, or until each wire is thoroughly cleaned. *Do not* allow the acid to reach the hemp center of the rope. Then rinse by immersing in boiling water that contains a small quantity of neutralizer (bicarbonate of soda). Use care not to allow acid to come into contact with any other part of the rope.

Bind the wires together and insert the end of the rope into the socket so that the ends of the wires are even with the top of the basket; then remove this temporary binding wire. Spread out the wires so as to occupy equally the entire space within the basket.

Clay should be applied to the annular space between the neck of the socket and the periphery of the rope to prevent the molten metal from running out. Hold the socket and several feet of rope in a vertical position, and pour in molten zinc until the basket of the socket is full. The zinc must be at the correct temperature. Before pouring, dip the end of a soft pine stick into the ladle for a few seconds and remove. If the metal adheres to the stick, the metal is too cold. If it chars the stick, the metal is too hot. Allow the socket to cool, then remove the seizings, except the one closest to the socket.

Note: *Never* use babbitt or lead for securing sockets on hoisting ropes. Both metals are far too soft for the exposure required, and efficiency can be as low as 25% of rope strength.

All hoisting ropes on elevators and mine hoists should be resocketed at frequent intervals, varying from monthly to semiannually according to the severity of service. Before resocketing, the socket should be annealed by heating it to a cherry-red color and then allowing it to cool. After the sixth annealing, the socket should be discarded.

Sometimes wedge-type sockets are used. In these, the end of the wire rope is fed through the socket, bent around the wedge, then reinserted through the socket again. Under tension, the wedge pulls tight against the bend in the wire rope, holding it solidly in place (see Fig. 9.2).

However, the efficiency of such sockets is low—only 70% of the strength of the rope. In using the wedge socket, care must be exercised

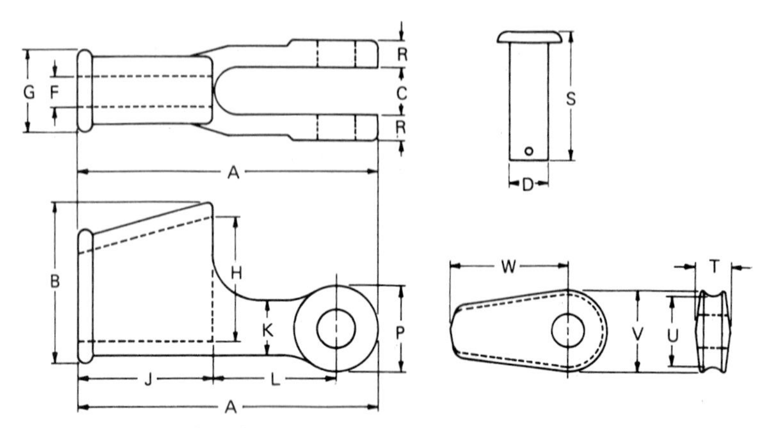

Figure 9.2 Typical wedge-type socket. *(Courtesy of Crosby Group/McKissick Products.)*

to install the rope so that the pulling part is directly in line with the clevis pin. Otherwise a sharp bend will be produced in the rope as it enters the socket.

Splicing

One of the most commonly used devices for making an end attachment is the eye splice, which should always be combined with a thimble to prevent excessive wear and distortion in the eye end. If a thimble is not used on a spliced eye, the efficiency of the connection can be reduced by as much as 10% because the rope flattens under loading.

The efficiency of a well-made eye splice with a heavy-duty thimble varies from about 95% for 5⁄16-in. rope to 88% for 3⁄4-in. rope, 76% for 1¼-in. rope, and 70% for 1½-in. rope. All splices should have at least four tucks, and the completed splice should be carefully wrapped with a wire serving to cover the protruding wire ends to eliminate the danger of lacerating the hands of riggers.

Rope splicing is a skill that requires training and proper tools to be done safely and effectively. Eye splicing should be done only by experienced splicers. Always refer to the wire rope manufacturer's handbooks and manuals for the proper splicing types, methods, and procedures to use.

There are three basic types of splices:

1. *Flemish eye (rolled eye).* With serving or with pressed metal sleeve
2. *Tucked eye (hand-splice eye).* With serving or with pressed metal sleeve
3. *Fold-back eye.* With pressed metal sleeve

Only the Flemish eye, with pressed metal sleeve, is recommended for all rigging and hoisting use. When properly made, this eye develops almost 100% of the catalog rope breaking strength.

Strand ends of the splice eye are secured against the live portion of the rope by means of a steel or aluminum sleeve set in place under pressure. This is the most dependable of the many types of mechanical splices available, since the basic strength of the splice is inherent in the hand splicing, regardless of the sleeve attachment.

Tucked eye splices develop only 70% of the strength of the rope and tend to come free as the rope unwinds.

Fold-back eye splices are made simply by bending the rope against the live portion to the free end by means of a steel or aluminum sleeve set in place under pressure. Never use this type of splice for overhead hoisting, since improper swaging or split sleeves will result in complete failure without warning.

Clips

The most common method of making an eye or attaching a wire rope to a piece of equipment is with cable or Crosby clips. These consist of the separate U-bolt and saddle clips and the double-saddle safety (fist-grip) clips (see Fig. 9.3).

Make sure when using clips that the corrugation in the saddles matches that of the rope lay being used—left-lay or right-lay clips for left-lay or right-lay ropes, respectively. Otherwise the ridges between the corrugation in the forging will run crossways, rather than parallel to and between the strands of the rope. This will result in cutting wire strands when the clip nuts are properly tightened.

U-bolt clips must be placed on the rope with the U bolts bearing on the short or dead end of the rope, and the saddle bearing on the long or live end of the rope. Make sure that all clips are attached correctly, for even one misapplied clip can reduce the efficiency of the connection to as low as 40%. These clips should be spaced not less than six rope diameters apart, and in no case should fewer clips be used than the number recommended in Table 9.2.

Clip efficiency will depend on the arrangement, care in tightening, and number of clips used:

Properly attached U bolts (all on dead end)	80%
Staggered clip attachment	75%
U bolts (all on load end)	70%
Improperly tightened nuts	Less than 50%

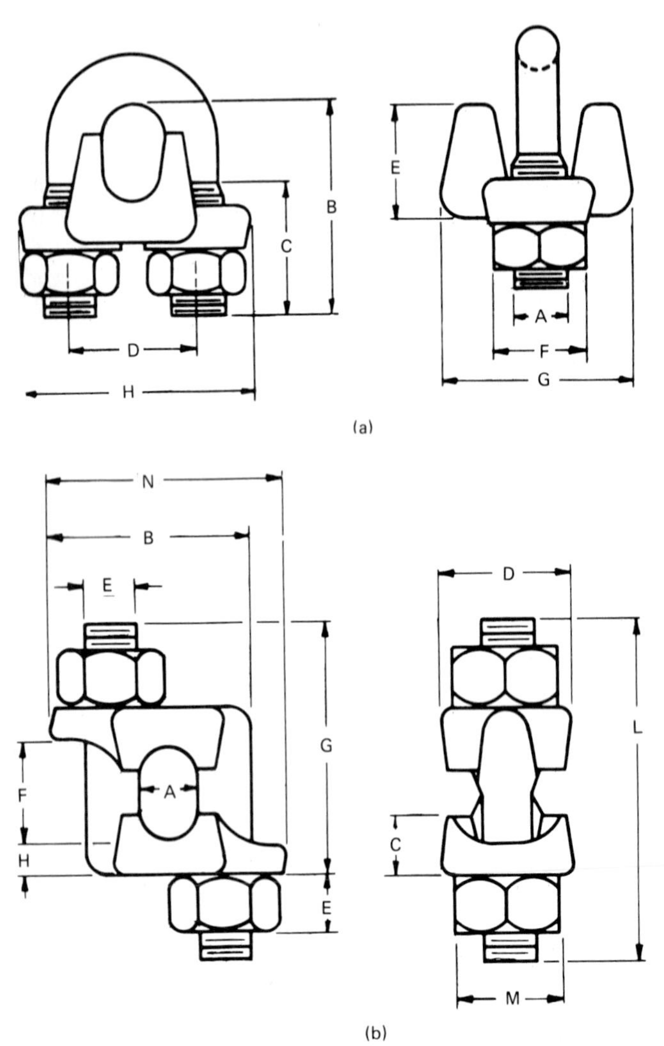

Figure 9.3 Typical cable clip attachments. (*a*) Regular. (*b*) Fist. *(Courtesy Bethlehem Steel Corp./Wire Rope Div.)*

Double-saddle clips have corrugated jaws that fit both parts of the rope and hence can be installed without regard to which part bears on the live or dead parts of the rope. This fist-type clip develops about 95% strength of the rope, thus one less clip is required than the number indicated in Table 9.2.

TABLE 9.2 Installation of Clips*

Rope diameter, in.	Number of Crosby clips for eye attachment	Spacing between clips, in.
1/4	2	1 1/2
5/16	2	2
3/8	2	2 1/4
7/16	2	2 1/2
1/2	3	3
5/8	3	4
3/4	4	4 1/2
7/8	4	5 1/4
1	4	6
1 1/8	5	7
1 1/4	5	8

*The proper number and spacing of Crosby clips are important.
SOURCE: Crosby Group/McKissick Products.

Clipped Eyes

When installing clips to form an eye in a wire rope, always use a thimble. If properly made, a clipped eye should develop about 80% of the strength of the rope.

Always use a heavy-duty thimble to form an eye in the end of a wire rope. Turn back the correct amount of rope for dead-ending to permit proper spacing of the clips. Bind the rope on itself at the toe of the thimble. Then apply the first clip (farthest from the thimble) one clip base width from the dead end of the rope. Attach nuts and tighten securely.

Apply the second clip nearest the thimble and apply the nuts handtight. Attach the remaining clips, leaving equal space between them (about six diameters).

Take up rope slack by applying tension to the eye, and while the rope is under strain, tighten the nuts on all loose clips. Take alternate turns on the two nuts of a clip so as to keep the saddle of the clip square.

After the rope has been in operation for a short time, check all clips and tighten nuts again. No slack should exist in either part of the rope between the clips.

Notes

1. *Never* connect two straight lengths of rope using clips. Always form an eye with a thimble in each length and connect the eyes together.
2. *Always* use new clips for connections since reused clips will not develop the proper efficiency.

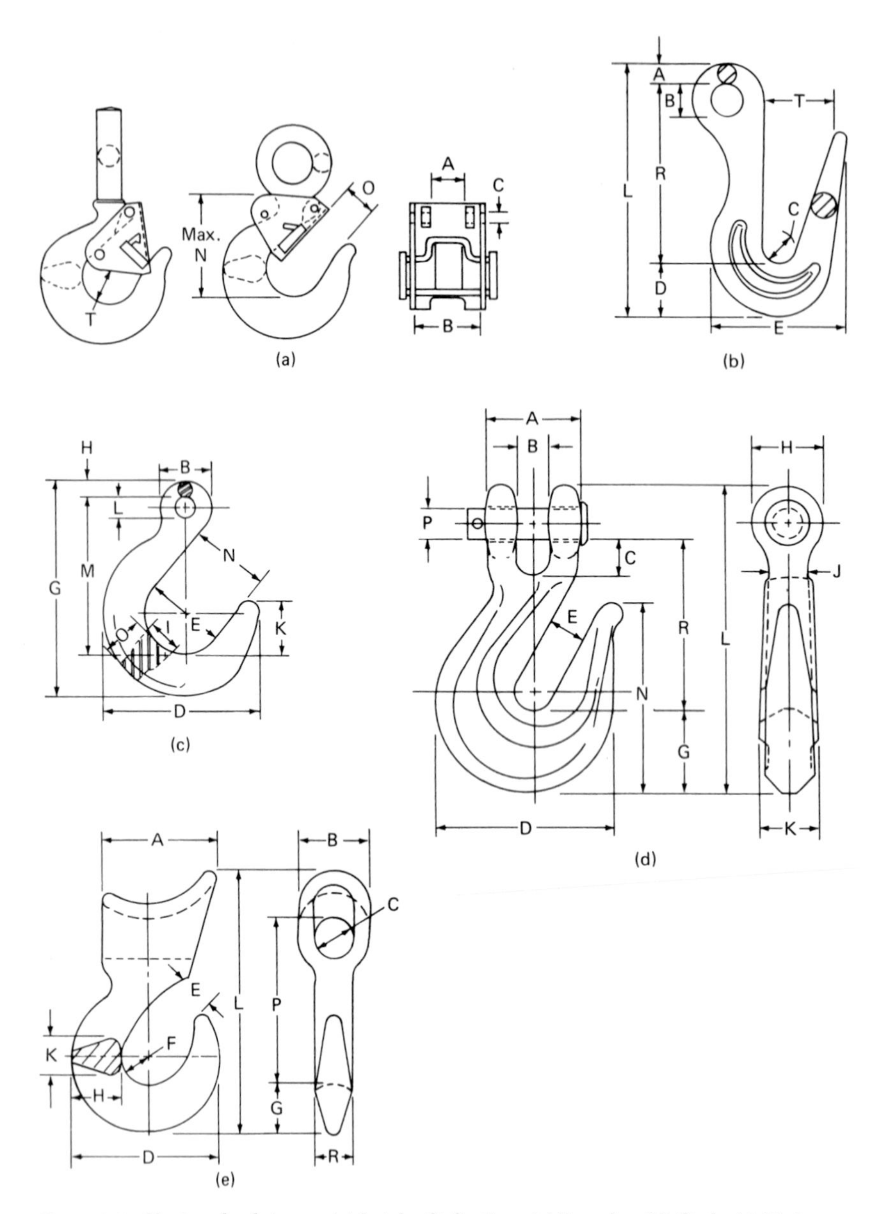

Figure 9.4 Various hook types. *(a)* Latch. *(b)* Sorting. *(c)* Foundry. *(d)* Grab. *(e)* Choker.

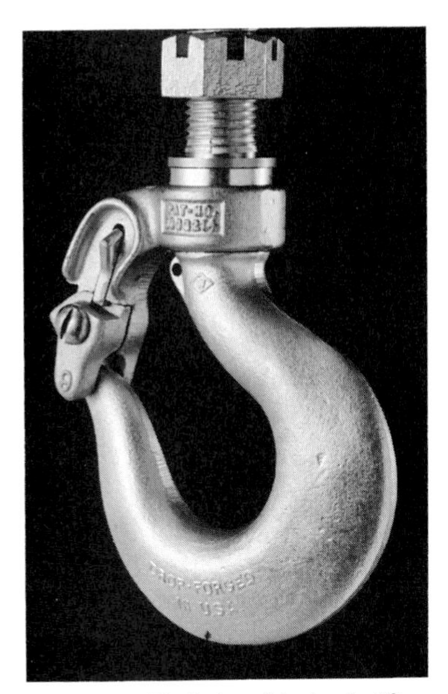

Figure 9.5 Tip-Lok safety hook. (*Courtesy of E. D. Bullard Co.*)

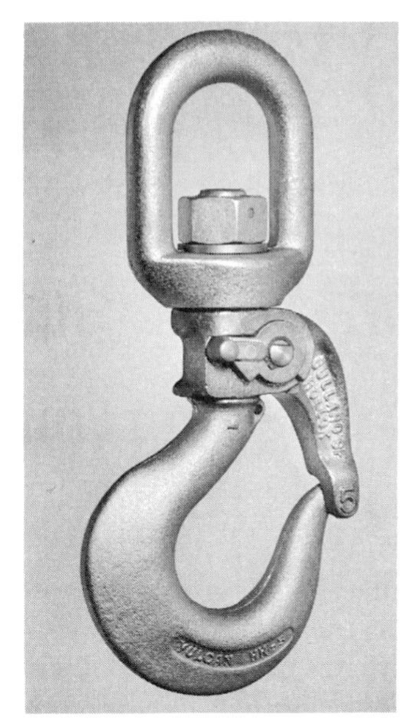

Figure 9.6 Safety hook. (*Courtesy of E. D. Bullard Co.*)

Hooks

It is extremely important to inspect all hooks frequently, checking their body for cracks (especially on neck and saddle), corrosion, and twisting; the saddle for wear; and the throat for possible opening up beyond new hook dimensions. Various hooks are illustrated in Figs. 9.4 to 9.8.

Make sure that all hoisting hooks, except grab and sorting types, are equipped with safety catches. Every hoisting hook should be equipped with swivel and headache ball. The ball should be securely attached to either hook or rope, thus preventing it from sliding on the load line (see Fig. 9.9). Remember a hook's safe working load applies only when the load is applied directly on the saddle of the hook. Eccentrically loading the hook, or applying a load anywhere between the saddle and the tip of the hook, will reduce the hook's rated safe working load considerably. Always refer to the manufacturer's load ratings for specific values for specific hooks.

Swivels

These coupling devices permit either half of a connection to rotate independently. They are available in a variety of designs, including eye

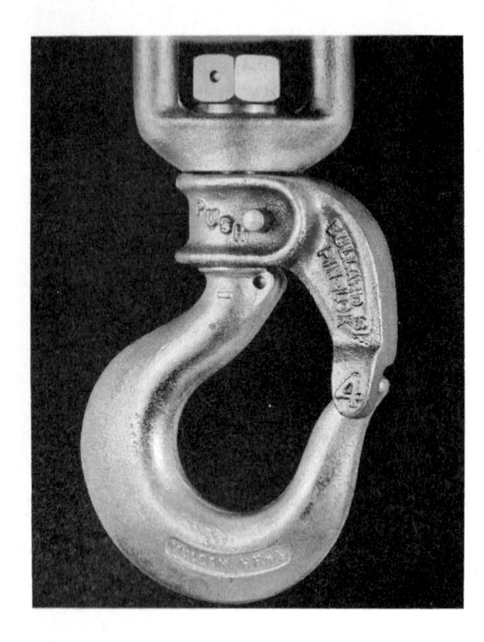

Figure 9.7 Pin-Lok safety hook. *(Courtesy of McKissick Products.)*

and eye, eye and clevis, clevis and clevis, hook and eye, hook and clevis, and chain type. Make sure that the swivel capacity is sufficient for loading application.

Shackles (Clevis)

Two types of shackles are commonly employed in rigging operations: anchor (bow-shaped) and chain (D-shaped). Both types of shackles

Figure 9.8 Various types of safety hooks. *(Courtesy of E. D. Bullard Co.)*

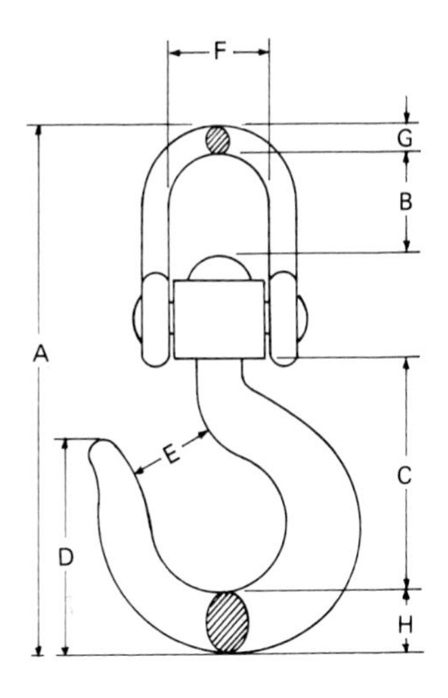

Figure 9.9 Typical rolling safety hook.

are available with screw pins, round pins, or bolt-type closures (see Fig. 9.10).

When using shackles, make sure that the load always pulls evenly. If the shackle is pulled at an angle, its capacity will be reduced considerably. If necessary, use washers to center the load on a shackle.

Never replace a shackle pin with anything but a properly fitting pin that is capable of carrying the loading normally applied to it. Always use cotter pins with any round pin shackle. If there is a chance that the pin might roll when loaded, *do not* use a screw-type pin shackle; the rolling action could loosen the pin. Destroy all shackles that show crown or pin wear greater than 10% of the fitting's original diameter.

Rings and Links

Most chain slings must be fitted with some type of end fitting, usually consisting of a master ring or link on the end that fits over a crane hook, and a hook or link at the other end that attaches to the load. These master rings and links have large inside dimensions that permit the fitting to slip easily onto the thicker crane hook section.

Master rings. These end fittings require a section diameter about 15% larger and an inside width about 33% greater than a comparable oblong link to withstand the same loading without being deformed.

Pear-shaped master links. These end fittings, once commonly used for chain end connections, are not as versatile as oblong links, and thus their use has declined considerably. The big danger of this type

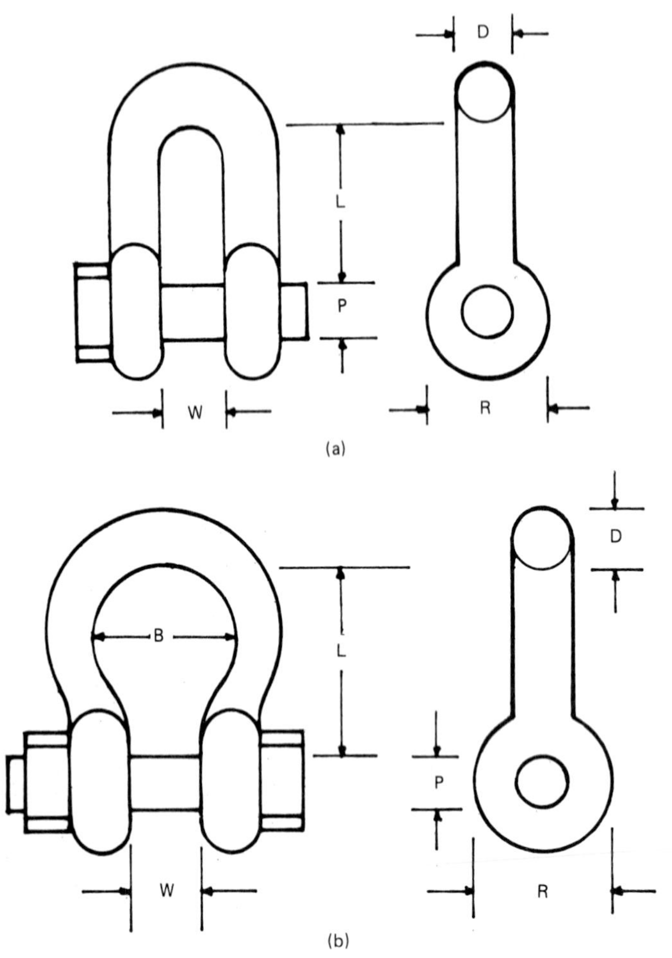

Figure 9.10 Typical shackles. (*a*) Chain pattern. (*b*) Safety-pin anchor.

of link occurs when the pear shape accidentally reverses in use, thus subjecting its narrow end to severe bending when the fitting is jammed down on the crane hook's thick saddle.

Oblong master links. These end fittings are preferred in modern rigging practice since they are less bulky and far more versatile than comparable other end shapes.

Connecting links. These connectors, either welded or mechanical, permit easy attachment of the sling chain to the hook or end links. Welded connecting links must be attached in plants by the manufacturer to assure the connection's compatibility with a chain sling's other components. However, mechanical couplers enable a rigger to assemble reliable, customized slings from component parts in the field.

Eye and Ring Bolts

All eye and ring bolts should be equipped with shoulders or collars. The shoulderless types are designed for vertical loading only, and will either bend or break when a load pulls on them at an angle (see Fig. 9.11).

However, even eye and ring bolts with shoulders have a reduced safe working load when subjected to angled loading. To assure the maximum safe working load of the bolt, its shoulder must be at a right angle to the axis of the hole, and the shoulder must be completely in contact with the bearing surface. To assure firm contact, washers may have to be used with the bolt.

Bolts secured by nuts must be properly torqued. Screw-type bolts should be set to a depth at least one and one-half times the diameter of the bolt.

Always apply loads in the plane of bolt eyes to minimize the possibility of bending, especially when using bridle slings, which always develop an angular pull in the eyebolts.

Always use a shackle when connecting a sling leg to an eyebolt. Never reeve slings through eyebolts for a lift.

Always turn eyebolts in line with each other, using washers under bolt collars, if necessary, to permit alignment when the bolt is tightened.

Turnbuckles

A variety of end fittings are used on turnbuckles: eye and eye, hook and hook, jaw and jaw, or stub and stub, as well as any combination of these types.

The rated load capacity of a turnbuckle depends on the type of end fitting as well as the outside diameter of the fitting's threaded section.

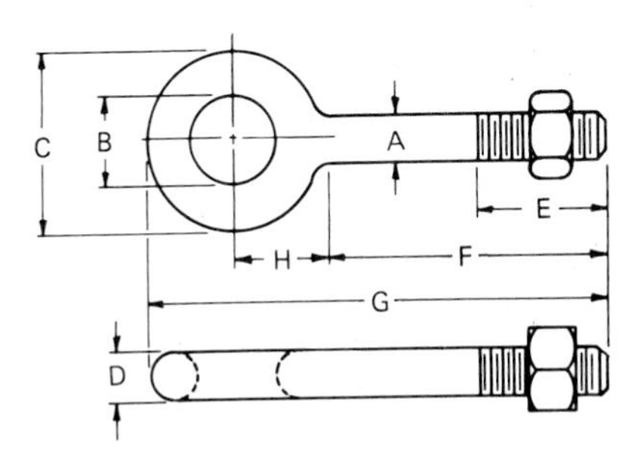

Figure 9.11 Regular nut eyebolt.

All turnbuckles of comparable size have equal, rated load capacities, except hook end types which, because of their configuration, have lower ratings. Make sure all hook end fittings are fitted with safety catches.

Always secure a turnbuckle frame to its end fitting if the turnbuckle is subject to vibrations, thus preventing the fitting from turning loose. Always use wire to lock the fitting in place, instead of heavier and less effective lock nuts or jam nuts.

Inspect all turnbuckles frequently. Check end fittings, rods, and bodies for cracks, bending, and deformation; the threaded portion, for thread damage.

Spreader Beams

A spreader beam is used to support long loads during lifts and to prevent the load from tipping, sliding, or bending. The beam also helps eliminate low sling angles and prevents a load from being crushed.

Equalizer Beams

An equalizer beam should be used to distribute a load equally between two sling legs or two hoist lines when making a lift in tandem.

Chapter

10

Blocks, Reeving, Sheaves, and Drums

Just as a rigger cannot do good work without proper tools, so the best wire rope cannot give the service expected of it unless the equipment on which it is used is properly designed and maintained. This is especially true of blocks, sheaves, and drums.

Blocks

The essential parts of any block are the shell, center pin, straps, sheaves, and block connections (see Fig. 10.1).

Shell. Protects the sheave and keeps the rope in the sheave groove. When made of steel, a shell also provides strength and rigidity to a block. Wood shells do not, and therefore they must only be reeved with fiber ropes.

Center pin. Transmits the sheave load to the straps. It should not be permitted to rotate in the block.

Straps and side plates. Add rigidity to the block and provide the structural means of transmitting the sheave load to the block's connections.

Sheaves. Transmit rope loads to the center pin, straps, and block connections. Sheaves on fiber rope blocks are of cast iron; on wire rope blocks they are of cast steel. Fiber rope sheaves must not be used with wire rope because their diameters are too small (see Fig. 10.2).

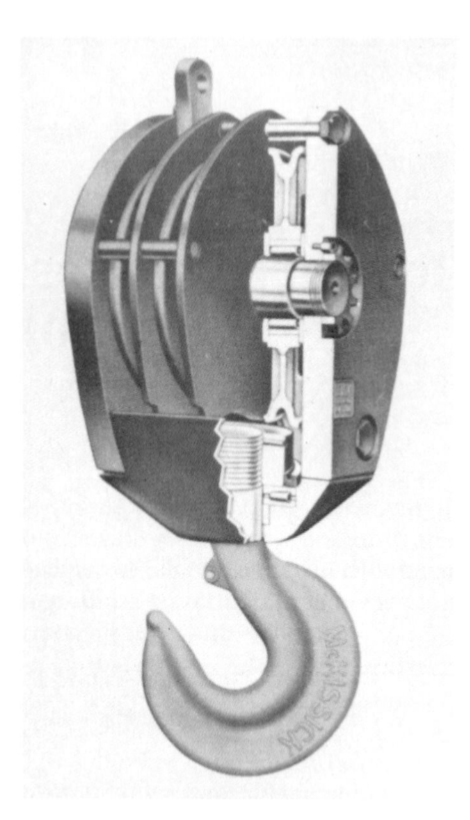

Figure 10.1 Cutaway of typical block.

Connections. Provide means for attaching the block to lifting devices and loads. These include a wide variety of hooks, wedge sockets, clevises, shackles, and swivels, as well as the becket (the rope anchorage point on a block).

All blocks can be designated by:

Application in rigging operations

Shape or type of construction

Number of sheaves

Position in a reeved system

Selection of the proper block and reeving for a particular loading requires a consideration of block size, rope size, and the number of sheaves per block in relation to the load to be lifted, as well as the pull that can be applied to the lead line. However, the load to be handled is the primary determining factor, rather than the diameter or the strength of

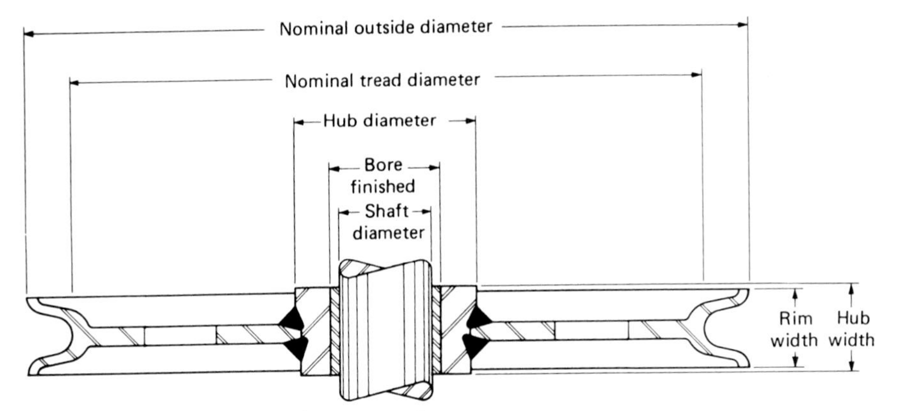

Figure 10.2 Basic sheave and its nomenclature.

the rope the blocks will carry. Other factors that must be considered include the possibility of overloading, excess friction, the angle of pull, the condition of the rope, sudden application of a load, and lubrication or the lack of it.

Regardless of the type or capacity of block selected, make sure that the block's sheaves have the maximum possible diameter for the size of rope being used. All blocks should be clearly and permanently marked with their weight, if it is significant, and with their rated capacity.

Types of blocks

Crane and hook. Equipped with heavy iron cheek weights; recommended for use in high-speed operations with heavy loads (see Fig. 10.3).

Wire rope. Lighter than crane and hook blocks. Cheek straps provide strength between end fittings and center pins. This type is *not recommended* for heavy-duty service and abuse.

Snatch. Opens on one side to permit the rope to be slipped over the sheave. It is available in many shapes for wire, manila, and synthetic-fiber ropes and is often used when necessary to change the line pull direction (see Fig. 10.4).

Tackle. Lighter than wire rope blocks; used with fiber ropes only. It is available with either wood or metal shells; with or without cheek straps, depending on capacity.

Furthermore there are diamond- and oval-type blocks (see Figs. 10.5

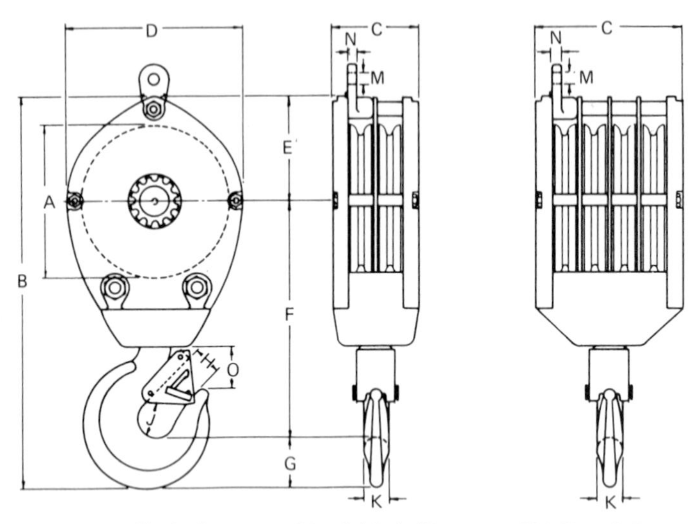

Figure 10.3 Typical crane and hook block.*(Courtesy of McKissick Products.)*

and 10.6). Blocks may be equipped with single sheaves or with multiple sheaves.

All block fittings should be made of forged alloy steel. Because shackles and eyes are inherently stronger than hooks, blocks equipped with them are rated at higher working loads. They are especially recommended for block connections where the block is to be mounted permanently, where standing blocks must support not only the load, but also the hoisting strain of the load line, and where it would be dangerous to have a hook connection become accidentally disengaged.

Remember: The anchorage point (becket) for the tackle or the blocks must be able to carry the total weight of the load plus the weight of the blocks, as well as the pull exerted on the lead line.

Inspection of blocks

Check beckets, end connections, sheave bearings, and center pins of blocks for excessive wear. Replace any defective parts.

Check for elongated links, eyes, or shackles; bent shackles, links, or center pins; and enlarged hook throats. Such defects are a sure indication of overloading and mean that the block should be replaced.

Check sheaves to be sure that they are lubricated and that they rotate properly and freely.

Check sheave grooves for smoothness. Excessive rope wear will occur if a wire sheave shows the imprint of the rope.

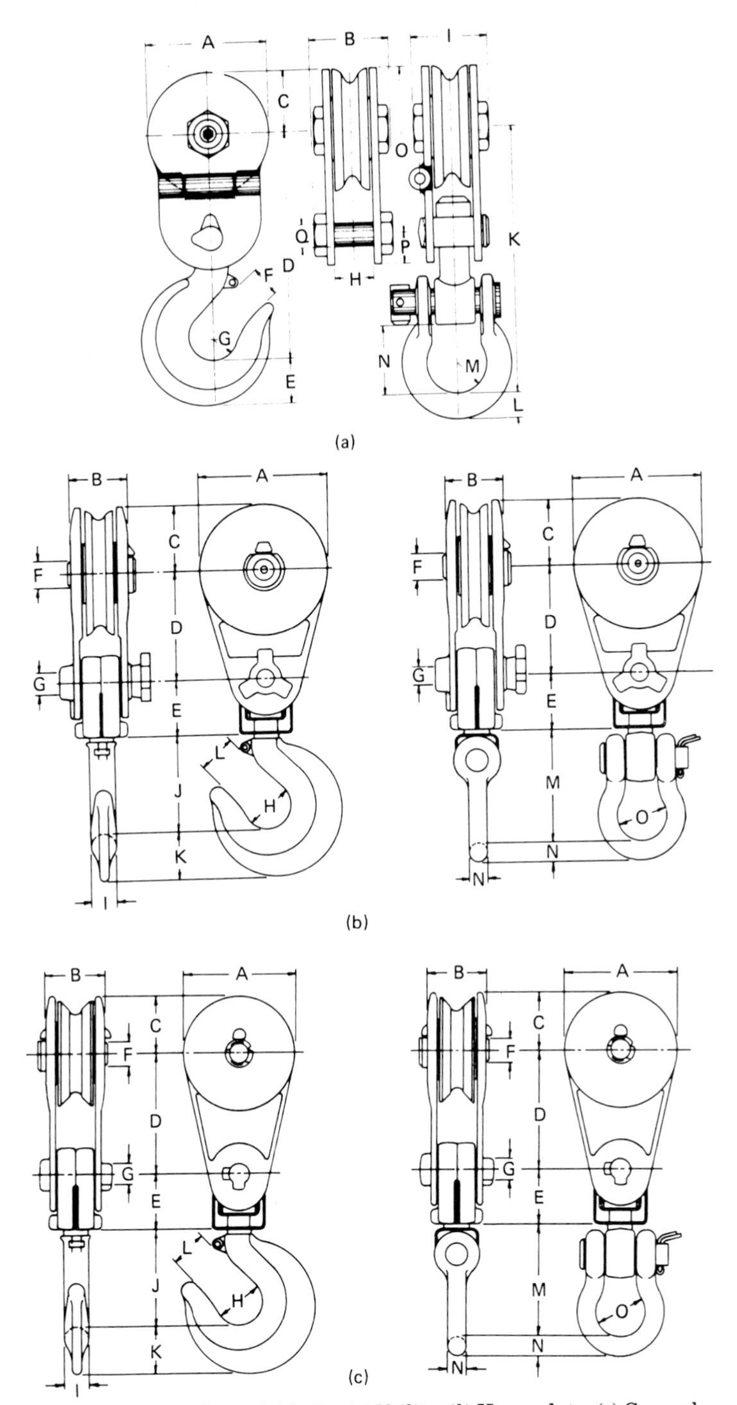

Figure 10.4 Typical snatch blocks. (*a*) Utility. (*b*) Heavy-duty. (*c*) General-purpose.

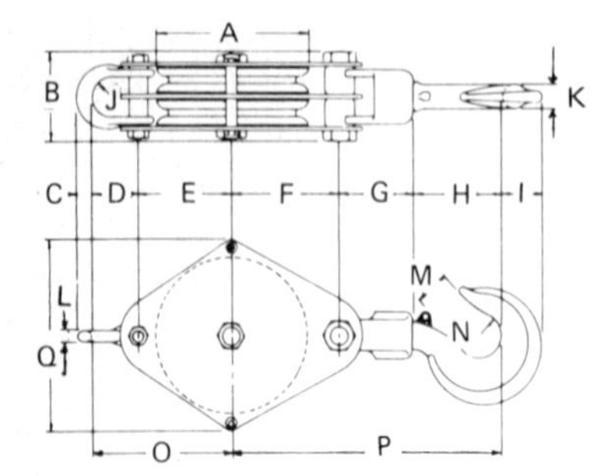

Figure 10.5 Diamond-type block.

Check the clearance between sheaves and cheek and partition plates to ensure that the rope will not slip between them while in use.

Reeving

Properly reeved blocks provide a mechanical advantage that permits lifting heavy loads with a reasonable pull of the haul line. The mechanical advantage of a multiple-part reeved system equals the number of parts of line actually supporting the running block and the load.

When a load is at rest, the tension on the rope is equal to the load divided by the number of the parts of rope supporting the lower block (running block). However, when the haul line is hoisting, the strain becomes somewhat greater due to sheave friction.

Each time a rope passes over a sheave, friction is produced, which reduces a block's efficiency (usually expressed in terms of a percentage). On well-maintained sheaves, friction losses are assumed to be

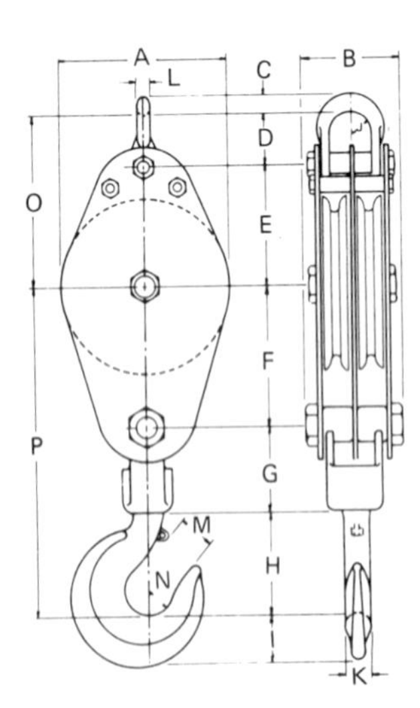

Figure 10.6 Oval-type block.

about 10% per sheave having plain bores used with manila-rope blocks, 5% per sheave having bronze bushings, and 3% per sheave having roller bearings.

Simply multiplying the number of sheaves by the appropriate percentage of friction and then adding this result to the lead-line static load produces a calculated load that is significantly lower than the actual lead-line stress. Thus, when calculating the effect of friction, it is imperative to calculate loads progressively, line by line.

For example, if a 10,000-lb load is to be raised using a five-part system, then:

Friction, sheave 1 = 2,000 × 10% = 200 lb

Load on line 2 = 2,000 + 200 = 2,200 lb

Friction, sheave 2 = 2,200 × 10% = 220 lb

Load on line 3 = 2,200 lb + 220 = 2,420 lb

Friction, sheave 3 = 2,420 × 10% = 242 lb

Load on line 4 = 2,420 lb + 242 = 2,662 lb

Friction, sheave 4 = 2,662 × 10% = 266 lb

Load on line 5 = 2,662 lb + 266 = 2,928 lb

However, it is not necessary for the rigger to repeat these calculations each time the lead-line pull must be determined. By using tables of multiplication factors, or ratios, for specific types of sheaves, the lead-line pull can be determined with one simple calculation (see Table 10.1).

To determine the lead-line pull when the load to be lifted and the number of parts of line are known,

$$\text{Lead-line pull} = \frac{\text{load to be lifted}}{\text{parts of line}} \times F$$

or

$$\text{Lead-line pull} = \frac{\text{load to be lifted}}{R}$$

where F = multiplication factor
R = ratio of line parts divided by F

To determine the maximum load that can be lifted with a particular reeving arrangement,

$$\text{Load to be lifted} = \text{lead-line pull} \times R$$

TABLE 10.1 Ratios R for Bronze-Bushed Sheaves

Number of parts of line	Ratio R
1	0.96
2	1.87
3	2.75
4	3.59
5	4.39
6	5.16
7	5.90
8	6.60
9	7.27
10	7.91
11	8.52
12	9.11
13	9.68
14	10.02
15	10.7
16	11.2
17	11.7
18	12.2
19	12.6
20	13.0
21	13.4
22	13.8
23	14.2
24	14.5

Use of Ratio Table

$$\frac{\text{Total load to be lifted in pounds}}{\text{Single-line pull in pounds}} = \text{ratio}$$

Example 1: Find the number of parts of line needed when weight of load and single-line pull are given:

$$\frac{\text{72,480 lb (load to be lifted)}}{\text{8,000 lb (single-line pull)}} = \text{9.06 (ratio)}$$

Refer to ratio 9.06 in table, which indicates 11 parts of line.

Example 2: Find the single-line pull needed when weight of load and number of parts of line are given:

$$\frac{\text{68,000 lb (load to be lifted)}}{\text{6.60 (ratio of 8-part line)}} = \text{10,303 lb (single-line pull)}$$

To determine the number of parts of line required to make a lift,

$$\text{Number of parts of line} = \frac{\text{load to be lifted}}{\text{lead-line pull}} \times F$$

Note that increasing the number of sheaves and line parts increases the mechanical advantage of a reeved system. But because friction losses will also be greater, the efficiency of the system will be lower. On the other hand, by reducing the number of sheaves and parts of line, friction losses can be reduced and the system's efficiency is increased. However, the mechanical advantage will then be lower. But beware. Sometimes increasing the number of parts on blocks does not necessarily mean an improved lifting capacity, since there is a limit beyond which the effort required to overcome friction becomes greater than that necessary to lift the load. For blocks having plain-bore sheaves, reeving beyond 6 or 7 parts of line provides little advantage; and reeving beyond 10 or 11 parts actually decreases the mechanical advantage due to the cumulative effect of sheave friction. Likewise, with blocks having bronze bushing sheaves, reeving beyond 9 or 10 parts has no particular advantage; and sheaves with roller bearings reach their practical advantage with 15 or 16 parts of line.

In addition, increased load-carrying ability means a slower travel speed. Thus, to maintain a mechanical advantage of 5, a system's lead line must be able to move five times faster than the load.

Important: The size and capacity of blocks and reeved systems must be sufficient to carry the loads to which they will be subjected.

Since a block hook will usually start bending when the block is loaded to 70% of its maximum strength, any such hooks showing signs of opening up are probably being overloaded. To avoid overloading, it is necessary to know exactly how a reeved system works and to understand how friction affects a system's capacity (see Fig. 10.7).

When blocks in a system have an equal number of sheaves, make sure the dead end of the rope is securely fastened to the becket of the standing (upper) block. When the system has an unequal number of sheaves, always fasten the rope to the becket of the block having the least number of sheaves.

When reeving a pair of tackle blocks, with one block having more than two sheaves, make sure the hoisting rope leads from a center sheave of the upper block. In this way, the hoisting strain is put on the center of the blocks, thus preventing them from twisting and toppling, which could injure the rope as it cuts across the edges of the block shell (see Fig. 10.8).

Reeve such blocks by placing them so that the sheaves in the upper block are at right angles to those in the lower one. Start reeving with the becket, or standing end, of the rope. Make sure that the becket connection is properly attached. Use a wedge socket or cable clip for wire rope blocks, a becket hitch for fiber rope blocks.

In a pair of reeved blocks, the upper one should be a shackle block; the lower one, a hook block. The shackle will prevent the block from becoming detached accidentally, should the block or its attaching point be jarred. The lower block's hook makes it easier to attach or detach the block from the load (see Fig. 10.8).

Do not lace blocks in reeving systems having five or more parts of line. Lacing will cause the traveling block to tilt, with resultant excessive wear and damage to both sheaves and rope.

Blocks having more sheaves than needed should be reeved symmetrically, with the rope distributed equally across the block so that loading on the system is balanced.

To prevent a block from toppling, when using a hook or line that is eccentrically reeved over the boom point of a crane, make sure reeving is in accordance with prescribed procedures.

Use new as well as very long tackles and reeved systems cautiously, since they tend to twist (usually because of the lay of rope, rather than the method of reeving). This tends to increase the power required to hoist, and causes damage to sheave and rope. Only braided fiber rope or nonrotating wire rope will help eliminate such problems.

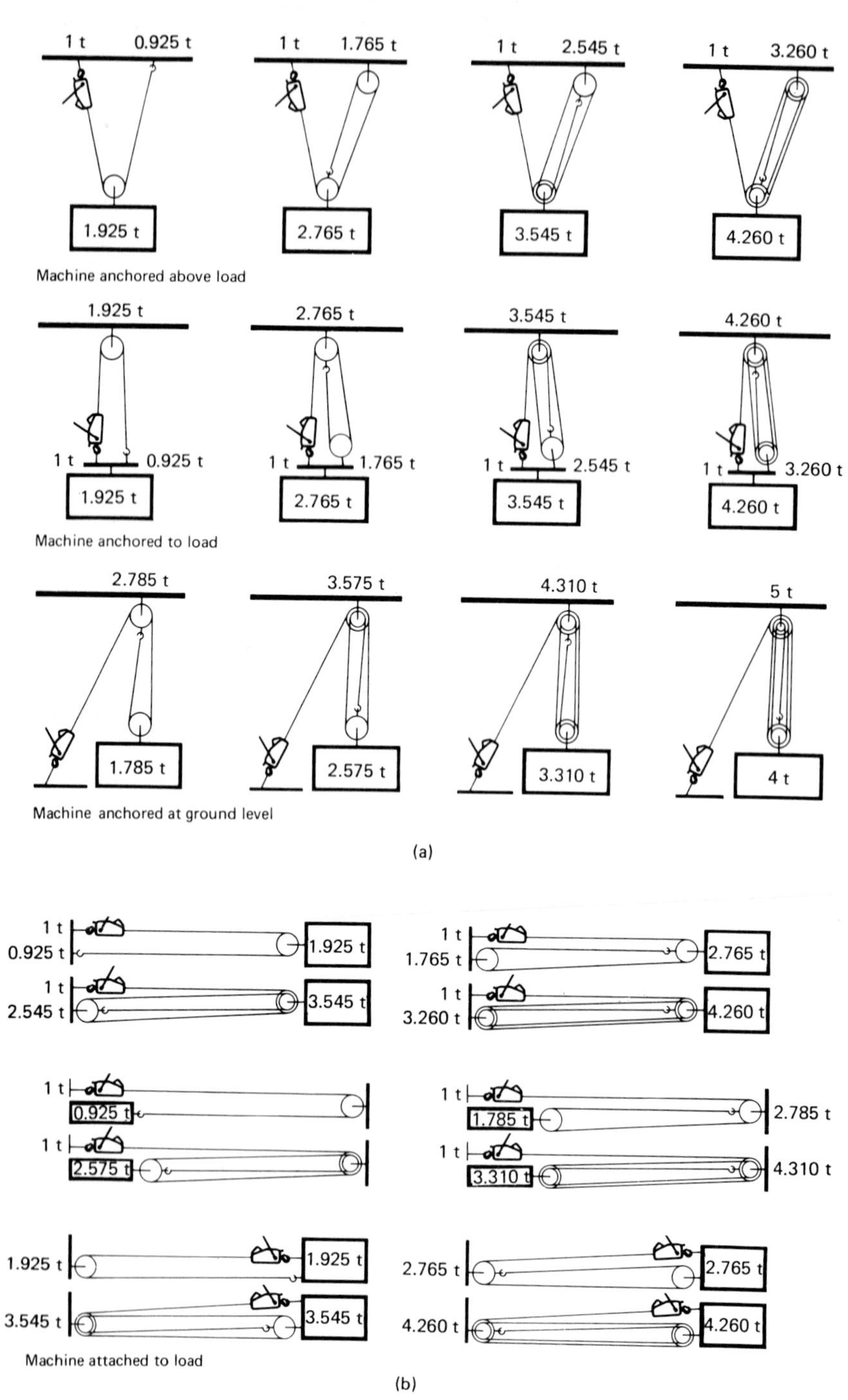

Figure 10.7 Groove gage.*(Courtesy of Griphoist, Inc.)*

Be careful when using single-rope falls, which tend to spin when the tension is removed from the hoist rope. Always allow rope and hooks to settle down before handling them.

Another consideration in using reeved systems is that all sheaves in a set of blocks revolve at different rates of speed.

To raise a load 1 ft, the lower block must be raised 1 ft. This means each working rope must be shortened 1 ft. Thus, assuming the sheave circumference is 1 ft, sheave 1 must make one revolution to shorten rope 1 by 1 ft. To raise each succeeding sheave 1 ft, that sheave must make one more revolution than the sheave preceding it. Each sheave revolves at a different rate of speed—sheave 2 rotating twice as fast as sheave 1; sheave 3, three times; sheave 4, four times; and sheave 5, five times as fast as sheave 1. As a result, sheaves nearest the lead line rotate faster and tend to wear out faster. All sheaves must be kept well lubricated to reduce friction and wear.

Sheaves

Sheaves are used in rigging operations to change the direction of travel of wire rope. When assembled in multiples, such as in blocks, sheaves can provide almost any required mechanical advantage.

In selecting the proper sheave for a standard rope application, three principal factors must be considered: sheave diameter, sheave material, and groove size and shape.

Sheave diameter

The tread diameter of a sheave determines the bending stress in the rope and the contact pressure between it and the sheave surface. The lives of both sheave and rope can be prolonged by using the properly sized sheave for the size and construction of a particular rope.

Types of sheaves

Plain bore. Cast iron; provides its own bearing; is used on fiber rope blocks; recommended for light intermittent use only. It must be oiled frequently (see Fig. 10.9).

Roller bushed. Unground rollers without races; recommended for light intermittent use only. It must be lubricated periodically with a heavy grease.

Bronze bushed. Self-lubricating; recommended for use where service requirements are severe or where block cannot be lubricated fre-

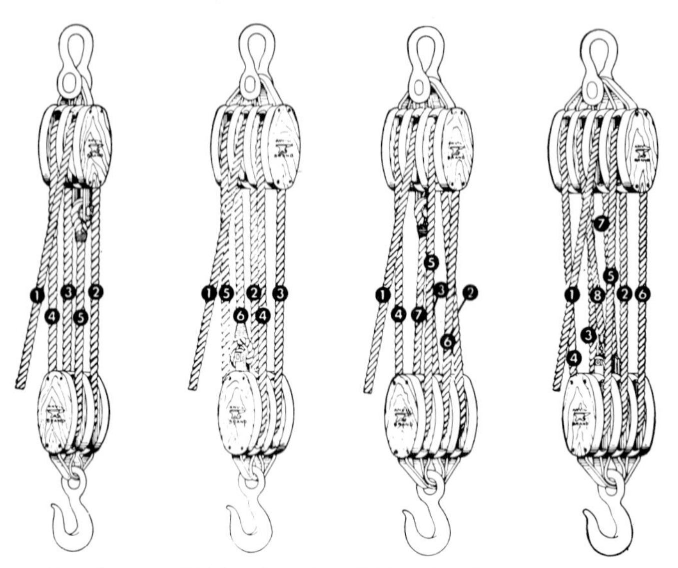

Figure 10.8 Various methods of reeving tackle blocks.*(Courtesy of Western Division/Crosby Group.)*

quently. It must not be subjected to high-speed or continuous operation. *Do not* lubricate or oil the plug-type bearing (see Fig. 10.10).

Bronze bearings. Pressure-lubricated; recommended for use where service requirements are extra heavy and continuous. They must be lubricated frequently.

Roller bearings. Ground rollers and full races; recommended for medium-duty use where operation is high-speed (see Fig. 10.11).

Antifriction. Precision bearings that combine long life with minimum maintenance; recommended for use in operations with continuous high speeds and heavy loads.

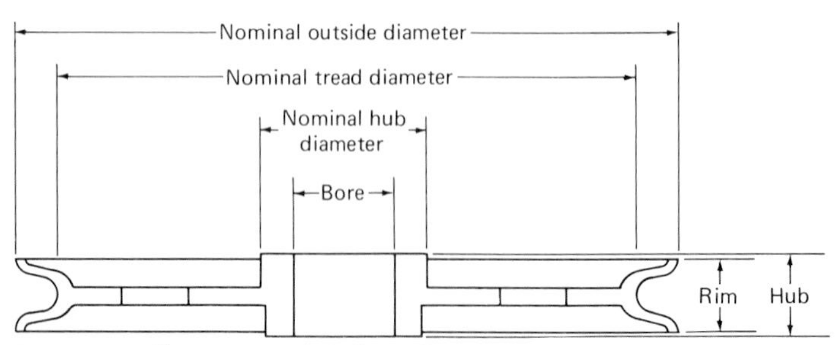

Figure 10.9 Common or plain bore sheave.

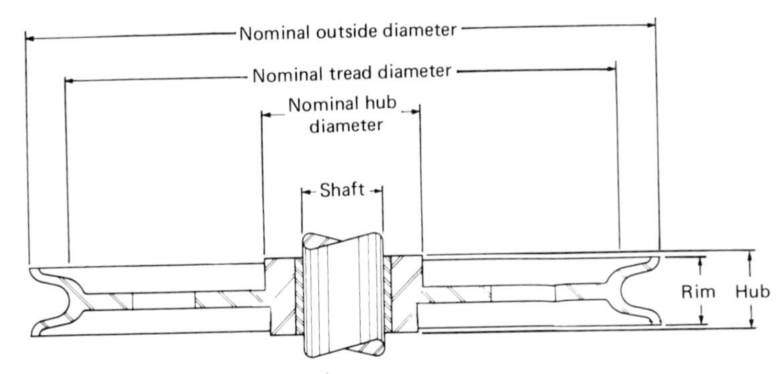

Figure 10.10 Bronze-bushed sheave.

If a wire rope is used with inadequately sized sheaves, the severe bending stresses imposed on the rope will cause wires to break from fatigue, even though actual wear may be slight. One of the fastest ways to destroy a rope is to run it over small sheaves. The excessive and repeated bending and straightening of the wires leads to premature failure from fatigue.

Small sheaves will also accelerate wear of both rope and sheave groove. Since the pressure per unit area of rope on a sheave groove for a given load is inversely proportional to the size of the sheave, the smaller the sheave diameter, the greater the rope pressure per unit area on the groove.

To determine the unit radial pressure between a rope and a sheave, use the formula

$$P = \frac{2L}{Dd}$$

where P = unit radial pressure, lb/in.^{2}
L = load on rope, pounds

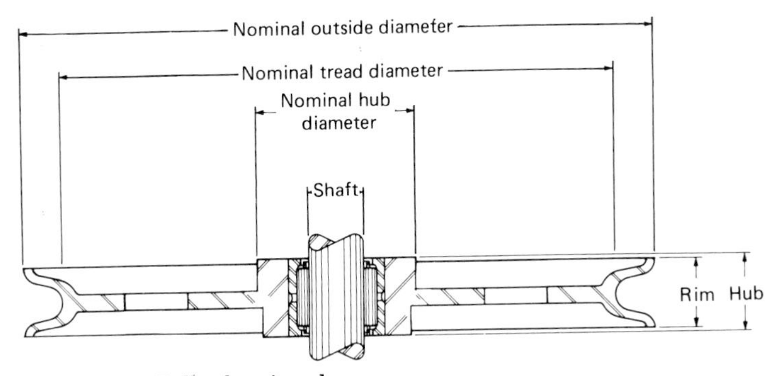

Figure 10.11 Roller bearing sheave.

TABLE 10.2 Allowable Unit Radial Bearing Pressures of Ropes on Various Sheave Materials

	Regular lay rope, lb/ in.^{2}				Lang-lay rope lb/in.^{2}			Flattened strand lang lay lb/in.^{2}	
Material	6×7	6×19	6×37	8×19	6×7	6×19	6×37		Remarks
Wood	150	250	300	350	165	275	330	400	On end grain of beech, hickory, gum
Cast iron	300	480	585	680	350	550	660	800	Average Brinell hardness of 125
Carbon steel casting	550	900	1,075	1,260	600	1,000	1,180	1,450	30-40 carbon; average Brinell hardness of 160
Chilled cast iron	650	1,100	1,325	1,550	715	1,210	1,450	1,780	Not advised unless surface is uniform in hardness
Manganese steel	1,470	2,400	3,000	3,500	1,650	2,750	3,300	4,000	Grooves must be ground and sheaves balanced for high-speed service

D = tread diameter of sheave, inches
d = nominal diameter of rope, inches

The allowable unit radial bearing pressures of various ropes on different sheave materials are given in Table 10.2.

If the unit radial pressure exceeds these maximum values, the material from which the groove is manufactured is too soft for the operating conditions. Therefore rapid wear of the grooves will result.

The sheave diameter can also influence the rope strength. When a wire rope is bent around a sheave, there is a loss of effective strength due to the inability of the individual strands and wires to adjust themselves entirely to their changed position. The rope strength efficiency decreases to a marked degree as the sheave diameter is reduced with respect to the diameter of the rope.

Wire rope manufacturers have established standards for sheave sizes that should be used with various rope constructions (see Table 10.3). Always use the maximum possible sheave diameter that the lifting equipment will carry.

Sheave material

Hard sheave surfaces offer the best bearing surface for wire rope, thus prolonging sheave and rope life. If a sheave is forged or cast from a

TABLE 10.3 Proper Sheave and Drum Sizes

Construction	Suggested D/d ratio*	Minimum D/d ratio*
6 × 7	72	42
19 × 7 or 18 × 7	51	34
6 × 19 Seale	51	34
6 × 27 H flattened strand	45	30
6 × 31 V flattened strand	45	30
6 × 21 filler wire	45	30
6 × 25 filler wire	39	26
6 × 31 Warrington Seale	39	26
6 × 36 Warrington Seale	35	23
8 × 19 Seale	41	27
8 × 25 filler wire	32	21
6 × 41 Warrington Seale	32	21
6 × 42 tiller	21	14

*D—tread diameter of sheave; d—nominal diameter of rope.

material softer than the wire rope, the sheave life and the wire rope life will be shortened. The sheave will have a tendency to take on the impression of the rope, causing scoring and corrugation of the line groove.

Cast manganese steel sheaves offer the ultimate in sheave material. This surface actually hardens to the use of wire rope and provides greatly extended sheave life and increased service time of the wire rope.

Groove size and shape

Sheave grooves must be smooth and slightly larger than the nominal rope diameter to provide maximum rope support.

Use of a groove gage is shown in Fig. 10.12. Newer gages, made to include the oversize tolerances for minimum groove conditions, are marked accordingly. Older gages include the standard oversize tolerances for new or regrooved sheaves, and are so marked.

If the groove diameter is too large, the rope will not be properly supported. It will tend to flatten and become distorted, thus accelerating the bending fatigue in individual wires and causing premature failure.

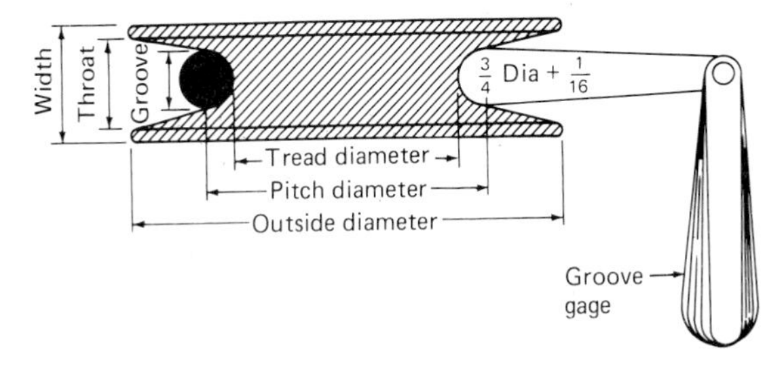

Figure 10.12 Groove gage.

Conversely, if the sheave groove is too narrow for the rope, the loading tension will pull the rope deeply into the groove, causing it to be pinched. Both rope and sheave will be subjected to severe abrasive wear.

Proper sheave design requires that the groove diameter have a maximum tolerance varying from 1/64 to 5/32 in. in excess of the nominal rope diameter, depending on the rope size. The depth of the sheave groove should be at least one and one-half times the rope's nominal diameter.

The wire rope in service will wear and decrease in diameter. The resulting distortion will cause poor distribution of the stress throughout the cross section of the rope, with a corresponding overloading of some portions.

As a rope is bent around a sheave, the strands and wires bind against each other and do not move freely. This increases internal and external abrasion and friction, as well as hindering the rope from readjusting to its load. As a result, some of the wire and strands are forced to carry more load than they should, thus shortening the rope's service life.

In addition, an undersized rope tends to wear an undersized groove in the sheave. Then when it is necessary to install a new rope, the sheave groove will not provide proper working clearances for the replacement rope. The resultant pinching or wedging of the rope causes binding that prevents proper load distribution and produces excessive internal and external friction.

To assure long and efficient rope life, the sheave grooves should be smoothly contoured, be free of surface defects, and have rounded edges.

Alignment

Sheaves should be mounted so that they are aligned exactly with each other. Since in actual use this is usually impossible, sheaves are manufactured with grooves that provide tolerance for misalignment and guide the rope into place.

A lead angle of only about two degrees can be accommodated without difficulty. Any appreciable misalignment will cause the rope to rub against the sides of the groove, resulting in both rope and sheave wear, thus shortening the useful life of both.

After every rope change, check that sheaves are properly aligned and running true; otherwise considerable wear in the rope and sheaves will result.

Always equip sheaves with cable keepers where an unloaded rope might possibly leave the sheave groove.

Bearings

Sheave bearings should be either permanently lubricated or equipped with a means for lubrication. Inadequate lubrication of a sheave that is too heavy for a load will cause the rope to slip in the sheave whenever the rope velocity changes. The momentum of the heavy sheave will cause it to continue turning after the rope has stopped, producing a grinding wheel action that can cause severe rope abrasion and wear flat spots in the sheave, further damaging the rope.

Friction losses in sheaves are a function of the style of rope, the ratio of sheave to rope diameter, and the type of bearing.

When computing friction losses at sheave bushings, it is reasonable to assume about a 4½% loss; for sheave bearings, the loss is about 1 to 2%, depending on the quality of the bearings. These losses are based on using a rope that makes a bend of 180° over the sheave. For smaller turning angles, these values can be reduced.

Inspection of sheaves

Common sheave troubles are illustrated in Fig. 10.13.

Check carefully for any sign of cracks in sheave flanges. Broken flanges can permit a rope to jump the sheave, or become badly cut as it rubs against the rough broken edge. Replace sheave.

Check groove contours for uneven wear. Sheaves that have worn out of round or have developed flat spots will set up vibrations in the rope that will result in premature wire fatigue. Remachine groove or replace sheave.

Check depth and flare of the groove to make sure that the rope does not rub against the sheave flange. Such rubbing may cause the rope

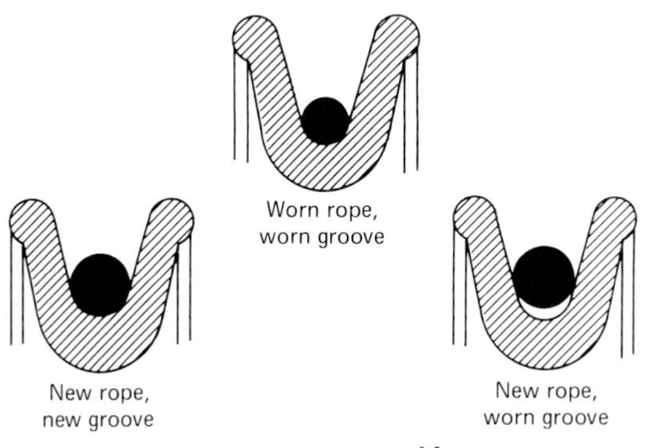

Figure 10.13 Common sheave troubles.

to fail prematurely. Remachine groove or replace sheave.

Check for wear on one side of groove, often due to excessive fleet angle or poor alignment. Move sheave to correct.

Check bearings for wear. Worn bearings will cause a sheave to wobble, putting vibration into the rope and increasing wire fatigue. Repair bearings or replace sheave.

Note: *Never* use wire rope on equipment designed for fiber rope. *Never* use fiber rope on sheaves that have been used for wire rope service.

Drums

Drums should be designed to conform to the rules for sheaves, although a drum can be made slightly smaller than the minimum recommended size for sheaves. This is because a rope is flexed only once by the drum, whereas it is flexed twice (bending and straightening) each time it passes over a sheave.

A drum should be designed, if possible, to handle all the rope in one smooth, even layer. Two and sometimes three layers are permitted; but more than three layers may cause crushing of the rope on the bottom layer as well as the end of any layer where pinching occurs.

The proper way to wind a rope on a drum will depend on the lay of the rope. Rope should wind in a helix that is opposite to its lay (see Fig. 10.14).

To locate the rope anchorage point on the drum, use the following rule of thumb. Stand behind the drum, looking toward the sheave. Since in most installations the rope winds onto the top of the drum, extend a

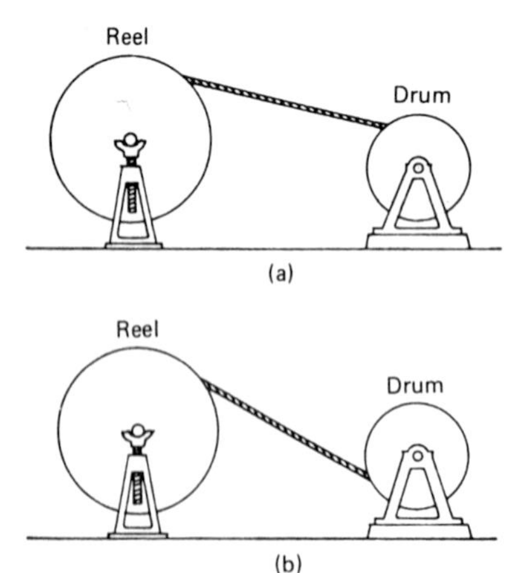

Figure 10.14 (*a*) Correct and (*b*) wrong way to wind wire rope from reel to drum. *(Courtesy of American Iron and Steel Institute.)*

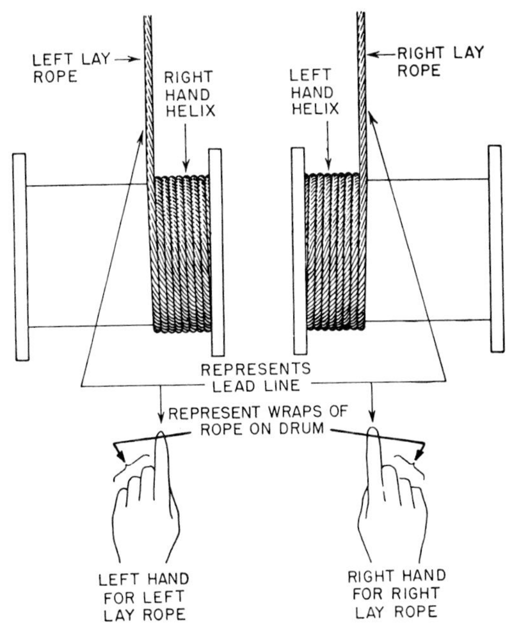

Figure 10.15 Hand rule for determining proper method of winding rope on drum for right- and left-lay wire rope.

clenched palm downward—right fist for right-lay rope, left fist for left-lay rope. Extending the index finger will point in the direction of the lead line; the thumb, to the drum anchorage point (see Fig. 10.15). Where the rope winds onto the bottom of the drum, simply turn the clenched fist over palm upward and extend the index finger and thumb to determine the direction of the lead line and the location of the anchorage point, respectively. The drum anchorage should be positioned to favor right-lay rope, since left-lay rope is not always available from stock.

The manner in which a rope spools on a drum also depends on:

Drum size and ratio of drum diameter to rope diameter

Rope speed

Amount of rope to be spooled

Type of hoisting equipment

Load on rope

To be certain that the rope spools evenly on the drum, either use a spooling device or keep the fleet angle to within the correct limits, maintaining tension on the rope at all times (approximately 10% of the working load).

To determine the approximate capacity of a given drum or reel for a particular rope diameter, use the following equation:

$$L = (A + B) \times A \times C \times F$$

where L = drum capacity, feet of rope
B = diameter of drum, inches
A = depth of rope layer, inches. $A = \frac{1}{2}(H - B) - M$ with H being the diameter of the flange and M the desired clearance, both in inches
C = distance between drum flanges, inches
F = drum or reel capacity factor

The drum capacity factor F applies to nominal rope size and level winding. Since new ropes are usually oversized by 1/32 in. per inch of rope diameter, the results obtained by the formula must be decreased to account for oversized new rope or random or uneven winding, or both, as follows:

Oversized ropes. Decrease calculated length from 0 to 6%.

Random wound ropes. Decrease calculated length from 0 to 8%.

Fleet angle

Where a wire rope passes over a sheave and onto a drum, the rope will remain in alignment with the sheave, but will deviate to either side of the drum, depending on its width and distance from the first fixed sheave. This angle between the centerline through the sheave and the centerline of the rope leading to the drum is called the fleet angle (see Fig. 10.16).

Because drums are usually offset in some installations, the fleet angle on one side of the drum could be larger than that on the other side. Always consider the larger of the two angles in making calculations.

To avoid excessive wear on the sheave and to prevent excessive chafing of the oncoming rope against previous wraps on the drum, it is desirable to keep the fleet angle as small as possible. If the fleet angle is too large, the rope will rub against the flanges of the sheave groove, or be crushed on the drum. If the angle is too small, it will cause the rope to pile up against the flange head, damaging both rope and equipment.

Too small an angle will also produce considerable vibration with sub-

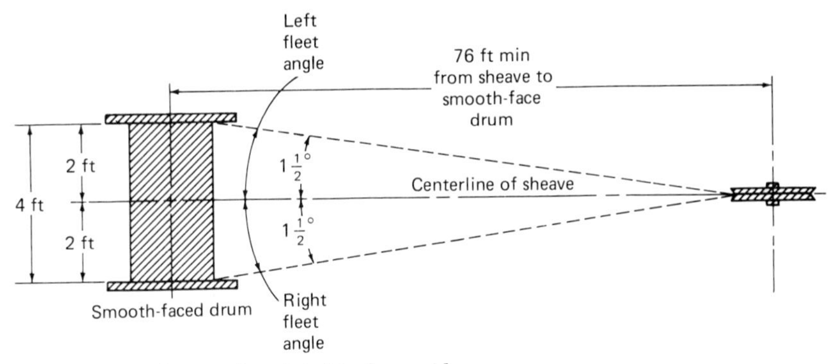

Figure 10.16 Fleet angle related to drum sides.

sequent deterioration of the rope. Intermediate idler sheaves can help eliminate this particular problem.

For average conditions, good practice would be to keep the fleet angle within 2° for a grooved drum, and within 1½° for a drum having a smooth face. This is equivalent to a distance of 29 and 38 ft, respectively, for each foot of drum width, on either side of the centerline of the sheave.

Flat-faced drum

When using a flat-faced drum, it is important that the rope be wound on the drum in a straight helix at its proper angle. Use a steel starting piece, cut in a tapered shape that fills the space between the first turn and the flange, to guide the rope into its correct position.

Make sure that all turns of the first layer of new rope on a drum are wound taut and correctly. If the winding is loose or nonuniform, serious damage will result to multiple windings. Tap each turn of the rope against the preceding one with a wooden mallet, but do not jam the strands together. Each succeeding layer of rope should lie in the groove formed by the preceding layer. *Never* permit wire rope to cross-wind on a drum.

Always fit the correct length of rope to a drum, for if the rope is too short, it will pay out and place the full load on the rope anchorage. Conversely, if the rope is too long, it might exceed the drum's spooling capacity, spilling over the flanges, and be damaged, crushed, or fail.

There should always be at least two or three full turns of rope left on a drum when the hook is at its highest point. Flanges on grooved drums should project either twice the rope diameter or 2 in. beyond the last layer of rope, whichever is greater. Flanges on ungrooved drums should project either twice the rope diameter or 2½ in. beyond the last layer of rope, whichever is greater.

Figure 10.17 Vertical unreeling stand. *(Courtesy of American Iron and Steel Institute.)*

Figure 10.18 Wrong method of unreeling wire rope. *(Courtesy of American Iron and Steel Institute.)*

Kinking

Kinking of wire rope can be avoided if ropes are properly handled and installed.

Kinking is caused by the rope taking a spiral shape as the result of unnatural twist in the rope. One of the most common causes for this twist is improper unreeling and uncoiling. Even though the kink may be straightened so that the damage appears to be slight, the relative adjustment between the strands has been disturbed so that the rope cannot give maximum service.

Unreeling and uncoiling

To unreel wire, the reel should be revolved and the rope taken off the same way it is put on the reel. One method is to put a shaft through the center of the reel and jack it up so that the reel will revolve freely. Pull the rope straight ahead, keeping it taut to prevent it from loosening up on the reel (see Figs. 10.17 to 10.19).

To uncoil wire, remove the ties and roll the coil along the ground so that the rope lies straight behind. There will be no twist or kink in the rope if these instructions are followed (see Figs. 10.20 and 10.21).

Figure 10.19 Wire rope reel mounted on a shaft supported by jacks can be rotated freely to unwind the rope either manually or by a powered mechanism. *(Courtesy of American Iron and Steel Institute.)*

Figure 10.20 Perhaps the most common and easiest uncurling method is to hold one end of the rope while the coil is rolled along the ground. *(Courtesy of American Iron and Steel Institute .)*

Figure 10.21 Wrong method of unreeling wire rope. *(Courtesy of American Iron and Steel Institute.)*

Section

3

Rigging Machinery

Chapter

11

Derricks and Cranes*

A derrick or crane is a piece of equipment designed to pick up a load, transport it a reasonable distance, and land it again through use of a hoisting mechanism using ropes. The basic difference between the two is that the derrick's hoisting engine is not part of the machine. There is almost no limit to the number of different designs of derricks and cranes that can be built.

Derricks

Chicago boom

A basic type of derrick is the Chicago boom (see Fig. 11.1). It is usually installed on a building frame, using a column as its mast and structural beam connections or bracing as the stiff legs. Columns should never be used as masts for Chicago booms.

Chicago boom lengths can range from 10 to 125 ft, lifting capacities from ¼ to about 35 tons. Most booms consist of trussed or latticed angle iron or steel tubing, or a combination of both.

The lower end of the boom is attached to the building column by means of a combination hinge pin and swivel pin to a heavy steel plate clamp secured to the column. Common hoisting blocks are used with this type of derrick; one is strapped at the boom tip, the other at the pivot fitting.

If heavy loads are to be lifted, steel guy ropes should be installed to carry the stress from the anchorage of the topping lift to a beam at the approximate elevation of the boom hinge or socket. Swing guys, run-

---

* Chapter 11 was developed from H. I. Shapiro, *Cranes and Derricks*, McGraw-Hill, 1980 with permission of the publisher

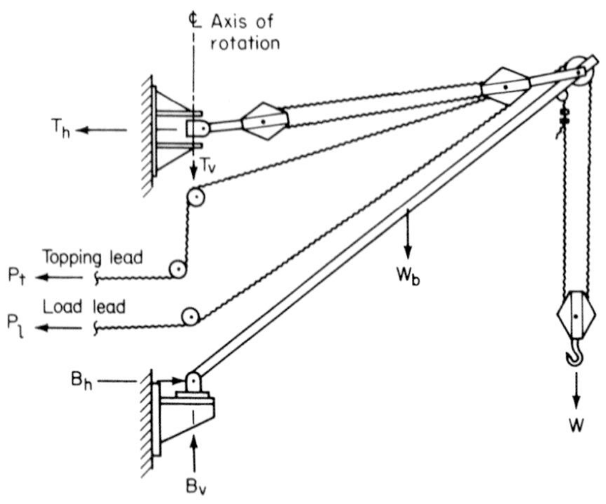

Figure 11.1 Chicago boom derrick. (*From H. I. Shapiro,* Cranes and Derricks, *McGraw-Hill, 1980.*)

ning from each side of the boom tip to a point of anchorage, enable the boom to swing.

Typically, a two-drum winch powers the hoisting and topping motions. When the boom is used only for lifting and swinging, the topping motion is not required, and a fixed rope guy line is used instead. The derrick's winch can be located at any level, but more often it is located at the boom foot's floor level. Heavy winches are usually located on the ground (see Fig. 11.2).

Direct and immediate communication is essential for landing a load accurately and safely. With the winch at the boom foot, the operator has direct communication with swing and load landing crews, and the boom, but not the load, is in view at all times. With the winch at ground level, the operator has direct communication with the sling loading crew, with boom and load in view at all times.

Because a Chicago boom is a low-production machine, it normally is used only where work is above the reach or capacity of mobile cranes.

When installing a Chicago boom, be sure the derrick is so located that its hook is directly over the center of gravity of the load it must lift. For light loads, the boom pivot can be located as close to the edge of the frame as one-half the width of the boom. For heavy loads, the boom pivot should be located as close to the building frame as possible.

Guy derrick

This type of derrick is basically a Chicago boom with its own integral mast held vertically by guy ropes (see Fig. 11.3). Standard practice is to use six guy wires spaced evenly around the derrick. Mast and boom

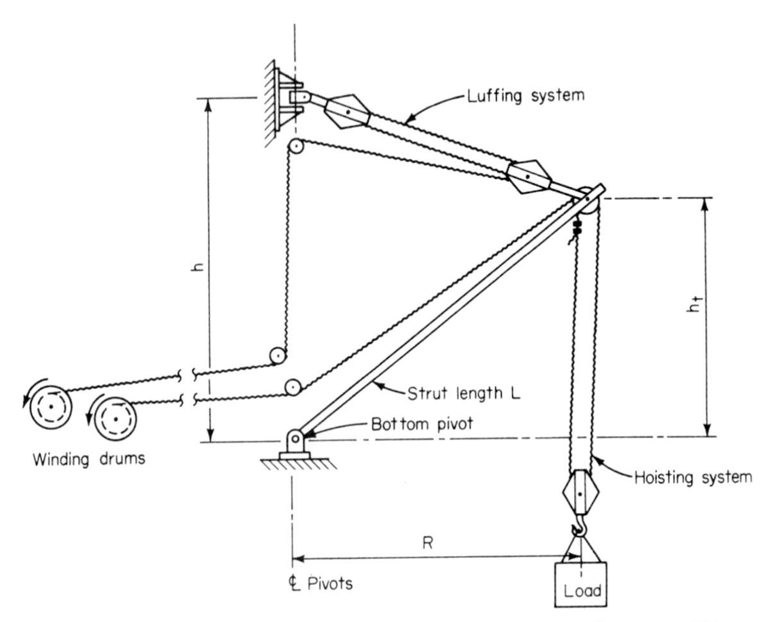

Figure 11.2 Basic derrick arrangement. (*From H. I. Shapiro,* Cranes and Derricks, *McGraw-Hill, 1980.*)

are usually constructed of latticed steel members. Common lengths of mast and boom are 125 and 100 ft, respectively, although these can be larger or smaller, depending on the particular application. Lifting capacities range up to 200 tons.

Mast and boom are mounted on a ball and socket joint that enables the mast to lean in the direction of loading. Because the boom is shorter than the mast, this derrick is able to swing through 360°. The mast top is fitted with a pivot pin (gudgeon) to which a ring fitting (spider) is attached from which the guy wires radiate. A large horizontal wheel (bull wheel) fitted to the bottom of the mast permits the system to be swung or slewed. The bull wheel is swung by wrapping rope from a winch around the wheel. A bull pole (or hand wheel with gearing) permits manual operation.

Topping and load lead lines are threaded through the mast, entering at a point above the boom foot and emerging from the mast top. Sheaves over which lead lines run to the winch are mounted in the derrick base.

When installing a guy derrick, the mast must be plumbed as closely as possible. The mast base is usually mounted on a steel grillage to transfer the loads to the building. Six to eight guys radiating in a horizontal circle from the mast are used to plumb the member. Guys are anchored to the building frame and hang in a predetermined catenary controlling the amount of mast lean when the system is under load.

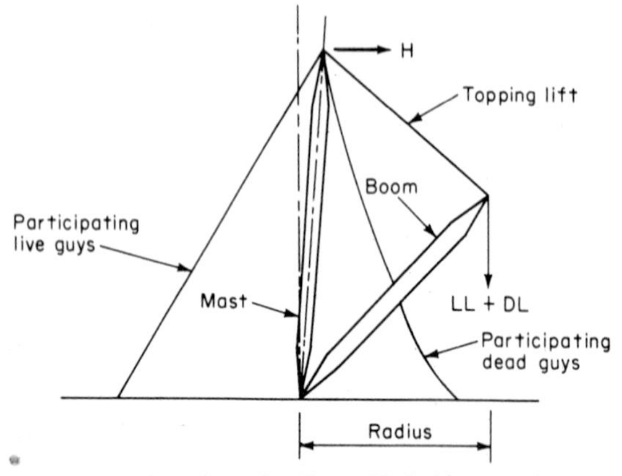

Figure 11.3 Guy derrick. (*From H. I. Shapiro,* Cranes and Derricks, *McGraw-Hill, 1980.*)

Because the mast is allowed to lean, design of the guying system is not as critical as that used for tower crane masts (see Fig. 11.4).

Since guys within 90° of either side of the boom do not support loading, they are considered *dead* guys. However, those within 75° of either side of the boom impose a dead-load moment on the system. The

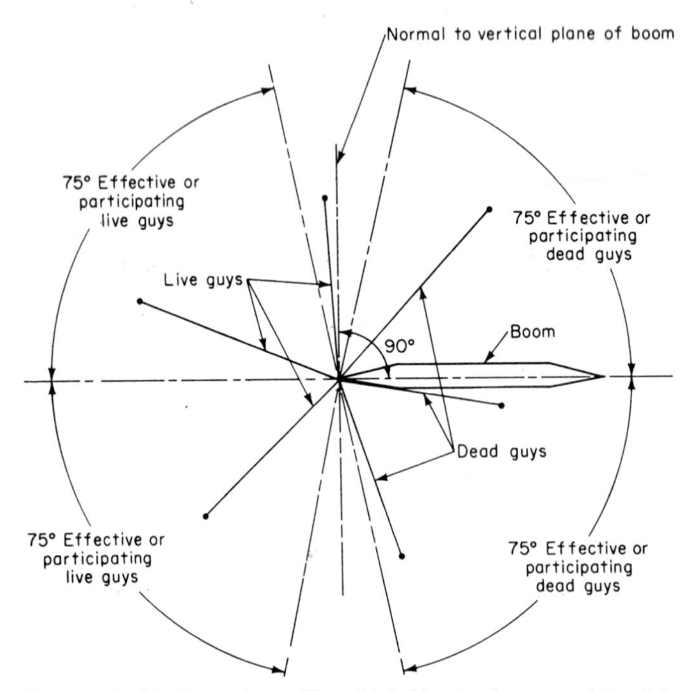

Figure 11.4 Guying systems. (*From H. I. Shapiro,* Cranes and Derricks, *McGraw-Hill, 1980.*)

remaining guys in the system are considered *live* guys, with those lying within a 150° area opposite the boom actually supporting the mast.

One of the most practical aspects of guy derricks is the ability of the derrick to jump itself to succeeding floors as frame erection proceeds. With the winch operated from the base floor, the boom is first lifted to the new floor level, where it is then used to lift the mast to a new grillage base. Initially a mobile crane must be used to erect the guy derrick. While installation costs are considerable, the two-person derrick crew provides an operational economy that more than offsets these preliminary costs.

Gin-pole derrick

This type of derrick resembles a guy derrick without a boom, held in position by guys fastened to its spider. Although only two guys, about 60 to 90° apart (in plan view) attached to the top of the gin pole are necessary to take the strain due to the load, normally four to six guys are installed to prevent the pole from falling over backward in the event of an unexpected jerk or a sudden release of the load (see Fig. 11.5).

Gin poles do not swing under load, but are used to lift loads, such as installed machinery or structural components, vertically at a fixed radius. Heavy gin poles lean only 5 or 10° from the vertical; lighter ones can lean more when fitted with a topping lift. Gin-pole lengths can be 250 ft or more, lifting capacities up to about 300 tons.

When installed, the gin pole should be plumbed nearly vertical (±10°). Because this type of derrick lifts without radius change, the pole height must be considerably greater than the height of the lift. Its guys should be heavily

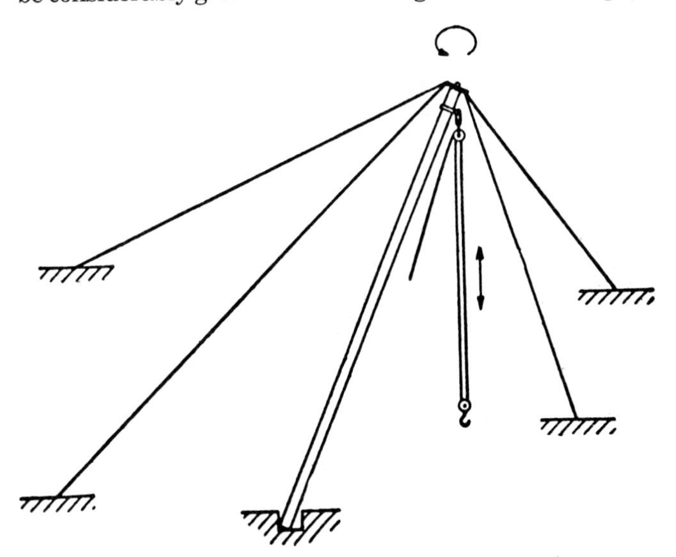

Figure 11.5 Gin pole. (*Courtesy of American Standards Institute.*)

preloaded to hold to a minimum the radius change resulting from loading. Care must be taken not to preload the guys to the point of overloading the pole in compression.

For erecting loads taller than a single gin-pole derrick, or where working space is confined, a pair of poles can be used. The poles are erected in tandem, leaning slightly away from each other. An open space is provided between guys at the top end of the load to provide clearance for the load.

Stiffleg derrick

This type of derrick has a short mast braced by two structural legs 60 and 90° apart horizontally, which limits the system's swing arc (see Fig. 11.6).

Stiffleg derricks are used primarily on sites where guys would not be feasible or would interfere with lifting. Components are usually of lattice construction, assembled from angle irons or tubular members in a wide range of sizes and capacities.

When necessary, stiffleg derricks can be mounted on towers or on fixed or traveling portal frames. In use, the mast of a stiffleg derrick rotates together with its boom. The mast has a top pivot where its legs join and a ball-and-socket joint at the bottom. Its base, in fixed position, houses lead sheaves.

As in a guy derrick, the lead lines are threaded down the mast's center and exit at the base sheaves. Positioning of the stiffleg mast on a building, and the location of its still anchorage points, are the most important installation considerations since each of these points is subject to both uplift and compressive loading when the derrick is in operation. These stresses often are far greater than the combined weight of the derrick plus the load being lifted.

The designer must find the position where the necessary derrick operations can be performed efficiently and the structure reinforcement is both practical and economically acceptable. For this reason, the mast must be positioned directly above a building column, and the sills should also be set above a column whenever possible.

Gallows frame

The configuration of two gin poles assembled with a horizontal beam across the pole heads is called a gallows frame. The beam supporting the upper load blocks is arranged to ensure concentric loading and to provide the gallows frame a greater capacity (about 50% more) than would the use of double gin poles. Gallows frames have lift heights of 200 ft with capacities to 600 tons. Shorter frames can lift up to 1200 tons.

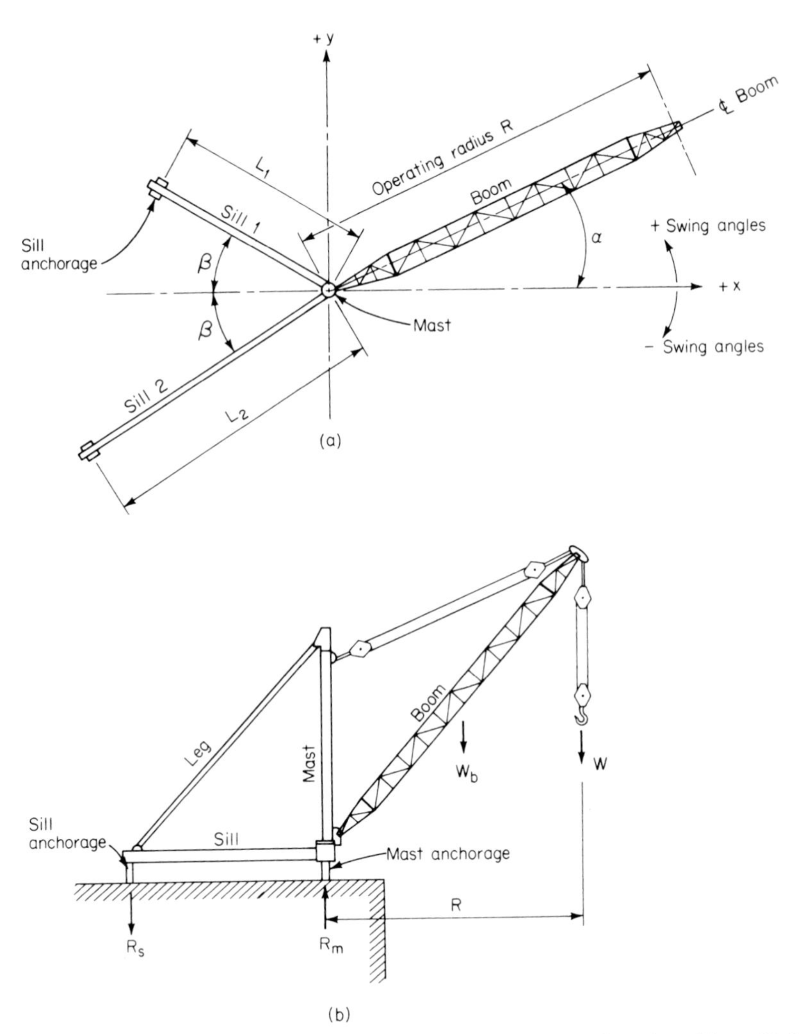

Figure 11.6 Installation of a stiffleg derrick. (*a*) Plan. (*b*) Elevation. (*From H. I. Shapiro,* Cranes and Derricks, *McGraw-Hill, 1980.*)

A-frame derrick

This type of derrick is similar to a Chicago boom, with the A frame taking the place of a column. Commonly mounted on barges for marine work, the A frame can be guyed or braced and can swing through a 180° arc.

Basket derrick

This derrick is another one in the Chicago boom family (see Fig. 11.7). Its boom is supported by four ropes tied off the adjacent frame mem-

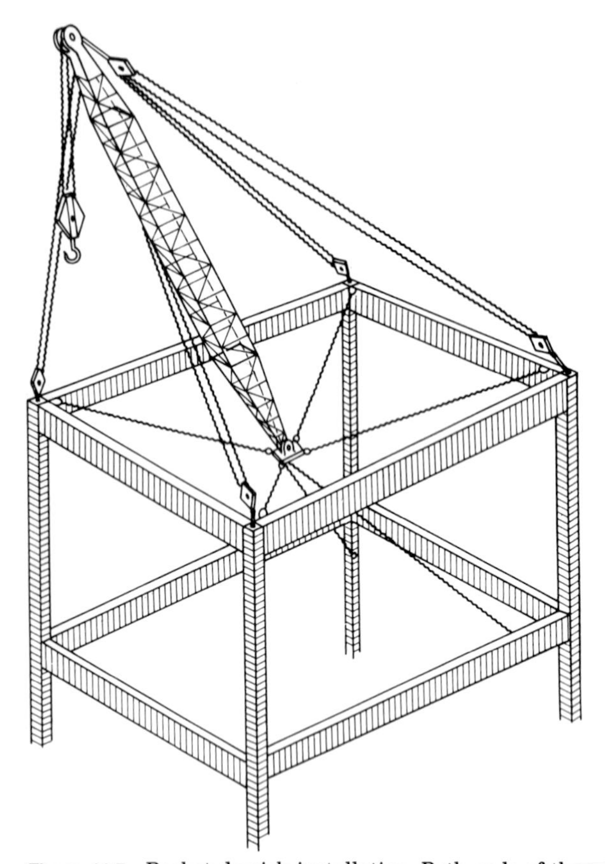

Figure 11.7 Basket derrick installation. Both ends of the boom are fully stabilized by ropes. The ropes going down to the lower level are called preventers. (*From H. I. Shapiro,* Cranes and Derricks, *McGraw-Hill, 1980.*)

bers. The derrick's boom foot is located below the support anchorages. Additional ropes provide stability.

The boom head and derrick can be turned by adjusting the guy lines, while the boom foot position can be changed by changing the lengths of the support ropes.

Mobile Cranes

A crane is basically a piece of hoisting equipment similar to a derrick, but having an integral hoisting engine. Any crane that moves from work area to work area under its own power is called a mobile crane. They can be mounted on crawlers or on rubber tires. These latter can be either truck cranes with separate hoisting and travel motors, or rough-terrain types with one motor performing both operations.

Crawler crane

Most crawler-mounted cranes have a single engine in the superstructure, which both propels the machine and lifts the loads. A system of shafts, sprockets, and drive chains working off the engine propel this type of crane. A few crawler cranes have hydraulic motors mounted on their side frames, which drive the tracks independently of the lifting mechanism.

Crawler cranes are capable of rotating through 360°. They have travel speeds ranging from about 0.5 to 1.5 mi/h, depending on the machine's size. Lifting capacities of standard crawler cranes range up to 300 tons. Various boom configurations and base modifications can increase this considerably.

Truck crane

A truck crane is mounted on a specially designed chassis that may have as many as nine sets of wheels (see Fig. 11.8). To accommodate the dual strain of operating around a job site's rough terrain as well as traveling on a highway, truck crane transmissions can have as many as 33 forward gears in addition to special high-ratio creeping gears.

Depending on the crane model, truck crane travel speeds can range from 35 to 50 mi/h over the road. They can easily negotiate job-site slopes of 20 to 40%.

Truck crane carrier widths range up to 13 ft; overall weights, to 350,000 lb. Heavier machines are shipped by rail because of load restrictions on highways.

Rough-terrain crane

This type of crane is mounted on a two-axle carrier with the operator cab either fixed or rotating—one that swings with the boom. All rough-terrain cranes are fitted with hydraulically telescoping booms.

Oversized tires make these cranes extremely maneuverable over a job site's irregular terrain. Over-the-road travel is limited to about 30

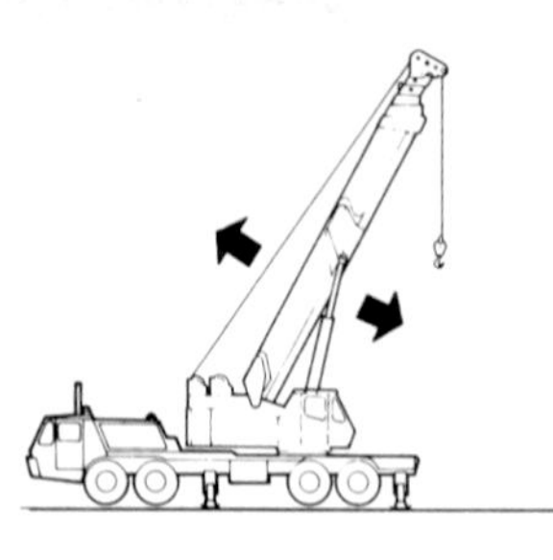

SETTING BOOM ANGLE. Position the boom at the proper angle for the load and working conditions. Refer to the rating plate for proper angle. Pull the boom hoist lever back, or depress the heel of the boom hoist pedal, to raise the boom. Push the boom hoist lever forward, or depress the toe of the boom hoist pedal, to lower the boom. Be sure to pay out line from the main and/or auxiliary winch, to prevent the hook blocks from coming in contact with the boom point.

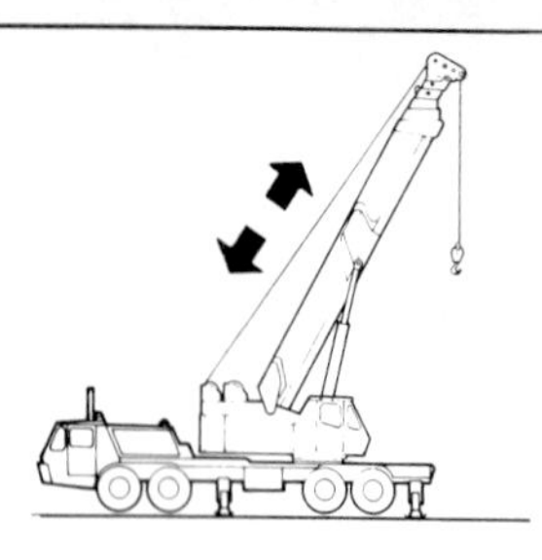

SETTING THE BOOM LENGTH. Push or pull the telescope lever to position the boom at the required length. Check to see that all boom sections are equally telescoped. If section lengths are uneven, depress and hold the individual button for the section to be telescoped. Move the telescope lever forward or back to control the movement in or out for the individual section. Be sure to pay out line from the main and/or auxiliary winch to prevent the hook block(s) from coming into contact with the boom point. Always refer to the rating plate for the proper boom length for the load being lifted.

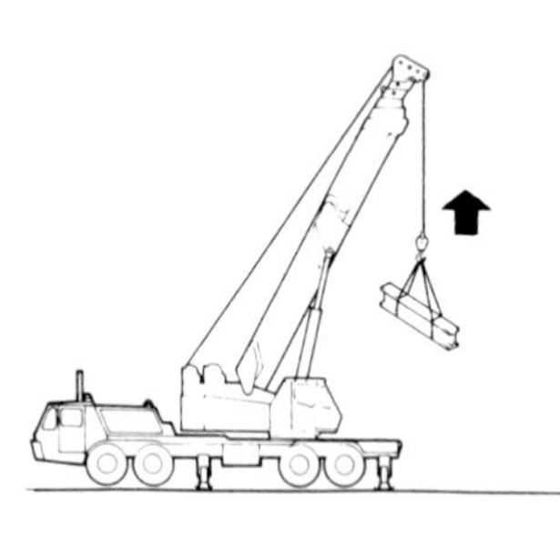

LIFTING THE LOAD. Push the main or auxiliary winch control lever forward to lower the hook block. Attach the load to the hook block. Then pull the main or auxiliary winch lever back to raise the load. Depress and hold the button on the side of the winch lever to operate the winch in high speed. The winch will operate in low speed when the button is not depressed.

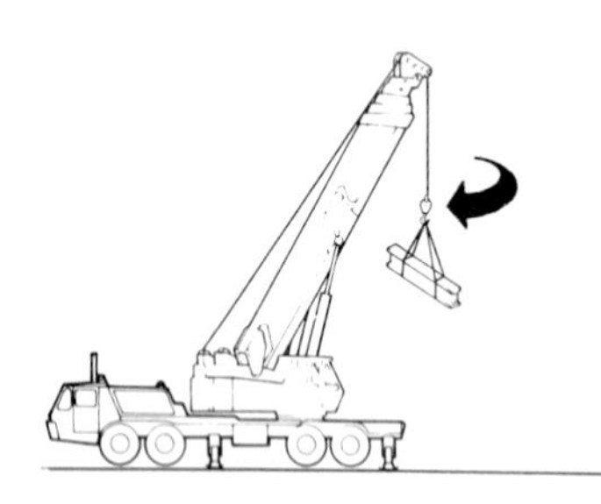

SWINGING. Push the swing lever forward to swing the upper to the left (toward the boom). Pull the swing lever back to swing the upper to the right (away from the boom). Plug the swing lever to bring the upper to a stop, and then depress the swing brake pedal to hold the upper. Place the swing brake lock, with the swing pedal depressed, in the ON position to hold the upper stationary without the use of your foot.

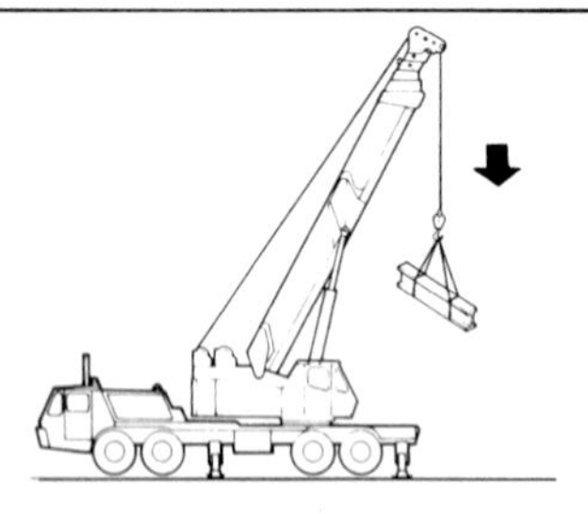

LOWERING THE LOAD. Push the main or auxiliary winch lever forward, to detent, to power down the boom. Push the winch lever past the detent to lower the load under controlled free-fall conditions.

CAUTION

Return the winch lever to the power down position slowly to avoid shock loading the winch and winch line.

Depress the button along side the winch lever to lower the load at high speed. The winch will lower at low speed when the button is not depressed.

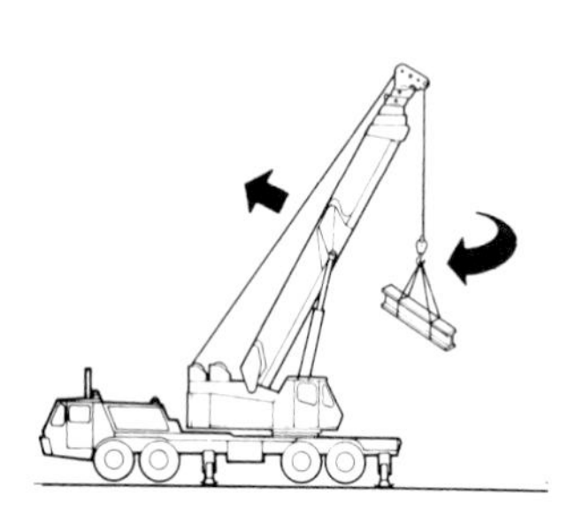

SPOTTING THE LOAD. Spotting the load requires accurate control of the boom and swing movements. It takes practice to locate the load at the exact spot without hunting or overshooting. Adjust the boom as required to accurately locate the load. Never extend the boom out so far that rating is exceeded. See the rating plate.

Figure 11.8 Telescopic crane operating cycles. (*From H. I. Shapiro,* Cranes and Derricks, *McGraw-Hill, 1980.*)

mi/h. Because of the machine's high center of gravity and mounting system, long-haul over-the-road travel is very uncomfortable. As a result, these machines are often transported by low-bed trailers.

Stability

Mobile crane stability is extremely critical since it determines the load rating. The rating itself is established by testing the machine in calm air with the crane in a level position on firm supports (see Figs. 11.9 to 11.12). A change in any one of these test conditions will change the crane's rated load capacity. Most tipping accidents can be attributed directly to the operator's failure to consider the effects of the machine or to operating conditions that vary from these test conditions.

Even though a crane is properly leveled, operating in calm air, and adequately supported with its unloaded hook set precisely to radius, it will exceed its rated capacity the instant it lifts the rated load for that radius. This can happen due to one of the following causes:

The machine deflecting under loading causes the frame to twist and the lifting radius to increase.

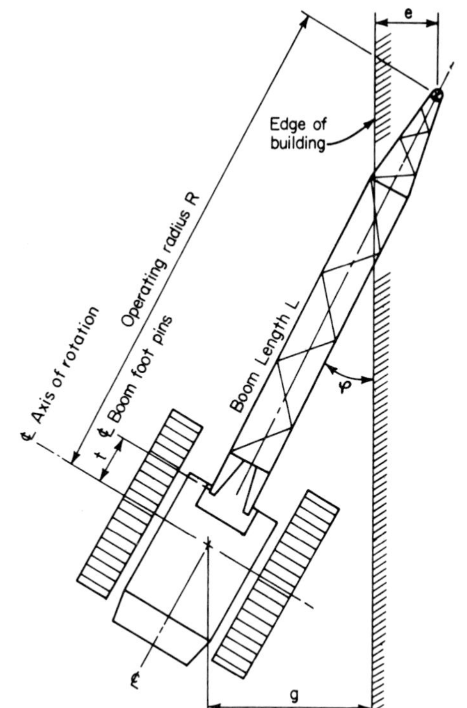

Figure 11.9 Plan view of a crawler crane showing the position parameters used for clearance calculations when the boom is at an angle to the wall. (*From H. I. Shapiro,* Cranes and Derricks, *McGraw-Hill, 1980.*)

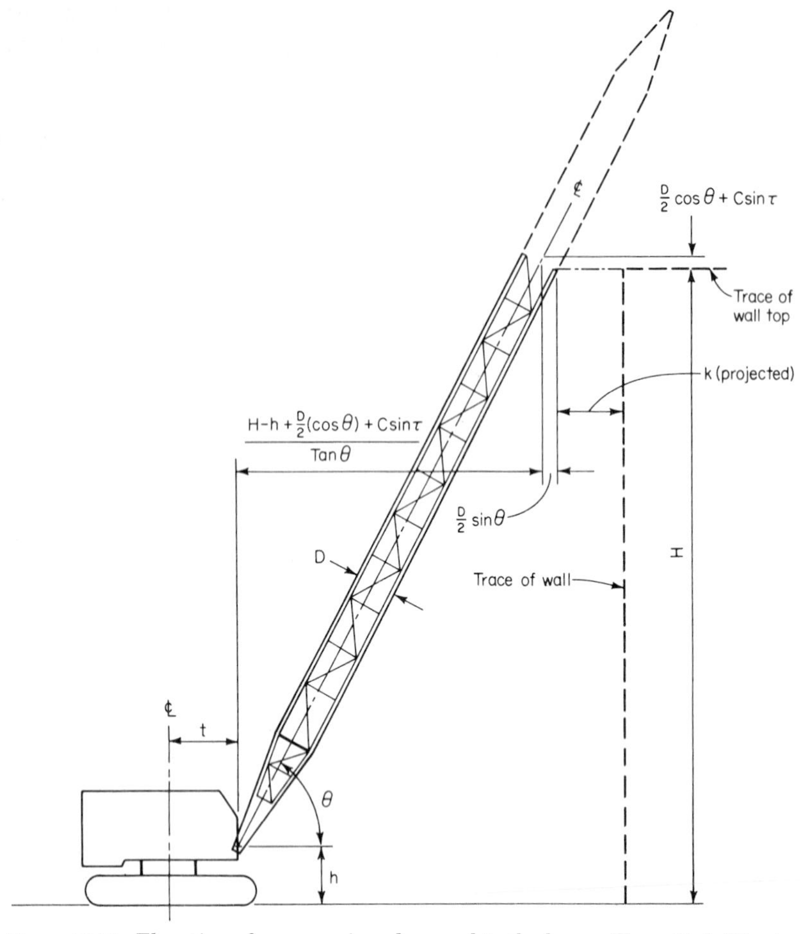

Figure 11.10 Elevation of a crane viewed normal to the boom. (*From H. I. Shapiro,* Cranes and Derricks, *McGraw-Hill, 1980.*)

The boom pendant lines and hoist ropes, elongating under load, further increase the lifting radius.

To allow for deflections and rope stretch, it is imperative that a lift at full rated load be made at a shorter radius. The load then can be boomed out to the required radius.

Wind, too, can be a critical factor in maintaining crane stability:

Blowing from behind the crane operator, wind is extremely critical since it applies force to the boom as well as to the load, adding to the overturning moment.

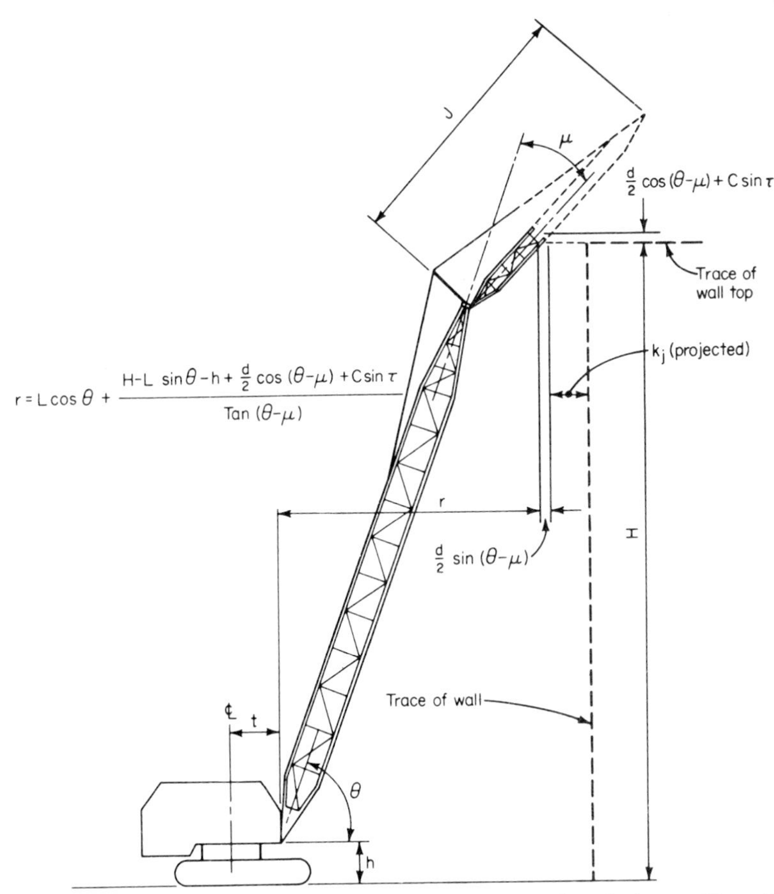

Figure 11.11 Elevation of a crane with jib viewed normal to the boom. (*From H. I. Shapiro,* Cranes and Derricks, *McGraw-Hill, 1980.*)

Blowing into the operator's face, wind tends to reduce the overturning moment.

Blowing from the side of the boom, wind becomes critical to boom strength.

Crawler crane. These cranes require a wide track spread to assure stability under loading. The wider the spread, the greater the crane's lifting capacity. Because track spread is limited by over-the-road restrictions, most manufacturers provide their machines with hydraulically extendable crawlers. These can increase the track spread by as much as 48 in., providing the stability required for lifting heavier loads (see Fig. 11.13).

Although crawler cranes have no leveling mechanism, stability very

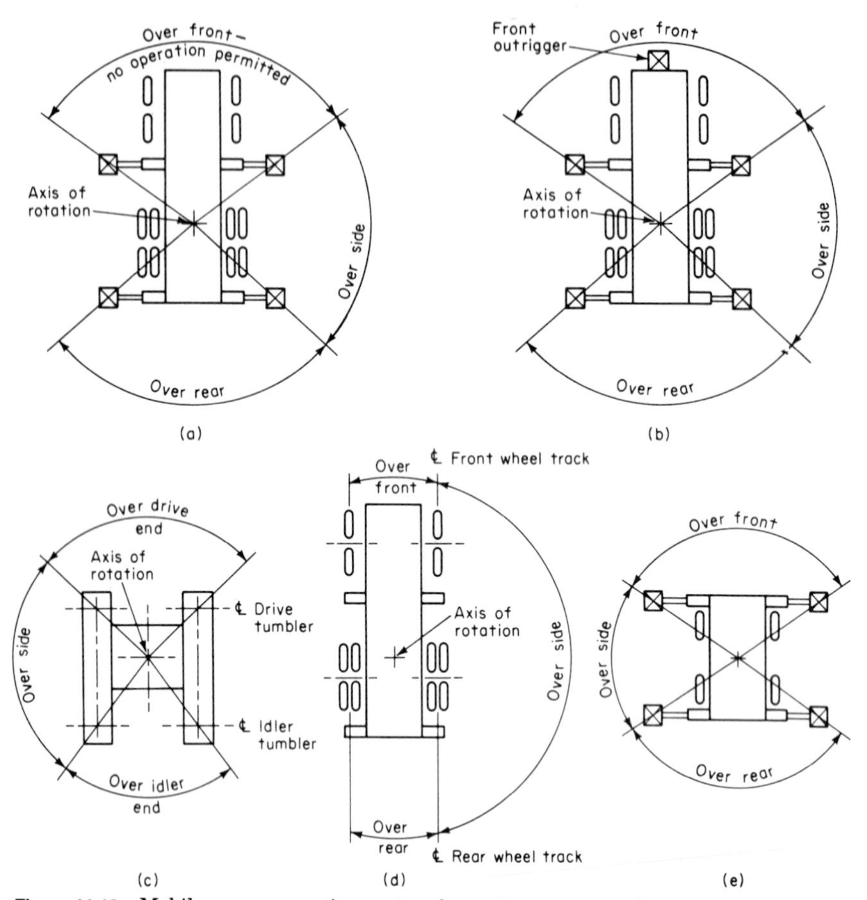

Figure 11.12 Mobile-crane operating sectors for various crane configurations. (*a*) Truck crane on outriggers, normal configuration. (*b*) Truck crane on outriggers with front outrigger. (*c*) Crawler-mounted crane. (*d*) Truck crane on tires. (*e*) Rough-terrain crane on outriggers. (*From H. I. Shapiro,* Cranes and Derricks, *McGraw-Hill, 1980.*)

much depends on their being operated on a level plane. This often requires laying timber planking to provide a level crane path.

A crawler crane's stability further depends on the position from which the load is lifted. When making a lift from the side of a crane, the side fulcrum line is located on the crane rollers that ride on the tracks. But because the tracks are loosely pinned and resting on the ground, the set opposite the tipping fulcrum cannot resist the machine's tendency to overturn during a side lift. Thus counterweights are necessary to gain the full capacity of a crawler crane. These are removed for travel or if the crane must be operated in a narrow configuration.

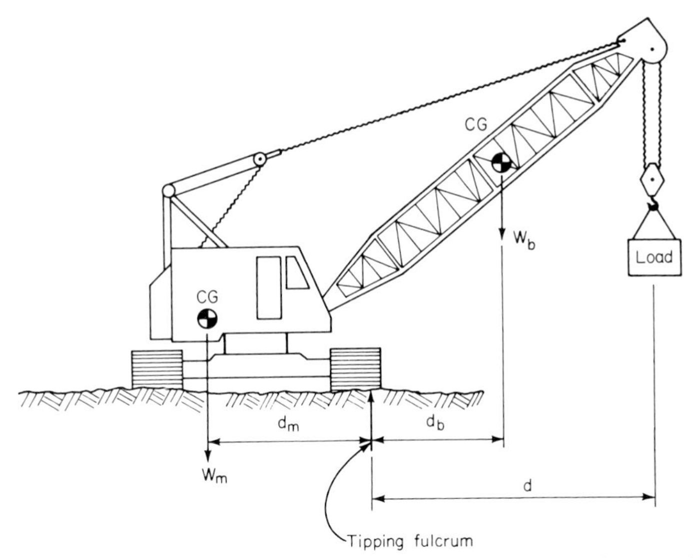

Figure 11.13 Crawler-mounted mobile crane annotated for consideration of stability. (*From H. I. Shapiro,* Cranes and Derricks, *McGraw-Hill, 1980.*)

Lifting a load over the end of a crane is more stable since the tipping fulcrum is located vertically below the centerline of the drive or idler tumbler shaft (see Fig. 11.14).

Tire-mounted crane. This type of crane is properly stabilized only if all weight is removed from the tires throughout the lifting operation. To accomplish this, manufacturers supply these cranes with hydraulically extendable outriggers that have knuckle-mounted floats on their ends. The crane is raised on its outriggers by hydraulic rams that either extend down from the outrigger ends, or push the outrigger beam downward from the chassis in a scissorlike manner (see Fig. 11.15).

To determine the location of the tipping fulcrum on a crane's outriggers, the machine should be raised so that outriggers carry the full weight. If the tires remain in contact with the ground, they reduce the outrigger beam's resisting moment and increase the crane's overturning moment.

With tire-mounted cranes having spring-mounted axles, the spring position determines the location of the fulcrum line (see Fig. 11.16). If, however, the two axles are mounted on (walking) beams that parallel the crane's centerline and pivot to the crane's frame (bogie), then the frame pivot controls the fulcrum location.

For cranes mounted on axles that are solidly fixed to the frame against rubber shock pads, the center of the tires or the center of a pair of dual tires denotes the location of the tipping line. This mount-

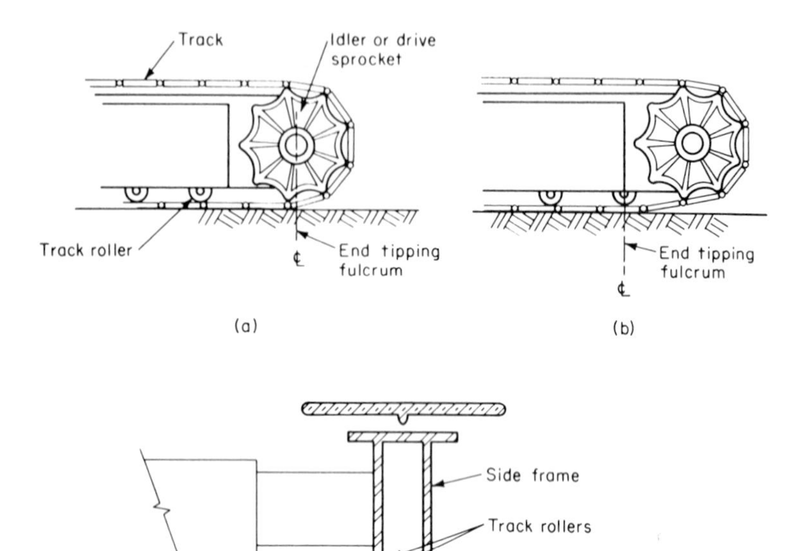

Figure 11.14 Crawler crane tipping lines. (*a*) Over-end tipping fulcrum at sprockets. (*b*) Over-end tipping fulcrum at first track roller. (*c*) Over-side tipping fulcrum. Any number of configurations of side frame, sprocket, track roller, and track are possible. (*From H. I. Shapiro,* Cranes and Derricks, *McGraw-Hill, 1980.*)

ing arrangement provides the greatest operating stability and permits the highest rated load of the three configurations.

Mechanical blocking can be used to increase the stability of rubber-tired cranes. Blocking locks a crane's axles to the frame so that the tires define the fulcrum line. This is done automatically on some newer cranes and manually on older machines.

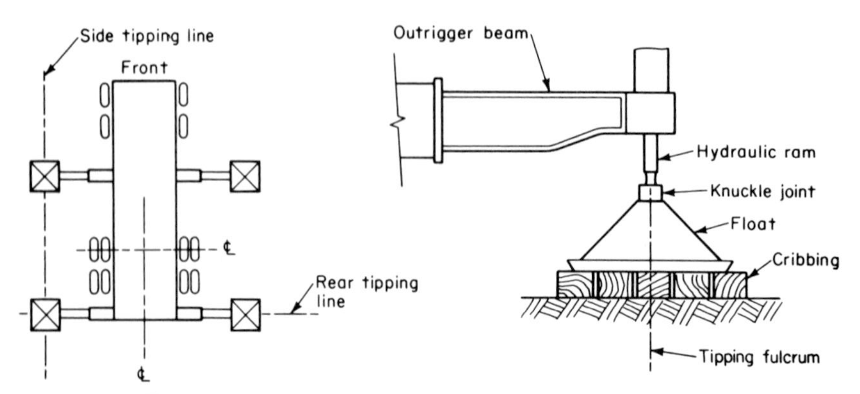

Figure 11.15 Tipping lines for a crane raised onto outriggers. (*From H. I. Shapiro,* Cranes and Derricks, *McGraw-Hill, 1980.*)

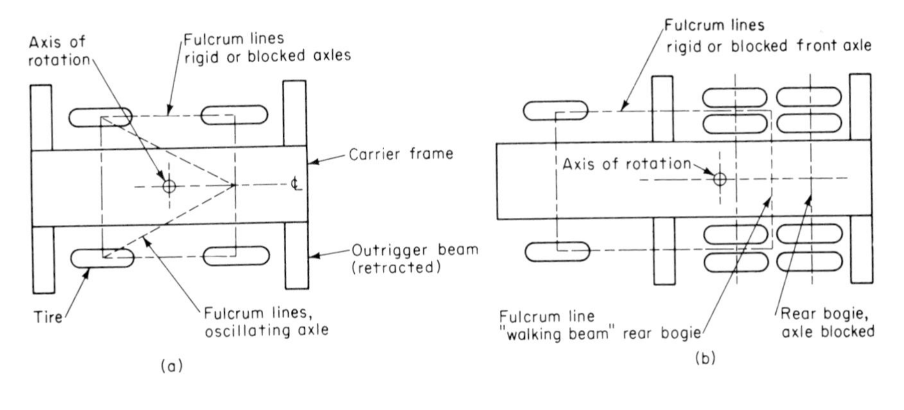

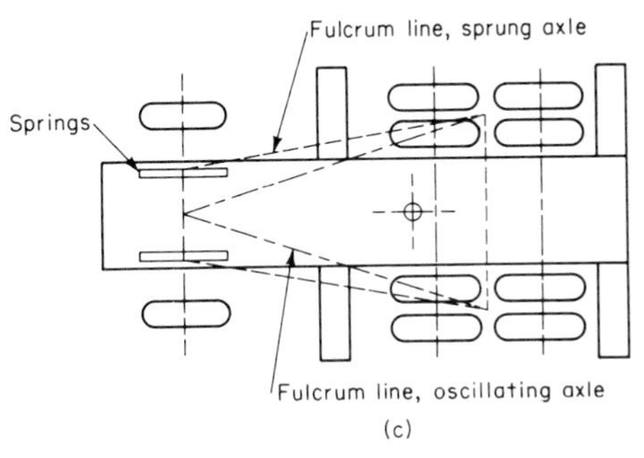

Figure 11.16 Tipping lines for cranes on tires with and without blocking. (*a*) Rough-terrain crane. (*b*) Truck crane with rigid front axle. (*c*) Truck crane with oscillating or sprung front axle. (*From H. I. Shapiro,* Cranes and Derricks, *McGraw-Hill, 1980.*)

Support

Ratings for mobile cranes are determined with the machine standing on a firm surface, capable of uniformly supporting the weight of the crane plus its load. Thus chart ratings should be used only if the outriggers, tracks, or tires are supported throughout the lifting operation to ensure maintaining the crane level to within ±1%.

For most job sites, once the crane is put in place, the initial settlement is the only settlement that need be considered. This is easily corrected by the first leveling of the crane. However, if the crane is to be installed in the same location for any length of time (weeks or months), releveling may be required periodically.

It is important to remember that the crane must be maintained level throughout all lifting operations. Crane support should be an essential part of planning all lifts in order to prevent tipping accidents.

Additional consideration must be given to outrigger floats that bear pressures on soil, because weight and stowage restrictions limit the size and area of such floats. Standard floats often produce high bearing pressures, which can trigger sudden soil shear failure with subsequent tipping of the crane. Cribbed floats often must be used to prevent such failure.

A rough-terrain crane should never be operated on soil with flat steel plate floats without first providing cribbing to support the floats. The sharp edge of these floats will cut into the soil and thus increase the possibility of crane tipping. Floats having turned up edges do not normally involve the same risk, since they tend to readjust themselves to a level position as the plate digs into the soil.

Booms and jibs

Crawler cranes are always fitted with lattice booms, while rubber-tired cranes may have booms that are either lattice or hydraulically telescoping.

The most common boom attachment consists of a base section plus inserts of varying lengths (normally 10-ft increments) and a tip section. Any boom length can be assembled using manufacturer specified combinations of inserts. When only the base and the tip section are used, the assembly is capable of lifting the maximum capacity.

Hammerhead tips are often used to increase the lifting height. In this configuration the upper block sheaves and the hook are located away from the boom centerline. Lightweight tips, having one or two sheaves, are used to provide even longer booms.

The basic boom for smaller cranes is about 30 ft long; for larger cranes, it runs from 70 to 100 ft. Maximum boom length will vary with the crane model. The longest booms currently in use are now in excess of 350 ft with cross sections up to 8 ft^2 or more.

A jib is a lightweight boomlike structure that is mounted at the boom tip to increase the height of the crane lift. Jibs have relatively low lifting capacities and usually have a single part line. The crane' boom tip usually carries the main hook, while the jib carries a light-duty hook (see Fig. 11.17).

When a jib is used, the backstay attachment is located on the boom, and fixed stay ropes are used to hold the jib in place. Jib mountings can be parallel to the boom centerline or offset up to 45° from the centerline.

A mobile crane will sometimes be fitted with a tower attachment, especially on sites where the crane must be located close to a structure yet be able to lift and place loads over the structure. The tower is pinned to the crane's upper structure where the boom is ordinarily mounted. A luffing boom is attached at the tower top, and a jib is sometimes fitted at the end of the boom (see Fig. 11.18).

Tower attachments usually are assembled and erected without the help of an auxiliary crane, although a small hydraulic unit often will speed unloading and placing boom and tower sections. Varying-length

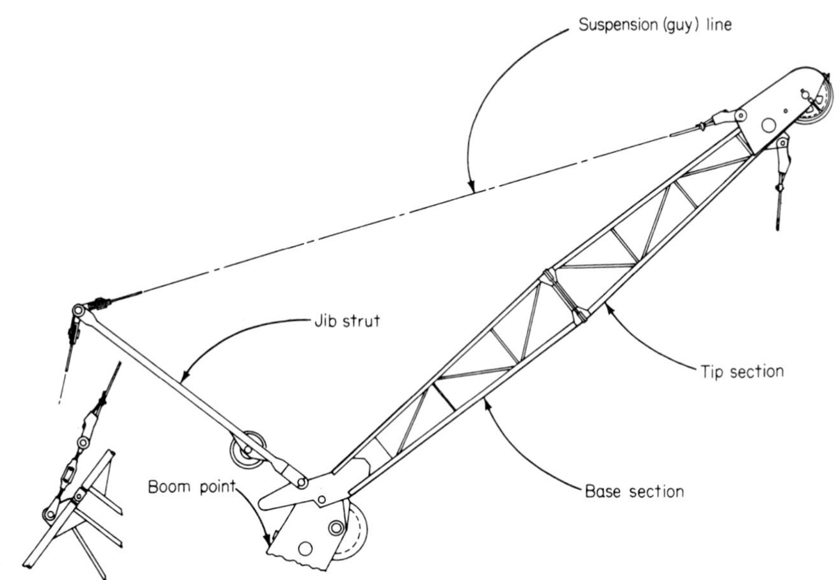

Figure 11.17 Typical jib mounting, showing the jib in its shortest configuration. The jib is offset 20° from the boom. (*From H. I. Shapiro,* Cranes and Derricks, *McGraw-Hill, 1980.*)

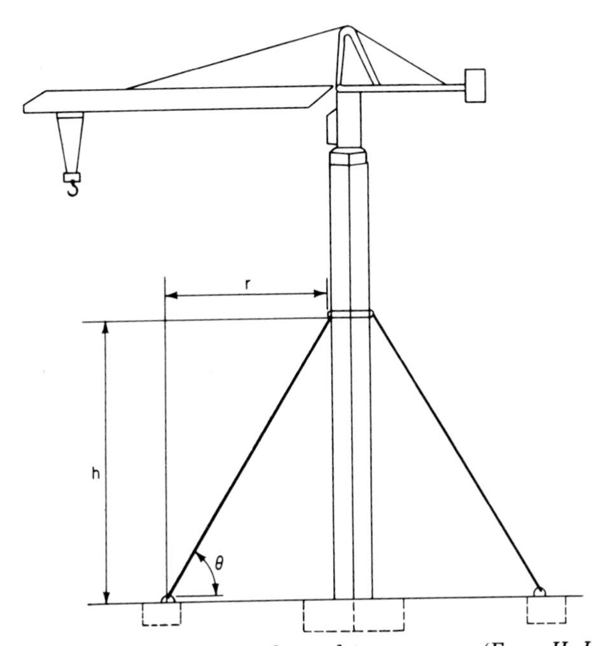

Figure 11.18 Elevation of guyed tower crane. (*From H. I. Shapiro,* Cranes and Derricks, *McGraw-Hill, 1980.*)

inserts permit the assembly of a number of tower-height and boom-length combinations.

Hydraulic telescoping booms are available on all rubber-tired cranes, for truck-mounted models extending to 200 ft with lifting capacities to 150 tons, and for rough-terrain models extending to 115 ft with lifting capacities to 80 tons.

Booms may be rectangular, trapezoidal, or any other symmetrically shaped cross sections that fit into each other. The largest section of a boom, at the bottom, is the base section; the smallest cross section, at the top of the boom, is the fly section. Sections in between are called the midsections. All sections, except the base, extend and retract by means of hydraulic cylinders that are mounted within the sections parallel to the boom centerline. The sections telescope, one within the other, on rollers.

Telescopic crane load ratings are determined by the machine's resistance to overturning, while the heavy boom resists the loads structurally by their cantilever bending action.

To increase the length of a telescopic boom, a nonpowered section can be added at the tip. The boom head is mounted on the insert, which is pinned in place to the fly section in either the retracted or the extended position.

An additional latticed extension can also be attached to the boom head. It acts as a cantilevered member and folds back sideways when not used for storage, latched to the boom. Solid or latticed jibs are also available for telescoping boom cranes and mount either on the boom head or at the tip of the latticed extension.

Tower Cranes

Even with an extremely long boom there is a practical limit to the height of buildings that can be erected using a mobile crane. To cope with this problem, crane manufacturers, over the past 20 years, have developed a variety of special cranes, known as tower cranes, for erecting tall buildings.

All functions on these cranes operate using electric motors that include high- and low-speed ranges, with stepped increments in each range. Some newer models have solid-state electronic stepless controls, with friction or eddy-current brakes, or both, which engage automatically as power is withdrawn.

The larger models of this type of crane have automatic acceleration and deceleration rate control devices for the various motions. Some industrial-type tower cranes even operate under remote control.

Mountings

Tower cranes have various base mountings, depending on crane model, size, or application including traveling, static, and climbing bases.

Traveling base. Tower cranes mounted on this type of base operate on wide-gage railroad tracks that run adjacent to a building's wall. Such cranes are used almost exclusively on construction buildings having limited height but considerable length.

The crane tower is rigidly fixed to a platform fitted with four sets of electrically driven wheels (bogies) that ride the steel rails. When installing a traveling-base crane, it is absolutely imperative that the rails be set on a level grade to maintain the tower plumbness throughout its operation, and that they be supported throughout to prevent possible bending or settling under loading.

With this type of crane, ballast weights are stacked on the tower platform.

The crane itself consists of a triangular boom that is raised to the vertical position and acts as the mast. At its top is an auxiliary boom, or jib, that can be raised or lowered to change the crane's working radius.

The operator's cab, in the form of an elevator car, is located within the telescoping mast and can be raised as construction progresses. This provides the crane operator with an unobstructed view of the erection floor at all times.

Static base. This mounting is used with either saddle or articulated jib cranes. The tower is set into a concrete foundation block, or bolted to it, throughout the life of the job.

Acting as a freestanding cantilevered structure, the tower must be able to resist both vertical and lateral forces, as well as over turning moments. The crane's total weight plus the load it lifts comprise the vertical force on the foundation block.

While wind load imposes only a minor lateral loading, a combination of wind and the load being lifted will add a significant overturning moment to the crane, which must be resisted.

Climbing base. This is perhaps the most versatile type of tower crane mounting. It is usually installed at the center of a building under construction, preferably in an elevator shaft. However, when this is not possible, a floor opening, to be closed later, is provided during construction (see Fig. 11.19).

The crane is usually mounted on a foundation of I beams bolted together to form a cross. A rotary crane is placed on top of the steel tower.

Initially the base beams are secured to a foundation block, or loaded on their outer ends with heavy counterweights. After the crane has erected several stories of the structure, its tower base is raised and bolted to a newly completed floor. When anchored to the floor frame, the tower base transmits the crane's vertical loads directly to the structure.

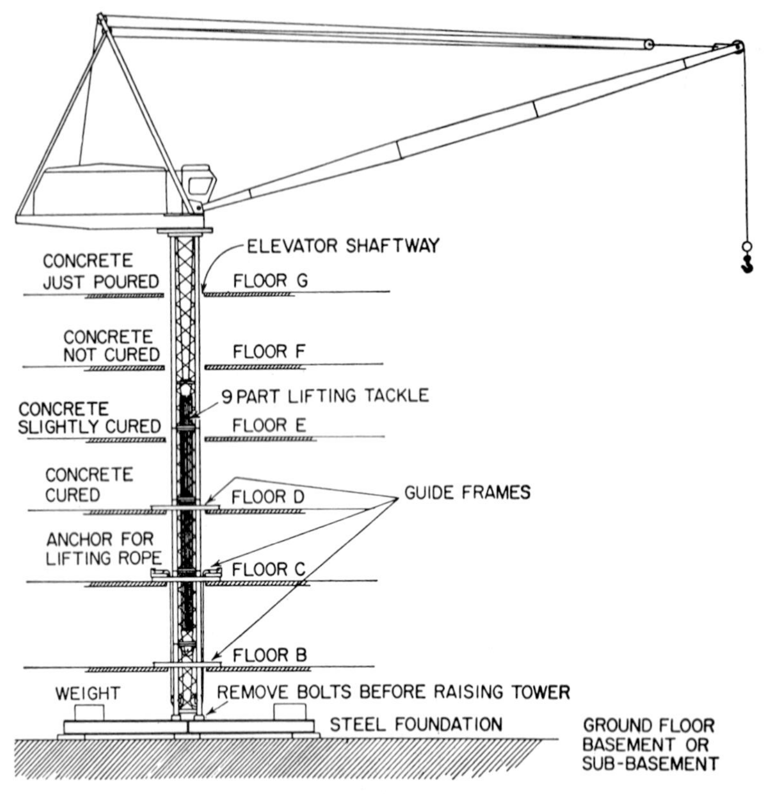

Figure 11.19 Raising a climbing tower derrick periodically. *(From H. I. Shapiro,* Cranes and Derricks, *McGraw-Hill, 1980.)*

Tower cranes are also classified as either fixed type—having their swing circle near the tower top-or slewing type—with their swing circle mounted near the base.

Jibs

Unlike booms on mobile cranes, tower crane booms are called jibs. When mounted in a horizontal or slightly upward sloping position from the tower, they are called saddle or hammerhead jibs. Those mounted with a pivot and derricklike system are called luffing jibs.

Saddle jibs. This type of jib has a trolley suspended from its bottom which travels the length of the jib. The crane's upper block is part of the trolley, and the load block is suspended from the upper block.

An independent winch and rope system controls trolley travel. By reeving the load hoist system with the dead end at the tip of the jib, the load block is maintained at a constant elevation as the trolley changes radius.

Projecting from the other side of the tower is a second jib, which serves as a counterweight. It carries the load winch, power plant, and control panel as well as counterweights.

A small tower projects above the jibs. Pendants supporting both jibs are anchored to the top of this tower. Steel bars instead of wire ropes are used for pendants on larger cranes. The tower also houses the head sheaves.

Just below the jib is the operator's cab, located to provide full view of the load at all times.

Larger jibs of this type can handle loads up to 264 tons at radii up to 145 ft, with maximum clearance under a hook as much as 260 ft.

It is important to remember that saddle-jib tower cranes require a 360° slewing capability, with no obstructions to impede free swing in the wind when the crane is idle.

Luffing jibs. This type of jib is used with towers that are mounted on a slewing platform. The platform also carries the power plant and counterweights.

Fixed pendant ropes supporting the luff jib run from the jib tip to struts that are pivoted at the tower top. From there they run down to a spreader that holds the sheaves through which the running ropes are reeved.

Because the luffing system's vertical ropes are installed nearly parallel to the tower, the tower actually acts as a compressive member and thus relieves most of the bending load.

A luffing jib can work at heights considerably greater than can a saddle jib that has the same tower height and jib length—at all but the longest radii.

Because of its configuration, the luffing crane can be used at sites where it is not possible to provide a 360° obstruction-free swing path.

Installation

The tower configuration found on most building job sites today is the static-mounted crane. This type of mounting also is the initial setup for all climbing, braced, and guyed crane systems.

A static crane is mounted on a mass concrete footing that serves as both its anchor base and its ballast. Vertical loads are transmitted to the ground through this footing block, which must also be designed to resist shear forces from wind and torsional effects from crane operation.

Although shear forces are generally small and seldom control the design of the footing block itself, shear must be considered when designing the connection between mast and block. Current design codes usually specify that the footing block be able to withstand from 1⅓ to 1½ times the applied overturning moment.

Footing settlement. Because uniform settling of the footing block will be small, it usually has no significant effect on crane height. However, differential settling can be a very critical design factor.

For differential settling, the crane must be monitored on a regular schedule, using a standard surveyor's level. Intervals between readings will depend on the soil characteristics of the foundation and any observed settlement of the tower.

A properly installed tower crane should not tilt more than 1:1000 in calm air when a load on the hook balances the crane's counterweight.

To correct for differential settling, the crane's jib should be pointed in the direction of the tower's lean during periods of shutdown (overnight). In this position the counterweight moment will act opposite to the beam and produce countersettlement.

If this procedure does not correct the settlement, it may be necessary to add ballast to the footing block. Make sure that settlement readings are taken frequently while trying to correct the tower tilt.

Wind load. Wind can be critical in designing a tower crane's footing. The resultant moment will vary with the amount of mast side exposed to the wind. The most critical exposures occur when the wind blows normal to the mast face and on its diagonal. Both values must be considered when designing an effective footing.

In areas where storm winds might produce greater than allowable loading for the required mast height, it will be necessary to install a guying system. In some cases, guy wires may have to be disconnected or left slack to permit crane operation. Under such conditions it is necessary to provide a system that permits crews to quickly secure the crane in the event of high winds.

Guys should be preloaded, using turnbuckles or other tensioning devices, to offset wind-imposed loading that will elongate the wires and permit the mast to lean with the wind. Extreme caution must be taken to maintain the crane's plumbness after the wires are preloaded.

Preload forces and mast plumbness should be monitored every few days after first installing a guyed system and after any signficant high wind. After the preload has stabilized, monitoring can be carried out less frequently. Check all fittings periodically for tightness.

If a static-mounted crane is installed alongside a building, it can be braced at intervals to the building frame, thereby increasing its working height considerably. Such bracing usually consists of two-dimensional triangular frames installed horizontally between building and mast, which adjust to variations in construction or distance from the building face (see Fig. 11.20). When installing bracing, make sure that there is sufficient clearance between crane and building.

Figure 11.20 Fixed or static type tower crane. (*Courtesy of Harnischfeger Corporation.*)

Tower cranelifts

Tower cranes that move up with the structure are called climbing cranes. They climb to a completed level in the structure in one of two ways: by climbing a pair of ladders or by lifting themselves on heavy steel rods or with a cable and winch system, depending on the crane model.

To provide for a lift, steel guide frames are first installed around the elevator shaft (or floor opening) at or above the new level of support. The guide frames support the tower laterally, once the base is unbolted from its anchorage. During the climb, the crane is suspended from the frames by a pair of climbing ladders (see Fig. 11.21).

The climbing mechanism, located in the lower portion of the crane mast, includes an upper cross beam fixed to the mast. Hydraulic rams under the beam are fitted with a floating beam having retractable dogs at its ends, which engage the climbing ladder rungs.

When the crane is ready to climb, ladders and ladder supports are moved up to the new level. The lower dogs are engaged in the ladder,

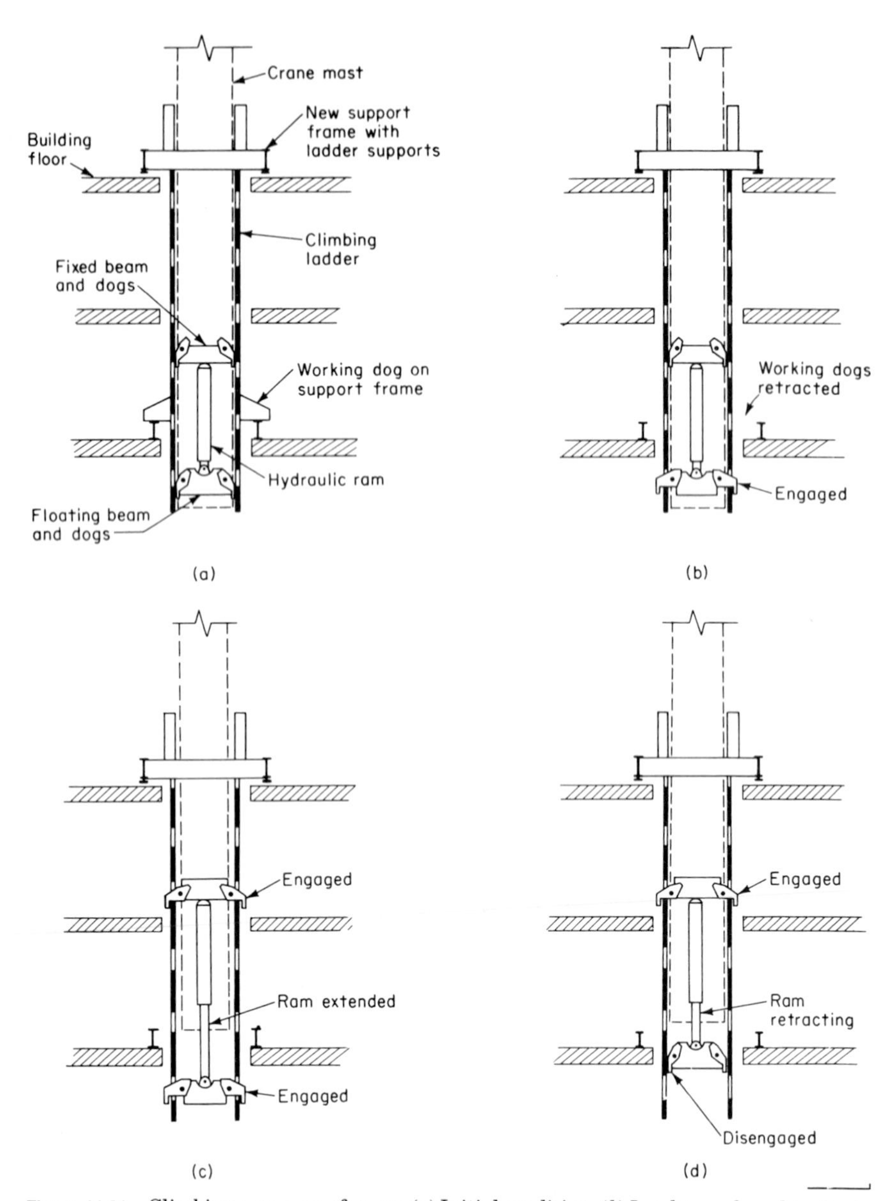

Figure 11.21 Climbing sequence of crane. (*a*) Initial condition. (*b*) Load transferred to climbing ladders. (*c*) Crane is pushed up by hydraulic ram. (*d*) Ram is retracted to start another cycle. (*From H. I. Shapiro,* Cranes and Derricks, *McGraw-Hill, 1980.*)

and the horizontal support wedges are loosened. Hydraulic pressure in the rams then lifts the crane. When the full ram stroke is reached, upper beam dogs are engaged to transfer the load to the ladders. When the rams are retracted, the floating beam is raised, the lower dogs are disengaged, and the cycle is repeated (see Fig. 11.22).

Cranes that lift themselves on rods use a pair of rams and a gripper mechanism mounted on opposite sides of a lifting frame. Rods, fixed at the base of the mast, run upward through the grippers.

With the rams retracted, the upper jaws lightly grip the rods and the working dogs are engaged. As the rams extend, the upper jaws grip the rods tightly and the crane rises. The working dogs are then retracted and the rams extended to full stroke. At the top of the stroke, the lower jaws grip the rod, while the upper jaws release it. The rams are then fully retracted for the repeat cycle.

With the cable-lifted crane, cables extend down the elevator shaft (or floor opening) alongside the guide frame and pass under a fixed

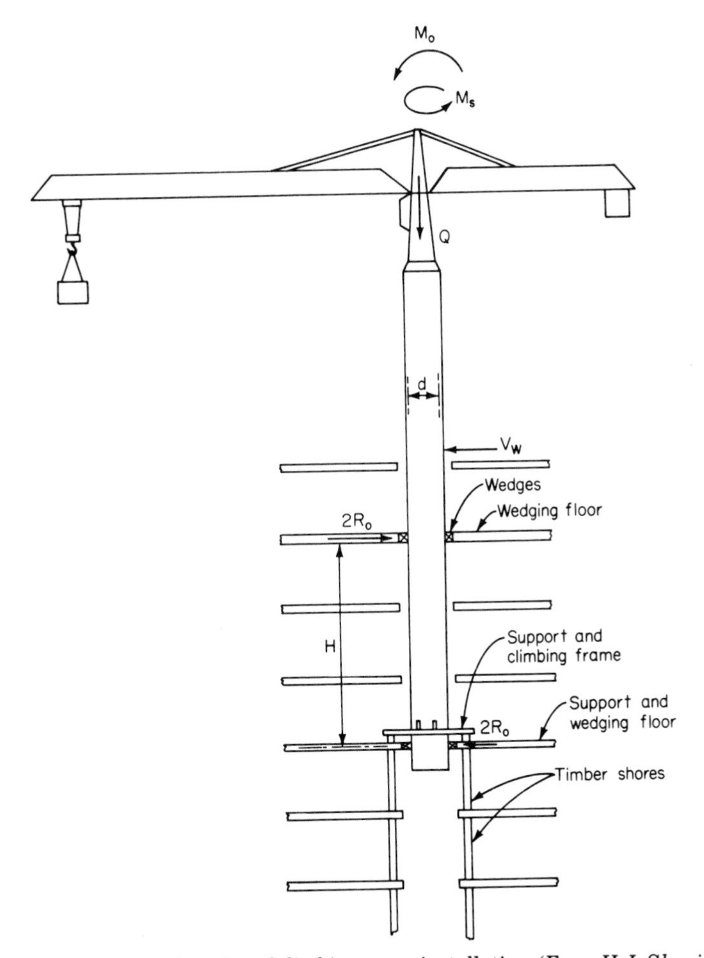

Figure 11.22 Elevation of climbing crane installation. (*From H. I. Shapiro,* Cranes and Derricks, *McGraw-Hill, 1980.*)

sheave at the bottom of the tower. The cable extends upward inside the tower and over a movable sheave, down and under another fixed sheave, and up the shaftway to a floor where its end is anchored.

A wire rope from the third drum on the hoist machine extends down the center of the tower through the pintle, and is reeved into a multipart tackle attached to the movable sheave. Using the hoist engine, the tower is raised until its lower end reaches the new anchorage level.

Hinged beams attached to the guide frames are swung into position and the tower is lowered to rest on them. The tower base then is bolted into place.

Once a climb is completed, the tower is wedged against the structure in four directions at two levels to transfer all horizontal forces and overturning moment to the structure's frame. Several thicknesses of exterior-grade plywood are usually used to shim a crane into its plumb position.

Since crane weights are considerable (as much as 200 tons), climbing crane reactions will invariably overload the building frame. These loads, therefore, must be distributed to the frame by using temporary shoring on floors below the crane tower's base.

Always make sure that there is extra shoring available before the crane climbs to another level. To prevent overloading the frame during the lift, the previously installed shores should remain in place until the new ones are set and the climb is completed.

Stability

When considering use of a tower crane, it is important to recognize that such cranes have a very high center of gravity in relation to their base width. This disproportionate ratio makes tower crane stability extremely sensitive. This is especially true of saddle and articulated types, which have their counterweights mounted at the very top of the crane. Cranes having their counterweight at their bases have a lower center of gravity and thus tend to be more stable in operation.

In addition, since tower crane masts act as cantilevered structures, they tend to deflect under unbalanced dead, live, wind, and dynamic loads.

Always consider and verify a tower crane's stability before making a lift. Make sure to check:

Basic stability. The crane under static loading, in calm air, with the rated load not exceeding two-thirds of the tipping load

Dynamic stability. The crane in service, under normal wind and dynamic loading, with the rated load not more than 75% of the tipping load

Extreme load stability. The crane out of service, subject to storm wind or to earthquake shock loading

Special mountings

Special mountings, used to raise the crane boom or jib to clear obstacles or to gain better access to work, include such types as pedestal, portal, portal/tower-mounted, revolver, overhead, and gantry.

Pedestal. Fixed mounting to which the crane's slewing ring is attached. The pedestal may be made of large steel tubes or framework, concrete tube or frame, or a concrete building core.

Portal. Steel-legged structure which raises the crane's working height while adding little support weight. This type of crane is used where passage beneath the crane is necessary, or where the crane travels on rails.

Portal/tower. Tower mounted on a portal frame provides still more working height for the crane. However the lifting capacity is limited by the combined support system's overturning resistance.

Revolver. Large lattice-boom cranes, usually barge-mounted with capacities up to 300 tons and more. These often are portal, pedestal, or tower-mounted on jobs that have long construction schedules, such as dam projects; or for more permanent installation, such as at shipyards.

Overhead. Electrically powered cranes, also called bridge cranes, that ride on a pair of tracks above the work floor and consist of a bridge spanning the tracks and a fixed or trolley-mounted hoisting system. These cranes, either riding on rails or hung from the bottom flange of the runway beams, have lifting capacities to 250 tons, with up to 100-ft spans.

Gantry. Leg-mounted cranes that ride on rails. They have a trussed or plate girder bridge that may be cantilevered on either or both ends. When the crane is supported on legs at one end and a runway beam at the other, it is called a half-gantry.

Chapter

12

Overhead Hoists

Portable overhead hoists, both hand-operated and motor-powered, are widely used in a variety of lifting and holding operations, especially where low overhead restricts the use of a crane or derrick.

Their simplicity, dependability, and relatively low cost have made them standard material-handling equipment for erection and maintenance work.

Selection

The most important consideration when selecting a hoist, either chain-operated or motor-powered, is its capacity for the heaviest load to be lifted (see Fig. 12.1). Motors may be electric or air-powered.

The performance and physical characteristics that must be considered when selecting a hoist for a given use and installation include headroom, height of lift, location, height of hand chain, type of suspension, and trolley clearances.

Basic considerations common to all types of hoists include initial cost, frequency of use, labor savings, safety, portability, and maintenance requirements. Other considerations are environmental considerations such as moisture, heat, chemicals, and foreign material in the atmosphere that may require weatherproofing or special protection.

Chain Hoists

The simplest and least expensive type of chain hoist is the rigger's ratchet type that has a mechanical advantage of approximately 15:1. A lightweight portable tool, it can be used for pulling at any angle.

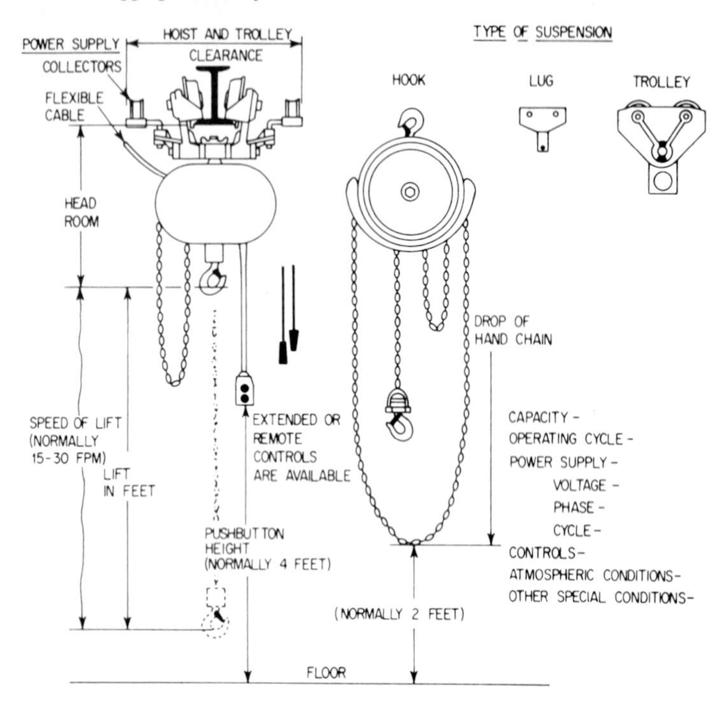

Figure 12.1 Hoist installation check diagram. (*From Higgins and Morrow,* Maintenance Engineering Handbook, *3d ed., McGraw-Hill, 1979.*)

When loading, a driving pawl engages the ratchet and, by turning the lift wheel, causes tension on the chain. By shifting the direction lever to the unloaded position, the tension can be released one tooth at a time (see Fig. 12.2).

The hand-chain hoist, suspended from a hook and operated by hand chain, is available in lift capacities to 50 tons, with almost no limit on its lift height. Although several types of hand-chain hoists may still be found on some job sites and in some maintenance shops, all but the high-speed spur-geared type are considered to be obsolete lifting devices.

Modern high-speed hand-chain hoists have relatively high mechanical efficiencies (65 to 80%) and mechanical advantages of approximately 22:1 (for low-capacity units). As hoist capacities increase, so does the mechanical efficiency as the handwheel size is changed and reduction gearing is used (see Fig. 12.3).

To suspend a load at rest and during lowering, these hoists are usually fitted with self-energizing brakes, which prevent the descending load from running away on the lowering motion. The brake enables the load to be lowered at a very slow rate and permits the operator to position it precisely.

Separate hand and load chains operate over pocket wheels that are con-

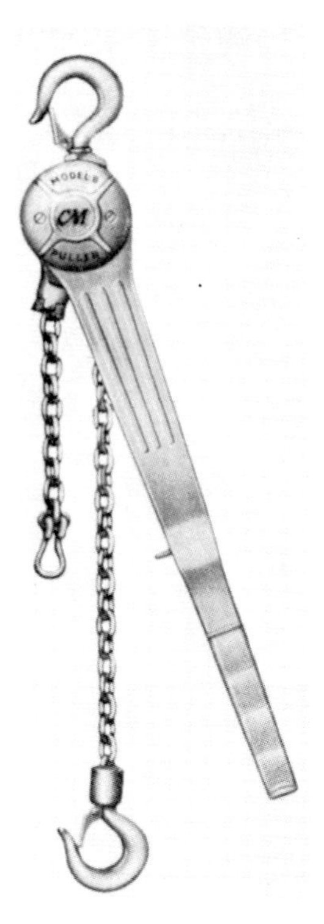

Figure 12.2 Puller or ratchet-lever hoist with link chain. (*From Higgins and Morrow,* Maintenance Engineering Handbook, *3d ed., McGraw-Hill, 1979.*)

nected by a gear train. A one-way ratchet mechanism disengages the brake during hoisting. Pulling the hand chain continuously in the reverse direction overcomes the brake torque and permits lowering the load.

All chain hoists are designed with their lower hooks as the weakest parts, the two hooks not being interchangeable. If the hoist is overloaded, the first indication is a spreading or opening of the lower hook. As designed, the inner contour of the hook is an arc of a simple circle, and any deviation from this circle is evidence of overloading. If sufficiently overloaded, the hook will gradually straighten out until it finally releases the load; yet no damage should have been done to other load-bearing parts of the hoist.

A distorted hook is positive evidence of overloading of great magnitude. All the hooks are rated at 1-ton capacity. Even an overload of 2,000 lb does not cause noticeable distortion. An overload of 3,800 lb on a 2,000-lb hook causes very little spreading. Yet it is not uncommon to see chain hoists in use with the hook opened up.

When a hook has been severely overloaded, it should be replaced by a new hook. Never attempt to force a spread hook back into shape. A new hook is too cheap to warrant taking any chances with an overloaded hook.

Figure 12.3 Typical chain hoists. (*a*) Differential. (*b*) Screw-geared. (*c*) Spur geared. (*From Higgins and Morrow,* Maintenance Engineering Handbook, *3d ed., McGraw-Hill, 1979.*)

If there is evidence of severe overloading, have the chain hoist sent to the maintenance shop for a complete internal examination and overhaul. Pay particular attention to the wear on the brake caused by excessive loading.

When inspecting a chain hoist, it is not only necessary to examine the hooks and the general appearance of the chain carefully, it is also extremely important to provide a thorough examination of the entire device. Among the many things to check, make sure that the load brake is receiving sufficient oil; otherwise the brake's holding power will be reduced. Make also sure that all hoist parts subject to stress, such as the hooks, swivels, chain, sprocket, gears, and similar parts, are made only of forged steel.

When using either a hand-chain hoist or a lever pull, it is almost impossible to estimate even roughly the load imposed on the hoist, and there is always a definite possibility of accidental overloading. However, if only one person pulls on the hand chain, or if only one person operates the lever pull (without lengthening it by means of a piece of pipe), then there is little danger of overloading the device.

Pullers, also known as come-alongs, are special lever-operated hoists that are much smaller and lighter than chain hoists of equal capacities. These devices are available in capacities from ¾ to 6 tons and have a mechanical advantage of approximately 25:1. This portable chain or wire rope hoist is used for short travel distances to lift or pull a load at any angle. A reversible ratchet mechanism in the lever permits short-stroke operation for both loading and unloading. An automatic friction brake, or releasable ratchet, holds the load securely. A load-sensing device is available that warns the operator when the puller is being loaded beyond its rated capacity.

Electric Hoists

Two types of electric hoists are currently used for high-speed repetitive lifting operations:

1. *Roller or link chain.* Lift capacities range from ⅛ to 5 tons (see Fig. 12.4).
2. *Wire rope.* Lift capacities range from ⅛ to 20 tons and lifting speeds from 2½ to 64 ft/min.

The electric hoist is essentially a drum or sprocket centered in a frame. Its drive motor is located at one end of the frame, with the motor shaft passing through or alongside the drum or sprocket. Reduction gearing at the other end of the frame connects to the drive shaft.

Such hoists are either suspended by an integral hook or bolt-type lug, or attached to a trolley that rolls on an I beam. The trolley may be the plain push type, geared hand-chain operated, or motor-driven.

Two independent braking devices—electric and spring-loaded—are used in the electric hoist. When the current is shut off, the electric brake is released, causing the spring-loaded disk brake to engage. When the current is turned on, a solenoid trips the spring system to release the brake.

Electric hoists equipped with a brake of the type used in a hand hoist rely on the motor to drive the load downward to release the brake. This type of brake generates considerable heat, which must be dissipated throughan oil-bath system. Because this brake does not act when the hoist is lifting a load, the hoist must also be provided with an auxiliaryhand-released or electrically released friction brake.

Current electric hoists have push-button-type controls instead of the now obsolete pendant control rope. Controls are of the dead-man type, stopping the hoist instantly upon release.

Safety limit switches for electric hoists control both up and down travel, as well as preventing the load hook from jamming against the bottom of the hoist or the chain from running out of the hoist. Upper-limit

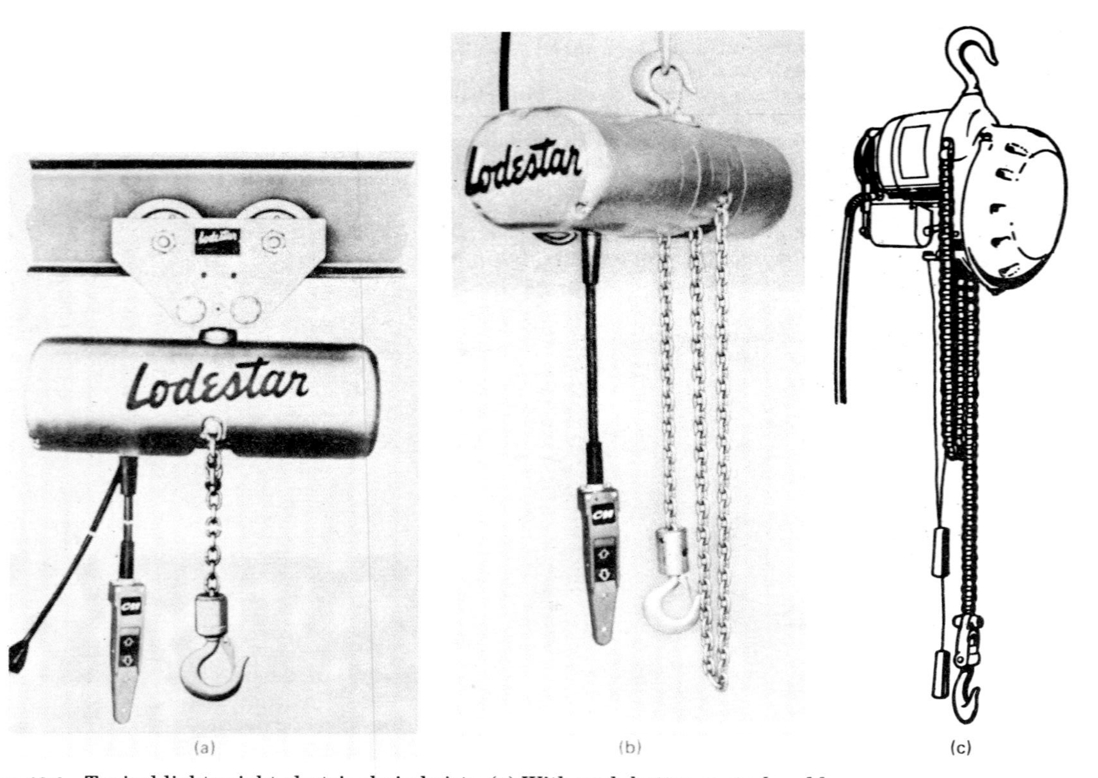

Figure 12.4 Typical lightweight electric chain hoists. (*a*) With push-button control and low headroom trolley. (*b*) With push-button control and hook suspension. (*c*) With roller chain and pendant rope control. (*From Higgins and Morrow,* Maintenance Engineering Handbook, *3d ed., McGraw-Hill, 1979.*)

switches are standard on all electric hoists, while lower-limit switches are standard on chain hoists and optional on wire rope hoists.

The hoist motor acts as a generator when lowering a load, putting current back into the line to control the lowering speed.

Electric motors for powering this type of hoist may be:

Single-phase, 115-V units that can be plugged into conventional three-prong receptacles

Three-phase, 230 V or 460 V, single-speed

Three-phase, 230 V or 460 V, two-speed

All new hoists have integral load protectors to prevent overloading.

Air Hoists

This type of hoist is similar to the electric hoist, except that it operates off an air-powered motor. Although smaller and lighter than electric hoists of equal capacity, air hoists are more expensive since they require mufflers for reasonably quiet operation and normally are fitted with automatic lubricators in the air supply. Similar to electric hoists, the air-operated devices are also available in two models:

1. *Roller or link chain.* Lift capacities to 1 and 3 tons
2. *Wire rope.* Lift capacities to 15 tons

They may also be suspended by hook or lug or attached to a trolley. An air hoist's horizontal movement usually is limited to about 25 ft because of its trailing air hose. For longer distances a runway can be fitted with a series of air ports along the system, which may be opened and closed as the hoist moves along.

An air hoist is equipped with a brake system that interlocks with the controls and automatically holds the load in neutral. Moving the control releases the brake, either mechanically or by air pressure.

Motors used for air hoists may be either rotary vane or piston types. Piston motors are more costly. However, they provide better starting and low-speed performance and are preferable for larger-capacity models.

Air hoists have infinitely variable speeds, controlled simply by an adjustment of the air valve. When severely overloaded, the air motor automatically stalls, thus preventing hoist damage. Because its motor is remotely located from the hoist itself, an air hoist is particularly suitable for safe use in an explosive atmosphere.

Preventive Maintenance, Lubrication, and Repairs

Preventive maintenance

Modern hand- and electric-chain hoists require a minimum of maintenance, provided they are used properly according to the manufacturer's recommendations. However, it is essential to provide correct and adequate routine maintenance for all types of lifting equipment to ensure safe, efficient use of such devices. Moreover, preventive maintenance is extremely important for reducing maintenance costs and improving equipment reliability and production output.

Initially the cost of a preventive maintenance program may seem high. But equipment downtime for periodic inspection only may be intolerable from a production standpoint. A key to holding down costs is a good record-keeping system for planning, scheduling, and training maintenance personnel.

Not only does preventive maintenance reduce the corrective maintenance workload, but as preventive maintenance takes over, the time of the corrective workload shifts from "when you must do it" to "when you want to do it." Thus work can be done more efficiently and at a considerably lower cost.

Preventive maintenance consists of inspecting equipment and keeping records to show wear or other deterioration of parts. Such records may include repetitive servicing, lubricating, painting, and cleaning, which form the basis of routine inspection and alert the hoist operator to the need for major overhaul before breakdown.

All hoist-operating mechanisms are subject to wear and must be inspected at regular intervals to assure that worn or damaged parts are removed from service before they become unsafe.

Preventive maintenance must take into account both the type of service for which the equipment is used, and whether the equipment is used regularly or is idle frequently. Inspection and testing procedures will depend upon these factors.

The types of service can be classified as follows:

1. *Normal.* Hoist operation with randomly distributed loads within capacity, or uniform loads up to 65% of capacity (not more than 15% for hand-operated hoists, 25% for electric- or air-powered hoists) during a single work shift.
2. *Heavy.* Hoist operation within rated capacity that exceeds normal service.
3. *Severe.* Hoist operation involving normal or heavy service in adverse environmental conditions.

Lubrication

It is essential to set up a regular lubrication schedule for all moving parts of a hoist for which lubrication is specified. Lubrication methods should be checked for proper delivery of lubricant.

Make sure to follow the manufacturer's recommendations as to the quantity and type of lubrication to be used, as well as the lubrication frequency. *Do not* substitute lubricants that are not recommended by the hoist manufacturer.

Repairs

Hoist manufacturers give assembly and disassembly instructions as well as the location and identification of all parts in their maintenance and parts manuals. When replacement parts are required, make sure to include name-plate information, especially the serial number of the hoist. Tables 12.1 to 12.3 are troubleshooting guides for various hoists.

Inspection

Inspection of hand- and electric-operated chain hoists is a critical part of preventive maintenance.

Initial inspection

Hoist users should inspect all new, altered, or repaired hoists prior to use. For hoists in regular service, inspections are classified as either frequent or periodic, with intervals of inspection dependent on the nature of the critical components of the hoist and the degree of their exposure to wear, deterioration, or malfunction. The maximum intervals between frequent and periodic inspections depend on hoist service.

Frequent inspection

This requires visual inspection by the operator or other designated person.

Normal. Monthly; no records required

Heavy. Weekly to monthly; no records required

Severe. Daily to weekly; no records required

Special (or infrequent service). Before and after operation; records required of each operation

Make sure you observe the hoist during operation for any damage that might appear between regular inspections. Carefully examine deficien-

TABLE 12.1 Troubleshooting Guide for Spur-Geared Hoists

Problem	Probable cause	Check/remedy
Hoist is hard to operate in either direction.	Load chain worn long to gauge, thus binding between liftwheel and chain guide.	Check gauge of chain. Replace if worn excessively.
	Load chain rusty, corroded, or clogged up with foreign matter such as cement or mud.	Clean by tumble polishing or solvent. Lubricate with penetrating oil and graphite. (SC-46 and SC-146)
	Load chain damaged.	Check chain for gouges, nicks, bent or twisted links. Replace if damaged.
	Liftwheel clogged with foreign matter or worn excessively, causing binding between the liftwheel and chain guide.	Clean out pockets. Replace if worn excessively.
	Hand-chain worn long to gauge, thus binding between handwheel and cover.	Check gauge of chain.
	Handwheel clogged with foreign matter or worn excessively, causing binding of chain between the handwheel and cover.	Clean out pockets. Replace if worn excessively.
	Liftwheel or gear teeth deformed.	Excessive overload has been applied. Replace damaged parts.
Hoist is hard to operate in the lowering direction.	Brake parts corroded or coated with foreign matter.	Disassemble brake and clean thoroughly. (By wiping with a cloth—not by washing in a solvent.) Replace washers if gummy, visibly worn, or coated with foreign matter. Keep washers and brake surfaces clean and dry.
	Chain binding.	See first three check/remedy items in this table.
Hoist is hard to operate in the hoisting direction.	Chain binding.	See first three check/remedy items in this table.
	Chain twisted. (3-ton capacity and larger)	Rereeve chain, or on 3- and 4-ton unit, if both chains are twisted, capsize hook block through loop in chain until twists are removed. Caution—do not operate unit in hoisting direction with twisted chain or serious damage will result.
	Overload.	Reduce load or use correct capacity unit.
Hoist will not operate in either direction.	Liftwheel gear key or friction hub key missing or sheared.	Install or replace key.
	Gears jammed.	Inspect for foreign material in gear teeth.

TABLE 12.1 Troubleshooting Guide for Spur-Geared Hoists. *(Continued)*

Problem	Probable cause	Check/remedy
Hoist will not operate in the lowering direction.	Locked brake due to a suddenly applied load, shock load, or load removed by other means than by operating unit in the lowering direction.	With hoist under load to keep chain taut, pull sharply on hand chain in the lowering direction to loosen brake.
	Chain binding.	See first three check/remedy items in this table.
	Lower hook all the way out. Load chain fully extended.	Chain taut between the lift wheel and loose end screw. Operate unit in hoisting direction only.
Hoist will not operate in the hoisting direction.	Chain binding.	See first three check/remedy items in this table.
Hoist will not hold load in suspension.	Lower hook or load side of chain on wrong side of liftwheel.	Lower hook must be on same side of liftwheel as upper hook. Refer to assembly. Rereeve chain.
	Ratchet assembled in reverse.	Ratchet must be assembled as shown.
	Pawl not engaging with ratchet.	Pawl spring missing or broken pawl binding on pawl stud. Replace spring and clean so pawl operates freely and engages properly with ratchet. Do not oil.
	Ratchet teeth or pawl worn or broken.	Replace pawl and/or ratchet.
	Worn brake parts.	Replace brake parts which are worn.
	Oily, dirty, or corroded brake friction surfaces.	Disassemble brake. Clean thoroughly. (By wiping with a cloth—not by washing in a solvent.) Replace washers if gummy, visibly, worn, or coated with foreign matter. Keep washers and brake surfaces clean and dry.

SOURCE: From L. R. Higgins, *Handbook of Construction Equipment Maintenance,* McGraw-Hill, 1979.

cies and determine whether they constitute a safety hazard. Inspect the following items for damage at the above specified intervals:

Braking mechanism. For evidence of slippage under load

Load chain. For lubrication, wear, or twists; broken, cracked, or otherwise damaged links; and deposits of foreign material that might be carried into the hoist mechanism

TABLE 12.2 Troubleshooting Guide for All-Electric Hoists

Problem	Probable cause	Check/remedy
Hook does not respond to the control station.	No voltage at hoist—main line or branch circuit switch open; branch line fuse blown or circuit breaker tripped.	Close switch, replace fuse or reset breaker.
	Phase failure (single phasing, three-phase unit only)—open circuit, grounded or faulty connection in one line of supply system, hoist wiring, reversing contactor, motor leads or windings.	Check for electrical continuity and repair or replace defective part.
	Upper or lower limit switch has opened the motor circuit.	Press the "other" control and the hook should respond. Adjust limit switches.
	Open control circuit—open or shorted winding in transformer, reversing contactor coil or speed selecting relay coil; loose connection or broken wire in circuit; mechanical binding in contactor or relay; control station contacts not closing or opening.	Check electrical continuity and repair or replace defective part.
	Wrong voltage or frequency.	Use the voltage and frequency indicated on hoist identification plate. For three-phase dual-voltage unit, make sure the connections at the conversion terminal board are for the proper voltage.
	Low voltage	Correct low-voltage condition.
	Brake not releasing—open or shorted coil winding; armature binding.	Check electrical continuity and connections. Check that correct coil has been installed. The coil for three-phase dual-voltage unit operates at 230 volts when the hoist is connected for either 230-volt or 460-volt operation. Check brake adjustment.
	Excessive load.	Reduce loading to the capacity limit of hoist as indicated on the identification plate.
Hook moves in the wrong direction.	Wiring connections reversed at either the control station or terminal board (single-phase unit only).	Check connections with the wiring diagram.
	Failure of the motor reversing switch to effect dynamic braking at time of reversal (single-phase unit only).	Check connections to switch. Replace a damaged switch or a faulty capacitor.
	Phase reversal (three-phase unit only).	Refer to installation instructions.

TABLE 12.2 Troubleshooting Guide for All-Electric Hoists (*Continued*)

Problem	Probable cause	Check/remedy
Hook lowers but will not raise.	Excessive load.	Reduce loading to capacity limit of hoist as indicated on the identification plate.
	Open hoisting circuit—open or shorted winding in reversing contactor coil or speed selecting relay coil; loose connection or broken wire in circuit; control station contacts not making; upper limit switch contacts open.	Check electrical continuity and repair or replace defective part. Check operation of limit switch.
	Motor reversing switch not operating (single-phase unit only).	Check switch connections and actuating finger and contacts for sticking or damage. Check centrifugal mechanism for loose or damaged components. Replace defective part.
	Phase failure (three-phase unit only).	Check for electrical continuity and repair or replace defective part.
Hook raises but will not lower.	Open lowering circuit—open or shorted winding in reversing contactor coil or speed selecting relay coil; loose connection or broken wire in circuit; control station contacts not making; lower limit switch contacts open.	Check electrical continuity and repair or replace defective part. Check operation of limit switch.
	Motor reversing switch not operating (single-phase unit only).	Check switch connections and actuating finger and contacts for sticking or damage. Check centrifugal mechanism for loose or damaged components. Replace defective part.
Hook lowers when hoisting control is operated.	Phase failure (three-phase unit only).	Check for electrical continuity and repair or replace defective part.
Hook does not stop promptly.	Brake slipping.	Check brake adjustment.
	Excessive load.	Reduce loading to the capacity limit of hoist as indicated on the identification plate.
Hoist operates sluggishly.	Excessive load.	Reduce loading to the capacity limit of hoist as indicated on the identification plate.
	Low voltage.	Correct low voltage condition.
	Phase failure or unbalanced current in the phases (three-phase unit only).	Check for electrical continuity and repair or replace defective part.
	Brake dragging.	Check brake adjustment.

TABLE 12.2 Troubleshooting Guide for All-Electric Hoists (*Continued*)

Problem	Probable cause	Check/remedy
Motor overheats.	Excessive load.	Reduce loading to the capacity limit of hoist as shown on the identification plate.
	Low voltage.	Correct low-voltage condition.
	Extreme external heating.	Above an ambient temperature of 104°F, the frequency of hoist operation must be limited to avoid overheating of motor. Special provisions should be made to ventilate the space or shield the hoist from radiation.
	Frequent starting or reversing.	Avoid excessive inching, jogging, or plugging. This type of operation drastically shortens the motor and contactor life and causes excessive brake wear.
	Phase failure or unbalanced current in the phases (three-phase unit only).	Check for electrical continuity and repair or replace defective part.
	Brake dragging.	Check brake adjustment.
	Motor reversing switch not opening start winding circuit. (Single-phase unit only.)	Check switch connections and actuating finger and contacts for sticking or damage. Check centrifugal mechanism for loose or damaged components. Replace defective part.
Hook fails to stop at either or both ends of travel.	Limit switches not opening circuits.	Check switch connections, electrical continuity, and mechanical operation. Check the switch adjustment. Check for a pinched wire.
	Shaft not rotating.	Check for damaged gears.
	Traveling nuts not moving along shaft—guide plate loose; shaft or nut threads damaged.	Tighten guide plate screws. Replace damaged part.
Hook stopping point varies.	Limit switch not holding adjustment.	Check switch connections, electrical continuity, and mechanical operation. Check switch adjustment. Check for pinched wire. Check for damaged gears. Tighten guide plate screws. Replace damaged part.
	Brake not holding.	Check the brake adjustment.

SOURCE: From L. R. Higgins, *Handbook of Construction Equipment Maintenance,* McGraw-Hill, 1979.

TABLE 12.3 Troubleshooting Guide for Two-Speed Electric Hoists

Problem	Probable cause	Check/remedy
Hoist will not operate at slow speed in either direction.	Open circuit.	Open or shorted motor winding loose or broken wire in circuit, speed-selecting contactor stuck in opposite speed mode. Replace motor, repair wire, and/or repair speed-selecting contactor.
	Phase failure.	Check for electrical continuity and repair or replace defective part.
Hoist will not operate at fast speed in either direction.	Open circuit.	Open or shorted motor winding, loose or broken wire in circuit, speed-selecting contactor stuck in opposite speed mode. Replace motor, repair wire, and/or repair speed-selecting contactor.
Hook will not raise at slow speed.	Open speed-selecting circuit.	Open or shorted winding in speed-selecting contactor coil. Loose connection or broken wire in circuit. Mechanical binding in contactor. Control station contacts not making or opening. Replace coil; repair connection, contactor, or control station.
Hook will not lower at slow speed.	Phase failure.	Check for electrical continuity and repair or replace defective part.
Hook will not raise at fast speed.	Excessive load.	Reduce loading to capacity limit of hoist as indicated on the identification plate.
	Phase failure.	Check for electrical continuity and repair or replace defective part.
	Brake not releasing.	Check electrical continuity and connections. Check that correct coil has been installed. The coil for three-phase dual-voltage unit operates at 230 volts when the hoist is connected for either 230- or 460-volt operation. Check brake adjustment.
Hook will not lower at fast speed.	Phase failure.	Check for electrical continuity and repair or replace defective part.
	Brake not releasing.	Check electrical continuity and connections. Check that correct coil has been installed. The coil for three-phase dual-voltage unit operates at 230 volts when the hoist is connected for either 230- or 460-volt operation. Check brake adjustment.
Hook moves in proper direction at one speed—wrong direction at other speed.	Phase reversal.	Wiring reconnected improperly. Interchange two leads of motor winding that is out of phase at the speed-selecting relay.

SOURCE: From L. R. Higgins, *Handbook of Construction Equipment Maintenance*, McGraw-Hill, 1979.

Load-bearing ropes. For wear, twists, distortion, or improper deadending to the hoisting drum and other attachments

Hooks. For deformation, chemical damage, or cracks (Replace such hooks if damage indicates that throat opening is more than 15% in excess of normal, or the plane of the hook has more than a 10° twist from the unbent position.)

Electric controls. For improper operation

Safety devices. For malfunction

Air hoist systems. For deterioration or leakage

Periodic inspection

This requires visual inspection by designated persons, keeping records of apparent external conditions to provide the basis for a continuing evaluation.

Normal. Equipment in place, yearly

Heavy. Same as for frequent inspections, unless external conditions indicate that disassembly should be done to permit detailed inspection semiannually

Severe. Same as for frequent inspection, except quarterly

Special (or infrequent service). Before the first such period of service, and as directed by the qualified individual for any subsequent special or infrequent period of service

Depending on hoist activity, severity of service, and environment, make a complete inspection of the hoist at the above specified intervals. Include all requirements of frequent inspections and, in addition:

Carefully examine any deficiencies and determine whether they constitute a safety hazard.

Check for external evidence of wear of chain, load sprockets, idler sprockets, and handwheel pockets for chain stretch.

Check rope, drums, and sheaves for wear.

Inspect for loose hook-retaining nuts, collars, pins, welds, or riveting that secure the retaining members.

Check for cracked hooks using dye penetrant, magnetic particles, or other suitable crack-detecting method at least once a year.

Examine brake mechanism for worn, glazed, or oil-contaminated friction disks; worn pawls, cams, or ratchets; and corroded, stretched, or broken pawl springs.

Look for worn, cracked, or distorted parts, such as hook blocks, suspension housing, outriggers, hand-chain attachments, clevises, yokes, suspension bolts, shafts, gears, bearings, drums or sheaves, pins, rollers, and locking and clamping devices.

Check for loose bolts, nuts, or rivets.

Inspect the supporting structure and trolley for continued ability to support the imposed loads.

Make sure the warning label on proper use of hoist is attached to the device and clearly legible from the operator's position.

Check for excessive wear on motor or load brakes, signs of pitting in electric apparatus, or any deterioration of controller contactors, limit switches, and push-button stations.

Idle-hoist inspection

Hoists not in regular use must be checked carefully before they are put back into service to meet the requirements of frequent and periodic inspections.

A hoist that has been idle for a period of 1 month or more, but less than 6 months, must be inspected by or under the direction of a designated person.

A hoist that has been idle for a period of 6 months and longer should be completely inspected.

Make sure that all inspection reports and records on critical items, such as brakes, hooks, and chains, are written, dated, and signed by the designated person, and that they include the time intervals specified in the inspection classification.

Operating Tests

All hoists

All new hoists should be tested by the manufacturer for proper operation and load-carrying capability. Before using any altered or repaired hoists, or those that have not been used within the preceding 12 months, make sure that they are tested for operation and load-carrying capability by or under the direction of a qualified person.

Test loading of all hoists and all load-sustaining parts should be 125% of the rated load. On hoists that have overload devices to prevent lifting 125% of rated load, apply at least 100% of rated load; then test the functions of the overload device.

For all hoists, hand-chain or electric-powered, check the following:

All hoist functions, including hoisting and lowering with the hoist suspended in the unloaded state

Operation of hooks by trip-setting of limit switches and limiting devices under no-load conditions; first by hand, if practical, then under the slowest speed obtainable, and finally with increasing speeds up to the hoist's maximum speed

Actuating mechanisms to be sure they will trip the switches or limiting devices sufficiently fast to stop motion without any part of the hoisting arrangement being damaged

All anchorages or suspension, or both, to ensure safe operation of hoist

Electric and air hoists

Make sure that every hoist is permanently marked with name and address of the manufacturer, manufacturer's unit identification number, and for:

Electric hoists. Voltage, frequency, and phase of power supply

Air hoists. Rated air pressure in pounds per square inch

Check the following:

Controls. Each element should clearly indicate the direction of resultant motion.

- Make sure that when controls are released, except when in automatic-cycling, they return to the OFF position and the hook motion stops instantly.
- Make sure controls interlock mechanically or electrically, or both, to prevent simultaneous actuation.
- Make sure that the voltage at pendant push buttons does not exceed 150 V for alternating current and 300 V for direct current.

Hoist rope. Use only hoist rope that is of a recommended construction for hoist service.

- Make sure that the rated load of the rope divided by the parts of rope used in the hoist does not exceed 20% of the nominal breaking strength of the rope. Make sure all rope ends are securely attached to the hoist to prevent their becoming detached during hook travel.
- Make sure the tension in rope parts is equalized when a load is supported by more than one part of the rope.

Braking system. A hoist's brakes should stop promptly and hold the load hook when supporting loads (up to 125% of rated load) upon release of the controls.

- Make sure hoist brakes can slow down load descent by at least 120% of the rated lowering speed for the load being handled.
- Make sure the braking system has ample thermal capacity depending on the operating frequency required by the service.
- Make sure the hoist brakes can be adjusted when necessary to compensate for wear.

Hooks. These should be sufficiently ductile so that they open noticeably before the hook fails under an overload.

Overtravel protection. Make sure the hoist is so designed and constructed that the load hook, whether loaded or empty, cannot exceed the upper limit of travel. Conversely, no hoist should be installed where the loaded hook can be lowered beyond the rated hook travel, unless the hoist is equipped with a lower-limit device.

Power failure protection. Make sure that if the power supply (air or electric) is interrupted, even partially, during operations, the load motion does not get out of control.

Power supply. Make sure all electric hoists are connected to the power system and are grounded in accordance with the National Electrical Code, ANSI C1.

- Make sure all air hoists are securely connected to air supply lines that have proper pressure and capacity.
- Make sure all fittings and devices are solidly secured and tested before operating the hoist.

Direction of motion. Make sure all electric hoist motors are connected to power lines so that hook motion agrees with control markings. DO NOT change internal connections in the hoist or control wiring to accomplish this.

- Make sure all air hoists are connected to the air supply lines so that travel motion agrees with control markings.

Operating Instructions

Prior to the start of normal operations, the hoist operator should test all limit switches, brakes, and other safety devices. Report any failure immediately and remove the equipment from service until repairs can be made. Each operator is responsible for performing the frequent inspection of a hoist.

The hoist operator is responsible for the safe operation of the equipment. During normal operation, if the equipment fails to respond properly, it is up to the operator to report the failure.

Remember: The hoist operator is the one responsible for the safe operation of the equipment.

- Make sure each operator is familiar with the hoisting equipment being used, and its proper care. *Allow only* qualified persons to operate electric hoists:

  Operators who have passed a practical operating examination on the specific equipment being used

  Maintenance and test crews in performance of their duties

  Hoist inspectors

- Make sure each operator reports to a designated supervisor any need for hoist adjustments or repairs.
- Make sure each operator notifies the next operator, upon changing shifts, of the hoist's condition.
- Test all hoist controls before beginning a shift, and adjust or repair any controls that do not work properly, before operating the hoist.
- Make sure every hoist is permanently marked with its rated load and, on hand hoists, that the markings are clearly legible from the operating position.
- Make sure the operator has safe access to the hoist and all persons are clear of the area before using a chain hoist or starting an electric or air hoist.
- NEVER load a hoist beyond its rated capacity, except for properly authorized and supervised tests.
- NEVER wrap hoisting chain or rope around the load.
- ALWAYS attach the load to the hook by means of slings or other approved devices.
- Make sure slings or lifting devices are properly seated in the saddle of the hook before operating the hoist.
- Make sure hoisting rope or chain is properly seated on the hoist drum, sheaves, or sprockets.

- NEVER move or lift a load more than a few inches until it is well balanced in a sling or lifting device.
- Make sure hoist ropes or chains are not kinked or twisted.
- DO NOT operate the hoist until it is centered over the load.
- Make sure the load does not touch any obstructions during the lifting operation.
- NEVER use an overhead hoist for handling people, unless the manufacturer specifically recommends such use and indicates this on a permanent nameplate attached to the hoist.
- Avoid making lifts or carrying loads over people.
- ALWAYS test hoist brakes each time you handle a load approaching the rated load. Raise the load sufficiently to clear the floor or its supports, then check brake action; continue with lift only when assured that the braking system is operating properly.
- NEVER leave a load suspended in air at the end of a work shift or for extended periods during operations, unless you take specific precautions to provide protection and properly secure the operating chain or rope.
- NEVER rotate a loaded hoist drum in the lowering direction beyond the point where less than two wraps of rope remain on the drum. If the drum is fitted with a lower-limit device, at least one wrap of rope should remain on the drum.
- NEVER use the hoist's upper-limit device as normal operating controls.

Inspection checklist

Make sure hoist or lift block displays a label that is clearly legible from the operating position, containing the following instructions, to bring label to attention of operator:

WARNING DO NOT

- Lift more than the rated load
- Operate hoist with twisted, kinked, or damaged chain or rope
- Operate damaged or malfunctioning hoist
- Lift people and loads over people
- Operate chain hoist with other than manual power
- Remove or obscure WARNING label.

•Make sure that:

Hand chain. Has proper shape and pitch to assure chain fits the hand-chain wheel without binding; has a guard to prevent disengagement of the chain from the hand-chain wheel; and is capable of withstanding, without distortion, a pull of three times the pull required to lift the rated load.

Load chain. Is accurately pitched so that it passes smoothly over all load sprockets without binding; has been proof-tested by the chain or hoist manufacturer with a load equivalent to at least 150% of the hoist's rated load divided by the number of chain parts supporting the load; and can be restrained in its fully extended position before it completely runs out of the hoist. The restraint should be such that the unloaded hoist can withstand a lowering hand-chain force of twice the pull required to lift the rated load.

Load sprockets. Have accurately formed pockets or teeth so that load chain is properly engaged; are guarded to minimize entrance of foreign objects; and permit chain to release freely and thus prevent jamming within the hoisting mechanism.

Lower hook. Can rotate freely through 360° when supporting a rated load.

Hooks. Are retained in their housing by positively secured lock nuts, collars, or other suitable devices that will prevent them from working loose.

Latch-type hooks. Are used, unless the use of such hooks increases the hazard of hoisting.

Bottom blocks. Are guarded against load chain jamming when used as recommended.

Braking system. Stops when actuating force is removed, and holds up to 125% of the rated load.

Supporting structure, including trolley or monorail. Has a load rating at least equal to that of the hoist.

Installation. Is such that the operator stands free of the load at all times, and that the location is not hazardous to the operator if grip or footing is lost while operating the hoist.

Chapter

13

Personnel/Material Hoists

Towers for platform-type personnel/material hoists are constructed of tubular steel similar to tubular scaffolding, but of much heavier members. These towers can be erected to great heights, alongside tall buildings. But they must be properly guyed or anchored to the building.

One point to keep in mind: diagonal bracing should be used to form triangles on all four sides of a tower. This includes both loading and unloading sides, where bracing is commonly omitted to permit passage or handling of material at the landings.

Where cross bracing is impractical, provide knee braces at the upper corners of the landing openings to afford at least some rigidity.

Two basic types of motor-driven hoists are in use: rack-and-pinion hoists and hydrostatic transmission hoists with wire lift rope.

Rack-and-pinion-type tower hoists are normally shipped as preassembled units consisting of base frame, cage, and all necessary power and control components. Hydrostatic types come with a separate power pack containing prime power (electric motor or gas engine), hydraulic pump with control valve, and the necessary electric controls.

For proper layout, a concrete foundation pad should be poured before the hoist is delivered to the site. Before erecting a hoist on site, carefully consider where to spot the tower, keeping in mind its accessibility as well as its position relative to the building. Make sure that there is sufficient clearance from the building, as well as from other obstructions such as at all landing levels and columns.

As the hoist is erected, attach the tower to the structure by special telescoping tie-in assemblies that lock into wall anchors. These should be placed up the side of the building at 25-ft intervals (20 ft for tower guide rails) (see Fig. 13.1).

Each tie-in requires only two mounting plates on the structure, located on either a vertical (wall) or a horizontal (floor) surface. The tie-in length is infinitely adjustable between minimum and maximum specifications.

If a crane is used to load or unload tower sections from their carrier, the pickup point is the hook eye, located inside and at the bottom of the base tower.

If a fork lift is used, pickup is at the same point, but a short cable should be looped over the tower section and the lift fork.

Important. Always check a tower hoist as soon as it is received at the site for any damage that may have occurred in shipment. Report any

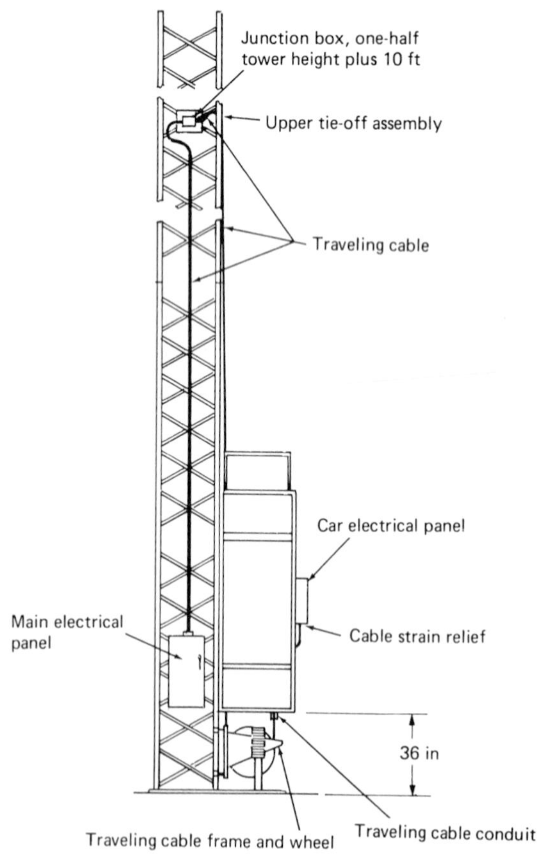

Figure 13.1 Typical hoist arrangement. (*Courtesy of Champion Hoist Company.*)

damage observed to the shipper immediately. Check also for any shortage of parts or components against the packing slip.

Rack-and-Pinion Hoist

If the tower is to be raised without a crane, using a stiffleg and hoist gin pole:

Place the tower base section on the pad and bolt it in place using only the anchor bolts located within the tower. Secure the anchor bolts under the cage later.

Level the tower, grouting it as required, and tighten the anchor bolt securely.

Connect the traveling cable to the main power panel and the cage control panel. Leave the wheel assembly with the cable on the ground so that the cable can be played out during erection.

Place the aligning pin in the hole at the top of the gear rack mounted on the base tower.

Bolt the lifting bale to the top of one tower section and, using a gin pole, hoist the first tower section atop the base section. Fasten the two sections with bolts.

Make the first wall tie at a point no farther than 25 ft from the ground. Position additional wall ties no farther apart than 25 ft.

Plumb the tower during erection, making necessary wall-tie adjustments to maintain a maximum $\pm\frac{1}{2}$ in. from the true vertical.

If the tower is to be raised with a crane, install the base section as described previously. With a crane three or four sections can be bolted together and then lifted into place as a unit. Exercise extreme caution when lifting bolted sections, especially if lifting more than four at one time, because the load can become too great for the lifting bale. Use slings attached to the lower legs when lifting more than four sections as a unit.

When jumping tower sections, make sure to follow proper wall-tie procedures. Never hoist the cage top more than 5 ft above the uppermost wall tie, except when jumping—and then never more than 12 ft above the uppermost wall tie. The tower hoist should never extend more than 24 ft above the uppermost wall tie.

Operation

After having erected the tower base section and connected the hoist's traveling cable, check to see that the oil level in the electric motor reducer is correct.

If a building gate interlock system or ground enclosure is used, it must be plugged into its proper receptacle. If such a system is not used, the jumper furnished must be plugged into its receptacle.

Connect the hoist to the proper electric supply and turn on the main power panel.

Check the disk of the phase-reversal relay and make sure that the electric hookup is in phase, allowing the motor to run in the proper direction. Detach the UP/DOWN control switch from its mounting plate on top of the cage so that the cage can be operated from the ground.

Pull the red mushroom switch to the ON position and close the gates. Check that all limit switches are electrically CLOSED. Standing outside the cage, run the cage up 3 or 4 ft. Turn off the main power switch.

Place a block under the cage, of sufficient size and strength to support the cage, so that it cannot possibly be lowered. Then place the buffer support bracket in position and the springs into holders. Place and tighten the remaining base anchor bolts. Finally remove the blocks, turn power ON, and lower the cage to a point several inches above the buffer springs.

Continue erecting the hoist tower sections. Carry the UP/DOWN control switch to the top of the cage and run the unit up high enough so that another section can be added to the tower. Be sure to insert aligning pins in each section. Repeat this procedure until the tower is erected to the desired height, with wall ties attached at proper intervals. Use a surveyor's transit to ensure true vertical alignment during erection.

Check out the travel limit switches and controls according to the manufacturer's instructions.

Safety check points

Check all safety points before and after the first day of operation and weekly thereafter.

Check motor safety.

Check running of unit (power, smoothness).

Check oil level in gear reducer.

Lubricate as necessary.

Instruct operator and service and maintenance personnel in proper hoist operation and maintenance.

Make sure each morning, or at the start of a new shift, that the operator makes a solo run before attempting to carry personnel, checking for safe and proper hoist operation.

TABLE 13.1 Safety Check Schedule

Check	Inspect for	Frequency
Electrical connectors	Loose connectors or clamp	Weekly
Electric cable	Severed insulation, broken wires	Daily
Traveling cables	Severed insulation, kinking, and free movement over sheaves and through guides	Daily
Limit-switch arms and rollers	Arms locked firmly to switch shafts, rollers move freely	Daily
Pinion gear and rack	Proper pinion and rack clearance	Weekly

Check the following items to make sure they are securely tightened: anchor bolts, tower bolts, wall-tie bolts, traveling cable, and guides and sheaves.

Table 13.1 is an example of a safety check schedule. Figures 13.2 and 13.3 show a hoist at work.

Overspeed or drop test

This test must be performed before placing the hoist in service and each time the tower height is increased. A drop test is a safe and simple

Figure 13.2 Hoist at work on tall building. (*Courtesy of Champion Hoist Company.*)

Figure 13.3 Arrangement of hoist cab and entry. (*Courtesy of Champion Hoist Company.*)

procedure, but it should only be performed by an erection supervisor or a qualified operator.

The overspeed device is designed to function when the cage reaches a downward speed in excess of 160 ft/min. To perform the test:

Load the cage with weight as prescribed by local safety codes. If no code exists, use a load equal to the specified capacity of the hoist.

Release the brakes slowly, allowing the cage to rest on the buffer springs. When the cage comes to rest on the springs, release the brakes fully.

Remove the UP/DOWN control switch from the top of the cage.

Turn power on.

While standing on the ground outside the cage, run the hoist up as far as the control cable will allow. Release the control to permit the cage to fall free. After the cage has fallen a few feet, the overspeed governor should become activated, catching the cage and locking it to the tower. If the governor does not catch after the cage has fallen 5 to 6 ft, *immediately* return the control to the UP position. The motors will stop the cage. As soon as the cage stops falling, release the control and allow the cage to fall a few more feet. Repeat this sequence until the cage rests on the buffer springs.

Check for, and locate, the cause of the governor malfunction before repeating the test.

Reset the hoist brakes before putting the hoist into operation.

Maintenance

Normal routine maintenance procedures apply to the hoist as they would to any similar piece of equipment. Make sure to check the motor pinion and cage rollers every week. Follow the manufacturer's instructions for the hoist lubricating schedule. An example is given in Table 13.2.

Troubleshooting

If problems arise, first isolate each problem area. Mechanical problems are usually accompanied by noise. No evidence of action or movement usually indicates electrical problems. After you isolate the problem area, refer to the manufacturer's troubleshooting chart or other available source data. Table 13.3 gives an example chart for electrical troubleshooting of rack-and-pinion hoists.

Hydrostatic Drive (Wire Rope)

Installation of this type of hoist tower follows the same procedures as those outlined for the rack-and-pinion type, except that after the tower base has been securely attached to the foundation pad, the hoist platform must be rigged up. Place the wire rope spool on the platform and reeve the cable according to the manufacturer's instructions.

Next install the power pack in an area that will not interfere with hoist operation. Run the hydraulic lines from the power pack to the base unit and make sure that quick disconnects are firmly seated.

TABLE 13.2 Lubrication Schedule

Item	Schedule
Gear reducer	Keep to proper level, change yearly
Rack	On erection and monthly, or as required by use
Erection jib gear box	Change yearly
Erection jib pulley, etc.	As required
Overspeed governor	Weekly (fitting provided)
Gate rollers	As required

TABLE 13.3 Electrical Troubleshooting of Rack-and-Pinion Hoists

Indication	Reason	Cause	Remedy
Green indicator lamp on cage does not light	Not receiving power	Check incoming power ON/OFF switch, and disconnect switch, fuses, and emergency stop switch; check gate limit switches; check contactor operation	
Motor will not stop	Not receiving power	Gate and overspeed switches open; final limit switch open	Check slide gate, split gate, overspeed switch
Car will not go up	Ac contactor not energized	Gate switch open; UP switch in hand control defective	Replace UP switch in hand control
Car will not go down	Ac contactor not energized	Gate switch open; DOWN switch in hand control defective	Replace DOWN switch in hand control
Safety mechanism activated	Governor switch tripped open	Reset safety mechanism after ascertaining that any danger that had activated the mechanism has been identified and remedied	

NEVER attempt to run the power pack unless all hydraulic hoses are properly connected. Failure to do so may cause a hose to burst at the extreme pressure that a totally confined pump is capable of producing.

If a gasoline engine powers the system, be sure the engine crank case is filled to its proper level with oil and that the fuel tank is filled with a good grade of gasoline, as specified in the manufacturer's engine manual.

Check the hydraulic fluid level in the upper tank. If it is not within 2 in. of the filler cap, fill with fluid as specified on the tank.

Check that both shut-off valves beneath the upper tank are fully opened.

Operation

Start up the engine and observe the vacuum gage. If the gage reading exceeds 10 in.Hg, allow the unit to idle until the reading drops below 10. If the reading is greater than 25 in.Hg, shut off power immediately and check for the cause.

Slowly wind the rope onto the drum, lifting the platform to take up slack. Be sure to straighten the rope on the drum, keeping each wrap held tightly against the previous wrap. As the platform rises, increase the drum speed by rotating the speed control clockwise.

NEVER drive the platform downward into the base structure. This will only cause the rope to unspool and tangle.

If there is a second platform on the hoist tower, repeat startup steps for the other platform.

Assemble the cage enclosure according to the manufacturer's instructions.

Testing

To run a broken-rope safety test:

Slowly run the platform down to the bottom of the tower and allow the lifting rope to go slack.

Visually check to see that the unit's safety shoes lift and engage tower legs. If the shoes do not engage, consult the manufacturer's troubleshooting chart and correct before proceeding.

Slowly and carefully take the slack out of the rope and run the platform sufficiently high to clear the bottom structure. Shoes should now drop free of the tower legs.

Place a test load equal to the hoist-lifting capacity on the platform.

Raise the load 15 to 20 ft and tie the platform to the tower, using the locking chain.

Ease the platform down until it is supported by the locking chain, then continue to run it down until you are able to pull a loop into the lifting rope.

Splice a short length of ½-in. rope in the lifting rope loop and tie securely with fist-grip clamps. DO NOT use regular cable clamps; USE ONLY fist-grip-type clamps.

Run the platform up until the locking chain can be removed and the load is supported by the lifting rope.

Hold the spliced loop of the rope and, using an oxyacetylene torch, cut the short length of the spliced rope. The platform should lock to the tower within a few inches of fall.

Remove ends of the cut rope from the lifting rope and being careful to keep layers straight, spool the rope onto the drum. Then lift the platform to release the shoes from the tower.

Lower the platform and remove the test load.

Maintenance

Hoist maintenance. Before beginning operation each day, inspect the hoist by easing the platform onto the bottom structure and allowing the lifting rope to go slack.

Check the following:

The unit's safety shoes to ensure that they engage the tower legs.

Shoes and linkage of brake rope safeties to ensure that they are clean, free, and properly lubricated.

Deflector and cathead sheaves with rope still slack, turning by hand to determine freedom of rotation, wear, and any condition that may damage the rope. Correct any of these conditions immediately.

Fist-grip clasps for tightness.

Rollers on upright carriage to determine freedom of rotation or wear. Frozen bearings should be replaced and signs of wear should be checked and corrected immediately.

Drum for proper rope winding, by running the unit up to remove slack from the lifting rope.

Rope for obvious defects, by running the platform to the top.

Repeat procedures for the other platform.

System maintenance

Fluid. In general, a fluid change performed annually or after 2,000 hours of operation, whichever occurs first, is adequate with a sealed reservoir system. A more frequent fluid change is required if the fluid has become contaminated by water or other foreign material or has been subjected to abnormal operating conditions.

Reservoir. The reservoir should be checked daily for the proper fluid level and the presence of water in the fluid. If fluid must be added to the reservoir, use only filtered or strained fluid. Drain any water as required.

Hydraulic lines and fittings. Visually check daily for any fluid leakage. Tighten, repair, or replace as required.

Heat exchanger. If used, the heat exchange core and cooling fins should be kept clean at all times for maximum cooling and system efficiency. Inspect daily for any external blockage and clean as required.

Dismantling

The dismantling of a hoist tower is essentially the reverse procedure of erecting the tower. However, when dismantling tower sections, it is extremely important to follow recommended safety procedures, such as maintaining proper wall ties throughout the operation. *Always* chain the platform to the tower while removing sections, and *never* ride the platform during tower dismantling.

Troubleshooting

Troubleshooting for a hydrostatic-drive-type hoist may be divided into three categories:electrical, hydraulic, and mechanical. If a problem arises, isolate the problem area and correct according to the manufacturer's instructions. Table 13.4 (pp. 328 to 334.) is a comprehensive troubleshooting chart.

TABLE 13.4 General Troubleshooting

Indication	Cause	Remedy
	A. General	
1. Electric motor will not start		
a. Not receiving power		Check incoming power ON/OFF switch, disconnect-switch fuses, and emergency stop switch.
b. Phase reversal relay not on	Improper phase	Switch two of incoming power lines.
c. Motor contactor not closed	Coil burned out	Replace.
	Transformer	Replace.
	Transformer fuse	Replace.
2. Gasoline engine will not start (Unit must never be run when filter gage reads above 10, except in extreme cold weather. Under such circumstances, do not allow unit to run more than 2 min before reading drops below 10. If necessary, adjustments should be made for cold-weather operation.)	Check emergency button. Consult engine manual furnished with hoist.	
a. High reading on filter gage	Shutoff valve closed	Open fully.
	Blockage in oil takeup circuit	Remove blockage.
	Excessively dirty oil filter	Close shutoff valve and replace filter cartridge. *BE SURE TO REOPEN VALVE BEFORE STARTING ENGINE.*
	Extreme cold weather	If prime mover is: i. *Gasoline engine,* remove fan blade. This will direct engine heat to oil heat exchanger to maintain warmer hydraulic oil temperatures. ii. *Electric motor,* disconnect fan (when used) and blanket oil cooler radiator to maintain warmer hydraulic oil temperatures.

3. Platform/carriage will not go up		
a. Pump not being activated	Control valve not electrified	See electrical troubleshooting in equipment manual.
	Valve electrified but not stroking (listen for whine in pump)	If no whine, pump is not stroking.
b. Hydraulic motor improperly fed	Quick disconnect not fully seated	Reseat firmly.
c. No output from pump	Blockage in oil uptake	Open any restriction and check pump for damage.
	Damaged motor	Repair or replace.
d. Platform overloaded	Pressure-relief valve activated to relieve overload	Reduce platform load.
4. Platform goes up too slowly or will not lift rated capacity		
a. Brake not releasing	Brake adjusted too tight	Readjust so that brake just holds full load.
b. Lifting cable is binding	Lifting rope off sheaves	Replace rope on sheave and check for damage to rope and sheave.
	Sheave frozen	Repair or replace sheave; check for damage to rope.
c. Insufficient pump output	Engine speed insufficient	See performance chart for proper speed.
	Throttle valve defective	Replace.
	Pump defective	Locate and repair cause.
	Pump being starved (should show high reading on filter gage)	Immediately shut down system. Be sure shutoff valve is fully opened. Problem is in takeup portion of system; continued operation will damage pump.
d. Pump not fully stroked	Speed control set too low	Readjust.
	Pump control valve receiving low voltage	Check voltage.
e. Brake dragging	Improper adjustment	Readjust with full load.
5. Platform/carriage will not come down		
a. Broken-rope safety mechanism activating	Spring adjusted too tight	Readjust only after it is definitely determined that improper adjustment is cause.

Source: Champion Hoist Company.

TABLE 13.4 General Troubleshooting (*Continued*)

Indication	Cause	Remedy
	Obstruction causing carriage to catch on tower. (Once weight of carriage is removed from rope, safety mechanism will activate.)	Remove obstruction.
	Lifting rope binding in sheave	Free rope and carefully check for damage to rope and sheave.
b. Pump not being activated	See item A3a.	
6. Platform comes down, but slowly		
a. Insufficient oil flow through return circuit	Obstruction in return circuit	Remove any obstruction.
7. Platform/carriage slides down, will not hold load		
a. Brake not holding	Brake improperly adjusted	Readjust with full load.
b. Brake partially released	Brake valve leaking	Check for foreign material on seats or replace.
	Brake-release linkage is binding	Repair cause.
	Brake lining is wet	Dry with air. Brief period of operation should dry out lining. In severe case, tighten brakes so that they drag, then operate a few times so that heat of friction dries lining. Readjust to proper setting.
	Brake lining worn out	Since brake is not used to stop load, but only to hold it, lining should not wear out. If it does, replace lining, but be sure to check for brake drag.
8. Engine will not return to idle		
a. Throttle is being activated	Throttle-valve linkage out of adjustment	Readjust.
	Cylinder displacement due to throttle-valve leakage	Repair or replace valve.

TABLE 13.4 General Troubleshooting (*Continued*)

	B. System Will Not Operate in Either Direction	
1. Low or zero charge pressure	System low on fluid	Locate and fix leaks causing the loss of fluid. Replenish fluid in reservoir to proper level. *Note: Use only recommended fluids.*
	Faulty control linkage	Check entire linkage, from control lever to pump arm, to make sure it is connected and free to operate as it should. Adjust linkage to pump arm. *Do not move pump arm to meet linkage.*
	Disconnected coupling	Check that coupling from engine to pump shaft or from motor shaft to driven mechanism is not slipping or broken.
2. High vacuum and low charge pressure	Filter or suction line from reservoir to charge pump plugged or collapsed	Replace or clean filter suction line.
3. Low or zero charge pressure. If problem is in charge pump, pressure will be low or zero when pump is in neutral; if in manifold, pressure will be low when in stroke.	Pressure-relief valve in charge pump or motor manifold damaged or open	Replace pressure-relief valve.
4. Zero charge pressure when pump is in neutral or when trying to go into stroke	Charge-pump drive key or shaft broken	Replace charge-pump assembly.
5. Pieces or flakes of brass in reservoir or filter; noisy unit (pump or motor)	Internal damage to pump or motor, or both	In fixed- or variable-displacement motors: remove charge pressure-relief-valve cap in motor manifold, remove charge-relief spring, insert solid shim of sufficient length in place of spring, and reinstall the cap. This will block relief valve in a fully closed position.

Source: Champion Hoist Company.

TABLE 13.4 General Troubleshooting (*Continued*)

Indication	Cause	Remedy
6. Neutral charge pressure maintained, but pump will not go into stroke; handle moves freely	Disconnected control valve internal	Disconnect control linkage at directional control arm, and move arm back and forth by hand. If it moves freely with no resistance, control valve should be removed and checked for broken or missing parts.
7. Neutral charge pressure normal, but pump will not go into stroke	Plugged control orifice	Remove bolts that hold control housing to pump and check orifice. *Caution: Do not allow orifice or O rings to fall into pump case.*
	C. System Operates in One Direction Only	
	Faulty control linkage	Check entire linkage to make sure it is connected and free to operate as it should. Adjust linkage to pump arm.
1. Loss of or lower than normal system pressure in one direction only	Faulty high-pressure relief valve	Switch two high-pressure relief valves. If system operates in direction it would not operate before, one of the relief valves is inoperative. Repair or replace damaged valve and reset system.
	Shuttle-valve spool jammed (located in motor manifold)	Remove and replace entire manifold assembly.
2. Loss of system pressure in one direction only; charge pressure might be higher than normal	One check valve faulty	Remove two check valves located in pump and cap under charge pump and check to see if: i. Poppet or ball is missing. ii. Valve seat is eroded or deformed. *Note:* If condition (i) exists, replace pump; if condition (ii) exists, replace both check valves.
3. Pump will not return to neutral	Control-valve-spool jammed, sticking	Replace control valve.

D. Neutral Difficult or Impossible to Find		
	Faulty linkage	Disconnect control linkage at directional control arm. If system now returns to neutral, linkage to control is out of adjustment or binding in some way.
	Control valve out of adjustment	Replace displacement control valve or readjust according to manufacturer's recommended instructions.
	Servo cylinder out of adjustment	Remove pump and return to factory for readjustment.
E. System Operating Hot (motor-case temperature above 120°F)		
1. Low charge pressure	Fluid level low	Replenish fluid supply.
	Fluid cooler clogged	Clean cooler air passage.
2. High vacuum and low charge pressure	Clogged filter or suction line	Replace filter. Clean or replace suction lines.
3. Lower than normal system pressure in one or both directions	Excessive internal leakage	Check high-pressure relief valves; one may be stuck partially open.
a. Lower than normal charge pressure may drop to or near zero when maximum obtainable system pressure is reached	Refer to B.	
b. Loss of acceleration and power		Replace pump, motor, or both.
	Case drain lines improperly plumbed	Check plumbing, reinstall to proper alignment.
	Continued operation at high-pressure relief-valve setting	Consult operator's manual for proper machine operation.
F. Acceleration and Deceleration Sluggish		
	Air in system	Check for low fluid level. Check inlet system, filter suction line, and so forth, for leaks allowing air to be drawn into system.

SOURCE: Champion Hoist Company.

TABLE 13.4 General Troubleshooting (*Continued*)

Indication	Cause	Remedy
	Control orifice plug partially blocked	Remove bolts that hold control housing to pump and check the orifice. *Caution: Do not allow O rings to fall into the pump case.* If orifice is clean, remove the charge pump and plug at charge-pressure-gage port and blow clean air through the passage between charge pump and control orifice port.
1. Pieces or flakes of brass in reservoir or filter; noisy unit (pump or motor)	Internal wear or damage in pump or motor, or both	In fixed- or variable-displacement motors, remove charge pressure-relief-valve cap in motor manifold, remove the charge-relief spring, insert solid shim in place of spring, and reinstall cap. This will block relief valve in a fully closed position.
	Engine lugs down	Consult engine manual.
	G. System Noisy	
1. Considerable amount of foam in reservoir; low and fluctuating charge pressure; spongy control	Air in system	Check for low fluid level. Check inlet system, filter, suction line, and so forth, for leaks allowing air to be drawn into system.
	Plumbing not properly insulated	Make sure base or tubing is not touching any metal that can act as a sounding board. Insulate base and tubing clamps with rubber to absorb noise.

SOURCE: Champion Hoist Company.

Chapter

14

Helicopters

One of the largest users of helicopters today is the construction industry, with all indications that rotorcraft application on future construction projects will increase considerably.

No longer considered merely a transportation vehicle, the helicopter has become an integral part of many construction operations that require a time-saving materials transport vehicle, or flying crane, for such operations as placement of concrete or structural members, construction of pipelines and transmission towers, and installation of equipment and machinery.

Commercial helicopters now coming on line differ considerably from their military counterparts. Technology and advanced design have combined to make the helicopter a tool that fulfills specific commercial needs at improved operating costs.

However, a decision to integrate the helicopter into construction operations will require careful analysis during the project-estimating stage and includes the required support system, operational costs, and the most economical method of procurement.

Support System

Once it has been decided that helicopter support is desired on a project, it will be necessary to determine the type, size, and number of units required to effectively maintain the construction schedule.

Helicopters are classified by gross weight (light, medium, or heavy), which has a direct correlation to their load-carrying capacity. Different tasks will require helicopters having different useful loads and different flying ranges. Even though a helicopter's useful load depends on its gross weight, the heavier the rotorcraft, the higher will be its operating costs.

Thus, it would not be economical to use a heavy helicopter on a job that could be done by a lighter rotorcraft at less cost; and on large complicated jobs that require the support of many helicopters, a mix of different sized helicopters will often prove to be more economical than a single sized craft.

If helicopter support on a project is to be only temporary, the contractor should probably hire a helicopter operator, rather than add rotorcraft to his equipment spread. But remember, not all operators do all types of work. Instead, they often specialize in contracts within a particular industry, such as offshore or pipeline construction. Others, of course, are limited in the kind and scope of work they do by the type (gross weight) of helicopters in their fleet. Then too, many small helicopter operators usually work only locally. Thus if a contractor's needs are great, a large operator might provide the best service.

Purchasing, Leasing, or Contracting

A contractor considering integration of rotorcraft into construction operations should analyze acquiring a helicopter in the same terms as any other capital investment. If contemplating purchasing a helicopter, the contractor should consider such factors as cost of capital, initial investment, revenue production, and salvage value (which for helicopters is very high).

On projects such as powerline construction, some contractors purchase support rotorcraft to use during the project and then sell them to the project owner or government agency for use in surveying and maintaining the line after completion.

Leasing is an alternative for contractors willing to hire the necessary operating and maintenance expertise. Several leasing sources are commonly used:

Airplane leasing groups.

Leasing corporations of the holding companies of large banks.

Major helicopter manufacturers that have lease/purchase agreements designed to assist the lessee in financing an eventual purchase; however, such agreements are usually unattractive if simple leases are considered.

Operators under the right circumstances sometimes will lease their aircraft.

However, by far, the most common approach to securing helicopter services is contracting with an operator of a rotorcraft fleet. The scale of the economics involved in helicopter operation often make this the most practical solution.

Helicopters require hangar facilities, maintenance and overhaul facilities and crews, spare parts inventory, and pilots. Training, if necessary, is expensive. More than one helicopter is usually required to support a contract that needs regularly scheduled flights, because if a helicopter is down, for scheduled or unscheduled maintenance, then another craft must be used.

In addition, far more complex operational efficiency can be achieved by supporting a larger helicopter with smaller craft, and if service is needed 24 h a day, then more than one crew and aircraft will be involved.

Helicopter operators can be located through civil aviation authorities, helicopter manufacturers, or the Helicopter Association of America.

Regardless of whether the rotorcraft is purchased, leased, or under contract, it must be certified by the Federal Aviation Administration (FAA) before it can be used. Make sure that each helicopter used displays its registration certificate, airworthiness certificate, and operating certificate in the craft, as required by the FAA.

Operational Costs and Insurance

Costs

Once equipment requirements have been determined, the contractor must then compute the cost of using rotorcraft on the project. This will require working closely with the helicopter operator to determine accurately both fixed and variable costs of operation.

Fixed costs. All items that increase with the number of helicopters in use, rather than with the number of hours flown. They cover:

Salary and administrative costs, including salaries of pilots, mechanics, and administrative personnel

Operating costs, including hangar rentals, fixed nonaircraft capital depreciation charges, local supplies, and local ground transportation

Acquisition costs of helicopters in the operator's fleet, including interest on loans for financing helicopters and spare parts, liability insurance, hull insurance, depreciation, and a reserve for contingencies

The first two costs vary little among different types of helicopters, while the last cost covers a wide range, depending on the type and size of the rotorcraft used.

Variable costs. All items that vary with flying time. As the number of flight hours increase, scheduled maintenance, overhauls, and the retirement of parts all occur with increasing frequency. Included as elements of variable costs are fuel, oil, and hydraulic fluids used.

The overall cost of helicopter service also is affected by many other factors, including:

Manufacturer's parts distribution and service support

Proximity and availability of spare parts and overhaul capabilities

Startup costs, which on shorter contracts cannot be spread over a longer period of time

Insurance

Insurance costs too must be included in a contractor's analysis of fixed and variable costs of operating a helicopter. Basically, two types of insurance are required—hull and liability.

Hull insurance is the responsibility of the operator. The contractor should ask the operator to furnish evidence of liability as well as the limits of the policy. Such insurance should cover passenger liability as well as damage to property.

Cargo liability insurance may be carried either by the operator or by the contractor. In either case, external-load liability will cost more than internal-load liability.

Although contractors sometimes are named as an additional insured under the operator's liability, this arrangement should be weighed cautiously since being so named could in effect reduce the contractor's overall coverage. Instead, it may be better for the contractor to require the operator to hold him harmless for damages resulting from helicopter operation.

Normally it is not considered necessary for the contractor to obtain a waiver of subrogation from the operator's underwriter, especially where the operator's personnel on the job are covered by workman's compensation. However, it may be that the operator will require the contractor to grant a waiver of subrogation for damages resulting from contractor negligence.

Regardless of the arrangement agreed upon, it is imperative that the contractor review general liability and comprehensive policies with his insurance underwriter if exposure to helicopter operations will be lengthy.

Planning Operations

To utilize rotorcraft successfully on a construction job one needs a sound project organization, a proper operational plan, and strict control, plus enforcement of operating procedures and on-site safety measures. Failure in any of these areas will produce a breakdown in coordinated efforts, increasing the amount of flight and ground time. These, in turn, will escalate costs and possibly delay the project's scheduled completion.

Close coordination between construction supervisor, contractor, and helicopter operator is essential for the efficient use of rotorcraft on a construction site.

The first thing a contractor must do is educate construction crews on the sequence of tasks and specific safety requirements before introducing the helicopter into operations. Make sure that every member of the construction team understands the importance of executing each step on time, so that there will be no delays to subsequent tasks. More than with most equipment on the contractor's spread, project operations must be built around the helicopter, not the helicopter around the construction operations. Cost and availability make this essential.

The helicopter operator and the contractor should have a prejob conference to coordinate all activities requiring use of the helicopter. This should include determining:

Precautions to be in effect at the helicopter landing area, pickup area, route to be flown, and delivery (setting) area; arrangements for compliance with any other mutual requirements.

Design and strength of rigging and how it will attach to the load.

Accuracy of weights, structural strength of the loads, size and number of loads, and number of ground crews required for the operation.

Assignment of responsibility for clearing and securing pickup and setting sites.

Maximum time that the helicopter can hover while ground crews are working beneath it.

Type and quantity of personal protection gear provided for the ground crews.

Type of scaffolding, if necessary, to be erected for ground crews to provide stable footing when attaching and unhooking the loads at elevated sites.

Working conditions that could be hazardous to ground crews, such as rotor wash, rain, static electricity, and gusty winds.

Operation of the rotorcraft in proximity to electric power lines, if applicable.

Clearance of nonessential personnel from the flight route.

Procedures for wetting down dusty and sandy areas.

Provision for a point of reference when the helicopter is hovering.

A helicopter crew should consist of one or two pilots and two mechanics, depending on the helicopter model used. The mechanics act as signal persons, one stationed at the pickup point on the ground and the other at the drop-off point on the structure.

The helicopter operator and the signal person should determine the signal person's position so that he or she can readily observe the load for pickup or delivery, and still be seen or heard by the operator. They should also agree upon the type of communications to be used for the specific lift—hand signals, ground-to-air radio, intercom, or relay signals, or a combination of them—and alternative procedures to be followed if communications or sight are lost between operator and signal person (see Fig. 14.1).

Note: Audible signals should be used as a warning to ground crews working with the helicopter during airlift operations.

Finally, both operator and contractor should agree upon any special identifying clothes or gloves that only the designated signal persons will wear. Signal persons should direct the pilot with hand signals to let the pilot know what control corrections are required.

Since union jurisdictional rules prevent helicopter crews from touching a load, the contractor should supply a minimum of two workers on the ground and four or more workers on the structure. For maximum accuracy in placing a load, the number of persons on the structure should be increased as the weight of the load increases.

Operational Procedures

Before putting a helicopter to work, the operator must first advise the FAA of the job. Frequently an FAA inspector will be on site to make sure that the work is done in accordance with FAA regulations.

The contractor, in turn, is responsible for notifying local police of the planned operation and requesting police help when necessary to keep traffic moving or to keep spectators out of the area of operation.

If the job is in a downtown area, it is usually necessary for the contractor to obtain also the approval of the city, and to have the help and cooperation of local police to block off streets around the area of operations.

Because of its unique vertical takeoff and landing capabilities, the helicopter requires no special landing facilities other than an open space with relatively level ground. With loads slung below the helicopter, it often is not necessary to land at all. The helicopter simply hovers over the desired location, releases the load, and flies off

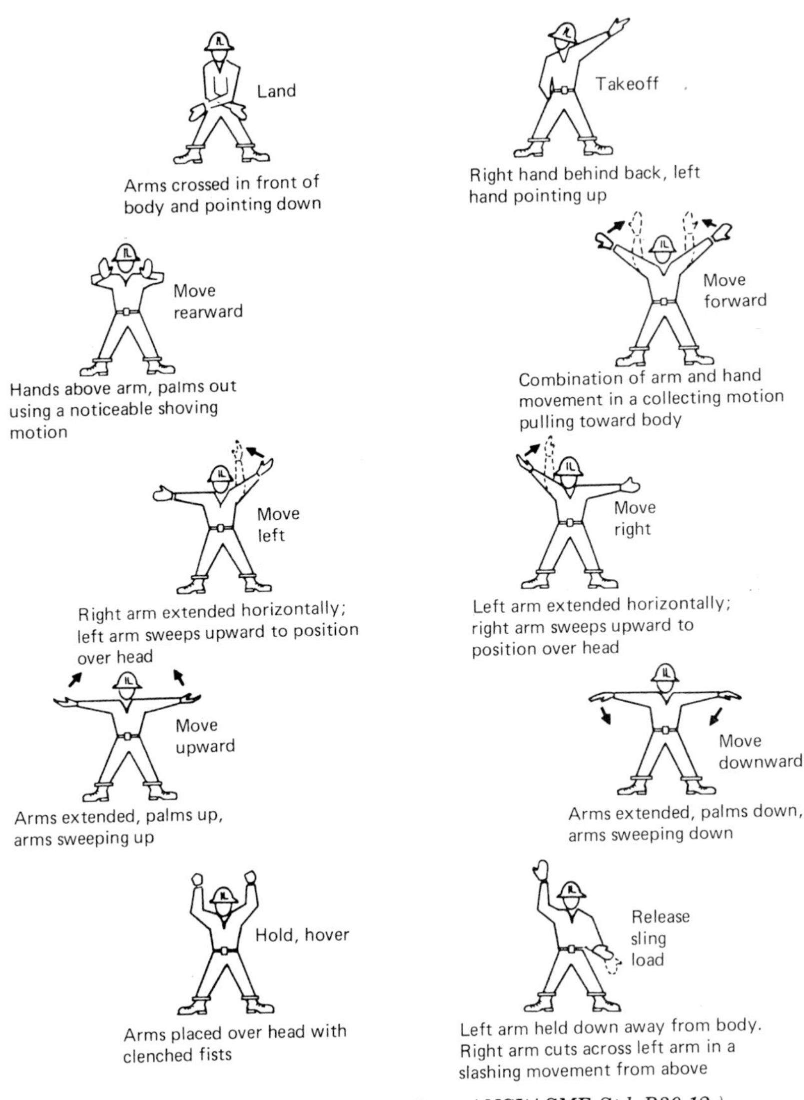

Figure 14.1 Recommended hand signals. *(From ANSI/ASME Std. B30.12.)*

to complete the task. In some cases it may be necessary during normal daily operations to land only to refuel.

However, a big concern of an operator, who will not be operating from home base, is where to base a helicopter during the support operation. A nearby airfield can often be used, or temporary hangar facilities can be set up on the construction site.

The length of contract and the amount of support needed will usually determine fleet maintenance requirements. Maintenance requirements, in turn, will determine facility needs.

Major considerations in determining flight operations are weather, altitude, and temperature, factors over which the pilot has no control, but which directly affect the craft's performance. The supervisor as well as the operator should be aware that engine efficiency is inversely proportional to temperature and altitude; as these two increase, the engine power is reduced.

Weather conditions must always be considered prior to flight planning. The supervisor and the operator must demand that the capabilities of both the operator and the machine be not overextended in marginal weather.

Airlifting loads by helicopter calls for some preparations that are not normally required for conventional crane lifting, especially because of winds (comparable to a strong wind storm) that are created by the helicopter's rotors.

Specific precautionary measures to be taken to prevent personnel injury and structural damage are:

Clear the entire operational area of all debris, such as loose tar paper, plywood, sheet metal, tarpaulins, polyethylene paper, roofing-tar pots, insulation, and all other objects that could move in a strong wind.

Properly secure all lumber and other large lightweight objects.

Water down thoroughly both the helicopter landing area and the load pickup area if they are dusty.

Make sure all wheel-type landing gear of helicopters is checked while on the ground or other supported area.

All workers must wear eye goggles and hard hats with chin straps (in accordance with OSHA regulations) when participating in the helicopter operation. This protective equipment should be supplied by the contractor.

Make sure the landing area set aside for the helicopter is at least 200 ft in diameter and is cleared of all obstacles such as poles, wires, trees, and sheds. Position a wind indicator near the landing site.

Make sure that, while the helicopter is on the landing site, the horizontal clearance between the main or auxiliary rotors and any obstacle is not less than 15 ft.

Clear the helicopter's flight route from the pickup area to the setting area of all people, vehicles, and structures.

Lifting Components

The primary cargo hook should be located as close to the helicopter as possible to prevent inadvertent entanglement of the hook and its suspension system with external fixed structures on the helicopter. The primary hook should include a quick-release device that enables the pilot to release the external load quickly during flight (see Fig. 14.2).

In addition to the primary control, an emergency release device (mechanical, electrical, hydraulic, pneumatic, explosive, or a combination of these) should be provided. The power source of this system must be independent of the normal primary hook release system power source.

The automatic hook release mechanism should be used only for specific operations where ground crews are not used, and should be activated only when actually placing a load.

Each lift should be made using a spreader bar that has four cables, with a safety hook on the end of each cable. The hooks should be attached to the suspended load by slings or by chokers attached to lifting points on the load (see Fig. 14.3).

Each load should have four pickup points on it. If the load does not have pickup points, and if it is designed so that eyebolts or clevises

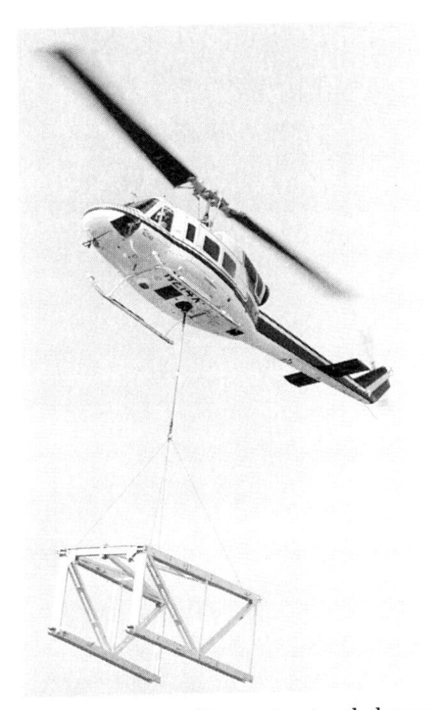

Figure 14.2 Helicopter handling a structural element. *(Courtesy Bell Helicopters.)*

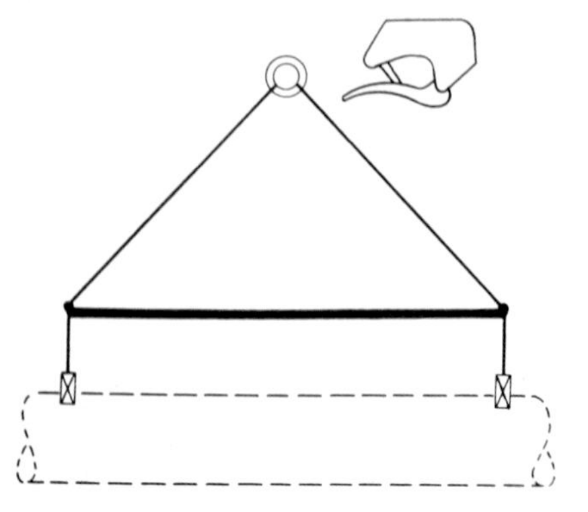

Figure 14.3 Two-legged spreader sling. *(From ANSI/ASME Std. B30.12.)*

cannot be installed or J hooks used, the lift can be made using nylon straps. If there is any sizable number of loads without pickup points, the contractor should provide special rigging for such loads (see Fig. 14.4).

To determine the service break strength S of each leg of a sling, the calculations are as follows:

1. Pendant (one-legged) sling or primary lifting cable,

$$S = 1.75nW$$

2. Two-legged sling,

$$S = 1.32nW$$

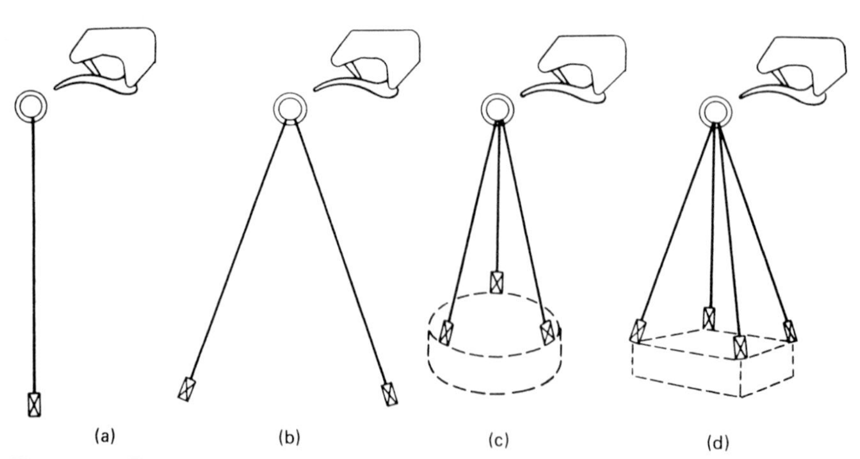

Figure 14.4 Types of slings. (*a*) Pendant, one-legged. (*b*) Two-legged. (*c*) Three-legged. (*d*) Four-legged. *(From ANSI/ASME Std. B30.12.)*

3. Three-legged sling,

$$S = 0.9nW$$

4. Four-legged sling,

$$S = 0.7nW$$

where n = design flight load factor for helicopter (see helicopter flight manual)
w = rated capacity of entire sling
S = service break strength of each leg

The service break strength is actually the strength of the sling after one year in service. The new break strength may be higher than the service break strength, depending on the amount of deterioration caused by environmental exposure on each sling material. The new break strength NS can be found by the formula

$$\text{New break strength} = \frac{\text{service break strength}}{\text{reduced strength factor}}$$

The reduced strength factors for sling materials are given in Table 14.1.

The length of sling legs should be such that no sling leg makes an angle from the vertical greater than 45°. The legs of multiple-leg slings should be connected at the top by an Apex fitting.

The rated load of the primary hook should be the maximum weight of load to be lifted, carried, pulled, or towed. This maximum weight should not exceed the weight approved by the FAA and specified in the helicopter flight manual, and is required by the FAA to be displayed on or adjacent to the hook. Rated loads for sling materials should be based on the catalog strength of the material used.

Safety Precautions

One of the most important arguments for proper planning where helicopters are used, is safety. Safety, always important in construction, is even more so around helicopters. The operator and the contractor should work closely on developing and implementing a safety program, and devising a set of emergency procedures that must be adhered to. Where the helicopter is part of the productive process, that process may require alterations to include provisions for a contingency in case of an emergency.

TABLE 14.1 Reduced Strength Factors for Sling Materials

				Pin diameter[d]		
Material	1 year [a]	Temperature variation [b]	Seawater immersion [c]	1¼ in. and up (31.8 mm)	1.00 in. (25.4 mm)	0.75 in. (19.1 mm)
Wire rope (IWRC)	1.00	1.00	1.00	Not applicable for thimbles		
Type X nylon	0.90	N.A.	0.85	1.00	0.90	0.81
Type XIX nylon	0.79	1.00	0.91	1.00	0.92	0.70
Type XXVI nylon	0.54	0.93	0.90	1.00	0.91	0.81
Type V polyester	0.62	1.00	1.00	1.00	0.91	0.81
Type VI polyester	0.90	0.94	1.00	1.00	0.91	0.81
Alloy chain						

[a]*Exposure* occurs when sling materials are subjected to weathering conditions. The single most destructive agent is ultraviolet light. When textile materials are exposed to the effects of ultraviolet light for long periods of time (up to a year), significant loss of strength occurs. Protection against the detrimental effects of ultraviolet light can be provided by the designer by using urethane or polymer coatings over the textile sling members. Sleeves over sling legs help screen out ultraviolet light. If the textile material is suitably protected against ultraviolet light by one or more of these methods, the exposure factor found in the table may be raised to 0.9. Slings should be stored away from strong sunlight when not in use.

[b]The material is subjected to temperatures from −65°F to 160°F (−53.9°C to 71.1°C) at 95% relative humidity.

[c]The material is soaked in seawater for periods up to 24 h and then washed off in freshwater. The use of waterproof coatings on textiles by the designer and the prompt and thorough washing by the user will reduce the detrimental effects of seawater.

[d]If the textile material is to be wrapped over a pin less than 1¼ in. (31.8 mm) in diameter, a corresponding strength reduction factor must be used. Wire rope slings should be IWRC (steel center). The end fittings should be thimbles with swaged eyes. If open choker eyes are used, a pin factor of 0.8 shall be used. The use of wire rope clips is not recommended; however, if they are used, precautions shall be taken prior to each use to ensure that they are properly tightened and installed. When wire rope clips are used, a strength reduction factor of 0.6 must be taken.

source: From ANSI/ASME Std. B30.12.

Caution must be exercised to ensure protection of the rotorcraft from objects being blown or drawn into the rotor system or engine intakes. All items susceptible of creating such hazards must be secured or removed from the operating sites.

Personnel should be instructed well in advance about safe behavior when working around a helicopter. When entering or leaving a helicopter, each person should do so toward the front of the helicopter, never the rear. No one other than the pilot should enter the rotorcraft, except qualified designated personnel considered essential for a specific lift operation.

Compliance with safety regulations, enforced by the civil aviation authorities, is clearly the responsibility of the operator. However, the contractor should determine how, if at all, these regulations will affect construction activities.

Smoking, open flames, or other sources of ignition must not be permitted within 50 ft of fueling operations or storage areas. No one other than authorized personnel should be allowed within 50 ft of fueling operations.

At least one 30-lb fire extinguisher, good for class A, B, and C fires, should be provided within 100 ft of the refueling operations.

While a rotorcraft is on a landing site, horizontal clearance between the main or auxiliary rotors and any obstacle should be not less than 15 ft.

Inspection

A prelift inspection should be performed prior to conducting each operation with the rotorcraft, and on at least a daily basis during construction operations. Unless the primary hook being utilized passes the following minimum requirements, it should not be used. Check the following:

Hook attachments for correct connections

Hook for manual release

Electrical connectors and wiring

Both normal and emergency release systems or circuits, functionally

Load beam for binding

Load beam and hook frame members, visually, for cracks, gouges, distortion, wear, and latch engagement

All other emergency release devices by testing them

Suspension members, if used, for alignment

Primary hook suspension ropes for broken wires

Manual release cable conduit for kinks; do not use if conduit is kinked

Visually inspect all ropes or chain used for slings each working day. If evidence of deterioration is found that could result in appreciable loss of original strength, carefully determine whether further use of the rope or chain would constitute a hazard.

Observe the following procedures and practices for the ground crew:

NEVER turn your back on the load.

Avoid getting directly under the load when it is airborne, except when necessary to the operation, such as to attach, detach, or guide the load into place.

Provide a means of access and egress when crews are required to work under hovering helicopters.

To avoid hand injury, keep hands clear and in view when steadying the load, and grasp cables from the side, not underneath or between the cables and the load.

Make certain that keepers work on the hooks.

Use all hooks the way they are intended to be used.

Make sure issued personal gear fits.

When using tag lines, make sure they are of such a length that contact with rotors is precluded.

Do not let tag lines wrap around any limbs or portions of the body.

Use railings, lifelines, or safety belts to avoid falling; watch your step when concentrating on the helicopter load.

Pull tight cinch cables on cargo nets prior to the airlift to prevent any objects from falling when the load is airborne.

Notify helicopter crew members immediately if you see a damaged sling.

Position the hooks so that when the strain is taken, cables will not break keepers and slip out.

Keep all sling legs free of any knots.

Hold hookup after detaching it from load so that it can be seen by signal person and will not snag objects on the ground when the helicopter is moving away.

Make signal person responsible for assuring that the load is free of hooks.

Make sure signal person is familiar with the emergency signals to be used and the need for rapid response.

Report any mishaps or near mishaps to helicopter crews immediately.

To prevent being pinned and injured, *do not* go or move between the load and another object.

Watch out for swinging rigging that is hanging from the helicopter.

Do not grab load or tag line if it is rotating or swinging too fast.

Note: If ground crews are transported by helicopter, they should be briefed by the helicopter crew on the method of entering and exiting the helicopter, and instructed in the use of safety belts and danger of turning rotors.

Section

4

Rigging Accessories

Chapter

15

Jacks, Rollers, and Skids

Portable jacks for heavy-duty use in rigging operations or in construction fall into three general categories:

1. Hand- or power-operated hydraulic jack
2. Mechanical ratchet
3. Hand- or power-operated mechanical screw

NEVER, under any circumstances, use an automobile-type jack for rigging or construction jobs. Regardless of the type of jack used, it is very important not to overload it.

Jack manufacturers provide each device with a lever or handle of a predetermined length to ensure against overloading. Although an extension pipe placed on the lever will make it possible to lift the rated load more easily, the same force on an extended lever will permit lifting an even greater load. Such a load, however, could easily be in excess of the rated or safe capacity of the jack. NEVER use a lever longer than the one the manufacturer furnishes with the jack.

Basic Types of Jacks

Hand- or power-operated hydraulic jack

This jack employs the tremendous power of hydraulics to lift heavy loads with little effort. The unit pressure exerted on a small area of incompressible fluid is transmitted undiminished to a larger area in the same closed circuit, thus multiplying the exerted force. It is not as

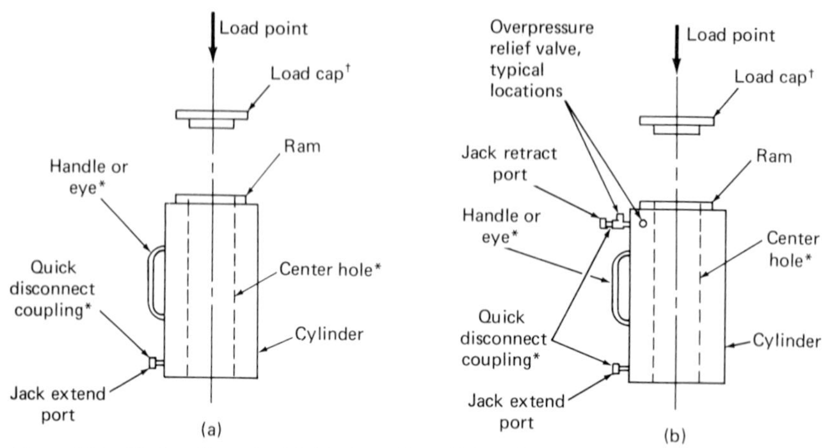

Figure 15.1 Hydraulic jacks (*a*) Single-acting. (*b*) Double-acting. *Optional. †If required by jack manufacturer. (*Courtesy of American National Standards Institute.*)

fast in operation as a ratchet-type jack, but it is excellent for easy lifting of heavy loads (see Figs. 15.1 and 15.2).

A variation of this type of pump is the hydraulic ram with separate hydraulic pump, operating on the same principle as the integral-pump-type jack. Rams enable users to operate the pump at a safe distance from the load being lifted.

Mechanical ratchet

This is the simplest jack construction (see Fig. 15.3). It employs the basic lever and fulcrum principle. A downward stroke of the lever raises the rack bar one notch at a time and the pawls spring into position, automatically holding the load and releasing the lever for the next lifting stroke. This type of jack should be used on lighter loads, usually

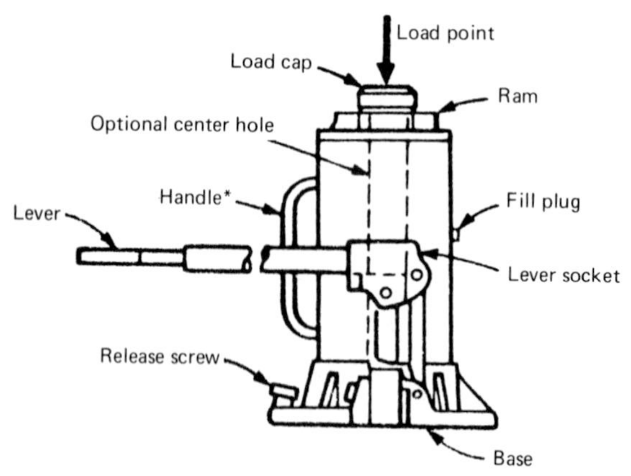

Figure 15.2 Self-contained hydraulic jack. *Optional. (*Courtesy of American National Standards Institute.*)

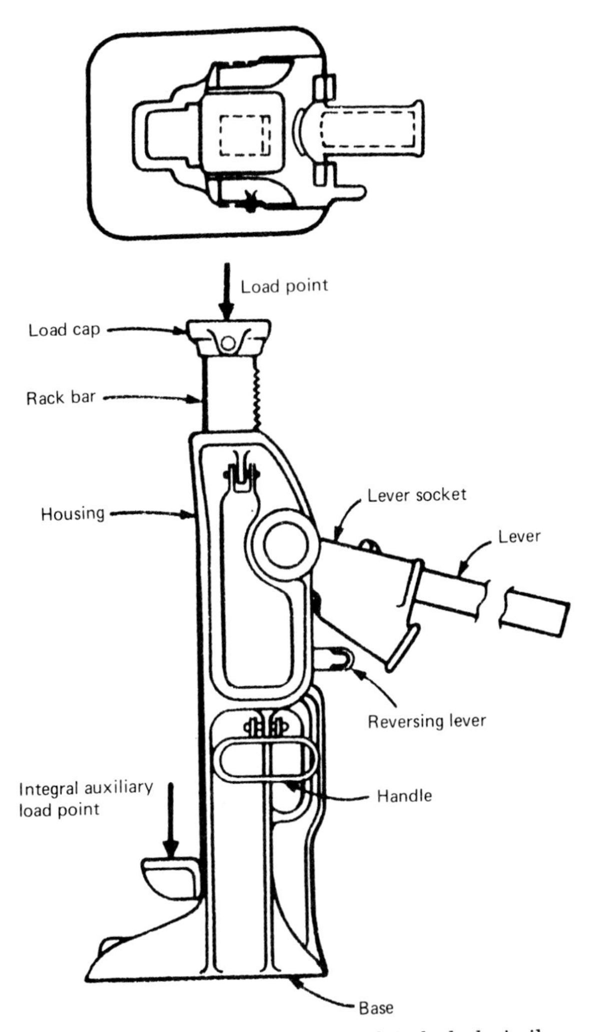

Figure 15.3 Typical ratchet jack. Track jacks look similar, but are trip-lowering and are not intended for use as general-purpose jacks. (*Courtesy of American National Standards Institute.*)

up to 20 tons, because of the physical effort required to operate it. These units are less expensive and operate quickly.

Hand-or power-operated mechanical screw

For heavier loads the screw and nut principle is employed. Two general types—regular and inverted—are available. For lighter loads a simple lever bar will apply sufficient power to turn the screw. As loads increase, gear reductions and ratchet devices serve to multiply the operator force.

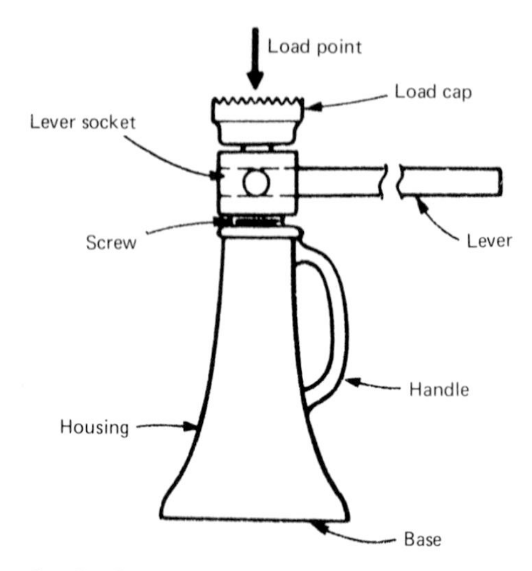

Figure 15.4 Basic screw jack. *(Courtesy of American National Standards Institute.)*

In the heaviest of jacks the screw is operated by an air motor for faster lifting or lowering with minimum effort (see Figs. 15.4 to 15.7).

Operation

When using jacks it is most important that the operator be familiar with the equipment and the manufacturer's instructions for operation, maintenance, and testing. All jacks should be visually inspected before each shift or each use, whichever is less frequent, as well as during operation for any deficiencies that might constitute a hazard.

Whenever there is evidence of any of the following deficiencies, the jack should be removed from service immediately and not used until the deficiency is corrected:

Improper engagement or extreme wear of pawl and rack

Cracked or broken rack teeth

Cracked or damaged housing

Excessive wear, bending, or other damage to threads

Leaking hydraulic fluid

Scored or damaged plunger

Improperly functioning swivel heads and caps

Loose bolts or rivets

Damaged or improperly assembled accessory equipment

Other items that may affect the jacking operation, as specified in the manufacturer's instructions

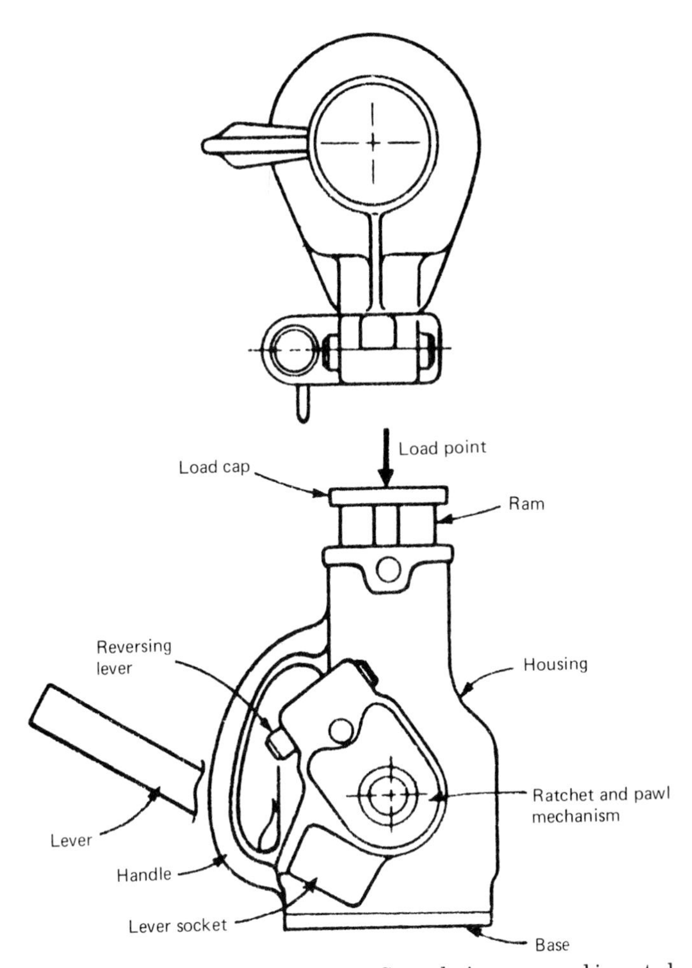

Figure 15.5 Screw jack with ratchet. Some designs are used inverted and have toe lifts, others are controlled self-lowering. (*Courtesy of American National Standards Institute.*)

Before operating a jack it is important to determine the load to be lifted and to assure that it is within the load rating of the jack.

All hydraulic jacks have the safe load indicated on the nameplate. Screw jacks occasionally are not labeled. Mechanical jacks sometimes have two ratings (sustaining and lifting) that should be indicated by the manufacturer. The hydraulic pressure or lever arm length and force to be applied should also be clearly marked on the jack.

A rough estimate of the safe load on a screw jack can be made using the following formulas and choosing the lowest value of W:

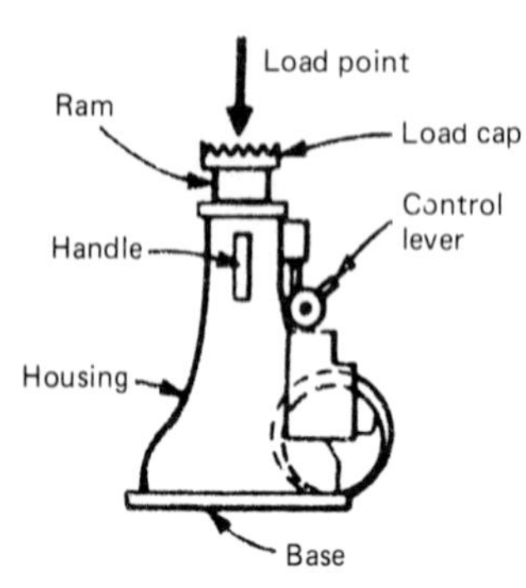

Figure 15.6 Power-driven screw jack. (*Courtesy of American National Standards Institute.*)

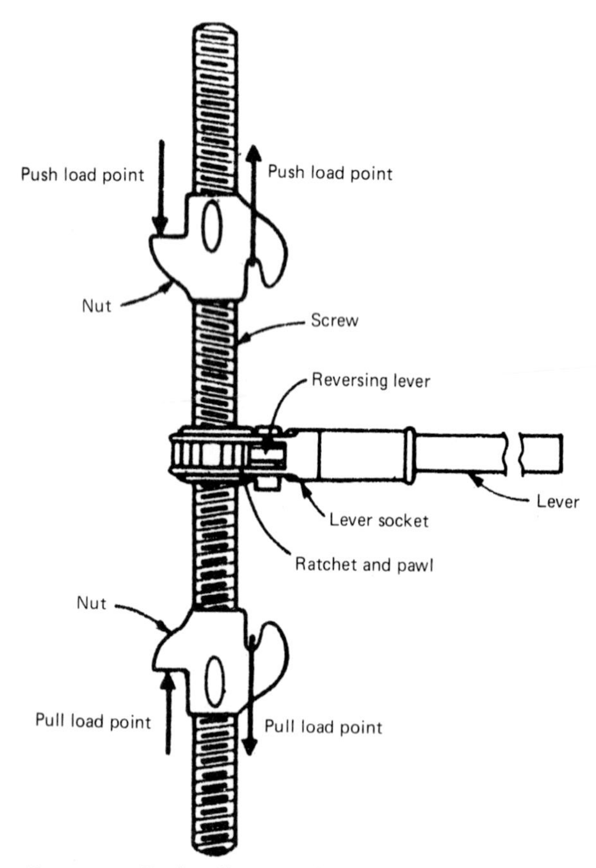

Figure 15.7 Push-pull-type jack. (*Courtesy of American National Standards Institute.*)

$$W = 31{,}400dt$$

or,

$$W = 14{,}000d^2$$

where W = safe load on jack, pounds
d = diameter of screw at root of threads, inches
t = thickness of nut or length of thread engaged, inches

The load that a screw jack should be able to lift is approximately

$$W = \frac{24PL}{r^2}$$

where W = load, pounds
L = length of lever, inches
P = pitch of screw threads, inches
r = average radius of screw thread, inches

The force on the lever can be assumed to be 120 lb for most cases.

When preparing for a lift, make sure that the jack is firmly supported at the base in such a manner that it cannot shift under the load. Always make certain that the jack is in a true vertical position (when lifting) and resting on good footing. Never place a jack directly on the ground, even though the soil appears to be firm.

If the load is to be raised in its entirety by several jacks, it should be secured laterally by struts to prevent all the jacks from upsetting in unison. When using a jack in a horizontal position to move an object, it should be lashed or blocked up to prevent it from falling should the object unexpectedly move faster than the jack.

After the load has been raised to the required height, shoring or cribbing blocks should be placed under the load and wedged to take the load off the jacks, or to be ready to take the load, should a jack fail. When required, use shims or constraints to prevent the base or load from slipping.

NEVER raise a load on a jack so high that the ram, screw, or ratchet runs out of the base, thereby dropping the load. Jacks designed in accordance with the ANSI safety requirements for jacks have positive stops or other methods to prevent overtravel. These will not alter the operating characteristics of the jack.

Hydraulic jacks are filled with a special hydraulic fluid that will not freeze. Under no circumstances should water be used in a jack.

A pressure-relief valve should be used and set at or below the rated pressure at the rated load where remotely operated hydraulic jacks are used. Double-acting hydraulic jacks should be marked to indicate the need for a relief valve as indicated.

Operating checklist

Make sure all operators are instructed in the proper use of jacks, including signals and other procedures for multiple jacks or other special lifts.

Make sure the recommended operating handle is used and properly seated in its socket.

Make sure operators do not straddle the operating lever of a mechanical jack.

Make sure there is sufficient swing area for the operating lever or levers.

Remove operating levers when not in use to avoid accidental dislocation of the jacks, thereby reducing the tripping hazard.

Make sure all personnel are clear of the load before lowering.

Follow every load with cribbing or blocking where practical.

DO NOT allow personnel to work or pass under a load until the load is secured by cribbing, blocking, or other means against accidental lowering.

Avoid off-center loading of jacks.

DO NOT use extenders unless authorized by a qualified person.

Inspection

Each jack, new or altered, should be thoroughly inspected prior to initial use, and periodically thereafter, depending on service conditions.

Inspections for jacks in regular service fall into two categories—frequent and periodic.

Frequent inspections. The operator or other designated person should examine the jacks visually, but need not keep records of the inspections.

Normal. One inspection a month

Severe. Before each use or daily, whichever is longer

Periodic inspections. An appointed person should examine the jacks visually, making records of apparent external conditions to provide the basis for a continuing evaluation.

Normal. One inspection a year

Severe. Before each use or daily, whichever is longer, unless external conditions indicate that the jack should be disassembled to permit more detailed inspection

Jacks should be disassembled for cleaning and examined for internal wear or damage if external appearances indicate that there may be internal deficiencies. This periodic inspection should be performed by an authorized service center or an appropriately trained authorized individual. Dated and signed inspection records should be kept on all periodic inspections.

Any jack that has been idle for one year or more should automatically be inspected before being put into use.

In addition, jacks should be inspected when:

Used constantly or intermittently at one location, once every six months

Sent out of the shop for special jobs, upon leaving and upon returning to the shop

Subjected to abnormal load or shock, immediately before and after making such a lift

Testing

After a jack has been repaired and before it is returned to service, it should be tested by:

Lifting a load equal to the rated capacity of the jack to within 1 in. of the full travel under the prescribed operating conditions for the particular type of jack being tested.

Adding to this original load an additional load equal to one-quarter of the rated load already applied to the jack.

Checking for any evidence that any part of the jack is stressed beyond the yield point of the material forming any part. Such evidence shall disqualify the jack for further service.

When testing a jack it is important that, as the load is lifted, blocks be placed under the load to hold it in the event the jack should fail.

Warning: Keep hands and fingers out from under the load during testing.

Maintenance

Lubrication. Regularly lubricate all moving parts of a jack requiring such service. Check lubricating systems for delivery of lubricant.

Follow the manufacturer's recommendations as to the points and frequency of lubricant, maintenance or lubricant levels, and types of lubricants to be used.

Hydraulic fluid. Use only hydraulic jack fluid that is fully compatible with the jack manufacturer's specifications.

Cleaning. Keep the threads of a screw jack free from grit and dirt. Lubricate the screw frequently. Clean all jacks exposed to rain, sand, or grit-laden air prior to use. Make sure the jack-operating lever and load-bearing surfaces are free of slippery material or fluid.

Storage. Jacks should be stored where they are protected from the elements, abrasive dust, and damage. Always store hydraulic jacks in a vertical position.

Repairs. Purchase repair parts only from the original manufacturer or authorized service center. If jack parts are other than those obtained from the original manufacturer, they must conform to ANSI specifications.

Specialized Hydraulic Jacking Systems

Today's heavy construction activity often involves considerable lifting and lowering of heavy loads. More and more extremely large and heavy components often are either prefabricated on site or delivered on special vehicles. They are subsequently installed as units, lifted, jacked horizontally, or lowered into their final position.

As soon as the weight of a load or its size exceeds the capacity of available cranes or lifting equipment, specialized hydraulic jacking systems must be used. These systems, developed by several manufacturers, combine hydraulic jacks with a regripping mechanism that allows the movement of almost unlimited loads quickly, over any distance, and in any direction. Loads of 1,000 tons or more can be lifted evenly, jacked horizontally, or lowered to their desired locations in situations where space is limited or ordinary cranes cannot reach.

This equipment is available on a hire basis, with installation, servicing, and maintenance handled exclusively by the manufacturer or its licensees.

The two basic systems in use are strand lifting and rod lifting. Both systems make use of similar techniques, but each has its own characteristics. For a certain application, one system is usually more suited than the other.

Strand-lifting system

The equipment consists of a moving unit with strand cables as tensile members. The moving unit is basically a hydraulic center-hole jack with two strand anchorages, an upper anchorage attached to the piston and a lower anchorage fixed to the support.

The system operates as follows: During the upward stroke of the piston, the strands are gripped by means of wedges seated in conical holes in the upper anchorage, which moves with the piston. During retraction of the piston, however, the strands are held at the lower anchorage. By this means a stage-by-stage advance of the cable and its attached load is achieved.

A special gripping mechanism assures an automatic and equal seating of the wedges. During lifting, the wedges of the lower anchorage remain in rubbing contact with the strand and will automatically grip, even if the oil pressure suddenly drops. This safety feature always ensures that the load is secure.

High tensile seven-wire prestressing strand with a nominal diameter of 0.6 in. is used to form the tensile member. The guaranteed breaking load per strand varies between 23.1 and 26.5 metric tons.

On the structure to be moved, the strands are anchored by means of compression fittings bearing against an anchor head. Both wedge and compression fitting anchorage develop the full breaking load of the strand.

The number of strands per moving unit can vary depending on the magnitude of the load. The same strand can be reused several times.

Each moving unit normally has its own hydraulic pump. The pumps can be either controlled directly or connected to a central console. Synchronized demand-regulated pumps ensure equal travel, irrespective of the load on individual moving units. Thus, large but relatively fragile loads, as well as hyperstatically supported structures, can be moved without introducing undue erection stresses.

Rod-lifting system

There are two basic types of this system, one employing cogged lifting rods, the other standard square rods, called jack rods.

Cogged jack-rod system. The equipment consists of a moving unit and a cogged lifting rod. During the upward stroke of the pistons, pawls on the upper yoke engage the lifting rod and advance it upward. As the lifting stroke ends, the pawls on the lower yoke engage the cogged lifting rod, supporting the load while the pawls on the upper yoke disengage and the jacks retract. This process is then repeated in a continuous automatic sequence, eliminating the need for resetting. Lowering a load is accomplished by reversing the operation.

Because of the inherent rigidity of the cogged lifting rods, the jacking system may be operated in either direction, pushing rather than pulling the load to its desired position. The lifting rods come in easily transportable lengths and may be coupled to give any desired length.

A large number of moving units may be connected to a central console, as with the strand-type lifting system, ensuring equal travel, irrespective of the load on the individual moving unit. Selective operation of the individual units allows corrections and adjustments to be made.

Spacer disks between the jack piston and the yoke provide mechanical security against possible oil leakage in the system.

The control console also contains automatic monitoring devices that survey such mechanical and hydraulic operations as the correct engagement of pawls and the right hydraulic pressure distribution. In the event of any deviation, automatic shutoffs stop the lifting procedure. Minor deviations between the jacks cannot accumulate since such disparities are automatically eliminated after each incremental advance.

Precise cog spacing on the lifting rods assures parallel movement, even over long distances, and provides at any time an accurate measurement of the distance covered.

Steel-bar jack-rod system. This system uses standard square steel bars and a grip-jaw arrangement that is essentially similar to that of a slip-form jack, except that the jaws are housed in thrust blocks above and below two parallel hydraulic cylinders. This allows the jack rod to be inserted or removed laterally from the lift-climbing device, or vice versa.

The lifting/lowering stroke is gage-set within the climber unit and is 4 in. (100 mm) for 12-, 16-, and 25-ton climbers; 8 in. (200 mm) for 42- and 100-ton units.

The most commonly used units in this system are hydraulically double-acting lift climbers. These climbers have a fully automatic lifting action and a semiautomatic lowering action. This means that at lifting, the operations are made only by the central hydraulic pump and at lowering the operations are made partly by the pump and partly manually by disengaging and engaging the grip jaws of the individual climbers.

Alternatively, if particular projects involve a considerable lowering distance, the climbers are provided with an up/down device. The changeover from lifting to lowering involves a simple manual operation with the up-down device at the point of directional change. Thereafter, the lowering operation is carried out fully automatically through the pump only.

A further device can be provided where the lowering of rigid loads is undertaken, thus avoiding differences in synchronization between particular lift climbers. The device is an equalizer that involves all the climbers being interconnected to a supplementary oil-cushion system. Equalizer devices are only used with lift climbers having the up-down arrangement fitted to allow for a considerable lowering distance.

A jacking system with the climbers can be arranged in one of two ways:

1. The climbers can be mounted on a static structure with the jack rods attached to the particular load. Thus the rod will pass through the climber.
2. The climber can be assembled under a load with the rods suspended from a supporting structure or fixed to a supporting trestle or mast structure.

For systems with lower lift capacity, the grip-jaw arrangement can be modified to allow using selected steel wire ropes with certain types of lifting operations.

A wide range of single- and double-acting hydraulic pumps are available with varying capacities to meet the range of requirements that may result from a particular lifting/lowering operation. In general, heavy lifting jobs require the use of high-pressure rubber hoses that will accommodate the normal working pressure of these pumps, which operate at between 100 and 170 kg/cm^2.

Rollers and Skids

For moving heavy loads across a floor or ground, hardwood (usually maple) rollers 7½ in. in diameter by 10 ft. long are commonly used. Pipe rollers may also be used, but never, under any condition, use an oxygen cylinder or any other high-pressure gas cylinder in place of a roller. When full, such cylinders may have a pressure in excess of 2,000 lb/in.^{2}

Timber skids are commonly used under heavy machinery or other equipment that is being moved on rollers. These are, for all practical purposes, simple wooden beams.

Definition of Terms

Double-Acting Jack lifting load on both upward and downward lever strokes.

Foot Lift Protruding toe or foot near base of jack for lifting low loads.

Pawls Devices, fastened to lever socket and frame, with teeth fitting into rack notches and providing the grip that raises, lowers, or holds the rack.

Rack Notched bar on ratchet jack that rises as lever is operated.

Single-acting Jack raised, only on downward stroke of lever, lowered only on upward stroke.

Chapter

16

Safety Belts, Lifelines, and Nets

Safety Belts, Lifelines, and Lanyards

A variety of personal restraining devices meant to be worn and used by employees working on the construction, repair, or demolition of structures are available to protect the wearer from falls and to minimize the risk of severe physical injury from a fall. These include safety belts, harnesses, lanyards, lifelines, and drop lines (see Figs. 16.1 to 16.5).

General-purpose belt

Designed to be worn on building projects, such as when erecting steel, the general-purpose safety belt consists of a waist belt, a restraining line, and metal components (buckles, D rings, and hooks). Its purpose is to catch the wearer smoothly in the event of a fall and to limit the distance of fall.

The safest design of a safety belt, however, consists of a set of straps, similar to parachute harness straps, which enable not only the trunk but also the legs of the wearer to absorb the forces exerted on the body during a fall. A number of factors determine which particular model of safety belt is applicable to a specific task. These include the method of work, the extent of exposure to a fall from a height, the length of time of such exposure, and so forth.

Figure 16.1 Typical safety-belt variations. In addition to the belt itself, there is a lanyard which is spliced to a D ring.

Proper use of these belts (and they must not only be worn but also hooked up) largely depends on the wearer's understanding of the need for safe working habits and willingness to act accordingly.

When using a safety belt, the hook attached to the end of the restraining line must be firmly anchored to a sufficiently strong fixture above the point of operation. On some projects, chains or steel-cable straps may have to be used to create the proper support. Such anchorages should be sufficiently high so that there is a minimum of slack line. Under no circumstances should it be lower than the belt attachment to the restraining line. The anchorage should be capable of supporting a minimum deadweight of 5,400 lb.

Because of its short restraining line, a general-purpose safety belt often restricts a wearer's movements. The drawback of restricted movement, however, can sometimes be obviated by fastening the hook of the restraining line to a horizontally stretched steel cable. Another possibility for extending the freedom of movement, especially in a vertical direction, is to use a rope-grab fall prevention device. This is a simple apparatus that is slid onto the restraining line and attached directly to the D ring of the waist belt.

When not under load, the device can easily be moved along the line (by depressing a catch) so that the wearer can move around. Once the catch is released, the device can no longer be moved.

Another aid in providing greater freedom of movement is to use a safety block in combination with the safety belt. The safety block, which must be suspended from a sufficiently rigid fixture, holds a steel wire rope mounted on a spring-loaded drum. The end of this rope is fastened to the safety belt.

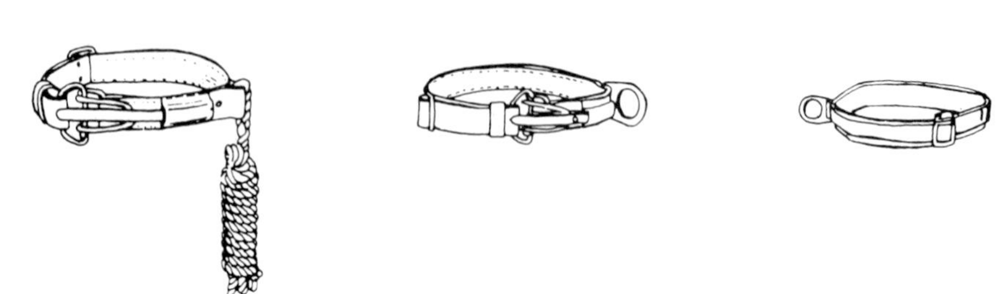

Figure 16.2 Some types of iron-worker quick-release safety belts.

Figure 16.3 Using lifelines and safety belt together.

During normal movement, the rope pulls out and winds up automatically, keeping the rope taut and giving the user freedom of movement. A sudden pull on the line engages a locking device and the rope is stopped. If the rope drum uses a friction brake, there is no jolt and the wearer is stopped smoothly from a fall.

Lineman's belt

These waist-worn restraining devices are used when climbing and working on towers, masts, and similar structures. They serve a dual function: they act as a support for the wearer, thereby freeing both hands for work; and they protect the wearer from the danger of falling from a height.

Construction of a lineman's belt is practically the same as that of a general-purpose safety belt, except that the lineman's waist belt is fitted at two points with a D ring. The length of the restraining line for such a device is sometimes adjustable by means of a friction-type buckle. A lineman's belt is sometimes also fitted with two restraining lines, such as those used by window cleaners.

In practice, the use of a lineman's belt presents few problems, because it is an important aid in the work the wearer is performing. Particular attention must be paid to the soundness of the anchorage points to which the restraining line or lines are secured.

Figure 16.4 Vertical and horizontal lifelines plus safety belt ensure worker safety.

A rope-grab fall protection device is occasionally used in combination with a lineman's belt. The restraining line used in such a case is a longer one and is fastened to one D ring. The rope-grab fall prevention device is attached to the other D ring.

Belts for enclosed spaces

These belts are worn for work in enclosed spaces, such as in tanks, where there may be potential danger from a lack of air or from existing vapors, fumes, or gases that might overcome a worker entering the space. With this type of belt it is possible for a worker on watch outside the confined space to rescue the wearer of the belt in the event that the wearer becomes injured or unconscious.

This type of belt consists of a set of straps to which a rescue line is securely connected. A wrist strap on the rescue line permits the wearer to be hauled up through an opening.

In practice, a few difficulties may be encountered with this type of belt. For example, a rescue line of sufficient length must be attached to the belt to permit the worker on watch to maintain contact with the belt wearer at all times.

Materials

The waist belt, the shoulder and leg strap harness, and the restraining line on all these safety devices should be made from synthetic fi-

Figure 16.5 Attaching lifeline to safety belt harness.

bers, which are strong, lightweight, and weatherproof. The older types of leather belts and harnesses should not be used, because the cross section of such devices is not always homogeneous, and thus not always strong at all points. In addition, leather is susceptible to weathering, deteriorating quite rapidly and thus requiring considerable maintenance.

Because synthetic fibers have a considerable amount of stretch before being permanently deformed, restraining lines made from them have a favorable effect on the size of the forces exerted on the body during a fall.

When in use such restraining devices are exposed to high temperatures or harsh environments, other materials may be used in place of synthetic fibers. For instance, lifelines used on rock-scaling operations or in areas where the lifeline may be subjected to cutting or abrasion, should be a minimum of $\frac{7}{8}$-in.-diameter wire core manila rope. For all other lifeline applications a minimum of $\frac{3}{4}$-in.-diameter manila, or equivalent synthetic-fiber, rope with a minimum breaking strength of 5,400 lb should be used.

Restraining lines subject to mechanical damage are sometimes protected by sleeving a PVC tubing over the line.

Safety-belt lanyards should be a minimum of $\frac{1}{2}$-in.-diameter nylon or equivalent, with a maximum length to provide for a fall of no greater than 6 ft. The rope should have a nominal breaking strength of 5,400 lb.

All hardware on safety belts and lanyards, except rivets, should be capable of withstanding a tensile loading factor of 4,000 lb without cracking, breaking, or permanently deforming.

Storage and maintenance

Safety belts must be kept in a cool, dry, and well-ventilated place, and should not be exposed to direct sunlight during storage. To satisfy these conditions and also to avoid unnecessary fouling of lines and danger of damage or attack by aggressive substances, these safety devices should be stored in a special room.

Safety belts returned after use should be cleaned, if necessary, according to the manufacturer's instructions. In addition to visual inspection upon issue and return, safety belts must be periodically examined by a qualified person for possible defects. If repairs are required, they should be made only by a competent person, preferably the manufacturer. Any lifeline, safety belt, or lanyard actually subjected to inservice loading, as distinguished from static load testing, must be immediately removed from service and should not be used again as a personal safety device.

Every safety belt should carry a serial number, so that records can be kept of repairs and results of examinations and testing, along with the appropriate dates of such actions.

Safety Nets

Safety nets are used on a wide variety of jobs, such as steel erection and bridge repairs, and must be provided when persons are working 25 ft or more above ground, water, machinery, or any other surface and are not otherwise protected by safety belts, lanyards, lifelines, or scaffolding (see Figs. 16.6 to 16.9).

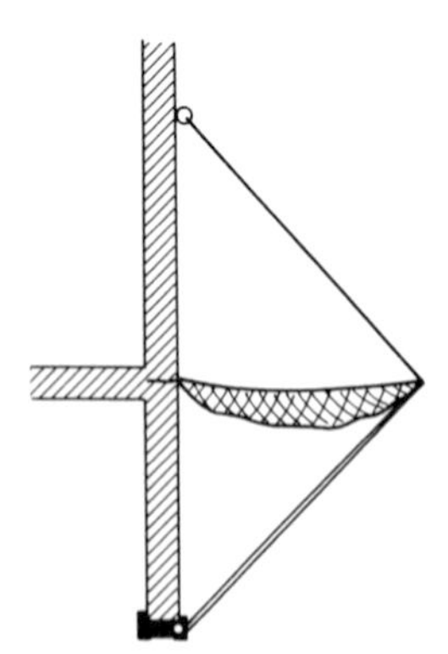

Figure 16.6 Perimeter nets protect building sides. *(Courtesy of Sinco Products Company.)*

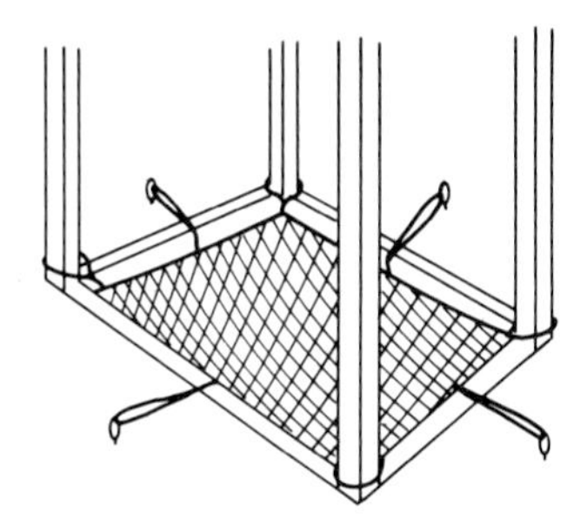

Figure 16.7 Deck opening with mechanical chases and shafts for vertical and horizontal steel work. *(Courtesy of Sinco Products Company.)*

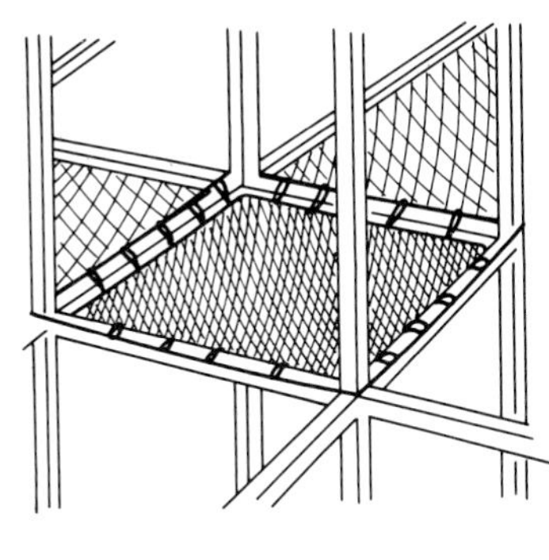

Figure 16.8 Decking nets hook to surrounding beams and girders. *(Courtesy of Sinco Products Company.)*

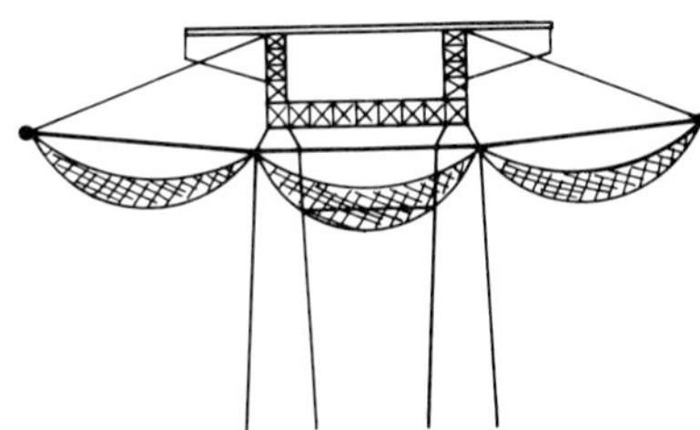

Figure 16.9 Beneath bridges nets can slide along as the work progresses. *(Courtesy of Sinco Products Company.)*

When using safety nets, not only the nets themselves, but also the structure to which the nets are attached, must be capable of catching and supporting at least two persons at a time.

Safety nets are usually made of synthetic fibers, which are strong, lightweight, weatherproof, and rotproof. Resistance against ultraviolet radiation is increased by impregnating the nets with a UV absorbent.

The maximum size mesh for safety nets must not exceed 6 by 6 in. The perimeter line to which the net is attached should have a minimum of 3⁄8-in.-diameter no. 1 grade pure manila rope, 1⁄4-in. nylon rope, 5⁄16-in. polypropylene rope, 1-in. nylon webbing, or their equivalents. Materials shall be compatible with each other.

All new nets shall meet accepted performance standards of 17,500-ft-lb minimum impact resistance as determined and verified by the manufacturer. Edge ropes shall provide a minimum breaking strength of 5,000 lb.

Eyes are often provided on the corners of safety nets thus strengthening the perimeter line (usually by increasing the diameter) to spread forces adequately.

Drop forge safety hooks or shackles (or other equivalent fastening means) that will support the design lead should be used to attach mats to supporting cables, structures, or beams projecting from the structure (see Fig. 16.10).

When two nets are used to cover a large work area, connections between net panels must develop the full strength of the net. This may be done by means of a cord made of artificial fibers similar to the cords of the net. The diameter of this connecting cord should be at least that of the perimeter line and it must be knotted every 3 ft. Another way of connecting nets is with self-closing hooks that should be attached directly against the perimeter line at 24-in. intervals (see Figs. 16.11 and 16.12).

If falling objects are also to be caught in the net, then the net lining should be of a mesh not more than 1 in. by 1 in., constructed of synthetic twine (or its equivalent) of a size not less than no. 18 nylon seine twine, or of wire not less than 22 gage (AWG). This reduction in mesh size can be met by using a net having such dimensions (not standard) or by using a net with standard size mesh together with an inner net of lighter construction having a smaller mesh diameter.

All nets should be cleared of debris and other objects at regular intervals during service, since continuous extended overloading can cause permanent elongation of meshes. Clearing a net also reduces the risk of injury in the event a worker falls into it (see Fig. 16.13).

When attaching nets:

The lowest point of the net should not be lower than 25 ft below the working level.

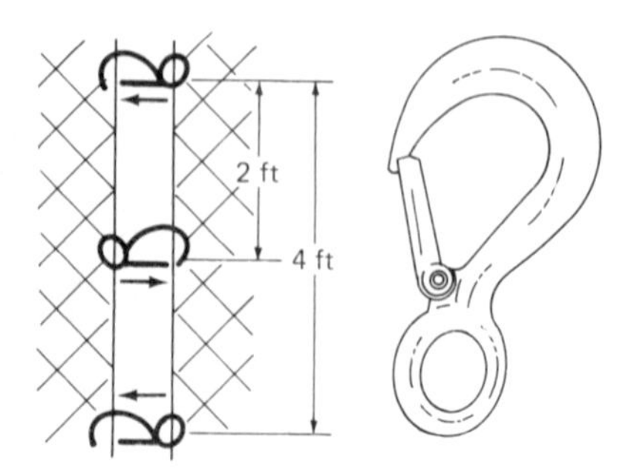

Figure 16.10 Simplified method of net hook attachment. *(Courtesy of Sinco Products Company.)*

Figure 16.11 Personnel nets cut hazards on bridge work. *(Courtesy of Sinco Products Company.)*

The distance between the lowest point of the unloaded net and an object below it should not be less than 12 ft.

The net should extend at least 8 ft beyond the edges of the working area.

To facilitate attachment of the net, it is necessary to make arrangements in the preconstruction stage. This may require welding steel bars to the steel structure or inserting fasteners in concrete columns, to which steel pipes or steel cables can be attached.

Maintenance and storage

Care, maintenance, and storage of safety nets should be in accordance with the manufacturer's recommendations.

Figure 16.12 Safety nets work with scaffolds during chimney construction. *(Courtesy of Sinco Products Company.)*

In addition to visual inspection upon issue and return, nets must be examined periodically for possible defects, which may be caused by sharp edges, sparks from welding or cutting tools, open flames, hot tubing, chemicals, and so forth.

Any repairs needed should be made preferably by the manufacturer. Every safety net should carry a serial number so that records can be kept of details such as repairs, results of examinations and testing, age in use, and appropriate dates of any action taken.

Before storage, nets should be completely dry and as far as possible clean. Net mesh must be stored in a dark, cool, dry, and well-ventilated place free from chemicals, hot tubing, and exposure to sunlight—preferably on wooden pallets.

Testing on the job

All safety nets should be tested on the job in a suspended position immediately following initial installation or major repair; and at 6-month intervals thereafter. Such a test consists of dropping a 400-lb bag of sand, not more than 30 in. in diameter, from a height of 25 ft above the net into the center of the net. No broken strands or significant distortion of the net pattern or the suspension system is permitted.

Figure 16.13 Nets make a good debris-collecting point. *(Courtesy of Sinco Products Company.)*

Inspection

During use, all safety nets, mesh ropes, perimeter lines, connectors, suspension systems, and the like should be completely inspected by a qualified person after each installation and not less than once a week thereafter. Additional inspections should be made after alterations, repairs, impact loading, and welding or cutting operations above the nets.

Nets that show mildew, wear, damage, or deterioration that substantially affects their strength should be removed from service immediately for complete inspection and repair or disposal.

Factors affecting net life

Sunlight. Special precautions must be taken to shield ropes ½ in. in diameter and smaller from the sun's rays. Ropes of natural or synthetic fibers can lose a significant amount of strength after prolonged exposure to direct sunlight's ultraviolet rays. Nylon safety netting should be dyed with a UV-absorbing dyestuff of known ability to increase outdoor durability significantly. All nets not in use should be protected from direct and indirect sunlight.

Abrasion. The adverse effects of abrasion should be constantly considered. Nets should not be dragged over the ground or other rough surfaces.

Sand. Embedded sand cuts fibers, reducing the strength of a net. Care should be taken to keep nets as clean and free of sand as possible.

Rust. Prolonged contact with rusting iron or steel can cause significant degradation and loss of strength. Nets should not be stored in metal containers that are rusty, but instead should be stored either on nonrusting hooks or on wooden pallets.

Airborne contaminants. Extremely high concentrations of many chemicals can adversely affect the strength of nets. Where such high concentrations may exist, the chemicals should be identified and the concentrations measured. The effect on the net material involved, if not known, should be determined by a test.

Working over or near water

Even with safety nets properly in place, when working over or near water where the danger of drowning exists, workers should be provided with U.S. Coast Guard approved life jackets or bouyant work vests.

Prior to and after each use such life preservers or buoyant vests should be inspected for defects that might alter their strength or buoyancy. Defective units must not be used.

Ring buoys with at least 90 ft of line should be provided and readily available for emergency rescue operations. The distance between ring buoys should not exceed 200 ft.

At least one lifesaving skiff should be immediately available at locations where employees are working over or adjacent to water.

Section

5

Scaffolding and Ladders

Chapter

17

Scaffolds and Ladders (General)

Practically all types of construction jobs require the use of scaffolds and ladders to enable workers to reach and work in locations that are otherwise inaccessible, and to do their jobs better and more efficiently. Scaffolds seldom are economical for work at heights of less than 30 ft, unless access to such work is impractical by ladder or any other means.

Scaffolds

The design of a scaffold depends on the feasibility of its erection, its economics, structural stability, and safety. Requirements of erection feasibility are mainly that the disposition, size, shape, and strength of a scaffold be capable of being erected by workers without the help of machinery. The economics of a scaffold, or scaffolding system, must take into consideration initial purchase, maintenance, durability, ability to reuse, ease of erection and dismantling, and structural characteristics. Above all, the scaffold must be capable of supporting itself and the applied load with an acceptable safety factor and without unacceptable deformation.

There are many different types of scaffolds used in construction to match a variety of job applications. In general, these are divided into two broad categories: support or falsework, and access. Only the access-type scaffold is covered in this handbook.

Access scaffolds consist of three main designs:

Swinging. Two-point suspension, masons' adjustable multipoint suspension, and one-person bosun's chairs and similar devices. These types of scaffolds are hoisted by hand-operated rope blocks and falls, by steel wire rope manual hoisting mechanisms, or by air-powered or electrically powered machines at platform level (see Fig. 17.1).

Stationary. Independent self-supporting structures erected adjacent to a permanent structure and tied to it for lateral stability; partially self-supporting structures (putlog) that rest partly on the permanent structure and obtain essential lateral stability from it; and truss-out, or cantilever, type structures that bear entirely on and cantilever out, or hang or span from the permanent structure (see Fig. 17.2).

Special. Truck-mounted hydraulic-arm platforms, telescopic mast, continuous masts, modular tubular metal systems, and tower-type cantilevered platforms (see Fig. 17.3).

Scaffolding materials include built-up timber structures, steel or aluminum tubing and couplers, and seamless steel or extruded aluminum tubing welded panels.

For the safety of workers, all scaffolds should comply with basic safety standards. Not only should scaffolds be provided with guardrails, midrails, and toeboards to protect the workers using the structures, but they also should have overhead protection for the workers and wire meshing on the scaffold's open side to prevent objects from falling and injuring those working or passing beneath the scaffold. Safety belts and separately attached lifelines should be provided for each worker on a swinging scaffold.

Figure 17.1 Safety swinging scaffold with two-point suspension. It utilizes steel wire rope with ratchet-action raising and crank-handle lowering.

Figure 17.2 Typical stationary scaffold. (*Courtesy of Patent Scaffolding Company.*)

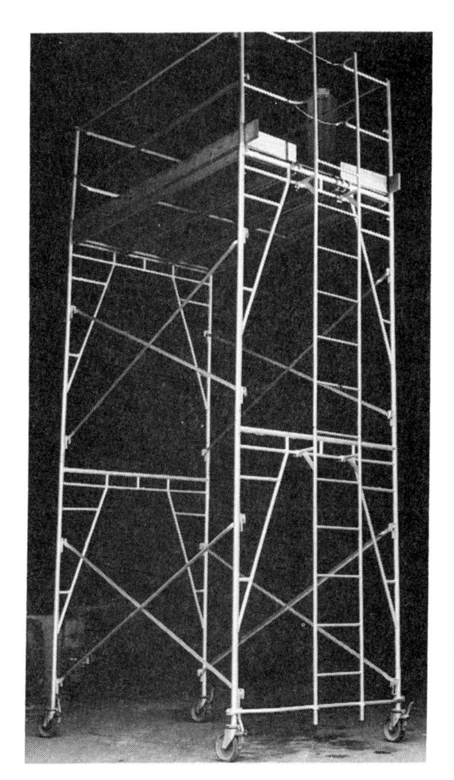

Figure 17.3 Sectional rolling scaffold. Steel frames are 5 ft wide and joined by pivoted diagonal braces 4 to 10 ft long. A hook-on type access ladder is used with a 3-ft-high grab rail at the top, guarded by steel wire rope with snap hooks. The complete lift of the ladder is set back 7½ in. from the scaffold frames.

Ladders

Portable ladders are available in a variety of designs and lengths, with varying duty ratings. Proper selection depends on the analysis of the particular job application and the service required of the ladder.

All ladders can be broadly classified as self-supporting or non-self-supporting:

1. *Self-supporting.* Step, trestle, extension trestle, and platform ladders
2. *Non-self-supporting.* Single, extension, and sectional ladders

Maximum allowable ladder lengths will depend on the material of construction: wood, metal, or plastic. Ladders are generally available in one of three duty ratings: heavy, medium, or light.

Scaffolding Glossary*

Bearer Horizontal member of a scaffold upon which the platform rests and which may be supported by ledgers.

Bosun's Chair Seat supported by slings attached to a suspended rope, designed to accommodate one worker in a sitting position.

Brace Tie that holds one scaffold member in a fixed position with respect to another member.

Bricklayer's Square Scaffold Scaffold composed of framed wood squares that support a platform; limited to light and medium duty.

Carpenter's Bracket Scaffold Scaffold consisting of wood or metal brackets supporting a platform.

Coupler Device for locking and aligning together the components of a tubular metal scaffold.

Crawling Board or Chicken Ladder Plank with cleats spaced and secured at equal intervals for use by a worker on roofs, not designed to carry any material.

Design Working Load (Maximum Intended Load) Total of all loads, including the weight of the people, materials, equipment, and platform.

Double-Pole (Independent-Pole) Scaffold Scaffold supported from the base by a double row of uprights, independent of support from the walls, and constructed of uprights, ledgers, horizontal platform bearers, and diagonal bracing.

Float (Ship) Scaffold Scaffold hung from overhead supports by means of ropes and consisting of a substantial platform having diagonal bracing underneath, resting on and securely fastened to two parallel plank bearers at right angles to the span.

Guardrail Barrier along the exposed (open) sides and ends of platform to prevent persons from falling.

Handrail Rail connected to a ladder stand running parallel to the slope or top step, or both, and serving the purpose of assisting the user in maintaining balance while climbing and descending.

---

* Based on American National Standard definitions as published in "Safety Requirements for Scaffolding," ANSI Std. A10.8.

Heavy-Duty Scaffold Scaffold designed and constructed to carry a working load not to exceed 75 lb/in.^{2}

Horse Scaffold Scaffold for light or medium duty, composed of horses supporting a work platform.

Interior-Hung Scaffold Scaffold suspended from the ceiling or roof of a structure.

Ladder Jack Scaffold Light-duty scaffold supported by brackets attached to ladders.

Ladder Stand Mobile fixed-size self-supporting ladder consisting of a wide flat tread ladder in the form of stairs. The assembly may include handrails.

Ledger (Stringer) Horizontal scaffold member that extends from post to post and which supports the putlogs or bearers, forming a tie between posts.

Light-Duty Scaffold Scaffold designed and constructed to carry a working load not to exceed 25 lb/in.^{2}

Manually Propelled Mobile Scaffold Portable rolling scaffold supported by casters.

Mason's Adjustable Multipoint Suspension Scaffold Scaffold having a continuous platform supported by bearers suspended by wire rope from overhead supports, so arranged and operated as to permit the raising or lowering of the platform to desired working positions.

Medium-Duty Scaffold Scaffold designed and constructed to carry a working load not to exceed 50 lb/in.^{2}

Mobile Scaffold (Tower) Light-, medium-, or heavy-duty scaffold mounted on casters or wheels.

Mobile Work Platform Fixed work level on casters or wheels, with bracing from platform to vertical frame.

Needle-Beam Scaffold Light-duty scaffold consisting of needle beams supporting a platform.

Outrigger Scaffold Scaffold supported by outriggers or thrust-outs projecting beyond the wall or face of a building or structure, the inboard ends of which are secured inside the building or structure.

Putlog Scaffold member upon which the platform rests.

Runner Lengthwise horizontal bracing or bearing members, or both.

Scaffold Temporary elevated platform and its necessary vertical, diagonal, and horizontal members used for supporting workers and materials, also known as scaffold tower.

Single-Point Adjustable-Suspension Scaffold Manually or power-operated unit designed for light-duty use, supported by a single wire rope from an overhead support, so arranged and operated as to permit the raising or lowering of the platform to desired positions.

Single-Pole Scaffold Scaffold with platforms resting on putlogs or cross beams, the outside ends of which are supported on ledgers secured to a single row of posts or uprights and the inner ends of which are supported on or in a wall.

Stone Setter's Adjustable Multiple-Point Suspension Scaffold Swinging-type scaffold having a platform supported by hangers suspended at four points so as to permit the raising or lowering of the platform to the desired working position by the use of hoisting machines.

Toeboard Barrier at platform level erected along the exposed sides and ends of a scaffold platform to prevent materials and equipment from falling.

Tube-and-Coupler Scaffold Assembly consisting of tubing that serves as posts, bearers, braces, ties, and runners, a base supporting the posts, and uprights and couplers that serve to join the various members, usually used in fixed locations.

Tubular Fabricated Sectional Folding Scaffold Sectional metal scaffold that folds. Either of ladder frame or of inside stairway design, substantially built of prefabricated sections consisting of end frames, platform frames, inside inclined stairway frame and braces, or hinge-connected diagonal and horizontal braces, capable of being folded into a flat package when not in use.

Tubular Welded-Frame Scaffold Sectional panel or frame metal scaffold substantially built up of prefabricated welded sections that consist of posts and horizontal bearers with intermediate members. Panels or frames are braced with diagonal or cross braces.

Two-Point Suspension Scaffold (Swinging) Scaffold whose platform is supported by hangers (stirrups) at two points, suspended from overhead supports so as to permit the raising or lowering of the platform to the desired working position by tackle or hoisting machines.

Window Jack Scaffold Scaffold whose platform is supported by a bracket or jack and projects through a window opening.

Ladder Glossary*

Ladder Appliance usually consisting of two side rails joined at regular intervals by cross pieces called steps, rungs, or cleats.

Step Ladder Self-supporting portable ladder, nonadjustable in length, having flat steps and a hinged back. Its size is designated by the overall length of the ladder measured along the front edge of the front side rail, including top cap and foot.

Single Ladder Non-self-supporting portable ladder, nonadjustable in length, consisting of but one section. Its size is designated by the overall length of the side rail, not including any foot or end cap.

Extension Ladder Non-self-supporting portable ladder, adjustable in length. It consists of two or more sections traveling in guides or brackets so arranged as to permit length adjustment. Its size is designated by the sum of the lengths of the sections measured along the side rails.

Combination Ladder Portable ladder capable of being used either as a step ladder or as a single or extension ladder. It may also be capable of being used as a trestle or stairwell ladder. Its components may be used as single ladders. Its size is designated by the overall length of the ladder measured along the front edge of the front side rail (step-ladder section), including the foot and the top cap, or the top step when no top cap is used. Combination-type ladders, where both components are constructed with rungs, shall be considered special-purpose ladders.

Sectional Ladder Non-self-supporting portable ladder, nonadjustable in length, consisting of two or more sections of ladder so constructed that the sections may be combined to function as a single ladder. Its size is designated by the overall length of the assembled sections. The size of any individual

---

* Based on American National Standard definitions as published in "Safety Requirements for Portable Metal Ladders," ANSI Std. A14.2.

section is designated by the length of the section, measured along a side rail, plus any material necessary for attachment.

Trestle Ladder Self-supporting portable ladder, nonadjustable in length, consisting of two sections hinged at the top to form equal angles with the base. The size is designated by the length of the side rails measured along the front edge, including foot or shoe.

Extension Trestle Ladder Self-supporting portable ladder, adjustable in length, consisting of a trestle ladder base and a vertically adjustable single ladder, with suitable means for locking the ladders together. The size is designated by the length of the trestle ladder base, along the front edge of the side rail, including the shoe plus the allowable extended length of the extension section measured along its side rail.

Special-Purpose Ladder Portable ladder that represents either a modification or a combination of design or construction features in one of the general-purpose types of ladders that will adapt the ladder to special or specific uses.

Platform Ladder Self-supporting ladder of fixed size with a platform provided at the working level. The size is designated by the length along the front edge of the front rail from the top of the platform to the base of the ladder, including any foot or shoe.

Chapter

18

Swinging and Suspended Scaffolds (Manual and Powered)

The swinging or suspended scaffold is available in a variety of shapes and sizes, ranging from the one-person bosun's chair to the articulated suspended-platform type scaffold. These units are used for such jobs as cleaning and painting, maintenance, repair, brickwork, and similar jobs on exterior walls, towers and tanks, especially where it is necessary to have access to large vertical surfaces and quick up-down mobility (see Fig. 18.1).

Swinging Scaffold

Most swinging scaffolds are two-person work platforms, of either wood or aluminum, suspended from lookouts that extend over the edge of flat-roofed buildings or from special U-shaped hangers on buildings that have a wide cornice.

Figure 18.1 Typical motorized swinging scaffold. (*Courtesy of Patent Scaffolding Company.*)

This type of scaffold may be hand-operated using simple ropes and block and tackle, or it may be suspended from high-tensile steel wire ropes, relying on hand-operated or power-driven remote-controlled winches to provide travel both vertically and horizontally (see Figs. 18.2 to 18.4).

In its most familiar form, the swinging scaffold consists of a frame similar in appearance to a ladder with a decking of wood slats. The structure is supported near each end by a steel stirrup to which the lower block of a set of manila rope falls is attached. Its frame should be constructed of clear straight-grained spruce rails and should have dimensions in accordance with those given in Table 18.1.

The rungs of a swinging scaffold should be made of straight-grained oak, ash, or hickory, at least 1⅛ in. in diameter with ⅞-in. tenons mortised into the side stringers at least ⅞ in. deep and spaced not more than 18 in. apart. The stringers should be tied together with tie-rods not less than ¼ in. in diameter, passing through the stringers and riveted up tight against washers on both ends.

The platform flooring should be ½-by-3-in. strips spaced not more than ⅝ in. apart, except at the side rails, where the space may be 1 in. The standard width of the scaffold is 28 in.

A 1-by-4-in. toeboard should be provided on the outboard side of such scaffolds. A guardrail not less than 2 by 4 in. should be located not less than 36 in., nor more than 42 in., above the platform floor, inserted into the sockets or loops in the stirrups provided for this purpose. When

Figure 18.2 Motorized scaffold going around king post at U.S. Naval shipyard. (*Courtesy of Patent Scaffolding Company.*)

Figure 18.3 Motorized scaffold used to move workers and materials. (*Courtesy of Patent Scaffolding Company.*)

Figure 18.4 Scaffold modules set above passageway so that work can progress while a train can operate on the track beneath. (*Courtesy of Patent Scaffolding Company.*)

TABLE 18.1 Frame Dimensions of Ladder-Type Platforms

	Length of platform ft				
	12	14 and 16	18 and 20	22 and 24	28 and 30
Side stringers, minimum cross section (finished sizes)					
At ends	1¾ × 2¾	1¾ × 2¾	1¾ × 3	1¾ × 3	1¾ × 3½
At middle	1¾ × 3¾	1¾ × 3¾	1¾ × 4¼	1¾ × 5	
Reinforcing strip, minimum	A ⅛ × ⅞-in. steel reinforcing strip or its equivalent shall be attached to the side or underside, full length.				
Rungs	Minimum diameter 1⅛-in. with at least ⅞-in.-diameter tenons; maximum spacing 12 in. center to center.				
Tie rods					
Number, minimum	3	4	4	5	6
Diameter, minimum, in.	¼	¼	¼	¼	¼
Flooring, minimum finished size, in.	½ × 2¾	½ × 2¾	½ × 2¾	½ × 2¾	½ × 2¾

SOURCE: From ANSI Std. A10.8.

Figure 18.5 Nonmotorized scaffold utilizes wooden planking and overhead protection. (*Courtesy of Patent Scaffolding Company.*)

required, a midrail of 1-by-4-in. lumber should be installed at all open sides on scaffolds more than 10 ft above the ground or floor (see Fig. 18.5).

Additional stanchions should be provided to keep the span of the guardrail to 10 ft or less. A screen of ½-in.-mesh rabbit wire is recommended between the guardrail and the toeboard.

When plank-type platforms are used, they should be constructed of not less than nominal 2-by-8-in. unspliced planks, properly cleated together on the underside, starting 6 in. from each end, with intervals not exceeding 4 ft. This type of platform should never extend more than 18 in. beyond the hangers. A bar, securely fastened to the platform at each end, will prevent its slipping off the hanger. Limit spans between hangers to 10 ft. For beam-type platforms, side stringers should be made of lumber not less than 2 by 6 in. set on edge.

The span between hangers on this type of platform should not exceed 12 ft. Platform flooring of 1-by-6-in. material, properly nailed and spaced not more than ½ in. apart, should be supported on 2-by-6-in. cross beams. These should be laid flat, set into the upper edge of the stringers with a snug fit at intervals of not more than 4 ft, and securely nailed in place.

If light-metal-type platforms are used, make sure that they are of tested design and approved for the particular scaffold duty intended. The steel stirrups that support the platform should be of wrought iron or mild steel design having a cross-sectional area capable of sustaining four times the maximum intended load. Stirrups should be placed between 6 and 18 in. from the ends of the scaffold and secured to it by U bolts of adequate size.

A set of ¾-in. no. 1 grade manila rope falls, consisting of a double- and a single-pulley 6-in. block, should be provided at each end of the scaffold. A safe means of supporting the upper blocks is absolutely necessary. This may be in the form of U-shaped roof or cornice hooks, ½-in. wire rope slings, or other approved design. Roof irons or hooks should be of wrought iron or mild steel of proper size and design, securely installed and anchored. A ¾-in. manila rope tieback should be used to secure the cornice hook to a fixed anchorage on the roof.

Good practice includes wiring the hook on the lower pulley block to the eye of the stirrup, thus preventing accidental detachment of the hook. A mousing would ordinarily be used for such a condition; however, its use here would interfere with the hitch of the rope.

Instead, a hitch is made by holding a strain on the rope with one hand and pushing a bight of the slack part of the rope through the inverted V or the stirrup. The bight is then given a 180° twist and is placed over the bill of the hook. The strain on the live part of the rope forces the dead part into the V, into which it jams. Although a very simple hitch, it is dependable.

To ensure against the scaffold falling if a worker were to lose grip on the rope accidentally while raising or lowering the scaffold, it is necessary to attach a special safety latch to the cheek of the lower block. The hauling part, or hand rope, is passed through the hole in the hinged plate, and the hitch is made in the manner described previously. If the worker accidentally lets go of the rope, the safety latch will then raise and grab the rope; the scaffold would be prevented from falling.

For special jobs a box-type scaffold is often used, which provides a two-member guardrail and toeboard on all sides.

When workers are using a swinging scaffold, make sure it is always secured to the building or structure to prevent it from moving away and allowing the worker to fall between. On building walls it is usually difficult to find something to lash a scaffold to, but it is often found practicable to provide a standard attachment for a window-cleaner's belt on a short length of manila rope. To hold the scaffold to the wall, the device is simply attached to the special bolt at the side of the windows and the rope is secured to the platform.

Safety belts for each worker on the scaffold, on separately attached lifelines, are essential for proper worker safety. Only new rope, ¾-in. or 1-in. diameter, should be used and it should be properly secured at the roof or upper part of the structure. Where the rope passes over sharp bends, such as copings or window sills, it should be padded to prevent abrasion.

Each worker should wear a 4-in. life belt of three-ply cotton (nylon) webbing with a ⅝-in. or ¾-in. rope tail line about 6 ft long. The tail line should be attached, as short as is practicable, to the hanging lifeline by a rolling hitch. This hitch can be readily slid up or down the hanging rope, yet if the worker falls, the hitch will jam and hold him. The rolling hitch is similar to the clove hitch, except that in tying the first part of the hitch, two wraps (instead of one) are made around the hanging line.

Nylon rope may be more suitable for lifelines because, in addition to being strong, it stretches and will stop a falling worker more gently. The lifeline must reach to the ground or other place of safety, and the worker must have the safety belt on and attached to the lifeline at all times, particularly when the scaffold is being raised or lowered.

On special hanging scaffolds, trolley cables run along the upper handrail members. The worker attaches the snap hook of the safety-belt tail line to the cable at the rear. This frees the worker to walk back and forth in his half of the scaffold. If a rope should fail, and one end of the scaffold swing downward, the workers probably would remain within the scaffold railings. In the event they were thrown out, their life belts would keep them from falling to the ground.

Although the preceding discusses manila rope falls for supporting the scaffolds, it is increasingly more common to suspend such scaffolds from wire ropes, using winches to raise or lower the scaffold. Where the lift of the scaffold exceeds 100 ft, wire rope and winches must be used. Also, where acid solutions are used, fiber ropes should never be permitted, unless they are approved as acidproof.

Among the many advantages claimed for using the wire rope winch suspension system are:

Greater safety due to more positive inspections

Ease of handling the scaffold since it can be inched up or down as desired

Lower headroom since the scaffold can be raised closer to the overhead supports

Less danger of failure when acid is used, such as in washing building walls

After a scaffold has been erected on a new job, it should be load-tested before workers risk their lives on it. Hoist the scaffold about 12 in. off the ground and apply a test weight equal to four workers for a period of 5 min. The test should be made every 10 days if the job continues for more than that time. A factor of safety of 4 should be used to calculate the strength of a swinging scaffold.

Where swinging scaffolds are suspended adjacent to each other, never place planks so as to form a bridge between scaffolds. Never permit more than two persons to work on a scaffold at any one time.

Occasionally it may be desirable to install wood bunters with rollers on a scaffold to hold it away from the building wall and to keep it from swinging or swaying.

Whenever a winch or a scaffold is to be removed to another location, always wind the cable on the winch properly. Never coil it up on the ground to save time. A kink in the cable will weaken and may even ruin it. The maximum load that can be suspended by one of these winches is 500 lb.

Suspended Scaffolds

Suspended scaffolds are most commonly used by bricklayers working on new buildings, but they are equally applicable for any heavy construction or repair work (see Figs. 18.6 and 18.7).

This type of scaffold consists of a number of outriggers, usually 7-in. steel or aluminum I beams located at roof level, from each of which two ½-in.-diameter wire ropes wind up on hand-operated winches on the scaffold platform.

The thrust-outs, or outriggers, of a suspended scaffold should not project more than 6 ft beyond the point where they bear on the support, unless used in pairs. The fulcrum point of the beam should rest on a secure bearing at least 6 in. in each horizontal dimension. Make sure that the beam is secured in place against movement and securely braced at the fulcrum point against tipping.

The inboard ends of the beams should be anchored to the roof steel by large U bolts and anchor plates. These ends should be measured from the fulcrum point to the extreme point of support, and should not be less than 1½ times the outboard end in length. They should be spaced not more than 10 ft apart. The suspension cables should be placed not more than 2 and 6 ft, respectively, out from the bearing point of the beam.

Figure 18.6 Bricklayer's scaffold shifts its position assisted by ratchets and cables. (*Courtesy of Patent Scaffolding Company*).

The platform of a suspended scaffold consists of planks resting on putlogs or bearers, which are supported by the winches. The width of the scaffold should not exceed 28 in. Planking should be 1¼ in. thick for spans up to 6 ft and 2 in. thick for spans up to 10 ft. These planks should be laid tightly and should extend to within 3 in. of the building. Planking should be nailed or bolted securely to the putlogs, which they should overlap by not more than 18 in. at each end. A standard guardrail not less than 36 in. nor more than 42 in. high and a 9-in. toeboard should be provided along the outer edge of the scaffold. A wire screen between them is recommended.

This type of scaffold equipment provides for a 2-in. plank decking or roofing above the workers. Special hooks should be provided to hold the scaffold close to the building wall. The allowable loading on an outrigger is approximately 2,000 lb.

To raise the scaffold, levers on the winches are operated up and down to rotate them. To lower the scaffold, depress the ratchet handle with the driving pawl held out of engagement. Then replace the driving pawl

Figure 18.7 Suspended scaffolding covers building facade. (*Courtesy of Patent Scaffolding Company.*)

and disengage the locking pawl so that the weight of the scaffold is sustained on the ratchet handle, which is allowed to rise, thus unwinding the cable from the drum.

Safety Cage and Bosun's Chair

When a relatively minor scaffold job is required, a safe and convenient work platform from which to work may be provided by a one-person aluminum safety cage. This lightweight rig is more comfortable for the worker than any bosun's chair, yet it is nearly as flexible in operation. These cages are often available with extensions attached to each side of the cage for use by two persons.

In general the cages are used with power-operated winches, while the bosun's chair uses powered winches or blocks and falls (see Fig. 18.8).

When using a bosun's chair, make sure that the seat is not less than a 12-by-24-in. board, at least 1 in. thick. The seat should be reinforced on the underside by cleats, which are securely fastened to prevent the board from splitting. Slings should be of ⅝-in.-diameter fiber rope, reeved through the four seat holes so as to cross each other on the underside of the seat.

If the bosun's chair is to be used by a worker handling a gas or arc welding torch, then the seat slings must be made of (minimum) 3/8-in.-diameter wire rope.

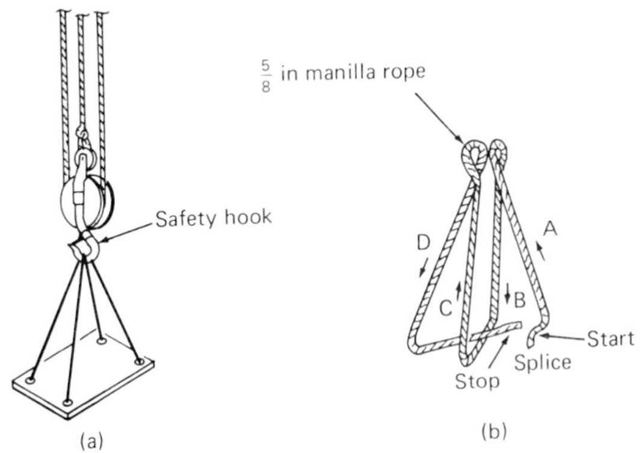

Figure 18.8 (*a*) Manual bosun's chair suspended from block. (*b*) Reeving of rope. (*Courtesy of Patent Scaffold Company.*)

Chapter

19

Stationary Scaffolds (Wood and Metal)

There are three basic types of stationary scaffolds: independent self-supporting, partially self-supporting, and cantilevered. They can be either built-up or prefabricated systems.

Built-up-type scaffolds are single- or double-pole structures made of wood or metal. The single-pole type is supported on one side by poles or uprights, on the other side by the wall or structure against which it is erected. The double-pole type, on the other hand, is erected independently of the building or structure (see Fig. 19.1).

These types of scaffolds must be built and inspected to conform with OSHA requirements as well as with state laws and local ordinances. If any portion of a wooden scaffold has been weakened or damaged by storm or accident, that scaffold should not be used until the necessary repairs have been made. Care must be exercised to prevent overloading any scaffold.

Wood-Pole Scaffolds

Construction

Woods most commonly used for scaffolds are spruce, fir, Douglas fir, and southern yellow pine. Because it has the highest ratio of strength to weight, spruce is the preferred material for wood scaffolds. All lumber used for scaffold members and planking should be reasonably free from serious defects.

When erecting wood-pole scaffolds, make sure that all poles, or uprights, are accurately plumbed. Poles should bear on a foundation of sufficient size and strength to spread the load from the pole over a suf-

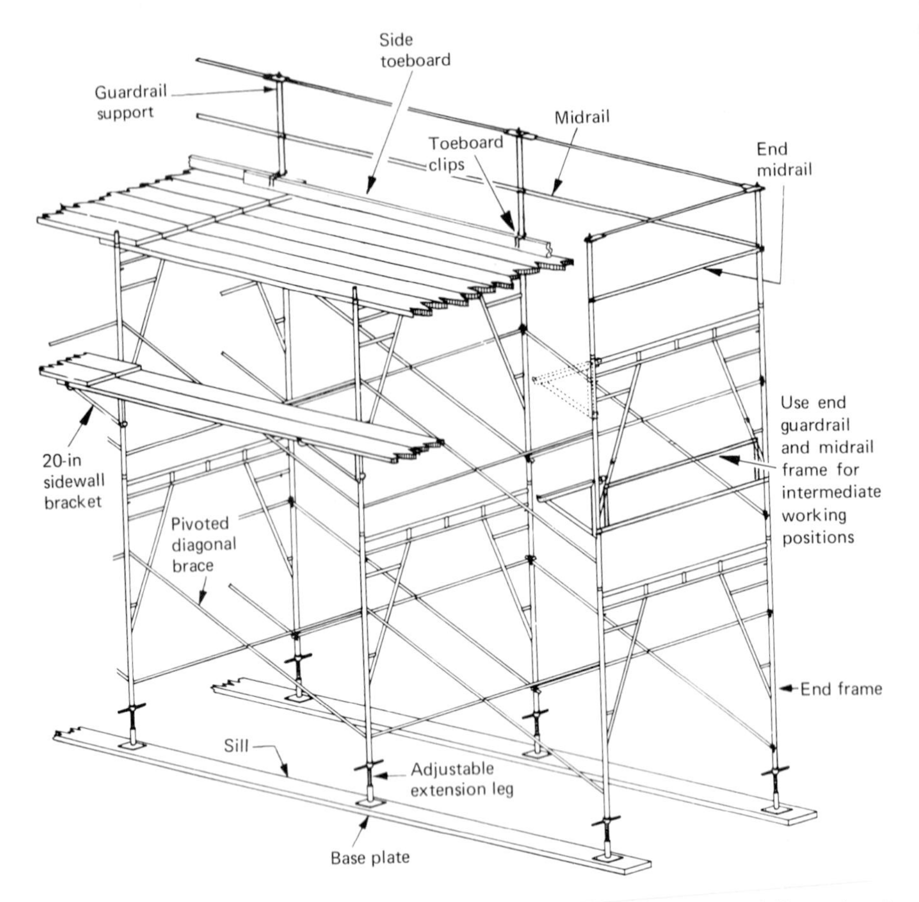

Figure 19.1 Stationary scaffolding. When scaffolds are to be partially or fully enclosed, specific precautions must be taken to assure the frequency and adequacy of the ties that attach the scaffolding to the building because of increased load conditions resulting from the effects of wind and weather. The scaffolding components to which the ties are attached must also be checked for additional lateral loads imparted thereto. *(Courtesy of Patent Scaffolding Company.)*

ficient area to prevent settlement. Independent scaffolds should be set as near to the wall of the building as practicable.

All pole scaffolds should be securely guyed or tied to the building or structure. Where height or length is greater than 25 ft, the scaffold should be secured at intervals not greater than 25 ft vertically or horizontally.

When poles must be spliced, make sure that abutting ends are squared, and that the bottom end of an upper member rests squarely upon the upper end of the lower member. Wood splice plates, not less than 4 ft long, should be used on at least two adjacent sides. Splices should overlap the abutted ends equally and should be of the same width and of a cross-sectional area not less than that of the pole.

Tables 19.1 and 19.2 give minimum sizes and maximum spacings for numbers of single-pole and independent-pole scaffolds.

Ledgers are horizontal members which extend along the length of the scaffold. They should be long enough to extend over two pole spaces. Ledgers should never be spliced between poles, but should always overlap the poles by at least 4 in. Wood blocks should be nailed to the pole below the overlapped ledgers to help support them.

Putlogs or bearers are horizontal members which extend across the width of the scaffold. They should be long enough to project over the ledgers beyond the inner and outer rows of poles by at least 3 in. for proper support. Putlogs should be set on the ledgers, with their greater dimension vertical, and nailed to the poles. Every wooden putlog on single-pole scaffolds should be reinforced with a 3⁄16-by-2-in. steel strip or its equivalent, secured to its lower edge throughout its entire length.

TABLE 19.1 Minimum Nominal Size and Maximum Spacing of Members of Single-Pole Scaffolds

	Light duty		Medium duty	Heavy duty
Uniformly distributed load	Not to exceed 25 lb/ft^2	Not to exceed 25 lb/ft^2	Not to exceed 50 lb/ft^2	Not to exceed 75 lb/ft^2
Maximum height of scaffold	20 ft	60 ft	60 ft	60 ft
Poles or uprights	2×4 in.	4×4 in.	4× 4 in.	4 × 4 in.
Pole spacing, longitudinal	6 ft	10 ft	8ft	6 ft
Maximum width of scaffold	5 ft	5 ft	5ft	5 ft
Bearers or putlogs			2× 9 in. or 3× 4 in.	2 × 9 in. or 3 × 5 in. (rough)
Bearers or putlogs to 30-ft width	2×4 in.	2 × 4 in.		
Bearers or putlogs to 5-ft width	2×6 in. or	2 × 6 in. or		
	3×4 in.	3 × 4 in. (rough)		
Spacing of bearers or putlogs			3ft 4 in.	6 ft
Ledgers	1×4 in.	1¼ × 9 in.	2× 9 in.	2 × 9 in.
Vertical spacing of horizontal members	7 ft	7 ft	9 ft	6 ft 6 in.
Bracing, horizontal	1×4 in.	1 × 4 in.	1× 6 in. or 1 ¼×4	2 × 4 in.
Bracing, diagonal	1×4 in.	1 × 4 in.	1×4 in.	2 × 4 in.
Tie-ins	1×4 in.	1 × 4 in.	1×4 in.	1 × 4 in.
Planking	1¼×9 in. (rough)	2 × 9 in.	2×9 in.	2 × 9 in.
Toeboards	4 in. high (minimum)	4 in. high (minimum)	4 in. high (minimum)	4 in. high (minimum)
Guardrail	2×4 in.	2×4 in.	2×4 in.	2×4 in.

*All members except planking are used on edge.

SOURCE: After ANSI A10.8.

Bracing

Pole scaffolds require full diagonal bracing across the entire face of the scaffold to prevent the poles from moving parallel to the wall of the building or from buckling. Independent-pole scaffolds require cross bracing between inner and outer sets of poles. The free ends of all pole scaffolds should be cross-braced. Make sure that all braces are spliced at the poles. All diagonal braces should be installed so that they create a number of true triangles. These triangles may all be arranged parallel, or they may be placed in a zigzag manner, or a long diagonal brace, extending two panels high, may be used, provided it is secured to the intervening horizontal members. The diagonal should start at the ground level, and not at a point one panel above the ground, as is occasionally done. If the scaffold is extra long with respect to its height,

TABLE 19.2 Minimum Nominal Size and Maximum Spacing of Members of Independent-Pole Scaffolds

	Light duty		Medium duty	Heavy duty
Uniformly distributed load	Not to exceed 25 lb/ft^2	Not to exceed 25 lb/ft^2	Not to exceed 50 lb/ft^2	Not to exceed 75 lb/ft^2
Maximum height of scaffold	20 ft	60 ft	60 ft	60 ft
Poles or uprights	2 × 4 in.	4 × 4 in.	4 × 4 in.	4 × 4 in.
Pole spacing, longitudinal	6 ft	10 ft	8 ft	6 ft
Pole spacing, transverse	6 ft	10 ft	8 ft	8 ft
Ledgers	1¼ × 4 in.	1¼ × 9 in.	2 × 9 in.	2 × 9 in.
Vertical spacing of horizontal members	7 ft	7 ft	6 ft	4 ft 6 in.
Spacing of bearers			8 ft	
Bearers			2 × 9 in. (rough) or 2 × 10 in.	2 × 9 in. (rough)
Bearers to 3-ft span	2 × 4 in.	2 × 4 in.		
Bearers to 10-ft span	2 × 6 in. or 3 × 4 in.	2 × 9 (rough) or 3 × 8 in.		
Bracing, horizontal	1 × 4 in.	1 × 4 in.	1 × 6 in. or 1¼ × 4 in.	2 × 4 in.
Bracing, diagonal	1 × 4 in.	1 × 4 in.	1 × 4 in.	2 × 4 in.
Tie-ins	1 × 4 in.	1 × 4 in.	1 × 4 in.	1 × 4 in.
Planking	1¼ × 9 in.	2 × 9 in.	2 × 9 in.	2 × 9 in.
Toeboards	4 in. high	4 in. high (minimum)	4 in. high (minimum)	4 in. high (minimum)
Guardrail	2 × 4 in.	2 × 4 in.	2 × 4 in.	2 × 4 in.

*All members except planking are used on edge.

SOURCE: After ANSI A10.8.

two or more diagonals may be used, placed either parallel or in an inverted V.

If a scaffold is higher than it is wide, bracing may be arranged so that the upper brace starts at the same elevation as the top of the lower brace. On all such scaffolds, diagonal bracing should be provided at every second pole and should continue from the ground to the top of the scaffold.

Scaffolds having a height greater than three times the width should be secured against overturning by means of ½-in. steel cables attached to the outer poles at two points. The cables should extend at about a 45° angle in a horizontal plane to the building columns or other adequate supports. Such tie-ins should be provided at every second or third panel vertically.

Where the scaffold is placed against or adjacent to an irregularly shaped wall, the vertical load from the upper sections of poles should be transferred to the lower sections by diagonal bracing. However, the primary system of bracing of such a scaffold should be kept continuous. If a balcony is bracketed out from a scaffold, then its bracing should be clamped to the projecting horizontal member.

A gantry is sometimes built across the entrance to a building to prevent obstruction by scaffold poles. In such a case, the entire structure must be subdivided into triangles to form the necessary truss.

Joining

Scaffolding lumber is usually joined with nails, although other fastening devices may also be used, depending on the duration of time the structure will be in use at a particular site.

Never use nails in tension, that is, when stress on members tends to pull them away from each other. Nails should be used where wood members are subject to shearing stress, such as when they tend to slide on each other. To develop full strength, at least one-half the length of a nail must be driven into the main member to which the secondary member is being nailed. The minimum number of proper size nails for securing a board or plank 4 in. wide is two; 6 in. wide, three; 8 in. wide, four; and 10 or 12 in. wide, five nails.

A common method of joining wood framework involves toenailing—driving a nail or a group of nails on a slant through the end or edge of an attached member and into a main member. Toenailing provides joints of greater strength and stability than ordinary end nailing.

Tests indicate that the maximum strength of toenailed joints under lateral and uplift loads is obtained by:

1. Using the largest nail that will not cause excessive splitting

2. Allowing an end distance (the distance from the end of the attached member to the point of initial nail entry) of approximately one-third the length of the nail
3. Driving the nail at a slope of 30° with the attached member
4. Burying the full shank of the nail, but avoiding excessive mutilation of the wood from hammer blows

For built-up scaffolding inside a room, extending from wall to wall, diagonal braces can be omitted, provided the scaffold is wedged against the walls to prevent its collapse. *Remember,* bracing must always be provided in both planes.

A scaffold member or members supporting concentrated loads should be properly reinforced or braced. When a load is applied, the horizontal member will bend under the weight and the feet of the poles will tend to spread. To overcome this, a tie member should be provided to hold the feet in place, or bracing should be added to prevent the bending of the horizontal member under load.

Independent-pole scaffolds require cross bracing between inner and outer sets of poles. The free ends of all pole scaffolds should be cross-braced. Nails driven into lead holes with a diameter slightly smaller than the nail have somewhat higher withdrawal resistance than nails driven without lead holes. Lead holes also prevent or reduce splitting of the wood, particularly for dense species.

Planking

Scaffold decking planks should be laid with their edges close together so that tools and material cannot fall through. Each plank should overlap its end supports by at least 12 in., and planks should overlap each other at the bearers by not less than 24 in.

Where the ends of planks abut each other to form a flush floor, the butt joint should be located at the centerline of a pole. Abutted ends should rest on separate bearers. Be sure to provide intermediate beams, where necessary, to prevent planks from becoming dislodged because of deflection. Nail or cleat plank ends to prevent dislodging.

When moving platforms to the next level, leave the old platform undisturbed until the new putlogs or bearers have been set in place, ready to receive the platform planks.

If a scaffold must turn a corner, make sure the platform planks are laid to prevent tipping. First lay planks that meet the corner putlog at an angle so that they extend over the diagonally placed putlog, far enough to have a good safe bearing without danger of tipping. Then place the planking that runs in the opposite direction at a right angle so as to extend over and rest on the first layer of planking.

Guardrails and toeboards

Every scaffold erected 10 ft or more above the ground must have a guardrail and a toeboard along the open sides and ends of the work level or platform.

Guardrails of 2 × 4 lumber or larger should be installed not less than 36 in., nor more than 42 in., high. Midrails and toeboards should be of 1 × 4 lumber, with supports set at intervals not greater than 10 ft.

Alternative guardrail material includes:

1¼-by-1¼-by-⅛-in. structural angle iron

1-by-0.070-in. wall steel tubing

1.990-by-0.058-in. wall aluminum tubing

When persons are required to work or pass under scaffolds, ½-in. wire mesh should be installed between toeboard and guardrail along the scaffold's entire opening. To protect workers on a scaffold from being struck by material or tools dropped by workers from above, erect a tight roof of 2-in. planks above the work platform.

Metal Tubular Scaffolds

Galvanized-steel or aluminum-alloy tubular scaffolding has all but displaced wood scaffolding for most operations. Metal scaffolding is not only stronger and safer to use than wood scaffolding, but it can be erected and dismantled more easily, is not subject to deterioration under exposure to harsh weather or handling, and provides a greater measure of safety from major fire when used with fireproofed planks (see Figs. 19.2 and 19.3).

Three general types of tubular metal scaffolding are generally used today:

1. *Tube and coupler.* Consists of straight tube members in steel or aluminum, in varying sizes and lengths, joined together by steel or aluminum couplers (see Figs. 19.4 and 19.5).
2. *Welded sectional steel.* Consists of prefabricated modular welded steel frames and accessories.
3. *Welded aluminum alloy.* Used in folding sectional stairway and ladder types.

Construction

The type of metal tube to be used for optimum efficiency will depend on the following factors:

Figure 19.2 Tubular scaffolding helps solve a difficult roof overhang condition. *(Courtesy of Patent Scaffolding Company.)*

Load-carrying ability. Not only the strength when resisting steadily applied stresses, but also deformations due to those stresses have to be considered, as well as temperature change, fatigue, and stress fluctuation.

Corrosive resistance. In some circumstances this is of major importance since the useful working life of a tube has considerable economic significance. Corrosion may perhaps reduce safety factors sufficiently to cause collapse.

Ease of handling. Most scaffolds are erected by manpower alone, and this factor governs the maximum acceptable weight of tubular components.

Availability. A certain type of tube may be greatly desirable for a specific purpose, but it may be in short supply or expensive, or both. The nominal 2-in.-diameter mild-steel tube is the most commonly used type.

Beam strength. In general, the safe UD load equals twice the safe central-point load based on safe bending stresses. Shear strength is not generally a criterion for tubes. Deflection is, however, definitely worthy of consideration. As bearers, 2-in. tubes are extremely flexible, and where light loads could permit long spans without exceeding the safe bending stresses, it may be found that excessive deflection is the limiting design factor. From the handling aspect, the 2-in. nominal mild-steel tube at its maximum length of about 21 ft, and mass of 62½ lb, represents about the maximum that one person can handle safely.

Figure 19.3 Metal walkways provide a safe, practical traveling surface. *(Courtesy of Patent Scaffolding Company.)*

Ties and lateral stability

The problem of providing efficient lateral ties applies to many types of scaffolds, but the most important application serves for putlog and independent scaffolds because of their great height-to-width ratio.

The two main criteria when considering the efficiency with which a scaffold is laterally tied are:

1. Efficiency of each tie
2. Location and number of ties used

With regard to the first criterion, all ties should be positive and capable of resisting a push or pull. Consequently, a normal load-bearing coupler should be of ample strength for connecting the tie tube to the scaffold. If, however, a positive horizontal force is required, it must be calculated and sufficient strength ensured through the use of safety couplers if necessary.

Figure 19.4 Tube-and-coupler scaffold permits working on this tall tower. *(Courtesy of Patent Scaffolding Company.)*

Ties formed by connecting to an adjustable shuttering prop wedged between floors are not reliable because of possible creep and shrinkage of the timber packing, unless they are checked at frequent intervals. If used in conjunction with a hard rubber pad, such as neoprene, which is not affected by temperature and humidity changes, then this type of tie could be reliable.

It is imperative that the maximum specified spacings for ties not be exceeded. Although many scaffolds are erected successfully with ties that are insufficient and inadequate according to the prescribed requirements, this would appear to indicate that the specifications are somewhat conservative. However, strict adherence to them will ensure safety of the system and the workers using the scaffolding. Nonadherence means gambling on uncalculated factors, which could pose serious dangers and problems. These factors include:

Limited but completely unknown tying effects obtained from putlog blades in the case of putlog scaffolds

Short life of most scaffold structures, reducing the risk of failure

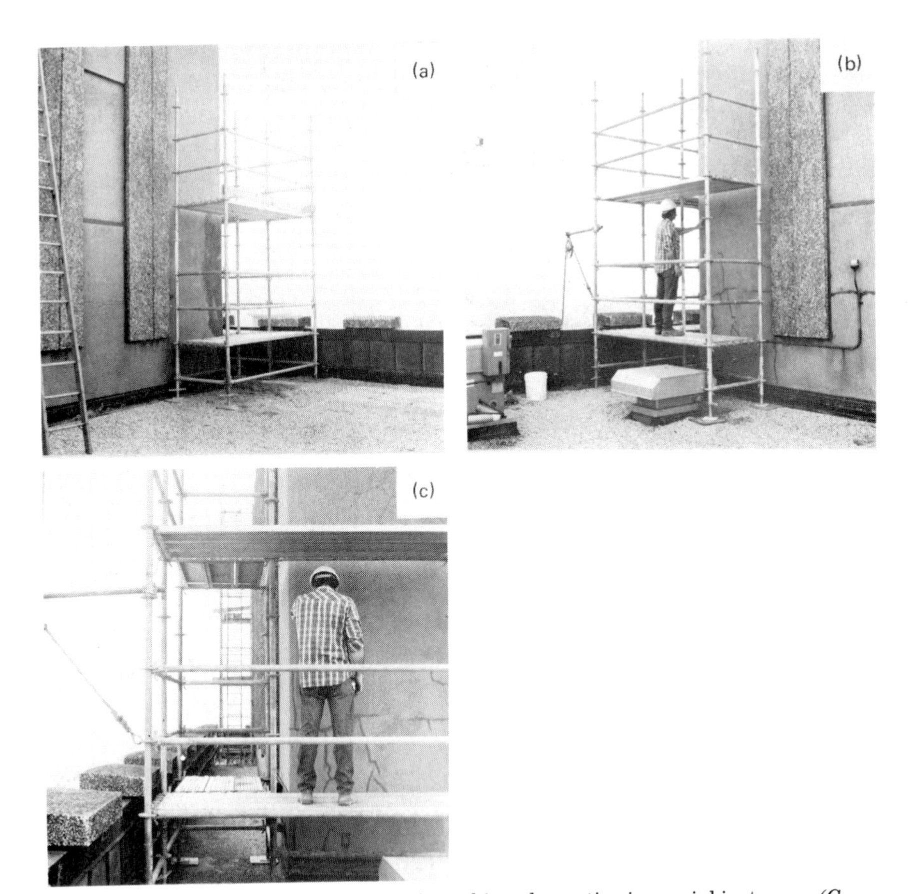

Figure 19.5 Modular scaffold offers a safe working alternative in special instances. *(Courtesy of Patent Scaffolding Company.)*

Reduced strength of standards because of increased flexibility between excessively spaced tie positions or tie positions not adjacent to node points

Ledger bracing increasing stiffness in the vertical direction

Protection fans and nets

These structures are intended to catch falling debris and workers. They are usually constructed as cantilevers of suitable width, slope, and vertical positioning all around high scaffolds or buildings that are either under construction or being demolished. In addition, they protect persons below from falling objects.

The actual catching surface must be securely attached to the fan structure, be capable of stopping the falling objects, and may consist of the following alternatives:

Scaffold boards, plywood, or heavier board if necessary

Corrugated iron sheeting

Industrial safety nets in accordance with ASTM specifications

The nets may consist of 4-in. mesh netting, over a 20-ft span, designed to catch a person falling up to 20 ft; or of ¾-in. mesh designed to catch debris.

Apart from the wind loading, fans should be designed for light access loads. At heights the uplift due to wind may be considerable and should be checked and allowed for where necessary.

It is customary for fans to slope down toward the building, and in the case of large corrugated sheeting fans, the possibility of large rainfall runoff may have to be dealt with.

Loading platforms

Where high independent scaffolds are in use, it is often convenient to raise building materials by crane and deposit them onto a platform attached to the scaffold at some convenient height. On large tower block construction, there may be dozens of such loading platforms at various positions. Also, on some buildings where no normal scaffolding is in use it may be required to provide a loading platform projecting from the building itself.

In either case the loading is usually determined in the form of an equivalent uniformly distributed load allowing for dynamic effects, and then designed by tracing the forces through a triangulated cantilever truss structure.

Corrosion

This problem is especially noted in steel tubes. The average useful life of a 2-in. nominal steel tube varies between 10 and 20 years. Several factors influence the life of a steel tube.

Degree of exposure to which the tube is subjected. This depends on atmospheric conditions and the time spent by a tube in those conditions. Although scaffolding structures are generally considered temporary in nature, it is possible that a particular scaffold may be required to stand for a year or more in an atmosphere exposed to bad corrosive conditions such as high rainfall or sulfurous or salty atmosphere. During this time, the corrosion rate could be so great as to reduce the structural safety of the scaffolding appreciably.

Variation of corrosion rate over tube surface. Assuming an exposed steel surface, the normal corrosion process will involve the formation of pits, which join up and gradually reduce the tube-wall thickness. This reduction does not occur uniformly over the surface, however. It is found that a tube tends to become more corroded at its ends, sometimes in extreme cases producing almost a knife edge. Not only the outside, but also the inside surface of the tube corrodes due to trapped moisture through rainfall or condensation.

In addition, the attachment of couplers on a tube may well result in increased local corrosion due to the local trapping of small pockets of moisture. There is little that can be done to counteract these variable corrosion rates over the tube surface, except that the inside surface may be protected by using tight rubber or plastic bungs in the tube ends.

Contact with other materials. This can also result in local corrosion. Galvanic corrosion occurs between steel and aluminum alloy where used in the same scaffold. Gypsum plaster, wood, brick, and concrete, combined with the presence of moisture, have been shown to cause local corrosion

Required load-carrying performance. As a tube deteriorates, it is possible that it could still be used safely in working conditions where the working stresses are adjusted to allow for the loss in strength. (Both vary with the moment of inertia.) However, for practical purposes it can be said that the reduction in beam and strut strength of a tube varies directly with the wall thickness. Hence, if a corroded tube has only three-quarters of its original wall thickness, it is only three-quarters as strong as when in new condition.

Alloy tubes are not subject to corrosion because the alloy forms a protective oxide skin.

Protective surface finishes

Unless a tube is to be sealed by bungs, it is essential that any protective surface be applied inside and outside. Otherwise only half of the required protection is obtained.

Hot-dip galvanizing. This is probably the most effective form of protection. Descaled tubes are dipped into a bath of molten zinc and given a uniform coating.

Zinc spraying. Fine zinc particles are sprayed onto the outside surface of a tube.

Painting. Because of the large amount of abrasive wear and tear and the handling to which tubes are subjected, the application of paint cannot be considered in the same light as for structural steelwork. If paint is used, however, probably one coat of paint applied frequently will provide better protection than several coats applied less frequently.

Tube-and-Coupler Scaffolding

Four basic components comprise the tube-and-coupler system: baseplates, interlocking tubing or pipes, bolt-activated couplers for making right-angle connections, and adjustable couplers for making connections other than at right angles. Special casters can be added to the baseplates to provide rolling scaffold systems.

With the tube-and-coupler system, horizontal runners can be placed at any point on the vertical posts and bearers can be placed at any point on the runners. Such scaffolds can be erected to any required dimension or height, on any terrain.

This type of scaffold provides maximum versatility for scaffolding around odd shapes, such as processing works and refineries, or other structures having uneven exteriors and projections. The same members can be used to build storage racks of any size or capacity.

To provide additional strength, posts on tube-and-coupling-type scaffolds can be spaced more closely, and more adequate bracing can be provided.

All members can be used interchangeably for posts, putlogs, ledgers, bracing, and handrails. Posts are joined end to end by placing the female end of a member down and locking it to the male upper end of the post below, giving a 90° twist. Rigid 90° couplers secure the various members together. Swivel couplers are available if necessary.

In erecting this type of scaffolding, the posts are placed on steel bases that distribute the load to footing planks. The ledgers then are clamped to the posts at the desired heights, with putlogs or bearers clamped to the ledgers close to the posts.

The braces on the outboard face are clamped to the projecting ends of the putlogs, close to the posts. Cross braces are installed at every fourth or fifth pair of posts. Usually the putlogs extend so as to bear against the wall of the building or other substantial support. Ties of ⅜-in. wire rope should be used to lash the scaffolding to the structure.

As the scaffold extends upward with construction progress, planking can be removed from the lower levels and placed higher up. Handrail members should remain in place at all times to provide added rigidity to the scaffold.

Post splices should be located a short distance above the ledger couplings and staggered so as to occur alternately at different levels. Where longitudinal diagonal bracing is made continuous from the ground to the top of the scaffold and braces do not meet the putlogs, be sure to clamp the braces to the post by means of swivel couplers.

To check the strength of tubular steel scaffolding of this type, first determine the live or movable load that may be applied to the plank decking. Do not count less than 25 lb/ft^2 of deck, preferably not less than 50 lb/ft^2. In a scaffold or staging more than one panel in length, each putlog carries a load on the decking equal to that on one-half the area of the panel on either side of it, or, usually the load on the area of one panel.

For example, if the posts are spaced 5 by 5 ft, the area supported is 25 ft^2. This area, multiplied by 25 lb/ft^2, gives a load of 625 lb uniformly distributed over the length of the span of the putlog. To this must be added the weight of the plank decking and toeboards, say 240 lb, which comes to a total load of 865 lb on the putlog.

Scaffold erection

Posts for tube-and-coupler scaffolds should be accurately erected on suitable bases and maintained plumb. They should be spaced, for the depth of the scaffold, no more than 6 ft apart; along its length, no further apart than 10 ft (light duty), 8 ft (intermediate duty), and 6 ft 6 in. (heavy duty).

Posts spaced not more than 6 ft by 8 ft apart along the length of the scaffold should have bearers of nominal 2½-in.-outside-diameter steel tube or pipe. Not more than 5 ft by 8 ft apart, they should have bearers of nominal 2-in.-outside-diameter tube or pipe. When other structural metals are used, they must be designed to carry an equivalent load.

All tube-and-coupler scaffolds should be limited in height and number of working levels and should be constructed and erected to support four times the maximum intended loads. Table 19.3 lists the data for light-, intermediate-, and heavy-duty tube-and-coupler scaffolds.

All tube-and-coupler scaffolds exceeding these limitations must be designed by a licensed professional engineer. Drawings and specifications should be prepared and copies made available to the contractor and for inspection purposes.

TABLE 19.3 Guidelines for Tube-and-Coupler Scaffolds

	Light Duty	
Uniformly distributed load	Not to exceed 25 lb/ft^2	
Post spacing, longitudinal	10 ft	
Post spacing, transverse	6 ft	
Working levels	Additional planked levels	Maximum height
1	8	125 ft
2	4	125 ft
3	0	91 ft
	Medium Duty	
Uniformly distributed load	Not to exceed 50 lb/ft^2	
Post spacing, longitudinal	8 ft	
Post spacing, transverse	6 ft	
Working levels	Additional planked levels	Maximum height
1	6	125 ft
2	0	78 ft
	Heavy Duty	
Uniformly distributed load	Not to exceed 75 lb/ft^2	
Post spacing, longitudinal	6 ft 6 in.	
Post spacing, transverse	6 ft	
Working levels	Additional planked levels	Maximum height
1	6	125 ft

SOURCE: ANSI A10.8.

Runners along the length of the scaffolds should be set not more than 6 ft 6 in. center-to-center, located on both the inside and the outside posts at even height. Make sure the runners are interlocked to form continuous lengths and are coupled to each post. Locate the bottom runners as close to the base as possible.

All bearers should be installed transversely between posts, coupling them securely to the posts bearing on the runner coupler. When coupled directly to runners, keep bearers as close to the posts as possible. Bearers should be at least 4 in., but not more than 12 in., longer than the post spacing or runner spacing. They may be cantilevered for use as brackets to carry not more than two planks.

Cross bracing should be installed across the width of the scaffold at least every third set of posts horizontally and every fourth runner vertically. Extend all such bracing diagonally from the inner and outer runners upward to the next outer and inner runners.

Longitudinal diagonal bracing should be installed at approximately a 45° angle from near the base of the first outer post upward to the extreme top of the scaffold. Where a long length of scaffold permits, duplicate such bracing at every fifth post. Install bracing also from the last post, extending back and upward to the first post, or to runners if conditions preclude attaching it to the posts.

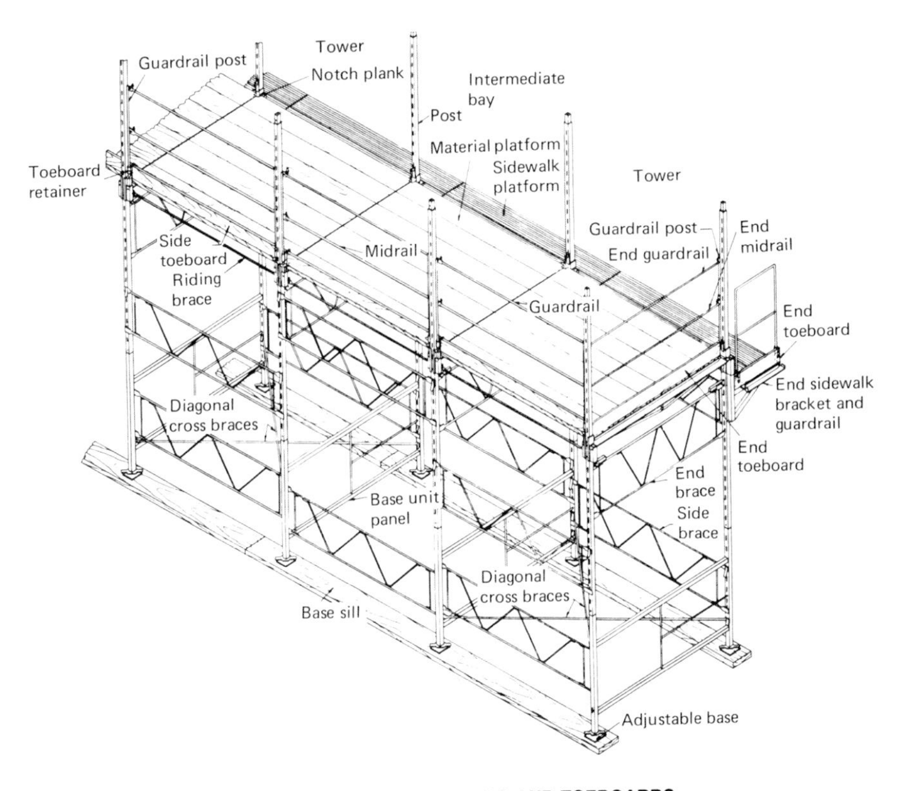

INSTRUCTIONS FOR PLANKING AND TOEBOARDS

1. Tie the two sidewalk brackets together in the first and last bays of a scaffold run by nailing the bracket planks to the bracket conversion plates.
2. Platform planking to be minimum 2 by 10 in. wide, full size, graded as scaffold plank, rough cut 7 ft.
3. Platform planking must be 6 ft 11½ in. long (6 ft 11¼ in. minimum). Ends must be square.
4. Notch the plank on material platform nearest to building at both ends, 1½ in. deep for 2 in. of its width.
5. Toeboard lumber must be minimum 1 in. nominal thickness and extend at least 4 in. above the platform.
6. Cut toeboard sections for the outside edge of the material platform, 6 ft 10 in. long.
7. Securely fasten end toeboards to the platform planking.

Figure 19.6 Typical guardrail, midrail, and toeboard installation. *(Courtesy of Patent Scaffolding Company.)*

Make sure to tie and securely brace the entire scaffold against the structure at intervals not exceeding 30 ft horizontally and 26 ft vertically. Guardrails, midrails, toeboards, and wire mesh should be installed in accordance with accepted standards (see Fig. 19.6).

Access to all built-up scaffolds must be by one of the following:

- Portable wood or metal ladders
- Access ladders positioned so that their use will not tend to tip the scaffold
- Hook-on or attachable metal ladders specifically designed for use in conjunction with proprietary types of scaffolds
- Step- or stair-type accessories designed specifically for use with proprietary types of scaffolds.
- DO NOT use cross braces as a means of access.

Welded Sectional Steel Scaffolding

This type of scaffolding consists of tubular steel frames—essentially two short poles, a putlog, and bracing members welded together as an integral unit.

To assemble, two frames are erected opposite each other and the re-

Figure 19.7 Sectional scaffolding supports overhanging top grandstand roof. *(Courtesy of Patent Scaffolding Company.)*

movable cross braces between them are placed on the stud bolts, which are welded to, and form an integral part of, the frame. The posts of the bottom frames rest on steel bases nailed to planks that distribute the load over the ground or floor surface. After the first tier of panels has been erected, the second tier of frames is socketed onto the upper ends of the first tier (see Figs. 19.7 to 19.10).

The prefabricated frames are standard 5 ft wide and range in height from 3 to 10 ft (5-ft-high frames are the most commonly used size). Because of the numerous designs of frames by various manufacturers, the allowable load uniformly distributed on the putlog (the top member of the frame) varies considerably. Likewise, because of the variation in the length of posts between the removable cross braces, there is a wide range in the load that these frames can support. Check the manufacturer's catalog before erecting such a system.

Because the frames are all interchangeable, unless the rigger knows exactly the load that each model will safely carry, the live loading should be limited to that permitted on the model having the lowest strength.

Figure 19.8 Sectional scaffolding provides access to ornate multiconfigured hotel facade. *(Courtesy of Patent Scaffolding Company.)*

Figure 19.9 Sectional scaffolding employed for building demolition work. *(Courtesy of Patent Scaffolding Company.)*

The safety factor on this type of scaffolding should be 4 to provide for possible overloading of the system by mistake. Because the dead load on the lower sections of a built-up scaffold varies almost directly with the height to which the scaffold is erected, the live load is generally a guess.

For example, when a scaffold is about to be erected, the rigger may assume that one pallet of bricks will be placed on the decking in each panel. But if through ignorance or carelessness two pallets are placed in the panels, this will almost double the live load. When the dolly used to move the pallets sets down the load, it may produce an additional impact load of as much as 100%.

The factor that usually limits the loading on a scaffold frame is the strength of the putlog or top horizontal member of the frame upon which the planks rest. The total load on any frame is one-half the combined live and dead loads on the two adjacent spans of decking.

The total loading on any frame should not exceed 2,000 lb on frames having diagonal bracing members, or 1,600 lb on those having no diagonal braces. These figures are based on a safety factor of 4.

The 2-in. scaffold planks come in 13-ft lengths. Hence regardless of the spacing of the frames, the same weight of planking will usually be employed, with the amount of overlap varying as required. Thus the

Figure 19.10 Sectional scaffolding having integral step units to scale this 330-ft-high curved stack. *(Courtesy of Patent Scaffolding Company.)*

weight of the planking on each frame putlog will be about 250 lb. The weight of the frames and their removable bracing may be estimated at about 10 lb (steel) and 5 lb (aluminum) per foot of scaffold.

Welded Aluminum-Alloy Scaffolds

Because of its lightness, mobility, and ease of erection, this type of scaffolding is most suitable for light-duty work, especially where the equipment requires frequent erection and dismantling.

These scaffolds are prefabricated from high-strength aluminum-alloy tubing and have internal stairways. They are often equipped with casters to permit easy movement for maintenance work.

The folding-ladder types are constructed in one-piece base sections that speed erection and dismantling. Ladder-type base sections are available in widths of 29 in. or 4 ft 6 in., with spans of 6, 8, or 10 ft between frames. This type of scaffold has two diagonal braces and one horizontal brace, which form integral parts of the folding unit. Intermediate extension and guardrail sections can be placed atop the folding unit, using individual end frames and braces.

A larger folding-type scaffold, with base dimensions of 4 ft 6 in. by 6 ft, has an external stairway, with its upper sections as well as the base section being one-piece folding units. When the scaffold must be erected higher than recommended for a base of this size, outriggers that clamp to the legs of the base section can be used. Leg equipment includes a leveling mechanism to level the scaffold on uneven ground. Casters on the legs are locked at both wheel and swivel. However, folding scaffold sections are heavier than individual components of demountable sectional scaffolding.

Sectional aluminum stairway scaffolds are designed with end frames of various heights to provide different working levels; adjustable bottom sections with casters, but without the folding feature; intermediate sections; half-sections; and guardrail sections. All components are demountable to reduce weight, and to make it easier to erect and dismantle. Internal stairways are used. Outriggers may be used to increase the base area.

Figure 19.11 Ladders integrated with scaffolding system provide multilevel access during this major boiler overhaul. *(Courtesy of Patent Scaffolding Company.)*

Folding sectional stairway and ladder scaffolds are used for outdoor cleaning and maintenance work, the folding or sectional-stairway types for high or heavy work and the ladder scaffolds for low to medium height and one-person jobs. Both types are especially suitable when the work is horizontal. When used indoors, they simplify work on walls and ceilings (see Fig. 19.11).

In general, the principles of sound design and construction practice referred to for mild-steel putlog and independent scaffolds are also applicable to scaffolding systems of aluminum alloys. However, the size and load-carrying capacity of alloy scaffolds is considerably less than that for steel scaffolding. Therefore when planning their use, the following factors must always be considered:

Flexibility. Compared with steel, aluminum-alloy tubes are three times as flexible, which reduces their efficiency as struts and beams.

Inconvenience. To avoid electrolytic corrosion, only alloy couplers can be used with alloy tubes, since it is essential to avoid mixing alloy and steel members on the same scaffolding system.

Convenience. Alloy tubes weigh one-third as much as steel ones, and are consequently easier to transport and handle.

Corrosion. Alloy tubes and fittings are less susceptible to corrosion than similar mild-steel tubes and fittings.

Chapter 20

Specialized Scaffolds

In addition to the swinging and the built-up scaffolds, a number of specialized scaffolds are commonly used for particular types of operations.

Outrigger Scaffold

This type of scaffold is frequently projected from a building window to enable workers to reach the upper portions of a wall or cornice.

Outrigger scaffolding consists of a pair of heavy timber beams, placed on edge, with their fulcrum points resting on a bearing plank on the window sill. Rough planking, which forms a working platform, is laid tight, to within 3 in. of the building wall, and nailed or bolted to the outrigger beams (see Fig. 20.1).

The outrigger beams should extend not more than 6 ft beyond the face of the wall. Their inboard ends, measured from the fulcrum point to the extreme point of support, should be not less than one and one-half times longer than their outboard ends.

To resist the upward thrust of their inboard ends, outrigger beams are usually nailed to vertical struts that are wedged against bearing blocks at the floor and ceiling of the building. Cross bridging between the outriggers at the fulcrum point resists any tendency of the planks to tip or roll over. The entire supporting structure should be securely braced in both directions to prevent any horizontal movement.

The bearing plank on the window sill should provide a surface of at least 6 in. in each horizontal direction to support each beam's fulcrum point. Guardrails, midrails, and toeboards should be installed on all open sides and ends of platforms more than 10 ft above ground or floor.

Guardrails of 2×4 timber, or structural angle iron, steel, or alumi-

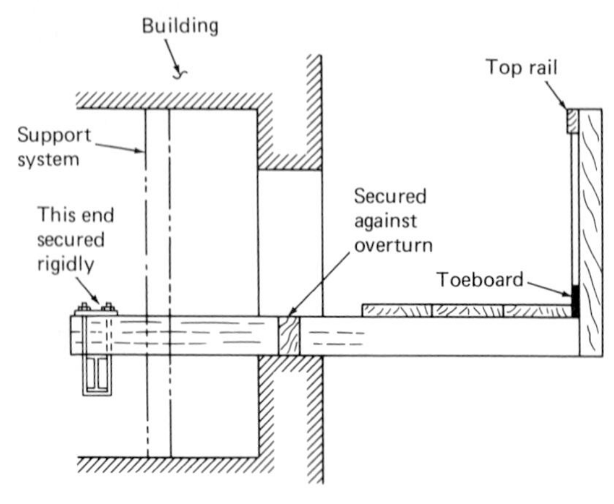

Figure 20.1 Outrigger scaffold. (*Courtesy of Patent Scaffolding Company.*)

num tubing of equivalent strength, should be installed not less than 36 in., nor more than 42 in., high with a midrail. Both guardrails and midrail supports should be installed not more than 10 ft apart along the platform. Toeboard and midrail should be of 1 × 4 lumber, or equivalent, with the toeboard to extend at least 3½ in. above the working surface. Where there is danger of material falling from the scaffold, a wire mesh or other enclosure should be provided between the guardrail and the toeboard. Outrigger scaffolding specifications are listed in Table 20.1.

The ends of planks should not extend more than 18 in. beyond outrigger beams. Where additional working levels are required to be supported by the outrigger method, the plans and specifications of the outrigger and scaffolding structure should be designed by a licensed professional engineer.

Mason's Adjustable Multiple-point Suspension Scaffold

This type of scaffold consists of a continuous wooden work platform supported by bearers suspended by wire rope from overhead outrigger

TABLE 20.1 Specifications for Outrigger Scaffolding

	Light duty	Medium duty
Maximum scaffold load, lb/ft^2	25	50
Outrigger size, in.	2×10	3×10
Maximum outrigger spacing, ft	10	6
Planking, rough, in.	2×9	2×9

beams, which are securely fastened or anchored to the structure's frame or floor system (see Fig. 20.2).

Each outrigger beam should be equivalent in strength to at least a standard 7-in., 15.3-lb steel I beam and at least 15 ft long. Beams should project not more than 6 ft 6 in. beyond their bearing point. If channel iron outrigger beams are used in place of I beams, they should be securely fastened together with the flanges turned out.

Where the overhead is greater than 6 ft 6 in., outrigger beams should be of stronger material, or multiple beams should be used, installed in accordance with approved design and instructions.

All outrigger beams should be set and maintained with their webs in a vertical position. They should rest on suitable wood bearing blocks.

The free end of the suspension wire ropes should be equipped with proper size thimbles and secured by splicing or other equivalent means, while the running ends should be securely attached to the drum of an approved hoisting machine. Make sure that at least four turns of rope remain on the drum at all times. Where a single outrigger beam is used, place the steel shackles of the clevis, with which the wire ropes

Figure 20.2 Aluminum sectional scaffold 4 ft 6 in. wide. Outrigger supports should be used when working at platform heights greater than four times the smallest dimension. (*Courtesy of Patent Scaffolding Company.*)

are attached, to the outrigger beams directly over the hoisting drum. The suspended rope should be capable of supporting at least six times the intended load. The scaffold work platform should be equivalent in strength to at least 2-in. planking capable of supporting a working load of at least 50 lb/ft^2. NEVER load the platform in excess of this figure. The maximum permissible span for 2 × 10 (nominal), 2 × 9 (rough) planks supporting this load is 8 ft.

All parts of the scaffold, such as bolts, nuts, fittings, clamps, wire ropes, and outrigger beams and their fastenings, should be maintained in sound and good working condition and should be inspected before each installation and periodically thereafter.

Guardrails and toeboards should be installed and protective wire mesh provided between guardrail and toeboard in accordance with specifications. Make sure that each scaffold is installed or relocated in accordance with approved designs and instructions, supervised by a designated and competent individual.

Needle-Beam Scaffold

This type of light-duty scaffolding, consisting of needle beams supporting a platform, is used only for very temporary jobs, in particular for riveting or bolting operations on steel structures. No material should be stored on this type of scaffold (see Fig. 20.3).

Two needle beams, 4 by 6 in. on edge, are placed parallel to each other in a horizontal or nearly horizontal plane, depending on circumstances, and no more than 10 ft apart. They are usually suspended by rope, or hangers, near each end. A center support is always required.

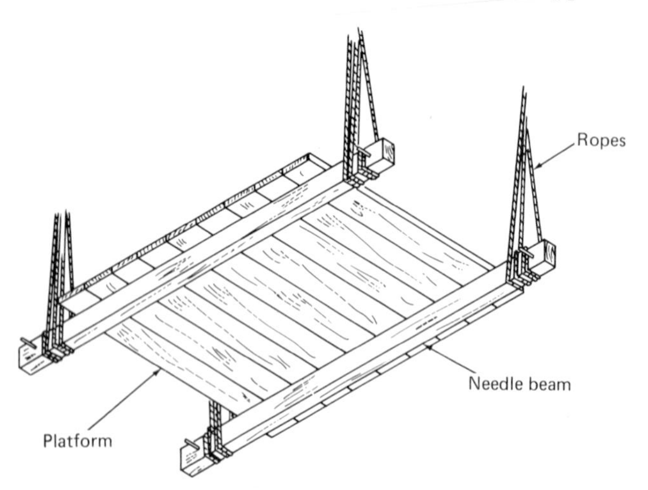

Figure 20.3 Needle-beam scaffold. (*Courtesy of Patent Scaffolding Company.*)

The rope supports should be equivalent in strength to 1-in.-diameter no. 1 grade manila rope. Attach the ropes to the needle beam by a scaffold hitch or a properly made eye splice in order to prevent the beams from rolling over on their sides under load.

Tie the loose ends of the rope with a bowline knot or a round turn and one-half hitch. Then extend the rope up over a structural member, down under the needle beam, and back over the structural member and secure it.

The rivet scaffold hitch is frequently used, although a clove hitch or rolling hitch is preferable. Make sure that you take proper precautions to prevent the ropes from slipping off the ends of the needle beams, particularly if the beams are inclined.

Decking should be of 2-in. planks having a length of at least 2 ft, and no more than 3 ft, longer than the span between the needle beams. Planking spans should be not less than 3 ft nor more than 8 ft when using 2-in. planks. For spans greater than 8 ft, platforms should be designed based on requirements for the special span. If the scaffold is not level, nail the planks to the beams or have cleats nailed on their undersides to engage the beams. One end of a needle-beam scaffold may be supported by a permanent structural member. *Make sure* that all unattached tools, bolts, and nuts used on needle-beam scaffolds are kept in suitable containers.

When working on a needle-beam scaffold 10 ft or more above the ground or floor, each worker should wear a safety belt with lanyard and fall-prevention device, which will limit a fall to no more than 6 ft. Make sure the lanyard and the fall-prevention device are secured to a structural member other than the scaffold, or attached to a lifeline suspended from a structural member other than the scaffold.

Interior Hung Scaffold

This type of scaffold, suspended from a ceiling or roof structure, consists of either a wood or a steel tube-and-coupler-type system (see Fig. 20.4).

For hanging wood scaffolds, the following minimum nominal-size material is recommended:

Supporting bearers, 2 × 9 in. on edge

Planking, 2 × 9 or 2 × 10 in., with a maximum span of 7 ft for heavy-duty and 10 ft for light- or medium-duty applications

Both wood and steel tube-and-coupler-type scaffolds should be designed to sustain a uniformly distributed working load up to heavy-duty

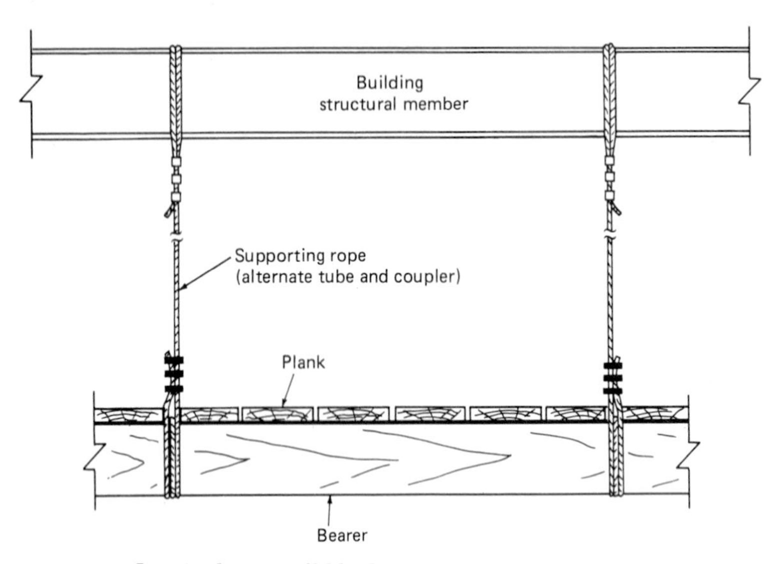

Figure 20.4 Interior hung scaffold. (*Courtesy of Patent Scaffolding Company.*)

scaffold loads, with a safety factor of 4. Install guardrails, midrails when required, and toeboards at all open sides on all scaffolds more than 10 ft above ground or floor. Provide wire mesh between guardrail and toeboard.

When wire rope is used to support such scaffolds, make sure the rope is wrapped at least twice around the supporting members and twice around the bearers of the scaffold. Secure each end of the wire rope with at least three standard wire rope clips. The suspension rope should be capable of supporting at least six times the intended load.

Inspect and check all overhead supporting members for proper strength.

Carpenter's Bracket Scaffold

These scaffolds consist of triangular frames of wood not less than 2 by 3 in. in cross section, or of equivalent-strength metal brackets (see Fig. 20.5). Each bracket should be attached to the structure by means of one of the following:

A bolt not less than ⅝ in. in diameter, which extends through the inside of the building wall

A metal stud attachment device

Welding to the structure

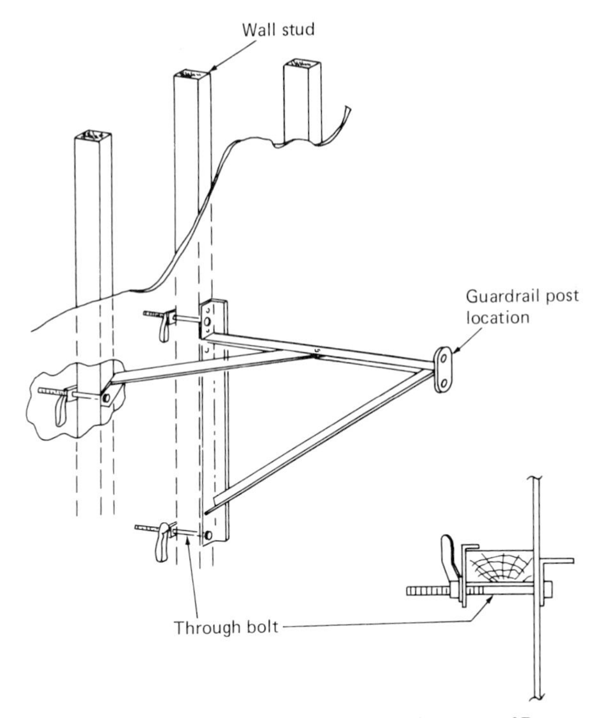

Figure 20.5 Carpenter's bracket scaffold. (*Courtesy of Patent Scaffolding Company.*)

Hooking over a well-secured and adequately strong supporting member

Make sure that brackets are spaced no more than 10 ft apart. Use at least two 2-by-9-in. (nominal size) wood planks for the work platform. Planks should extend not more than 18 in., nor less than 6 in., beyond each end support. Install guardrails, midrails when required, and toeboards on all open sides when a scaffold is more than 10 ft above ground or floor. Provide wire mesh between guardrail and toeboard.

NEVER allow more than two persons to occupy any given 10 ft of a bracket scaffold at any one time. Make sure that tools and materials do not exceed 75 lb in addition to workers.

Bricklayer's Square Scaffolds

This type of scaffold consists of square shapes of framed structural wood, which support a wooden platform. The wood frames should be stacked no higher than three tiers, one square resting directly above the other (see Fig. 20.6). The dimensions for this scaffold are given in Table 20.2.

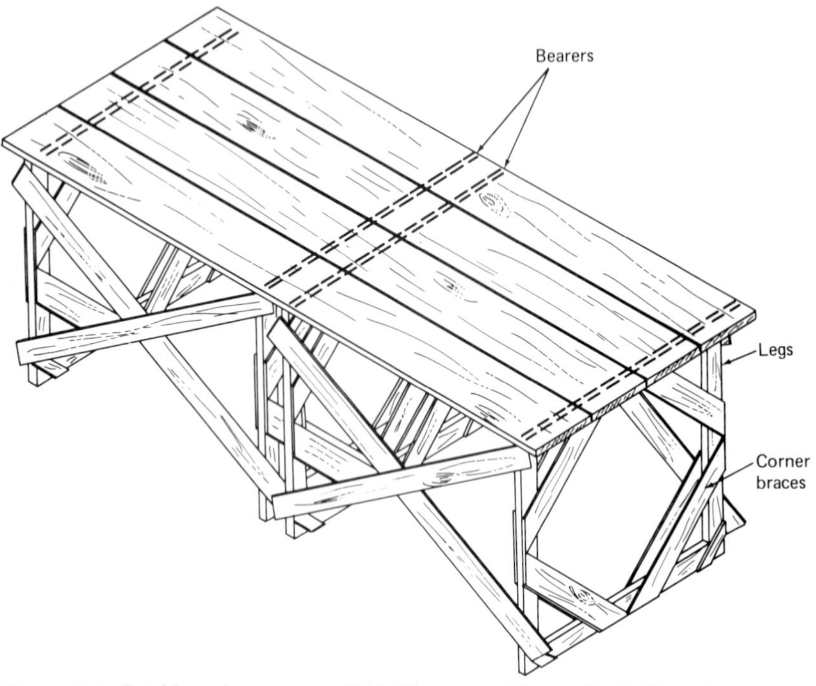

Figure 20.6 Bricklayer's square scaffold. (*Courtesy of Patent Scaffolding Company.*)

Each upper tier stands on a continuous row of planks laid across the next lower tier and is nailed down or otherwise secured to prevent displacement. The squares should be no wider, nor higher, than 5 ft. They should be set not more than 5 ft apart for medium-duty scaffolds and not more than 8 ft apart for light-duty scaffolds.

Make sure each square is reinforced on both sides of each corner with 1-by-6-in. gusset pieces. Place bracing of 1-by-8-in. lumber on both sides, running from center to center of each member, to ensure scaffold strength and rigidity. Extend 1-by-8-in. bracing from the bottom of each square to the top of the next square on both the front and the rear sides of the scaffold to provide sufficient strength and rigidity. Before use, check that the scaffold is level and resting upon a firm foundation. Platform planks should be at least 2 by 9 in. nominal size, with their ends overlapping the bearers of the squares. Support each plank on at least three squares.

TABLE 20.2 Dimensions for Bricklayers' Scaffolds

Members	Dimensions, in.
Bearers or horizontal members	2 × 6
Legs	2 × 6
Braces at corners	1 × 6
Braces diagonally from center frame	1 × 8

Horse Scaffold

Wood horses may be used to construct light- or medium-duty scaffolds, provided they are spaced not more than 5 ft apart, for medium duty, and not more than 8 ft apart for light duty. The horses support the work platform and may be arranged in tiers not more than two tiers high, nor more than 10 ft above ground (see Fig. 20.7).

When arranged in tiers, each horse should be placed directly over the horse in the tier below, with the legs of the upper horses nailed down to the platform planks of the tier below to prevent displacement or thrust. Make sure each tier is cross-braced substantially. The dimensions for a horse scaffold are given in Table 20.3.

NEVER use horses or parts that have become weak or defective. Immediately remove such items from service and dismantle the horse or replace damaged parts.

Ladder Jack Scaffold

This is a light-duty type scaffold consisting of a work platform supported by brackets attached to ladders. Ladder jack scaffolds should be limited to light duty and a height of 20 ft above floor or ground.

Make sure the ladder jack is so designed and constructed that it will bear on the side rails in addition to the ladder rungs. If it bears on rungs only, make sure to provide at least 10 in. of bearing area on each rung.

Ladders used in conjunction with ladder jacks should be placed, fastened, held, or equipped with devices so as to prevent slipping. The

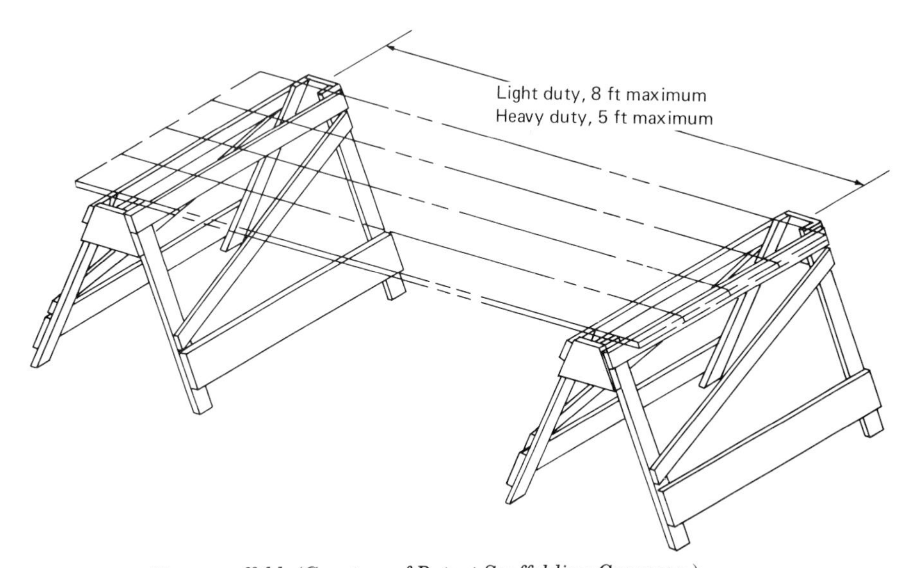

Figure 20.7 Horse scaffold. (*Courtesy of Patent Scaffolding Company.*)

TABLE 20.3 Dimensions for Horse Scaffolds

Members	Dimensions, in.
Horizontal members or bearers	3×4
Legs	1¼×4½
Longitudinal brace between legs	1×6
Gusset brace at top of legs	1×8
Half-diagonal braces	1¼×4½

ladders must be of the heavy-duty type, designed and constructed in accordance with ANSI standards A14.1 and A14.2.

Wood platform planks should be not less than 2-in. nominal thickness, with support spans no greater than 8 ft. Make sure that platform planks, whether metal or wood, overlap the bearing surface by not less than 12 in. Platforms should be not less than 18 in. wide. *Do not* permit more than two persons to occupy a ladder jack scaffold at any one time.

When working on a ladder jack scaffold 10 ft or more above the ground or floor, each worker should wear a safety belt tied to a lanyard and a fall-prevention device that will limit the fall to 6 ft. Make sure the lanyard and a fall-prevention device are secured to a structural member, or attached to a lifeline suspended from a structural member, other than the scaffold.

Window Jack Scaffold

A window jack scaffold consists of a work platform supported by a bracket or jack that projects through a window opening. The window jacks should be designed and constructed so as to provide a secure grip on the window opening and be capable of supporting the design load (see Fig. 20.8).

This type of scaffold should be used only for the purpose of working at the window opening through which the jack is placed. *Never* support planks between one window jack and another, nor use window jacks to support other elements of scaffolding. Make sure window jacks have suitable guardrails, or provide each worker with approved safety belts and attached lifelines. *No more* than one person should work on a window jack scaffold at a time.

Bosun's Chair

The bosun's chair is a suspended seat designed to accommodate one worker in a sitting position (see Fig. 20.9). The chair consists of wood planking not less than 12 by 24 in. and 1 in. thick, reinforced on the underside by cleats securely fastened to prevent the board from splitting. Seat slings of ⅝-in.-diameter fiber ropes are reeved through the four seat holes so as to cross each other on the underside of the seat.

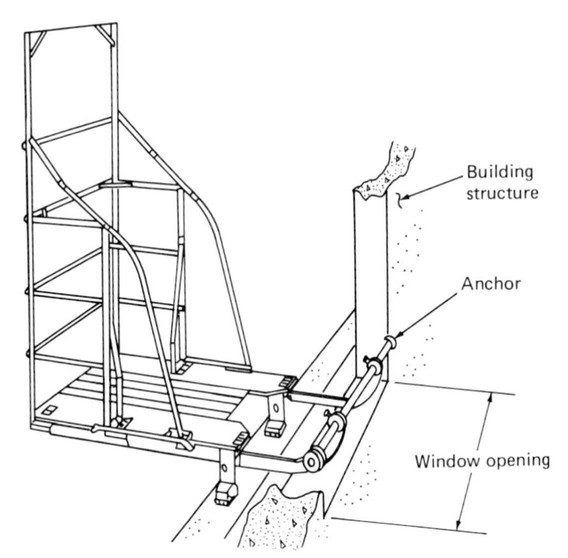

Figure 20.8 Window jack scaffold. (*Courtesy of Patent Scaffolding Company.*)

Tackle should consist of the correct size ball bearing or bushed blocks and properly spliced no. 1 grade manila rope. When a worker in a bosun's chair is welding or working with an open flame, seat slings should be of at least ⅜-in. wire rope.

Make sure that roof irons, hooks, or the objects to which the tackle is anchored are securely installed. When tiebacks are used, they should be installed at right angles to the face of the building and securely fastened to a chimney, standpipe, skylight, or other part of the structure.

Manually Propelled Mobile Scaffolds

These portable rolling scaffolds constructed of tubular welded metal members are supported by casters that permit ease of travel (see Fig. 20.10).

The casters of the mobile tower are designed to have the strength and dimensions to support four times the maximum intended load, and are provided with positive locking devices to hold the scaffold in position. They should have rubber or similar resilient tires with a minimum of 5-in.-diameter wheels, unless specific design requirements dictate the use of other materials. If the casters have plain stems, they should be secured in the scaffold to prevent them from falling out.

Free-standing mobile scaffold towers should not be erected any higher than four times the minimum base dimensions, including any outrigger frames that may be used to increase stability. Make sure that the towers are properly braced with cross bracing as well as horizontal bracing, in accordance with standard practice.

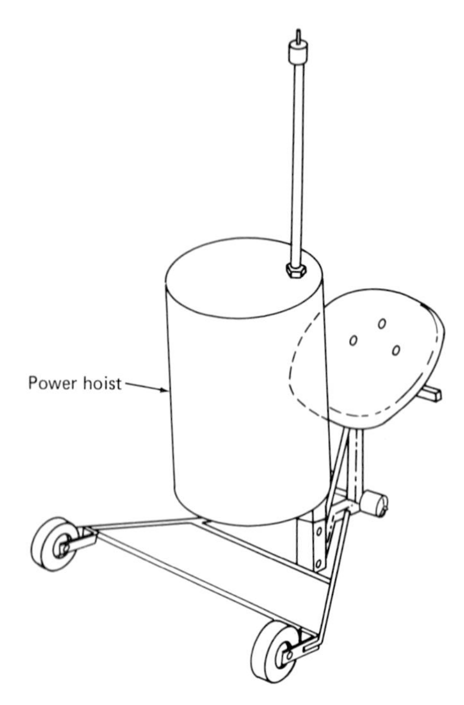

Figure 20.9 Powered bosun's chair. (*Courtesy of Patent Scaffolding Company.*)

Platforms for mobile scaffolding should be tightly planked for the full width of the scaffold, except for necessary entrance openings. Make sure that platforms are securely fixed in place to prevent any horizontal movement. For scaffolds more than 10 ft above the ground or floor, provide guardrails, midrails, and toeboards, with wire mesh between guardrail and toeboard.

For proper access and egress, a ladder or stairway must be affixed or built into the scaffold and so located that when in use it will not have a tendency to tip the scaffold. These can be:

Portable wood or metal ladders

Separate attachable or built-in ladders with regularly spaced steps or rungs, having maximum variations between adjacent rungs of 2 in. Spacing between rungs may be up to 16½ in. if such spacing is necessitated by practical limitations of the equipment in conjunction with which the ladder is being used

Step- or stair-type accessories specifically designed for use with proprietary types of scaffolds

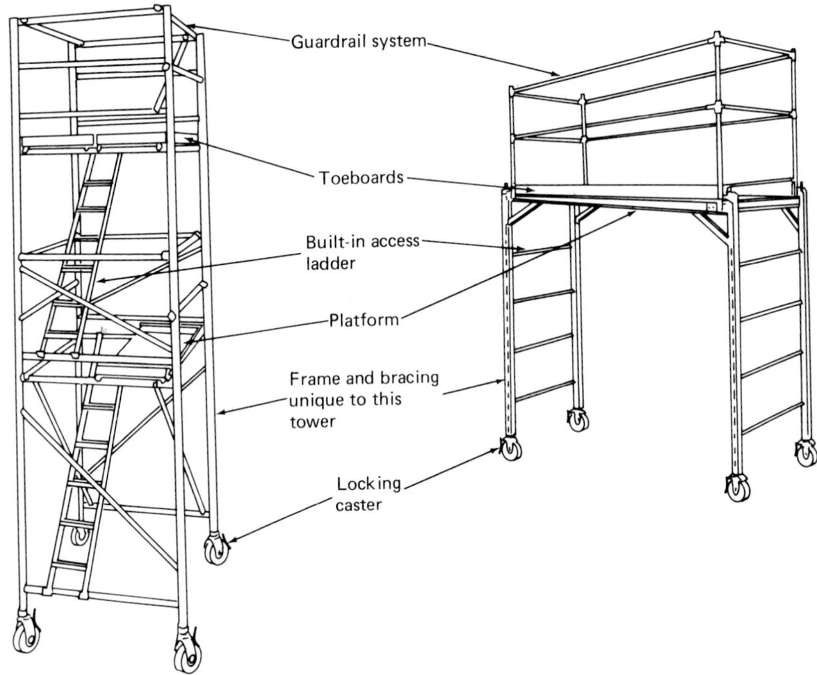

Figure 20.10 Manually propelled mobile scaffold. (*Courtesy of Patent Scaffolding Company.*)

Landing platform intervals should not exceed 35 ft.

When mobile scaffolds are used, make sure that they rest on a suitable footing and that they stand plumb. Always lock casters or wheels before using a mobile scaffold to prevent any movement.

When it is necessary to level the mobile scaffold platform, use screw jacks or other means in each leg section for adjusting the height. Make sure that at least 6 in. of the screw jacks are in each scaffold leg, with no more than 12 in. extended.

When moving a mobile scaffold, always apply the force as close to the base as practical. Make sure that the tower is stabilized during movement from one location to another. Move scaffolds only on floors free of obstructions and openings.

Workers are permitted to ride on manually propelled mobile scaffolds only if guardrails, midrails, and toeboards are installed on all open sides of the platform, and only when:

The floor or surface is within 1½° of level concrete, free from pits, holes, or obstructions

The minimum dimensions of the scaffold base, ready for rolling, are at least one-half its height. Outriggers, if used, should be installed on both sides of the staging and then can be calculated as part of the base dimension

All tools and materials are secured, or removed from the platform, before the mobile scaffold is moved

Workers on the mobile scaffold are advised and aware of each movement in advance

Section

6

Procedures and Precautions

Chapter

21

Portable Ladders

Although most portable ladders used today are still made of wood, both metal and reinforced-plastic ladders have gained wide acceptance in recent years for a variety of applications. This chapter covers the requirements for the selection, proper care, and use of the common types of portable ladders: step, extension, sectional, trestle, extension-trestle, and special-purpose, considering all three materials (see Fig. 21.1).

Many industrial organizations that use a large number of ladders have an inspector carefully examine each ladder when received to make sure that it meets the purchase specifications. Such inspection procedures should be part of the purchase requirements for every organization, including construction contracting, utilities, and others whose operations require the use of portable ladders. The purpose of such specifications is to provide reasonable safety for life, limb, and property.

Most specifications are based primarily upon data contained in the safety requirements for portable ladders published by the American National Standards Institute. These codes, regarding specific uses for ladders, include:

ANSI A14.1, "Safety Requirements for Portable Wood Ladders"

ANSI A14.2, "Safety Requirements for Portable Metal Ladders"

ANSI A14.5, "Safety Requirements for Portable Reinforced Plastic Ladders"

ANSI A14.3, "Safety Requirements for Fixed Ladders"

Not only should these standards serve as a basis for purchase require-

Figure 21.1 Typical extension ladder, designed to meet ANSI and UL requirements. (*From L. R. Higgins,* Handbook of Construction Equipment Maintenance, *McGraw-Hill, 1979.*)

ments, they should also be used for instructing workers in ladder safety practices.

Before putting any new ladder into use, the inspector should first check all the dimensions of the ladder, including the sizes of the side rails and rungs to make sure they are not less than the minimum allowed by the code. Well-built ladders usually have side rails larger than the code dimensions.

Next the inspector should examine the ladder's hardware to determine whether it is of ample strength. Malleable iron and cast iron should be avoided for parts subject to bending or tensile stress. All hardware and fittings should be securely attached. Rungs and steps should be tightly fitted into the side rails and secured against turning.

Special, or trick, ladders that can quickly convert into an extension or stepladder should never be used for industrial or commercial purposes.

The spreader bars on stepladders should be designed to prevent serious finger-pinching hazards, as well as to prevent the ladder from folding up accidentally if pushed along the floor. The moving parts should operate freely without binding or undue play.

Where bucket shelves are an integral part of a stepladder, they should be fastened so that they can be folded up when the ladder is closed.

TABLE 21.1 Classification of Various Species of Wood Acceptable for Use in Ladders

Group 1		
White ash	Hickory	Red oak
Beech	Locust*	White oak
Birch	Hard maple	Pecan
Rock elm	Red maple	Persimmon

The allowable fiber stress in bending for these species, when used for side rails, must not exceed 2,150 lb/in.^{2} Group 1 woods may be substituted for those in group 3 provided the dimensions are not more than 10% smaller for each cross section; or, if the thickness remains unchanged, the width is not more than 15% smaller when used edgewise (as in a rail), nor 25% smaller when used flatwise (as in a tread).

Group 2		
Douglas fir (coastal)	Western larch	Southern yellow pine

The allowable fiber stress in bending for these species, when used for side rails, must not exceed 2,000 lb/in.^{2} Group 2 woods may be substituted for those in group 3 provided the dimensions are not more than 7½% smaller for each cross section; or, if the thickness remains unchanged, the width is not more than 11% smaller when used edgewise (as in a rail), nor 20% smaller when used flatwise (as in a tread).

Group 3		
Red alder	Douglas fir (Rocky Mountain)	Poplar
Oregon ash	Noble fir	Redwood*
Alaskan cedar*	Gum	Eastern spruce
Port Orford cedar*	West coast hemlock	Sitka spruce
Cypress*	Oregon maple	Sycamore
Soft elm	Norway pine	Tamarack

The allowable fiber stress in bending for these species, when used for side rails, must not exceed 1,600 lb/in.^{2}

Group 4		
Aspen	Idaho white pine	Eastern hemlock
Basswood	Ponderosa pine	Soft maple
Butternut	Cottonwood	Lodgepole pine
Incense cedar*	White fir	Northern white pine
Western red cedar*	Sugar pine	

The allowable fiber stress in bending for the species, when used for side rails, must not exceed 1,375 lb/in.^{2} Group 4 woods may be substituted for those in group 3 provided the dimensions are at least 5% greater for each cross section; or, if the thickness remains unchanged, the width is at least 7½% greater when used edgewise (as in a rail), or 15% greater when used flatwise (as in a tread).

Notes 1. Where resistance to decay is required, the species having the most durable heartwood, as indicated by an asterisk(*), should be used; or the wood should be given a treatment with a preservative before using a ladder under conditions favorable to decay.

2. All minimum dimensions and specifications for side rails and flat steps are based on group 3 species. Substitution of species in other groups is permitted only when used in sizes that provide at least equivalent strength, as specified.

3. All minimum dimensions and specifications for rungs and cleats are based on group 1 species. However, cleats may be made of wood from any other group, provided that the cross-sectional dimensions specified for group 1 species are increased by the following factors:

Species	Each dimension	Width only (thickness unchanged)
Group 1	1.00	1.00
Group 2	1.03	1.05
Group 3	1.11	1.19
Group 4	1.17	1.26

Extension ladders should have their guide brackets long enough to engage the full width of the side rails on the other section of the ladder. Also, the locks should be of proper design, and the rope and pulley of ample strength. Near the lower end of the upper section of extension ladders a rung is often omitted at the location of the locks. However, a special offset rung may be necessary, because the sections of an extension ladder are frequently separated and used independently.

Finally, the inspector should make a thorough examination of the ladder material itself, especially with wood ladders. This is one part of the inspection that is usually omitted in part or in its entirety. All parts of a wood ladder should be free from splinters and defects. All parts of metal and plastic ladders should be free from burrs, cracks, and defects.

ANSI A14.1, "Safety Requirements for Portable Wood Ladders," classifies the various species of wood into four groups on the basis of the mechanical properties acceptable for use in ladder construction (see Table 21.1).

In addition to these general considerations, the inspector should assure a ladder's conformity to individual specifications for particular types of ladders before allowing it to be used.

Stepladders

By definition, a stepladder is a self-supporting portable ladder, nonadjustable in length, having flat steps and a hinged back (see Fig. 21.2). The length of such a ladder is measured along the front edge of its side rails. Stepladders longer than 20 ft DO NOT meet standard safety requirements.

All stepladders are classified by use:

Type I. Industrial; heavy-duty, for use in utilities, construction, and industrial applications; 3 to 20 ft long; capable of supporting up to 250 lb

Type II. Commercial; medium-duty, for use in painting, office, and light maintenance applications; 3 to 12 ft long; capable of supporting up to 225 lb

Type III. Household; light-duty; 3 to 6 ft long; capable of supporting up to 200 lb

Type III ladders should never be used with ladder jacks and scaffold planks.

Stepladder steps should be spaced no more than 12 in. apart. They should be parallel and level when the ladder is open. The minimum width between side rails at the top, inside to inside, should be not less

Figure 21.2 Platform stepladder, useful because the worker is fully supported while having both hands free. (*From L. R. Higgins,* Handbook of Construction Equipment Maintenance, *McGraw-Hill, 1979.*)

than 12 in., with the side rails spreading from top to bottom at least 1 in. for each 12 in. of ladder length.

Standard dimensions for type I and type II wood stepladders are listed in Table 21.2.

When a stepladder is in the open position, the front section should slope not less than 3½ in., the back section not less than 2 in., for each 12-in. length of side rail.

A metal spreader or locking device of sufficient size and strength should securely hold the front and back sections in the open position. The spreader should not be more than 6½ ft above the ladder base.

Portable Rung Ladders

There are four principal types of portable rung ladders: single, extension, sectional, and trestle or extension trestle.

TABLE 21.2 Dimensions for Wood Step Ladders

	Type I						Type II					
	12-ft length or less		14- and 16-ft length		18- and 20-ft length		3–8-ft length		10-ft length		12-ft length	
	Thickness, in.	Depth, in.	Thickness, in.	Depth, in.	Thickness, in.	Depth, in.	Thickness, in.	Depth, in.	Thickness, in.	Depth, in.	Thickness, in.	Depth, in.
Side rails	25/32	3¼	25/32	3½	1 1/16	3½	¾	2⅝	¾	2⅝	¾	3
Back legs	25/32	2¼	25/32	2⅝	1 1/16	2¼	¾	1⅝	¾	1¾	¾	2
Steps	25/32	3⅝	25/32	4¼	25/32	4¼	¾	3½	¾	3½	¾	3⅝
Tops	25/32	5½	25/32	5½	25/32	5½	¾	5	¾	5	¾	5

Notes 1. Flat steps should be secured with at least two 6-d nails at each end. Each step should be reinforced by a steel rod not less than 0.160 in. in diameter, which is fastened through metal washers on each end. A truss block should be fitted between the rod and the center of each step. As an alternative, a metal angle firmly secured to the steps and side rails may be used to provide the required bracing.

2. All steps 3⅝ in. wide and 27 in. or more long, and all steps 4¼ in. wide and 32 in. or more long, should be braced at each end with a metal angle securely attached to the step and side rails.

3. The back legs should be reinforced by a rivet that passes through the depth of the leg above the hinge point to prevent splitting, or by metal plates or collars at the hinge point.

4. Minimum width of step or tread for metal and plastic ladders:

Type I ladders 3 in.

Type II ladders 3 in.

Type III ladders 2½ in.

Note Metal and plastic ladders should be fitted with slip-resistant material at the bottom of the four rails.

As with the steps on stepladders, the rungs on these ladders should be parallel, level, and uniformly spaced. Wood ladder rungs should be bored through the side rails and should extend at least flush with the outside rail surface. The shoulder of each rung should be forced firmly against the side rails and the tenon secured in place with a nail or its equivalent to prevent the rung from turning and to maintain its position on the side rail. With metal and plastic ladders, rungs should be rigidly connected to side rails to ensure specified strength.

Round rungs of group 1 wood should be not less than 1⅛ in. in diameter for lengths up to 36 in. between side rails, and 1¼ in. in diameter for lengths over 36 in., up to and including 72 in. Tenons should be not less than ⅞ in. in diameter. When wood rungs between side rails are 28 in. or longer, they should also be provided with center bearing.

Oval rungs or rungs of any other cross section may be used on wood ladders, provided they are secured by a nail at each end, and have at least the same strength and bearing as round rungs of the same length.

When measured along the side rails, no wood rung ladder or section should be more than 4 in. shorter than the specified length.

All metal parts and fittings, including hooks and nonslip bases, should be securely bolted, riveted, or attached by equivalent fasteners.

Single ladders

A single portable rung ladder is a non-self-supporting type, nonadjustable in length, consisting of but one section. Its size is designated by the overall length of the side rail. Single ladders longer than 30 ft DO NOT meet standard safety code requirements.

The minimum dimensions of single-ladder side rails made of group 2 or group 3 wood are given in Table 21.3.

Extension ladders

An extension ladder is a non-self-supporting type, consisting of two or more sections that travel in guides or brackets so arranged as to permit the user to adjust the overall length. The size of an extension ladder is designated by the sum of the lengths of the sections, measured along the side rails.

Two-section wood and metal extension ladders longer than 60 ft DO NOT meet standard safety requirements; three-section metal ladders longer than 72 ft DO NOT meet standard safety requirements. Dimensions for wood extension ladders are given in Table 21.4.

The nominal size of a metal or plastic extension ladder is designated by the sum of the lengths of one side rail of each section, measured along the side rails exclusive of any foot or end cap. Dimensions are given in Table 21.5.

Extension ladders should be equipped in such a manner that the ladder cannot be used with an overlap less than that specified in the

TABLE 21.3 Dimensions for Single Ladders

Ladder length, ft	Thickness, in.	Depth, in.
Up to and including 16	1⅛	2½
Over 16 up to and including 22	1¼	2¾
Over 22 up to and including 30	1¼	3

Notes 1. Smaller side rails are acceptable in all ladders of this type if they are reinforced by a steel wire, rod, or strap that runs the length of the side rails and is adequately secured to them.

2. The width between side rails at the base, inside to inside, should be at least 11½ in. for single wood ladders up to and including 10 ft. This minimum width should increase by at least ¼ in. for each additional 2 ft of ladder length.

3. Metal or plastic single ladders should not exceed the following lengths:

Type I	30 ft
Type II	24 ft
Type III	16 ft

TABLE 21.4 Dimension for Wood Extension Ladders

Overall length of ladder, ft	Side rails for two-section ladders, in.	
	Thickness × depth, group 2 wood	Thickness depth, group 3 wood
16	1¹⁄₁₆ × 2	1⅛ × 2
20	1¹⁄₁₆ × 2	1⅛ × 2¼
24	1¹⁄₁₆ × 2½	1⅛ × 2½
28	1¹⁄₁₆ × 2¾	1⅛ × 2¾
32	1⅛ × 2¾	1⁵⁄₁₆ × 3
40	1⁵⁄₁₆ × 2¾	1⅜ × 3
44	1⁵⁄₁₆ × 3	1⅜ × 3¼
48–52		1⅜ × 3¾
56–60		1⅝ × 3¾

Notes 1. Smaller side rails are permitted when reinforced by a steel wire, rod, or strap running the length of the side rails, adequately secured to the rails. Minimum dimensions of side rails are based on the extension ladder's working length, which is the size of the ladder less the minimum overlap.

2. Minimum distance between side rails of the bottom section (inside to inside):

Ladder length, ft	Minimum distance, in.
Up to and including 28	14½
Over 28 up to and including 40	16
Over 40 up to and including 60	18

3. Minimum overlap of wood extension ladder side rails:

Ladder length, ft	Minimum overlap
Up to and including 36	3
Over 36 up to and including 48	4
Over 48 up to and including 60	5

tables. The fly section of metal or plastic extension ladders should contain the following warning in letters not less than ⅛ in. high: "CAUTION—This Ladder Section Is Not Designed for Separate Use."

All locks and guide irons for extension ladders should be so constructed so as to develop the full strength of the side rails. The metal locks should be positive-gravity or spring-action types, which engage the rung of the base section or the base section rung and the mating fly section rung.

If the fly section of a metal extension ladder incorporates a stationary lock or locks, which by their location eliminate a rung in the section, then the section should be permanently marked: "CAUTION—This Ladder Section Is Not Designed for Separate Use."

Guide irons should be securely attached and so placed on the base section of the ladder as to prevent the upper section from tipping or falling out while it is being raised or lowered, or while the extension is in use.

TABLE 21.5 Dimensions for Metal Extension Ladders

Ladder duty rating	Two-section ladder, ft	Three-section ladder, ft
Type I, heavy duty	Up to and including 60	Up to and including 72
Type II, medium duty	Up to and including 48	Up to and including 60
Type III, light duty	Up to and including 32	

Notes 1. All metal extension ladders should be clearly marked to indicate both the total nominal length of the sections and the maximum nominal extended length or maximum nominal working length.

2. The minimum clear width between side rails of the uppermost (fly) section of metal extension ladders shall be not less than 12 in. The minimum clear width between side rails at the bottom of the base section shall be not less than the following dimensions:

Ladder length, ft	Minimum clear width, in.
Up to and including 16	12½
Over 16 up to and including 28	14
Over 28 up to and including 40	15
Over 40 up to and including 72	18

3. Minimum overlap of metal or plastic extension ladder side rail when fully extended:

Ladder length, ft	Two-section ladder total overlap, ft	Three section ladder total overlap, ft
Up to and including 36	3	6
Over 36 up to and including 48	4	8
Over 48 up to and including 72	5	10

An extension ladder's rope and pulley should be securely attached to the ladder so as not to weaken either the rungs or the side rails. Minimum pulley and rope specifications are given in Table 21.6.

Rope should be of sufficient length for proper operation. Where cable is used on three-section metal ladders, the cable should be at least 3⁄16 in. in diameter.

Sectional ladders

These ladders are not as common as the other types; they are used in particular by window cleaners. Sectional ladders are defined as non-self-supporting portable ladders, nonadjustable in length, and consist of two or more sections of ladder so constructed that the sections may be combined to function as a single ladder. The size of such ladders is designated by the overall length of the assembled sections. Minimum dimensions are listed in Table 21.7.

TABLE 21.6 Minimum Pulley and Rope Specifications

Ladder material	Pulley diameter, in.	Rope diameter, in.	Breaking strength, lb
Wood	1¼	5⁄16	560
Metal or plastic	1	¼	560

TABLE 21.7 Minimum Dimensions of Side Rails, Group 2 or Group 3 Woods

Length, ft	Thickness, in.	Depth, in.
Up to and including 21	1⅛	2¾
Over 21 up to and including 31	1⅛	3⅛

Ladders of this type should have either straight sides that converge slightly toward the top of each section, or flaring sides at the bottom of the bottom section, with the top section having side rails that converge to a width not less than 4 in. The minimum width between side rails of the bottom section should be 11 in.

Individual sections are joined by means of a groove in the bottom end of the upper section's rails, which sets firmly over extensions outside the side rails of the next lower section's topmost rung. A groove in the top end of the lower section's rails sets firmly under the bottom rung, inside the side rails of the section immediately above.

The distance between the topmost rung of one section and the bottom rung of the next section above should be not less than 12 in. The grooved ends of the sections should be reinforced with a metal plate (not less than 18 gage) secured to the rail by a rivet through the depth of the rail. Rungs and grooves should fit securely, without binding or unnecessary play.

Trestle and extension-trestle ladders

A trestle ladder is a self-supporting portable ladder, nonadjustable in length, consisting of two sections hinged at the top to form two equal angles with the base. Its size is designated by the length of the side rails measured along the front edge of the ladder.

The combination extension-trestle ladder, on the other hand, is an adjustable self-supporting portable ladder that consists of a trestle ladder base and a vertically adjustable single ladder, with suitable means for locking the ladders together. The size of this type of ladder is designated by the length of the trestle ladder base.

Trestle ladders, or extension sections, or base sections of extension-trestle ladders longer than 20 ft *do not* meet standard safety requirements. Minimum dimensions are given in Tables 21.8 and 21.9.

TABLE 21.8 Minimum Dimensions for Trestle Ladder Side Rails or Extension-Trestle Ladder Base Sections, Group 2 or Group 3 Woods

Length, ft	Thickness, in.	Depth, in.
Up to and including 16	1$\frac{5}{16}$	2¾
Over 16 up to and including 20	1$\frac{5}{16}$	3

TABLE 21.9 Minimum Dimensions for Trestle Ladder Side Rails or Extension-Trestle Ladders with Parallel Sides, Group 2 or Group 3 Woods

Length, ft	Thickness, in.	Depth, in.
Up to and including 12	1 5/16	2 1/4
Over 12 up to and including 16	1 5/16	2 1/2
Over 16 up to and including 20	1 5/16	2 3/4

Trestle ladders and base sections of extension-trestle ladders should open so that the spread of the trestle at the bottom, inside to inside, is at least 5½ in. per foot of ladder length. Side rail width at the base of these ladders should be at least 21 in. for all ladders and sections up to and including 6 ft. For longer lengths, side rails should be at least 1 in. greater for each additional foot of ladder length. The width between side rails of the extension sections should be not less than 12 in.

All rungs should be parallel and level, spaced not less than 8 in., nor more than 18 in., apart on trestle ladders or base sections of extension-trestle ladders; and not less than 6 in., nor more than 12 in., apart on extension sections.

Special ladders

Two special ladders commonly used on construction sites are the mason's ladder and the cleat-type ladder.

The mason's ladder is a single ladder that should *not* be supplied in sizes longer than 40 ft. The width between side rails at the bottom rung, inside to inside, should be not less than 12 in. for all mason's ladders up to and including 10 ft in length. This should be increased by at least ¼ in. for each additional 2 ft of ladder length. Table 21.10 lists the minimum dimensions for mason's ladders.

Rungs of mason's ladders should be spaced not less than 8 in. nor more than 12 in. apart.

Cleat ladders, also of the single type, should not be supplied in sizes longer than 30 ft. Wood side rails of these ladders should be not less than 1⅝ in. thick and 3⅝ in. deep (nominal 2 × 4 in.). Minimum dimensions of wood cleats are given in Table 21.11.

TABLE 21.10 Minimum Dimensions for Mason's Ladder Side Rails

	Side rails		Diameter	
Length, ft	Thickness, in.	Depth, in.	Rung, in.	Tenon, in.
Up to and including 12	1 5/8	3 5/8	1 3/8	1
Over 22 up to and including 40	1 5/8	4 1/2	1 3/8	1

TABLE 21.11 Minimum Dimensions for Wood Cleats, Group 1 Woods

Length, ft	Thickness, in.	Width, in.
Up to and including 20	25/32	3
Over 20 up to and including 30	25/32	3¾

Wood cleats should be inset into the side rails not less than ½ in. If the cleats are attached directly to the side rails, they should be fastened by three 10-d wire nails to each rail, and filler blocks of the same thickness as the cleats should be securely attached to the edge of the rail for the full length between cleats.

Care and Maintenance

Proper care and maintenance of ladders will ensure their continued safety and serviceability. The following checklist is designed to help maintain ladders in good usable condition at all times.

Never subject ladders to unnecessary dropping, jarring, or misuse. Ladders, like any tool, are designed for a specific purpose or use. Therefore, any variation from this use is a mishandling of equipment.

Inspect ladders periodically—initially upon receipt, as well as each time before use—and remove from service, for repair or destruction, any ladder that has developed defects. Tag all ladders that are to be repaired: "DANGEROUS—DO NOT USE."

Make sure that the joint between steps or rungs and side rails is tight. On metal ladders, check for side rail dents or bends, or excessively dented rungs. Never attempt to straighten or use a bent ladder.

Keep steps and rungs free of grease, oil, and other slippery substances.

Make sure all bolts and rivets are in place and tight before using a ladder. Never use a ladder if any bolts or rivets are missing.

Check hardware, fittings, and accessories frequently. Make sure that they are in proper working condition and that all moving parts operate freely without binding or undue play. Lubricate metal bearings of locks, wheels, and pulleys regularly.

Inspect ropes or cable frequently and replace any that are worn or frayed.

Check the ladder's safety shoes or padded feet. If they are excessively worn, do not use the ladder until shoes or feet are repaired.

Tie down ladders securely to a truck rack for transporting. Make sure such ladders are supported adequately to prevent sagging. Secure the feet from pivoting about the ladder while the vehicle is in motion. The ladder truck rack should be designed to clamp the ladder into a fixed position and prevent any chafing that might be caused by horizontal and vertical movement of the ladder with respect to the truck rack, the truck, or other ladders or individual sections. Keep to a minimum any ladder overhang beyond supporting points. These points should be of hardwood or rubber-covered iron pipe to reduce chafing and the effects of road shock. Tying the ladder to each support point will greatly reduce any damage that may be caused by road shock.

Store ladders in such a manner as to provide easy access or inspection, and to prevent danger of accident when withdrawing a ladder for use. Ladders, stored in a horizontal position or on edge on racks, should be supported at a sufficient number of points to avoid sagging and permanent set. At no time should material be placed on the ladder while it is in storage.

Store wood ladders at a location where they will not be exposed to the elements, but where there is good ventilation. Never store wood ladders near radiators, stoves, steam pipes, or other places subjected to excessive heat or dampness.

Use a suitable protective coating on wood ladders. Painting is satisfactory, provided that prior to painting the ladders are carefully inspected by competent and experienced inspectors.

Inspect all metal ladders that are exposed to excessive heat visually for damage and then test them for reduced deflection and possible loss of strength characteristics. (In doubtful cases, refer to the manufacturer.)

Protect plastic reinforced ladders from excessive exposure to ultraviolet rays (sunlight), which may cause damage to glass fiber. Where this condition is evident, wash down the ladder with a suitable commercial solvent, or liquid detergent solution compatible with the ladder's composition, such as acetone, isopropyl alcohol, methyl ethyl ketone, or the like, and allow the ladder to air dry. Provide adequate ventilation, prohibit smoking, and make sure workers wear the proper protective equipment when using such solvents. After the ladder has been washed, handle it as little as possible. Brush or spray the ladder rail with a good commercially available grade of acrylic lacquer, epoxy coating, or other suitable material. One or two coats may be required depending on the degree of glass fiber prominence. If a plastic ladder must be stored outdoors for any extended period, apply a coat of good commercially available nonslip paste wax or other suitable material on the fiberglass side rails of the ladders to help reduce the possibility of glass fiber prominence. A semiannual coating program should be carried out to ensure the ladder's longevity.

Maintain plastic ladders in a clean condition at all times to ensure the retention of the original electric insulating characteristics. Remove all surface buildup of dirt, dust, grease, grime, and carbonaceous and other conductive materials that will provide a ready path for electric currents and thus create a potential danger for the user.

Never use a metal ladder near exposed electric equipment, including overhead distribution wires, crane ways, trolley conductors, or switchboards.

Never load any ladder beyond its rated capacity.

Chapter 22

General Site Safety

Accidents don't just happen. They are caused by someone or something working unsafely or protected improperly.

Controlling accidents on a work site is the result of management and employees working together for a common objective. However, the leadership for such an effort must originate with management.

While individual workers have a basic responsibility for their own and their fellow workers' safety, management and supervision have the overriding responsibility for ensuring that the workers who prepare the equipment, use the equipment, and work with or around it are well trained in both safety and operating procedures.

The degree to which management accepts this responsibility will determine the success or failure of the entire effort. Unless management is firmly convinced of the need to control accident losses, there is little chance that any safety program it implements will succeed.

Policy

Accident control must begin with a written policy—management's commitment to accident prevention objectives.

Lead off the written policy with a statement that affirms the company's commitment to safety and its intentions, through the medium of the policy, to take positive action. This statement is vital to the success of both policy and program, because safety, more than any other facet of an operation, demands individual participation by every member of the organization—and at every level. Participation by subordinates will occur only when each individual believes that management really means what it says.

Key management personnel should meet and agree that they sincerely want an effective accident prevention program and are willing to commit every effort toward its attainment. Top management should then decide who is to be in charge of this program, how it is to be done, and to whom the supervisory staff is to report.

Although executive management has the ultimate responsibility for safety, specific responsibilities and authority must be assigned to each level of supervision. Management's policy should be passed down through the supervisory level to all personnel involved in the overall operations.

Most construction safety programs that fail do so because the management policy was expressed in terms under which the company was subsequently unwilling or unable to operate. Make sure the policy initially contains only what can be supported by existing intentions and resources. Then, as capabilities grow, that policy may be expanded to conform with any changes.

A management policy should include:

Objectives for accident prevention and loss control

Authority for the safety program

Assignment of responsibilities

Designation of functions

Support role of company resources

Objectives

Above and beyond a statement of intentions to provide a safe working environment, it is necessary to conform to various safety laws and regulations at municipal, county, state, and federal levels. Furthermore, certain goals must be kept in mind, such as the following:

Curbing the general cost of insurance coverage, the incidental one of compensation, and the ever-present expense of lost time on work in progress

Providing a means of making workers and supervisors aware of sound safety practices and providing a time and a place for this learning process to occur

Establishing a back-up for a full-time safety program by setting up the organization, the reporting, and the record-keeping system required

Once these basic goals are set, the extent of the program must be determined. As with any management program, safety implementation costs money—strictly in accordance with its depth and its degree of sophistication.

The scope of the safety program will involve situations in which there is cause for management concern, and situations in which there is both cause for concern and a directed preventive action. These concerns include:

On-site injury. Any injury to owner's employees, subcontractor's employees, vendor's employees, tenants, trespassers

Off-site injury. Any injury involving contractor operations, equipment, or employee on duty

On-site damage. Caused by own equipment or operations, by action of others to own supplies and equipment, or damage to work in place, not accepted by owner

Off-site damage. Any damage involving contractor operations, equipment, or employee on duty

Authority

The authority to enforce a safety program must be meted out carefully and in strict accordance with its form. A good practice in assigning line authority is to transmit it along the same channels used for other company directions. In this way it flows down to the workers in a familiar and expected manner.

The amount and nature of staff authority will depend on the particular needs and the sophistication of the safety program. But the person working for the program's goals must be independent of the demands for safety record upkeep and the day-to-day need to produce specific results.

The true function of the staff safety specialist is to use his or her professional skills to develop meaningful rules that the line can put into use. This individual can make recommendations to management and should secure voluntary compliance from line supervisors, but may not direct conformance.

Responsibility

Everyone in the organization must at all times be made responsible for all safety practice. There are no exceptions, and the policy must say so firmly and clearly. Because this is true, each member of the organization must receive a copy of the complete policy, for whose contents everyone is individually responsible.

A management representative, preferably someone with basic training in accident prevention, should be assigned the responsibility for coordinating all loss control activities. This person should maintain the liaison between top management and field supervision.

Management and supervisory responsibilities should ensure that:

Proper equipment and materials are available for a particular application

Correct load ratings are determined for the equipment and materials being used

Equipment and materials are maintained in proper working condition

Rigging and scaffolding crews have proper work supervision

Loads are rigged and scaffolding is erected correctly

Equipment and materials used on the job have the necessary capacity and are in safe working condition

Equipment and materials are correctly assembled and erected as required during operations

Work crews and other personnel are afforded adequate safety as they are affected by particular rigging and scaffolding operations

Functions

Duties should include preplanning, employee and supervisory training, setting up minimum safety standards, medical facilities, accident investigation, and record keeping.

Preplanning

Since safety program implementation is essentially a management endeavor, the present dimensions of safety within the organization must be examined. Begin by reviewing such safety history as is available. Time spans for looking backward vary from company to company, but a 3- to 5-year period is perhaps the best general coverage. Of particular importance are:

1. An exact and clear statement about the accident cause
2. An accurate assignment of the responsible area or element of equipment

Armed with this information, some understanding of the reasons for accidents will emerge.

In addition to a review of the past, the accident potential should be considered. This involves inspecting every aspect of the operation and pinpointing those conditions and methods that appear to threaten safety. It means getting out on every job and observing every operation and every arrangement and hookup. It also means taking an up-close look at worker practices in an effort to discover which ones contain the seeds of future trouble. It also involves taking the measure of supervisory personnel in terms of their safety training, latent knowledge, and their awareness of the essential nature of 100% job safety.

Training

Preventing accidents necessitates creating a safe environment and then teaching people to respect it. Aside from some effort in apprenticeship classes, training workers in safety practices is the prime responsibility of the employer. Company safety programs must provide for the safety education of every supervisor, as well as the opportunity and facilities to train workers under supervision.

The extent and type of training to be provided will depend on the type of work, the geographic area, the number of employees involved, and their assignments. Nevertheless, there are certain basic elements common to nearly all effective safety efforts.

Training needs should be considered at the time the accident prevention personnel are studying the job and formulating plans. Begin safety instruction when a worker is first hired. A new worker forms attitudes about an employer immediately and should be acquainted at the start with the safety program and its importance to the company.

New employee indoctrination should cover at least the following: hazards involved in department or job, disciplinary action involved for disregarding safety rules, importance of good housekeeping, and worker's responsibility for it. Make sure new employees understand any safety rules that are a condition of employment, such as wearing eye protection or a hard hat, and their enforcement.

The campaign for safety is a never-ending one. Regular meetings should be scheduled to reinforce an awareness of the subject and keep the workers updated on accident prevention and safety training measures. Supervisors should meet weekly to discuss and review any accidents of the preceding week, to consider upcoming work and any training needed to overcome the hazards it presents, and to develop subjects for toolbox meetings—five- to fifteen-minute weekly sessions.

An insurance carrier can provide advice, assistance, films, literature, and other aids for these safety meetings. Use posters and bulletin boards to provide visual reminders.

To make sure a training program retains its effectiveness in the field, monitor it. Have members of management look for and report any safety violations they see when visiting a site. Make it known that they are not spies, but worker protectors charged with remedying faults that may cause injury or death.

Safety standards

Safety standards usually are the minimum control to make a hazard tolerable. If the hazard and the conditions surrounding it are understood, the protection and other benefits can be substantially increased with a little additional input.

Medical facilities

The availability of first-aid services and medical personnel, plus an ongoing health maintenance program, are important considerations in any loss control program. Immediate and competent attention to a construction injury is often the difference between a minor disability and death or permanent physical impairment.

The program should include rescue and resuscitation training, as well as emergency treatment of wounds, burns, and sickness. The local office of the American Red Cross offers such training programs.

It is the employer's responsibility to ensure that arrangements are made to provide immediate attention to an injured employee. Management must provide for first-aid treatment of the injured, first-aid equipment, and trained personnel to handle injuries on every job. At least one employee on each shift should have first-aid training and maintain a current first-aid card issued by an agency such as the American Red Cross or the U.S. Bureau of Mines. Whenever possible, all members of the supervisory staff should attend a class in first aid conducted by qualified instructors.

At the start of each job, obtain the names and locations of qualified physicians in the immediate area. Contact these doctors to explain the nature of the work and the type of injuries that might occur. In addition, contact an ambulance service and a hospital before the work starts. Post the telephone numbers and addresses of doctors, ambulance services, and hospitals near all job telephones.

Instruct every employee to report all injuries, regardless of apparent severity, to their immediate supervisors as soon as accidents occur. These should be logged by time and date for future reference.

Accident investigation

A project supervisor should investigate every accident that occurs, regardless of its apparent severity at the time. More serious accidents should be detailed in a report to management, with inclusion of the steps necessary to prevent future similar types of accidents.

To assure the success of any safety program it is essential that supervisory personnel root out hazards on a daily basis. Accidents prevented are also losses prevented. Most accident hazards are also production hazards, and their existence costs money.

The practice of considering hazard control as the objectives for a safety program, rather than as a means to the objective, has generally obscured the opportunity to exploit the production economics of accident prevention. Management should take into consideration just how much of its labor cost is claimed by a broken ladder, missing step, burned-out light bulb, clogged passageway, makeshift piece of equipment, or a multitude of other similar hazards.

It is also important to consider the cost involved in getting a worker to the point of operation, as well as what that individual works from and with, after arriving at the work station. A negligent effort is worse than no effort at all. A negligent effort is frequently one that serves to camouflage a hazard, to give a dangerous situation a false appearance of safety.

Record keeping

To assure a successful loss control program, supervisors must take the lead role in accident investigation and maintain complete accident records in order to keep management informed of safety progress and to identify areas of strength and weakness in the program. Calculate the frequency and severity rates and publish them on a monthly basis. These should be developed for each job, as well as for the company as a whole.

Where accidents have a potential for future claims, the availability of a report covering details of the investigation of the accident is of great importance. Accident reports and related data will have to be supplied to the insurance carrier, as well as to state, federal, and other officials.

General Site Safety

For the most part, accidents result from one of two sources: unsafe acts or unsafe conditions. On every job it should be standard practice for supervisory personnel to look for unsafe acts and unsafe conditions and demand that they be immediately corrected, or obtain a definite commitment as to when the correction will be made.

Housekeeping

A basic incentive in developing good attitudes is maintaining the working areas in a clean, orderly fashion. Poor housekeeping is not only a direct contributor to many injuries, but is a primary cause of fires.

The essential element of good housekeeping is planning—at the beginning of each job—for equipment arrangement, material storage, safe environmental conditions, and final cleanup.

Storage areas. All materials should be maintained in organized stockpiles for ease of access. Keep walkways clear of loose materials and tools.

Work areas. Clean up loose materials, waste, and other supplies immediately. This is especially important in the vicinity of ladders, ramps, stairs, and machinery. Tools and loose materials should be removed immediately if they threaten to become a hazard.

Areas used by personnel. Empty bottles, containers, and papers should not be allowed to accumulate where lunches are eaten on the job site.

Oil and grease. Spills of oil, grease, or other liquids should be removed immediately or sprinkled with sand.

Disposal of waste. An effective means of preventing litter is the provision of suitable receptacles for waste, scrap, and other materials. Combustible waste, such as oily rags, paper, and the like should be stored in a safe place in covered metal containers and disposed of regularly.

Protruding nails. Such nails must be removed. This should be done as this hazard develops. Cleaned lumber should be stacked in orderly piles. Workers performing this task should wear heavy gloves and puncture-proof insoles.

Lighting. Lighting at an adequate level should be provided in or around all work areas, passageways, stairs, ladders, and other areas used by workers.

Personal protective gear

All levels of supervision should be held responsible for seeing that workers wear proper protective gear and that it is kept in good repair.

Hard hats. All construction workers and all others entering the site must wear hard hats approved for the particular exposure at all times on the job. They should not be permitted to wear hard hats that have the shell altered by drilling or cutting, since these alterations weaken the shell, and the hat will not provide adequate protection.

Gloves. Where needed, workers should wear work gloves in good condition that are suited to the type of work involved. Chrome-tanned leather gloves, due to their resistance to abrasion and sparks, are recommended for general use and materials-handling jobs. Neoprene or rubber gloves should be worn for protection from oils, acids, or chemicals. Electrically insulated gloves, coupled with leather protection and tested for the maximum voltage encountered, must be used on power lines and other points of electrical exposure. These gloves should be inspected prior to use and dielectrically tested periodically by a reputable laboratory. Defective gloves must be destroyed.

Shoes. Safety shoes are recommended for use by all construction workers. All safety shoes should meet nationally recognized standards. In addition to safety shoes, canvas or leather leggings and spats should be worn by welders. Workers should be encouraged to keep their shoes in good repair. Do not permit workers to wear shoes with worn heels or thin and worn soles, nor sneakers or shoes that have been slit or have holes cut in them.

Glasses. Workers should wear safety glasses in all circumstances where there is exposure to flying particles. Side shields offer additional protection.

Plastic face shields. These should be worn for guarding against spraying liquids or corrosives, flying particles, or similar work; coverall acid goggles when washing masonry walls, fluxing metals, or handling corrosives; coverall chipping goggles when caulking, drilling, sawing, or chipping, and for other dust-producing operations; and burning goggles or a welder's hood with lenses that have the correct color density for the type of welding (gas or electric arc) involved.

In summary, then, an attitude of safe practice and procedures should be developed and maintained throughout an organization so that everyone is committed to the idea of performing accident-free work. Management and supervisory personnel should set the right example by following established safety rules without exception.

ANSI Standards

A detailed set of specifications, prepared by the American National Standards Institute (ANSI), has been included in the Appendix.

Inspection Checklist

As part of any safety program, management should develop a general inspection checklist that will apply to most of the problems that are likely to be encountered. In addition, special checklists should be tailored for specific requirements of individual operations.

The general checklist given in Table 22.1, developed from the *Manual of Accident Prevention in Construction* by the Associated General Contractors of America, can serve as a guide for preparing a checklist for rigging and scaffolding operations. It is not intended to be a comprehensive checklist, nor to be followed in form, if the job requirements would be better served otherwise.

TABLE 22.1 Safety Checklist

Checklist	Yes	No
Accident Prevention Organization		
Schedule for posting safety material		
Hard hat requirements		
Safety meetings scheduled and posted		
Housekeeping and Sanitation		
General neatness of working areas		
Regular disposal of waste and trash		
Passageways and walkways clear		
Adequate lighting		
Projecting nails removed		
Oil and grease removed		
Waste containers provided and used		
Sanitary facilities adequate and clean		
Drinking water tested and approved		
Adequate supply of water		
Salt tablets provided		
Drinking cups or sterilized bubblers supplied		
First Aid		
First-aid station set up		
First-aid supplies available		
First-aid instruction on the job		
Telephone numbers and locations of nearby physicians posted		
Telephone number and location of nearest hospital posted		
Injuries reported promptly and to proper persons		
Fire Protection and Prevention		
Fire instruction to personnel		
Fire extinguishers identified, checked, lighted		
Phone number of fire department posted		
Hydrants clear, access to public thoroughfare open		
Good housekeeping		
NO SMOKING posted and enforced where needed		
Handling and Storing Materials		
Orderly storage areas, clear passageways		
Materials stacked neatly		
Stacks on firm footings, not too high		

TABLE 22.1 **Safety Checklist (*Continued*)**

Checklist	Yes	No
Handling and Storing Materials		
Proper number of personnel for each operation		
Materials protected from heat and moisture		
Dust protection observed		
Extinguishers and other fire protection on hand		
Traffic routing and control planning		
Barricades		
Floor openings planked over or barricaded		
Roadways and sidewalks effectively protected		
Adequate lighting provided		
Traffic controlled		
Hoists, Cranes, Derricks		
Inspect cables and sheaves		
Check slings and chains, hooks and eyes		
Secure equipment firmly or be sure it is supported		
Use outriggers as needed		
Be sure all power lines are inactivated, removed, or at safe distance		
Always use proper loading for capacity at lifting radius		
Keep all equipment properly lubricated and maintained		
Employ signal persons wherever needed		
Make certain that signals are understood and observed		
Equipment Maintenance		
Install a planned maintenance and inspection program		
Maintain adequate equipment records		
Use only proper oils, fuels and lubricants		
Scaffolding		
Erection under proper supervision		
All structural members adequate for use		
All connections adequate		
Safe tie-in to structure		
Ladders and working areas free of debris, snow, ice, grease		
Proper footings provided		
Passerby protection from falling objects		
Supports plumb, adequate cross-bracing provided		
Guardrails and toeboards in place		
Scaffold machines in working order		
Ropes and cables in good condition		
Frequent inspection		
Scaffolding Erection		
Safety nets		
Hardhats, safety shoes, gloves		
Taglines for tools		
Fire hazards controlled at welding operations		
Floor openings covered and barricaded		

TABLE 22.1 Safety Checklist (*Continued*)

Checklist	Yes	No
Scaffolding Erection (*Continued*)		
Ladders, stairs, or other access provided		
Hoisting apparatus checked		
Ladders		
Stock ladders inspected and in good condition		
Properly secured top and bottom		
Side rails on fixed ladders extending above top landing		

Safety Program Review*

A. Management organization

1. Does the company have a written policy on safety?
2. Can you draw a clear organization chart to show the line and staff relationships? (If so, draw one.)
3. To what extent does executive management accept its responsibility for safety?
   a. To what extent does management participate in the safety effort?
   b. To what extent does it assist in administering the safety effort?
4. To what extent does executive management delegate safety responsibility? How is it accepted by:
   a. The superintendent or top production people?
   b. The foremen or supervisors?
   c. The staff safety people?
   d. The employees?
5. How is the company organized for programmed safety?
   a. Are there staff safety personnel?
   b. If so, are the duties clear?
   c. Are responsibilities and authorities clear?
   d. Where is staff safety located?
   e. How can it be reached, and whom can it reach?
   f. What influence does it have?
   g. To whom does it report?
   h. Are there safety committees?
      i. What is the makeup of the committees?
      ii. Are the duties clearly defined?
      iii. Do the committees seem effective?
   i. What type of responsibility is delegated to the employees?
6. Does the company have written operating rules or procedures?

* Adapted from D. Petersen, *The OSHA Compliance Manual*, McGraw-Hill, 1979.

a. Is safety covered in these rules?
b. Is it built into each rule, or are there separate safety rules?

B. Accountability for safety:
1. Does management hold line personnel accountable for accident prevention?
2. What techniques are used to fix accountability?
   a. Accountability for results
      i. Are accidents charged against departments?
      ii. Are claim costs charged?
      iii. Are premiums prorated by losses?
      iv. Does supervisory appraisal include looking at the supervisors' accident records?
      v. Are bonuses influenced by accident records?
   b. Accountability for activities
      i. How does management ensure that supervisors conduct toolbox meetings, inspections, accident investigations, regular safety supervision, and coaching?
      ii. Other means?

C. Training and supervision
1. Is there safety indoctrination for new employees?
   a. Who conducts it?
   b. What does it consist of?
2. What is the usual procedure in training a new employee for a job?
   a. Who does the training?
   b. How is it done?
   c. Are written job instructions based on job analysis used?
   d. Is safety included?
3. What training is given to an older employee who has transferred to a new job?
4. What methods are used for training the supervisory staff?
   a. For new supervisors?
   b. Continuous training for the entire supervisory force?
   c. Who does the training?
   d. Is safety included?
5. After completion of training, what is an employee's status?
   a. What is the quality of the supervision?
   b. What use is made of the probation period?

D. Medical program
1. What first-aid facilities, equipment, supplies, and personnel are available to all shifts?
2. What are the qualifications of the people responsible for the first-aid program?
3. Is there medical direction for the first-aid program?

4. What procedure is followed in obtaining first aid?
5. What emergency first-aid training and facilities are provided when regular first-aid people are not available?
6. Are there any catastrophe plans or disaster plans?
7. What facilities are available for transportation of the injured to a hospital?
8. Is a directory of qualified physicians, hospitals, and ambulances available?
9. Does the company have any special preventive-medicine program?
10. Does the company engage in any activities in the health education field?

E. Systems to identify problems and hazards

1. Are routine inspections made?
   a. Who is responsible for inspections?
   b. By whom are the inspections made?
   c. How often are they made?
   d. What types are they?
   e. To whom are the results reported?
   f. What type of follow-up action is taken?
   g. By whom?
2. Are any special inspections made of elevators, hoists, overhead cranes, chains and slings, ropes, hooks, electrical insulation and grounding, ladders, scaffolding and planks, lighting, ventilation, materials-handling equipment, fire and other catastrophe hazards, noise and toxic controls?
3. Are any special systems set up?
   a. Job safety analysis?
   b. Critical-incident technique?
   c. High-potential accident analysis?
   d. Fault-free analysis?
   e. Safety sampling?
4. What procedure is followed to ensure the safety of new equipment, materials, processes, or operations?
5. Is safety considered when purchasing equipment, materials, and supplies?
6. When corrective action is needed, how is it initiated and followed up?
7. When faced with special or unusual jobs, how does management ensure that they are done safely?
   a. Is there adequate job and equipment planning?
   b. Is safety a part of the overall consideration?

8. What are the normal exposures for which protective equipment is needed?
   a. What are the special or unusual exposures for which personal protective equipment is needed?
   b. What personal protective equipment is provided?
   c. How is such equipment initially fitted?
   d. What type of care maintenance program is instituted for such equipment?
   e. Who enforces the wearing of such equipment?

F. Accident records and analysis
1. What injury records are kept? By whom?
2. Are standard methods of frequency and severity recording used?
3. Who sees and uses the records?
4. What type of analysis is applied to the records?
   a. Daily, weekly, monthly, annual?
   b. By department, cost, other?
5. What is the accident investigation procedure?
   a. What circumstances and conditions determine which accidents will be investigated?
   b. Who does the investigating?
   c. When is it done?
   d. What types of reports are submitted?
   e. To whom do they go?
   f. What follow-up action is taken?
   g. By whom?
6. Are any special techniques used?
   a. Estimated costs?
   b. Safe-T-Scores?
   c. Statistical control charts?

Chapter 23

OSHA: Safety and Health Provisions

> ...No contractor or subcontractor...shall require any (employees) in the performance of the contract to work in surroundings or under working conditions that are unsanitary, hazardous, or dangerous to (the employee's) health or safety.—*Federal Occupational Safety and Health Act* (1970).

Under OSHA it is the legal responsibility of the contractor to furnish its employees employment, and places of employment, free from recognized hazards that might cause, or would be likely to cause, serious physical harm or death; and to comply with the occupational safety and health standards issued in the Act.

The contractor's employees, in turn, have an obligation to familiarize themselves with those standards that apply to them and to observe them at all times.

OSHA also requires that each contractor initiate and maintain an accident prevention program to facilitate compliance with the Act. Such programs should provide for frequent and regular inspections of job sites, materials, and equipment. And, the contractor must post OSHA's "Safety and Health Protection on the Job Site" poster in an area where all employees are exposed to it daily.

A brief summary of the major points of the general safety and health provisions, as they apply to all businesses and industry, follows.

Accident Prevention (1926.20)

The employer is responsible for frequent and regular inspections of the job sites, materials, and equipment—to be made by competent persons designated by the employer.

Use of any machinery, tool, material, or equipment that is not in compliance with any applicable requirement of the standards is prohibited.

Any such items should either be identified as unsafe, by tagging or locking the controls to render them inoperable, or be physically removed from their place of operation.

It is the employer's responsibility to make sure that only those employees qualified by training or experience operate equipment and machinery.

Signs and Tags (1926.200)

Signs and symbols required to prevent accidents must be visible at all times when work is being performed, and should be removed or covered promptly when the hazard no longer exists.

Danger signs. Shall have red as the predominant color for the upper panel, black outline on the borders, and a white lower panel for additional wording. These signs should be used only where an immediate hazard exists.

Caution signs. Shall have yellow as the predominant color, black upper panel and borders, yellow lettering of "caution" on the black panel, and the lower yellow panel for additional wording in black. These signs should be used only to warn against potential hazards or to caution against unsafe practices.

Exit signs. Shall be lettered in legible red letters, not less than 6 in. high, on a white field, and the principal stroke of the letters should be at least three-fourths of an inch wide.

Safety instruction signs. Shall be white with a green upper panel having white letters to convey the principal message. Additional wording on the sign shall be in black on the white background.

Directional signs. Shall be white with a black panel and a white directional symbol. Additional wording on the sign shall be in black on the white background.

Traffic signs. Shall conform to ANSI D6.1c-1984, "Manual on Uniform Traffic Control Devices for Streets and Highways." All points of traffic hazard on construction sites or within construction areas must be posted with legible traffic signs.

Accident prevention tags. Shall be used as a temporary means of warning employees of an existing hazard, such as defective tools, equipment, and so forth. They must not be used as a substitute for accident prevention signs (see Table 23.1).

In addition to the OSHA standards, ANSI Z35.1-1972, "Specifications for Accident Prevention Signs," and ANSI Z35.2-1968 (R-1974), "Specifications for Accident Prevention Tags," contain certain rules not specifically prescribed by OSHA but with which employers must comply.

Barricades (1926.202)

Barricades for the protection of employees shall conform to ANSI D6.1c-1984, "Manual on Uniform Traffic Control Devices for Streets and Highways—Barricades."

TABLE 23.1 Accident Prevention Tags and Specifications

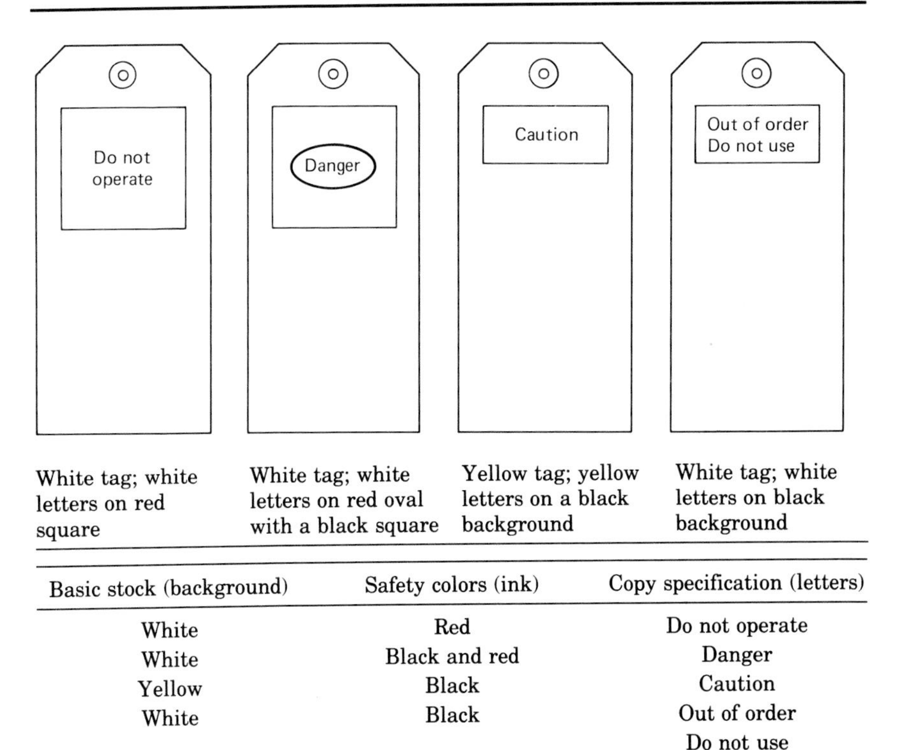

White tag; white letters on red square	White tag; white letters on red oval with a black square	Yellow tag; yellow letters on a black background	White tag; white letters on black background

Basic stock (background)	Safety colors (ink)	Copy specification (letters)
White	Red	Do not operate
White	Black and red	Danger
Yellow	Black	Caution
White	Black	Out of order Do not use

Signaling (1926.201)

When construction operations are such that signs, signals, and barricades do not provide the necessary protection on or adjacent to a highway or street, flagmen or other appropriate traffic controls shall be provided by the contractor.

Flagmen shall use red flags or sign paddles at least 18 in. square, and in period of darkness, red lights. They shall be provided with and shall wear a red or orange warning garment while flagging. Warning garments worn at night shall be of reflectorized material.

A flagman's signaling directions shall conform to ANSI D6.1c-1984, "Manual on Uniform Traffic Control Devices for Streets and Highways."

Storage Requirements (1926.250)

All materials stored in tiers shall be stacked, racked, blocked, interlocked, or otherwise secured to prevent sliding, falling, or collapse.

When storing materials within buildings and structures, maximum safe load limits of floors shall not be exceeded. Such maximum safe loads shall be conspicuously posted in all storage areas (except when storage is on a floor or slab at grade).

Aisles and passageways shall be kept clear to provide for the free and safe movement of material-handling equipment or employees. When a difference in road or working levels exists, ramps, blocking, or grading shall be used to ensure the safe movement of vehicles between the two levels.

Materials stored inside buildings under construction shall not be placed within 6 ft of any hoistway or inside floor openings, nor within 10 ft of an exterior wall that does not extend above the top of the material stored.

If employees are required to work on stored material in silos, hoppers, tanks, and similar storage vessels, they shall be equipped with lifelines and safety belts that meet OSHA requirements.

Materials in excess of those needed for immediate operations shall not be stored on scaffolds or runways.

Bagged materials. Shall be stacked by stepping back the layers and cross-keying the bags at least every 10 bags high.

Bricks. Shall not be stacked more than 7 ft high. When a loose brick stack reaches 4 ft high, it shall be tapered back 2 in. in every 1 ft height above the 4-ft level.

Masonry blocks. Shall be tapered back one-half block per tier above the 6-ft level.

Lumber. Shall be stacked no higher than 20 ft, on level and solidly supported sills, so as to be stable and self-supporting. Lumber to be handled manually shall be stacked no more than 16 ft high. Used lumber shall have all nails withdrawn before stacking.

Structural steel (poles, pipe, bar stock, and other cylindrical materials). Shall be stacked and blocked, unless racked, to prevent spreading or tilting.

Lighting Requirements (1926.56)

Construction areas, ramps, runways, corridors, offices, shops, and storage areas shall be lighted not less than the minimum illumination intensities listed in Table 23.2 while any work is in progress.

For areas or operations not covered in the table, refer to ANSI/IES RP7-1983, "Practices for Industrial Lighting," for recommended values of illumination.

Fire Protection (1926.150)

The employer shall be responsible for developing a fire protection program to be followed throughout all phases of the construction project, and for providing the firefighting equipment as specified by OSHA.

The program shall include a trained and equipped firefighting organization, as warranted by the project, to assure adequate protection to life.

TABLE 23.2 Minimum Illumination Intensities, Footcandles

Footcandles	Area of operation
5	General construction area lighting
3	General construction areas, concrete placement, excavation and waste areas, accessways, active storage areas, loading platforms, refueling, and field maintenance areas
5	Indoors: warehouses, corridors, hallways, and exitways
5	Tunnels, shafts, and general underground work areas (Exception: minimum of 10 footcandles is required at tunnel and shaft heading during drilling, mucking, and scaling. U.S. Bureau of Mines approved cap lights shall be acceptable for use in the tunnel heading.)
10	General construction plant and shops (batch plants, screening plants, mechanical and electric equipment rooms, carpenter shops, rigging lofts and active storerooms, barracks or living quarters, locker or dressing rooms, mess halls, and indoor toilets and workrooms)
30	First-aid stations, infirmaries, and offices

All firefighting equipment shall be conspicuously located and access to such equipment shall be maintained at all times. Such equipment shall be periodically inspected and continually maintained in operating condition, with defective equipment immediately replaced.

Water supply

A temporary or permanent water supply, of sufficient volume, duration, and pressure, required to properly operate the firefighting equipment, shall be made available as soon as combustible materials accumulate.

Portable firefighting equipment

A fire extinguisher, rated not less than 2A, shall be provided for each 3,000 ft^2 of the protected building area, or major fraction thereof. Distance from any point of the protected area to the nearest fire extinguisher is not to exceed 100 ft. One or more fire extinguishers, rated not less than 2A, shall be provided on each floor of a multistoried building, with at least one extinguisher located adjacent to stairways.

Approved substitutions for a fire extinguisher having a 2A rating include:

> One 55-gal open drum of water with two fire pails.
>
> A ½-in.-diameter garden-type hose, not longer than 100 ft and equipped with a nozzle provided it is capable of discharging a minimum of 5 gal/min with a minimum hose stream of 30 ft horizontally. Hoses shall be mounted on conventional racks or reels, with number and location such that at least one hose stream can be applied to all points in the area.

A fire extinguisher, rated not less than 10B, shall be provided within 50 ft of wherever more than 5 gal of flammable or combustible liquids or 5 lb of flammable gas are being used on the jobsite.

Extinguishers and water drums shall be protected from freezing.

Carbon tetrachloride and other toxic vaporizing liquid fire extinguishers are prohibited.

Portable fire extinguishers shall be inspected periodically and maintained in accordance with ANSI/NFPA 10-1981, "Portable Fire Extinguishers."

Fixed firefighting equipment

In facilities that include the installation of automatic sprinkler protection, the installation shall closely follow the construction and be placed in service as soon as applicable laws permit completion of each story.

A hose, 100 ft or less, of 1½-in.-diameter, with a nozzle capable of discharging water at 25 gal/min, or more, may be substituted for a fire extinguisher rated not more than 2A in the designated area, provided the hose can reach all points in the area. If hose connections are not compatible with local firefighting equipment, the contractor shall provide adapters, or equivalent, to permit connections.

An alarm system (telephone, siren, and so forth) shall be established by the contractor whereby employees on the site and the local fire department can be alerted for an emergency. The alarm code and reporting instructions shall be conspicuously posted at phones and at employee entrances.

Fire walls and exit stairways, required for the completed building, shall be given construction priority. Fire doors, with automatic closing devices, shall be hung on openings as soon as practicable.

Fire Prevention (1926.151)

Electric wiring and equipment for light, heat, or power shall be installed in compliance with ANSI/NFPA 70-1984, "National Electrical Code."

Internal combustion-engine-powered equipment shall be so located that the exhausts are well away from combustible materials.

Smoking shall be prohibited at or in the vicinity of operations that constitute a fire hazard, and shall be conspicuously posted: "No Smoking or Open Flame."

Portable battery-powered lighting equipment, used in connection with the storage, handling, or use of flammable gases or liquids, shall be of the type approved for hazardous locations.

The nozzle of air, inert gas, and steam lines or hoses, when used in the cleaning or ventilation of tanks and vessels that contain hazardous concentrations of flammable gases or vapors, shall be bonded to the tank or vessel shell.

Open-yard storage

Sites shall be kept free from accumulation of unnecessary combustible materials. Stored combustible materials shall be piled with regard to the stability of piles, and in no case higher than 20 ft. No combustible material shall be stored outdoors within 10 ft of a building or structure.

Driveways between and around combustible storage piles shall be at least 15 ft wide and maintained free from accumulation of rubbish, equipment, or other articles or materials.

Portable fire-extinguishing equipment, suitable for the fire hazard involved, shall be provided at convenient, conspicuously accessible locations in the yard area. Such extinguishers shall be rated not less than 2A and so placed that the maximum distance to the nearest unit does not exceed 100 ft.

Indoor storage

All materials shall be stored, handled, and piled with regard to their fire characteristics, with noncompatible materials (which may create a fire hazard) segregated by a barrier having a fire resistance of at least 1 hr.

Material shall be piled to minimize the spread of fire internally and to permit convenient access for firefighting. Aisle space shall be maintained to safely accommodate the widest vehicle that may be used within the building for firefighting. Storage shall not obstruct, or adversely affect, means of exit.

Personal Protective Equipment (1926.28)

The employer is responsible for requiring employees to wear appropriate personal protective equipment in all operations where there is an exposure to hazardous conditions; or where the need for using such equipment is to reduce the hazards to employees.

Where employees provide their own protective equipment, the employer shall be responsible to assure its adequacy, including proper maintenance and sanitation of such equipment.

All personal protective equipment shall be of safe design and construction for the work to be performed.

Foot protection

Safety-toe footwear shall meet the requirements and specifications of ANSI Z41-1983, "Safety-Toe Footwear."

Head protection

Helmets for the protection against impact and penetration of falling and flying objects shall meet the specifications of ANSI Z89.1-1981, "Protective Headwear for Industrial Workers."

Helmets for protection against exposure to high-voltage electric shock and burns shall meet the specifications of ANSI Z89.1-1981, "Protective Headwear for Industrial Workers."

Eye and face protection

Equipment to protect eye and face shall be provided when machines or operations present potential eye or face injury from physical, chemical, or radiation agents. Such eye and face protection equipment shall meet the requirements specified in ANSI Z87.1-1979, "Practice for Occupational and Educational Eye and Face Protection."

Face and eye protection equipment shall be kept clean and in good repair; use of this type of equipment with structural or optical defects shall be prohibited (see Table 23.3).

Protection against radiant energy in welding operations requires the selection of the proper shade numbers of filter lenses or plates in accordance with specifications listed in Table 23.4.

Noise Exposure (1926.52)

Protection against the effects of noise exposure shall be provided when the sound levels exceed those shown in Table 23.5 as measured on the A scale of a standard sound level meter at slow response.

When employees are subjected to sound levels exceeding those listed, feasible administrative or engineering controls shall be utilized. If such controls fail to reduce sound levels within the specified acceptable levels, personal protective equipment shall be provided and used to reduce sound levels within the levels specified.

In all levels where the sound levels exceed the values specified, a continuing, effective hearing conservation program shall be administered.

Hearing Protection (1926.101)

Whenever it is not feasible to reduce the noise levels or the duration of exposures to those specified, ear protection devices shall be provided and used. Ear protective devices inserted in the ear shall be fitted or determined individually by competent persons. Plain cotton is NOT an acceptable protective device.

Gases, Vapors, Fumes, Dusts, and Mists (1926.55)

Exposure to inhalation, ingestion, skin absorption, or contact with any material or substance at a concentration above those specified in "Threshold Limit Values of Airborne Contaminants for 1970" of the American Conference of Governmental Industrial Hygienists shall be avoided.

To achieve compliance with these specifications, administrative or engineering controls must first be implemented whenever feasible. When such controls are not feasible to achieve full compliance, protective equipment or other protective measures shall be used to keep the exposure of employees to air contaminants within the limits prescribed by OSHA.

Any equipment and technical measures used for this purpose must first be approved for each particular use by a competent industrial hygienist or other technically qualified person.

TABLE 23.3 Eye and Face Protector Selection Guide and Applications

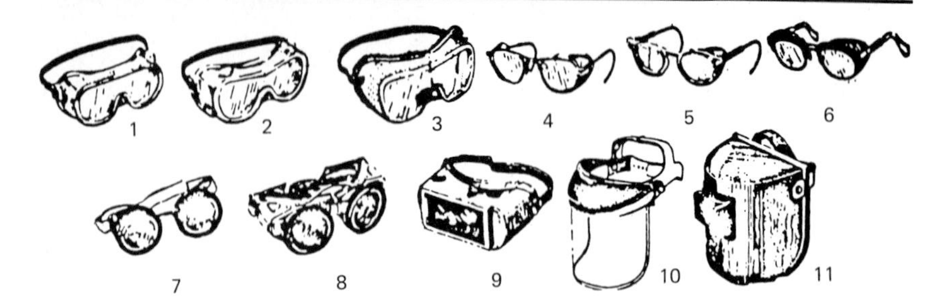

1. Goggles, flexible fitting, regular ventilation
2. Goggles, flexible fitting, hooded ventilation
3. Goggles, cushioned fitting, rigid body
4.* Spectacles, metal frame with sideshields
5.* Spectacles, plastic frame, with sideshields
6.* Spectacles, metal-plastic frame, with sideshields
7.† Welding goggles, eyecup type, tinted lenses (illustrated)
7*a*. Chipping goggles, eyecup type, clear safety lenses (not illustrated)
8.† Welding goggles, coverspec type tinted lenses (illustrated)
8*a*. Chipping goggles, coverspec type, clear safety lenses (not illustrated)
9.† Welding goggles, coverspec type, tinted plate lenses
10. Face shield (available with plastic or mesh window
11.† Welding helmets

Applications		
Operation	Hazard	Recommended protectors
Acetylene burning, cutting, welding	Sparks, harmful rays, molten metal, flying particles	7, 8, 9
Chemical handling	Splash, acid burns, fumes	2, 10 (For severe exposure add 10 over 2.)
Chipping	Flying particles	1, 3, 4, 5, 6, 7a, 8a
Electric (arc) welding	Sparks, intense rays, molten metal	9, 11 (11 in combination with 4, 5, 6 in tinted lenses advisable.)
Furnace operations	Glare, heat, molten metal	7, 8, 9 (For severe exposure add 10.)
Grinding, light	Flying particles	1, 3, 4, 5, 6, 10
Grinding, heavy	Flying particles	1, 3, 7a, 8a (For severe exposure add 10.)
Laboratory	Chemical splash, glass breakage	2 (10 when in combination with 4, 5, 6.)
Machining	Flying particles	1, 3, 4, 5, 6, 10
Molten metals	Heat, glare, sparks, splash	7, 8, (10 in combination with 4, 5, 6 in tinted lenses.)
Spot welding	Flying particles, sparks	1, 3, 4, 5, 6, 10

* Non-side-shield spectacles are available for limited hazard use requiring only frontal protection.

† See Table 23.4.

TABLE 23.4 Filter Lens Shade Numbers for Protection against Radiant Energy

Welding operation	Shade number
Shielded metal-arc welding, 1/16-, 3/32-, 1/8-, 5/32–in. diameter electrodes	10
Gas-shielded arc welding (nonferrous), 1/16-, 3/32-, 1/8-, 5/32-in. diameter electrodes	11
Gas-shielded arc welding (ferrous) 1/16-, 3/32-, 1/8-, 5/32-in. diameter electrodes	12
Shielded metal-arc welding, 3/16-, 7/32-, 1/4-in. diameter electrodes	12
5/16-, 3/8-in. diameter electrodes	14
Atomic hydrogen welding	10–14
Carbon-arc welding	14
Soldering	2
Torch brazing	3 or 4
Light cutting, up to 1 in.	3 or 4
Medium cutting, 1 to 6 in.	4 or 5
Heavy cutting, over 6 in.	5 or 6
Gas welding (light), up to 1/8 in.	4 or 5
Gas welding (medium), 1/8 to 1/2 in.	5 or 6

Ventilation

When ventilation is used as an engineering control method, the system shall be installed and operated according to the following requirements:

Local exhaust ventilation shall be designed to prevent dispersion into the air of dusts, fumes, mists, vapors, and gases in concentrations

TABLE 23.5 Permissible Noise Exposures

Duration per day, hours	Sound level dBA, slow response
8	90
6	92
4	95
3	97
2	100
1½	102
1	105
½	110
¼ or less	115

causing harmful exposure. Such exhaust systems shall be so designed that hazardous airborne substances are not drawn through the work area of employees.

Exhaust fans, jets, ducts, hoods, separators, and all necessary appurtenances, including refuse receptacles, shall be so designed, constructed, maintained, and operated as to ensure the required protection by maintaining a volume and velocity of exhaust air sufficient to gather hazardous airborne substances from said equipment or process, and to convey them to suitable points of safe disposal, thereby preventing their dispersion in harmful quantities into the atmosphere where employees work.

The exhaust system shall be maintained in operation continually during all operations that it is designed to serve; and shall continue to be operated after the cessation of such operations if the employee remains in the contaminated zone. The length of time shall depend on the individual circumstances and effectiveness of the general ventilation system.

The air outlet from every dust separator and the airborne hazardous materials collected by an exhaust or ventilating system shall discharge to the outside atmosphere. Dust and refuse discharged from an exhaust system shall be disposed of in such a manner that it will not result in harmful exposure to employees.

Collection systems that return air to work areas may be used if concentrations, which accumulate in the work area, do not result in any harmful exposure to employees.

Respiratory protection

Appropriate respiratory protective devices shall be provided by the employer and shall be used by employees in emergencies or when controls required to rid an area of airborne hazardous substances either fail or are inadequate to prevent harmful exposure to employees.

Respiratory protective devices shall be approved by the U.S. Bureau of Mines or acceptable to the U.S. Department of Labor for the specific contaminant to which the employee is exposed. The chemical and physical properties of the contaminant, as well as the toxicity and concentration of the hazardous substance, shall be considered in selecting the proper respirators (see Table 23.6).

TABLE 23.6 Selection of Respirators

Hazard	Respirator*
Oxygen deficiency	Self-contained breathing apparatus. Hose mask with blower. Combination air-line respirator with auxiliary self-contained air supply or an air-storage receiver with alarm.
Gas and vapor contaminants, immediately dangerous to life and health	Self-contained breathing apparatus. Hose mask with blower. Air-purifying, full face piece respirator with chemical canister (gas mask). Self-rescue mouthpiece respirator (for escape only). Combination air-line respirator with auxiliary self-contained air supply or an air-storage receiver with alarm.
Not immediately dangerous to life and health	Air-line respirator. Hose mask without blower. Air-purifying, half-mask or mouthpiece respirator with chemical cartridge.
Particulate contaminants, immediately dangerous to life and health	Self-contained breathing apparatus. Hose mask with blower. Air-purifying, full face piece respirator with appropriate filter. Self-rescue mouthpiece respirator (for escape only). Combination air-line respirator with auxiliary self-contained air-supply or an air-storage receiver with alarm.
Not immediately dangerous to life and health	Air-purifying, half-mask or mouthpiece respirator with filter pad or cartridge. Air-line respirator. Air-line abrasive-blasting respirator. Hose-mask without blower.
Combination gas, vapor, and particulate contaminants. Immediately dangerous to life and health	Self-contained breathing apparatus. Hose mask with blower. Air-purifying, full face piece respirator with chemical canister and appropriate filter (gas mask with filter). Self-rescue mouthpiece respirator (for escape only). Combination air-line respirator with auxiliary self-contained air supply or an air-storage receiver with alarm.
Not immediately dangerous to life and health	Air-line respirator. Hose mask without blower. Air-purifying, half-mask or mouthpiece respirator with chemical cartridge and appropriate filter.

* For the purpose of this part, "immediately dangerous to life and health" is defined as a condition that either poses an immediate threat to life and health or an immediate threat of severe exposure to contaminants, such as radioactive materials, which are likely to have adverse delayed effects on health.

Employees required to use respiratory protective equipment approved for use in atmospheres immediately dangerous to life shall be thoroughly trained in its use.

Respiratory protective equipment shall be inspected regularly and maintained in good condition. Gas mask canisters and chemical cartridges shall be replaced as necessary to provide complete protection. Mechanical filters shall be cleaned or replaced as necessary to avoid undue resistance to breathing.

Respiratory protective equipment that has been previously used shall be cleaned and disinfected before it is issued by the employer to another employee. Emergency rescue equipment shall be cleaned and disinfected immediately after each use.

Chapter

24

OSHA: Rigging Equipment and Accessories

Safe operation of hoisting equipment to lift a given load is not the only factor the equipment operator must consider. The load must be secured to a hook by means of an attachment, and both hook and attachment must be of adequate strength to lift the load safely. A brief summary of the major points of the OSHA safety practices for rigging equipment and accessories follows.

General

Rigging equipment for handling material shall not be loaded in excess of its recommended safe working load as specified by OSHA 1926.251, Tables H-1 through H-20.

All such equipment shall be inspected prior to its use on each shift, and as necessary during its use, to ensure that it is safe. Remove from service immediately any rigging equipment found to be defective.

When not in use, rigging equipment is to be removed from the immediate work area so as not to present a hazard to employees.

The employer shall mark all special custom-design grabs, hooks, clamps, or other lifting accessories to indicate their safe working loads; and proof test each to 125% of its rated load before allowing its use.

Whenever a sling is used, the following practices shall be observed:

Slings shall not be shortened with knots or bolts or other makeshift devices.

Sling legs shall not be kinked.

Slings used in a basket hitch shall have the loads balanced to prevent slippage.

Slings shall be padded or protected from the sharp edges of their loads.

Shock loading is prohibited.

A sling shall not be pulled from under a load when the load is resting on the sling.

Hands or fingers shall not be placed between the sling and its load while the sling is being tightened around the load.

Chains and Ropes (1926.251)

Alloy steel chains

Rated capacity (working load limit) for welded alloy steel chain slings shall conform to the values listed in Table 24.1.

Whenever wear at any point of any chain link exceeds that specified in Table 24.2, the assembly shall be removed from service.

All such slings shall have permanently affixed durable identification, stating size, grade, rated capacity, and sling manufacturer.

Hooks, rings, oblong links, or makeshift fasteners formed from bolts, rods, and so forth, or other such job-fabricated attachments shall not be used.

Wire rope

Safe working loads of various sizes and classifications of improved plow-steel wire rope and wire rope slings, with various types of terminals, shall be determined from Tables 24.3 and 24.4. For sizes, classifications, and grades not included in these tables, the safe working load recommended by the manufacturer for specific, identifiable products shall be followed, provided that a safety factor of not less than 5 is maintained.

Wire rope shall not be used in hoisting or lowering, or in pulling loads, if in any length of eight diameters the total number of visible broken wires exceeds 10% of the total number of wires; or if the rope shows other signs of excessive wear, corrosion, or defect.

Never secure wire ropes with knots.

Each wire rope used in hoisting or lowering, or in pulling loads, shall consist of one continuous piece without knots or splices, except for eye splices in the ends of wires and for endless rope slings.

TABLE 24.1 Rated Capacity (Working Load Limit) for Alloy Steel Chain Sling,* (Pounds)

		Double sling			Triple and quadruple sling		
	Single-branch sling,	Vertical angle†			Vertical angle†		
		30°	45°	60°	30°	45°	60°
Chain	90°	Horizontal angle‡			Horizontal angle‡		
size, in.	loading	60°	45°	30°	30°	45°	30°
¼	3,250	5,650	4,550	3,250	8,400	6,800	4,900
⅜	6,600	11,400	9,300	6,600	17,000	14,000	9,900
½	11,250	19,500	15,900	11,250	29,000	24,000	17,000
⅝	16,500	28,500	23,300	16,500	43,000	35,000	24,500
¾	23,000	39,800	32,500	23,000	59,500	48,500	34,500
⅞	28,750	49,800	40,600	28,750	74,500	61,000	43,000
1	38,750	67,100	54,800	38,750	101,000	82,000	58,000
1⅛	44,500	77,000	63,000	44,500	115,500	94,500	66,500
1¼	57,500	99,500	81,000	57,500	149,000	121,500	86,000
1¾	67,000	116,000	94,000	67,000	174,000	141,000	100,500
1½	80,000	138,000	112,500	80,000	207,000	169,000	119,500
1⅜	100,000	172,000	140,000	100,000	258,000	210,000	150,000

* Other grades of proof-tested steel chain include Proof Coil, BBB Coil, and Hi-Test Chain. These grades are not recommended for overhead lifting and therefore are not covered by this code.

† Rating of multileg slings adjusted for angle of loading measured as the included angle between the inclined leg and the vertical.

‡ Rating of multileg slings adjusted for angle of loading between the inclined leg and the horizontal plane of the load.

An eye splice made in the wire rope shall have not less than three full tucks. (Another form of splice or connection may be used if it can be shown to be as efficient and not otherwise prohibited.)

Eyes in wire rope bridles, slings, or bull wires shall not be formed by wire rope clips or knots. When U-bolt wire rope clips are used to form eyes, the number and spacing of U-bolts shall be determined from Table 24.5. The U-bolt shall be attached for an eye splice so that the U section is in contact with the dead end of the rope.

TABLE 24.2 Maximum Allowable Wear at Any Point of Link

Chain size, in.	Maximum allowable wear, in.	Chain size, in.	Maximum allowable wear, in.
¼	3/64	1	3/16
⅜	5/64	1⅛	7/32
½	7/64	1¼	¼
⅝	9/64	1⅜	9/32
¾	5/32	1½	5/16
⅞	11/64	1¾	11/32

TABLE 24.3 Rated Capacities for Single-Leg Slings, Tons (2,000 lb) 6 × 19 and 6 × 37 Classification Improved Plow-Steel Grade Rope with Fiber Core (FC)

Rope		Vertical			Choker			Vertical basket*		
Diameter, in.	Construction	HT	MS	S	HT	MS	S	HT	MS	S
¼	6 × 19	0.49	0.51	0.55	0.37	0.38	0.41	0.99	1.0	1.1
5⁄16	6 × 19	0.76	0.79	0.85	0.57	0.59	0.64	1.5	1.6	1.7
⅜	6 × 19	1.1	1.1	1.2	0.80	0.85	0.91	2.1	2.2	2.4
7⁄16	6 × 19	1.4	1.5	1.6	1.1	1.1	1.2	2.9	3.0	3.3
½	6 × 19	1.8	2.0	2.1	1.4	1.5	1.6	3.7	3.9	4.3
9⁄16	6 × 19	2.3	2.5	2.7	1.7	1.9	2.0	4.6	5.0	5.4
⅝	6 × 19	2.8	3.1	3.3	2.1	2.3	2.5	5.6	6.2	6.7
¾	6 × 19	3.9	4.4	4.8	2.9	3.3	3.6	7.8	8.8	9.5
⅞	6 × 19	5.1	5.9	6.4	3.9	4.5	4.8	10.0	12.0	13.0
1	6 × 19	6.7	7.7	8.4	5.0	5.8	6.3	13.0	15.0	17.0
1⅛	6 × 19	8.4	9.5	10.0	6.3	7.1	7.9	17.0	19.0	21.0
1¼	6 × 37	9.8	11.0	12.0	7.4	8.3	9.2	20.0	22.0	25.0
1⅜	6 × 37	12.0	13.0	15.0	8.9	10.0	11.0	24.0	27.0	30.0
1½	6 × 37	14.0	16.0	17.0	10.0	12.0	13.0	28.0	32.0	35.0
1⅜	6 × 37	16.0	18.0	21.0	12.0	14.0	15.0	33.0	37.0	41.0
1¾	6 × 37	19.0	21.0	24.0	14.0	16.0	18.0	38.0	43.0	48.0
2	6 × 37	25.0	28.0	31.0	18.0	21.0	23.0	49.0	55.0	62.0

* These values only apply when the D/d ratio for HT slings is 10 or greater, and for MS and S slings, 20 or greater, where D—diameter of curvature around which the body of the sling is bent, d—diameter of rope.

HT—hand-tucked splice; MS—mechanical splice; S—swaged or zinc-poured socket.

Fiber rope (natural and synthetic)

Safe working loads of various sizes and classifications of natural- and synthetic-fiber ropes shall be determined from Tables 24.6 to 24.9.

All splices in rope slings provided by the employer shall be made in accordance with the fiber rope manufacturer's recommendations.

Manila rope. Eye splices shall contain at least three full tucks; short splices shall contain at least six full tucks (three on each side of the centerline of the splice).

Layed synthetic fiber rope. Eye splices shall contain at least four full tucks; short splices shall contain at least eight full tucks (four on each side of the centerline of the splice).

Strand end tails of splices shall not be trimmed short (flush with the surface of the rope) immediately adjacent to the full tucks. (This applies to both eye and short splices and all types of fiber rope.)

For fiber ropes under 1 in. in diameter the tails shall project at least 6 in. beyond the last full tuck. In applications where projecting tails may be objectionable, tails shall be tapered and spliced into the body

TABLE 24.4 Rated Capacities for Single-Leg Slings, Tons (2,000 lb) 6 × 19 and 6 × 37 Classification Improved Plow-Steel Grade Rope with Independent Wire Rope Core (IWRC)

Rope		Vertical			Choker			Vertical basket*		
Diameter, in.	Construction	HT	MS	S	HT	MS	S	HT	MS	S
¼	6 × 19	0.53	0.56	0.59	0.40	0.42	0.44	1.0	1.1	1.2
5⁄16	6 × 19	0.81	0.87	0.92	0.61	0.65	0.69	1.6	1.7	1.8
⅜	6 × 19	1.1	1.2	1.3	0.86	0.93	0.98	2.3	2.5	2.6
7⁄16	6 × 19	1.5	1.7	1.8	1.2	1.3	1.3	3.1	3.4	3.5
½	6 × 19	2.0	2.2	2.3	1.5	1.6	1.7	3.9	4.4	4.6
9⁄16	6 × 19	2.5	2.7	2.9	1.8	2.1	2.2	4.9	5.5	5.8
⅝	6 × 19	3.0	3.4	3.6	2.2	2.5	2.7	6.0	6.8	7.2
¾	6 × 19	4.2	4.9	5.1	3.1	3.6	3.8	8.4	9.7	10.0
⅞	6 × 19	5.5	6.6	6.9	4.1	4.9	5.2	11.0	13.0	14.0
1	6 × 19	7.2	8.5	9.0	5.4	6.4	6.7	14.0	17.0	18.0
1⅛	6 × 19	9.0	10.0	11.0	6.8	7.8	8.5	18.0	21.0	23.0
1¼	6 × 37	10.0	12.0	13.0	7.9	9.2	9.9	21.0	24.0	26.0
1⅜	6 × 37	13.0	15.0	16.0	9.6	11.0	12.0	25.0	29.0	32.0
1½	6 × 37	15.0	17.0	19.0	11.0	13.0	14.0	30.0	35.0	38.0
1⅝	6 × 37	18.0	20.0	22.0	13.0	15.0	17.0	35.0	41.0	44.0
1¾	6 × 37	20.0	24.0	26.0	15.0	18.0	19.0	41.0	47.0	51.0
2	6 × 37	26.0	30.0	33.0	20.0	23.0	25.0	53.0	61.0	66.0

* These values only apply when the *D/d* ratio for HT slings is 10 or greater, and for MS and S slings, 20 or greater, where *D*—diameter of curvature around which the body of the sling is bent, *d*—diameter of rope.

HT—hand-tucked splice; MS—mechanical splice; S—swaged or zinc-poured socket.

of the rope using at least two additional tucks (which will require a tail length of approximately six rope diameters beyond the last full tuck).

For all eye splices, the eye shall be sufficiently large to provide an included angle of not greater than 60° at the splice when the eye is placed over the load or support.

Knots shall not be used in place of splices.

TABLE 24.5 Number and Spacing of U-Bolt Wire Rope Clips

Improved plow-steel rope diameter, in.	Number of clips		Minimum spacing, in.
	Drop-forged	Other material	
½	3	4	3
⅓	3	4	3¼
¼	4	5	4½
⅞	4	5	5¼
1	5	6	5
1⅓	6	6	6¼
1¼	6	7	7½
1⅛	7	7	8¼
1½	7	8	9

TABLE 24.6 Rated Capacity of Manila Rope Slings, Pounds (Safety Factor = 5)

			Eye and eye sling						Endless sling					
					Basket hitch						Basket hitch			
					Angle of rope to horizontal						Angle of rope to horizontal			
					90°	60°	45°	30°			90°	60°	45°	30°
					Angle of rope to vertical						Angle of rope to vertical			
Nominal rope diameter, in.	Nominal weight per 100 ft, lb	Minimum breaking strength, lb	Vertical hitch	Choker hitch	0°	30°	45°	60°	Vertical hitch	Choker hitch	0°	30°	45°	60°
½	7.5	2,650	550	250	1,100	900	750	550	950	500	1,900	1,700	1,400	950
9/16	10.4	3,450	700	350	1,400	1,200	1,000	700	1,200	600	2,500	2,200	1,800	1,200
⅝	13.3	4,400	900	450	1,800	1,500	1,200	900	1,600	800	3,200	2,700	2,200	1,600
¾	16.7	5,400	1,100	550	2,200	1,900	1,500	1,100	2,000	950	3,900	3,400	2,800	2,000
13/16	19.5	6,500	1,300	650	2,600	2,300	1,800	1,300	2,300	1,200	4,700	4,100	3,300	2,300
⅞	22.5	7,700	1,500	750	3,100	2,700	2,200	1,500	2,800	1,400	5,600	4,800	3,900	2,800
1	27.0	9,000	1,800	900	3,600	3,100	2,600	1,800	3,200	1,600	6,500	5,600	4,600	3,200
1 1/16	31.3	10,500	2,100	1,100	4,200	3,600	3,000	2,100	3,800	1,900	7,600	6,600	5,400	3,800
1⅛	36.0	12,000	2,400	1,200	4,800	4,200	3,400	2,400	4,300	2,200	8,600	7,600	6,100	4,300
1¼	41.7	13,500	2,700	1,400	5,400	4,700	3,800	2,700	4,900	2,400	9,700	8,400	6,900	4,900
1 5/16	47.9	15,000	3,000	1,500	6,000	5,200	4,300	3,000	5,400	2,700	11,000	9,400	7,700	5,400
1½	59.9	18,500	3,700	1,850	7,400	6,400	5,200	3,700	6,700	3,300	13,500	11,500	9,400	6,700
1⅝	74.6	22,500	4,500	2,300	9,000	7,800	6,400	4,500	8,100	4,100	16,000	14,000	11,500	8,000
1¾	89.3	26,500	5,300	2,700	10,500	9,200	7,500	5,300	9,500	4,800	19,000	16,500	13,500	9,500
2	107.5	31,000	6,200	3,100	12,500	10,500	8,800	6,200	11,000	5,600	22,500	19,500	16,000	11,000
2⅛	125.0	36,000	7,200	3,600	14,500	12,500	10,000	7,200	13,000	6,500	26,000	22,500	18,500	13,000
2¼	146.0	41,000	8,200	4,100	16,500	14,000	11,500	8,200	15,000	7,400	29,500	25,500	21,000	15,000
2½	166.7	46,500	9,300	4,700	18,500	16,000	13,000	9,300	16,500	8,400	33,500	29,000	23,500	16,500
2⅝	190.8	52,000	10,500	5,200	21,000	18,000	14,500	10,500	18,500	9,500	37,500	32,500	26,500	18,500

TABLE 24.7 Rated Capacity of Nylon Rope Slings, Pounds (Safety Factor = 9)

			Eye and eye sling						Endless sling					
					Basket hitch						Basket hitch			
					Angle of rope to horizontal						Angle of rope to horizontal			
					90°	60°	45°	30°			90°	60°	45°	30°
					Angle of rope to vertical						Angle of rope to vertical			
Nominal rope diameter, in.	Nominal weight per 100 ft, lb	Minimum breaking strength, lb	Vertical hitch	Choker hitch	0°	30°	45°	60°	Vertical hitch	Choker hitch	0°	30°	45°	60°
½	6.5	6,080	700	350	1,400	1,200	950	700	1,200	600	2,400	2,100	1,700	1,200
9/16	8.3	7,600	850	400	1,700	1,500	1,200	850	1,500	750	3,000	2,600	2,200	1,500
5/8	10.5	9,880	1,100	550	2,200	1,900	1,600	1,100	2,000	1,000	4,000	3,400	2,800	2,000
¾	14.5	13,490	1,500	750	3,000	2,600	2,100	1,500	2,700	1,400	5,400	4,700	3,800	2,700
13/16	17.0	16,150	1,800	900	3,000	3,100	2,600	1,800	3,200	1,600	6,400	5,600	4,600	3,200
7/8	20.0	19,000	2,100	1,100	4,200	3,700	3,000	2,100	3,800	1,900	7,600	6,600	5,400	3,800
1	26.0	23,750	2,600	1,300	5,300	4,600	3,700	2,600	4,800	2,400	9,500	8,200	6,700	4,800
1 1/16	29.0	27,360	3,000	1,300	5,100	5,300	4,300	3,000	5,500	2,700	11,000	9,500	7,700	5,500
1 1/8	34.0	31,350	3,500	1,700	7,000	6,000	5,000	3,500	6,300	3,100	12,500	11,000	8,900	6,300
1¼	40.0	35,625	4,000	2,000	7,900	6,900	5,600	4,000	7,100	3,600	14,500	12,500	10,000	7,100
1 5/16	45.0	40,850	4,500	2,300	9,100	7,900	6,400	4,500	8,200	4,100	16,500	14,000	12,000	8,200
1½	55.0	50,350	5,600	2,800	11,000	9,700	7,900	5,600	10,000	5,000	20,000	17,500	14,000	10,000
1 5/8	68.0	61,750	6,900	3,400	13,500	12,000	9,700	6,900	12,500	6,200	24,500	21,500	17,500	12,500
1¾	83.0	74,100	8,200	4,100	16,500	14,500	11,500	8,200	15,000	7,400	29,500	27,500	21,000	15,000
2	95.0	87,400	9,700	4,900	19,500	17,000	13,500	9,700	17,500	8,700	35,000	30,500	24,500	17,500
2 1/8	109.0	100,700	11,000	5,600	22,500	19,500	16,000	11,000	20,000	10,000	40,500	35,000	28,500	20,000
2¼	129.0	118,750	13,000	5,600	26,500	23,000	18,500	13,000	24,000	12,000	47,500	41,000	33,500	24,000
2½	149.0	132,000	15,000	7,400	29,500	25,500	21,000	15,000	24,500	11,500	53,000	46,000	37,500	26,500
2 5/8	160.0	133,900	17,100	8,600	34,000	29,500	24,000	17,000	31,000	15,500	61,500	53,500	43,500	31,000

TABLE 24.8 Rated Capacity of Polyesther Rope Slings, Pounds (Safety Factor = 9)

			Eye and eye sling						Endless sling					
					Basket hitch						Basket hitch			
					Angle of rope to horizontal						Angle of rope to horizontal			
					90°	60°	45°	30°			90°	60°	45°	30°
					Angle of rope to vertical						Angle of rope to vertical			
Nominal rope diameter, in.	Nominal weight per 100 ft, lb	Minimum breaking strength, lb	Vertical hitch	Choker hitch	0°	30°	45°	60°	Vertical hitch	Choker hitch	0°	30°	45°	60°
½	8.0	6,080	700	350	1,400	1,200	950	700	1,200	600	2,400	2,100	1,700	1,200
9/16	10.2	7,600	850	400	1,700	1,500	1,200	850	1,500	750	3,000	2,600	2,200	1,500
5/8	13.0	9,500	1,100	550	2,100	1,800	1,500	1,100	1,900	950	3,800	3,300	2,700	1,900
¾	17.5	11,875	1,300	650	2,600	2,300	1,900	1,300	2,400	1,200	4,800	4,100	3,400	2,400
13/16	21.0	14,725	1,600	800	3,300	2,800	2,300	1,600	2,900	1,500	5,900	5,100	4,200	2,900
7/8	25.0	17,100	1,900	950	3,800	3,300	2,700	1,900	3,400	1,700	6,800	5,900	4,800	3,400
1	30.5	20,900	2,300	1,200	4,600	4,000	3,300	2,300	4,200	2,100	8,400	7,200	5,900	4,200
1 1/16	34.5	24,225	2,700	1,300	5,400	4,700	3,800	2,700	4,800	2,400	9,700	8,400	6,900	4,800
1 1/8	40.0	28,025	3,100	1,600	6,200	5,400	4,400	3,100	5,600	2,800	11,000	9,700	7,900	5,600
1 ¼	46.3	31,540	3,500	1,800	7,000	6,100	5,000	3,500	6,300	3,200	12,500	11,000	8,900	6,300
1 5/16	52.5	35,625	4,000	2,000	7,900	6,900	5,600	4,000	7,100	3,600	14,500	12,500	10,000	7,100
1 ½	66.8	44,460	4,900	2,500	9,900	8,600	7,000	4,900	8,900	4,400	18,000	15,500	12,500	8,900
1 5/8	82.0	54,150	6,000	3,000	12,000	10,400	8,500	6,000	11,000	5,400	21,500	19,000	15,500	11,000
1 ¾	98.0	64,410	7,200	3,600	14,500	12,500	10,000	7,200	13,000	6,400	26,000	22,500	18,000	13,000
2	118.0	76,000	8,400	4,200	17,000	14,500	12,000	8,400	15,000	7,600	30,500	26,500	21,500	15,000
2 1/8	135.0	87,400	9,700	4,900	19,500	17,000	13,500	9,700	17,500	8,700	35,000	30,500	24,500	17,500
2 ¼	157.0	101,650	11,500	5,700	22,500	19,500	16,000	11,500	20,500	10,000	40,500	35,000	29,000	20,500
2 ½	181.0	115,900	13,000	6,400	26,000	22,500	18,000	13,000	23,000	11,500	46,500	40,000	33,000	23,000
2 5/8	205.0	130,150	14,500	7,200	29,000	25,000	20,500	14,500	26,000	13,000	52,000	45,000	37,000	26,000

TABLE 24.9 Rated Capacity of Polypropylene Rope Slings, Pounds (Safety Factor = 6)

			Eye and eye sling						Endless sling					
					Basket hitch						Basket hitch			
					Angle of rope to horizontal						Angle of rope to horizontal			
					90°	60°	45°	30°			90°	60°	45°	30°
					Angle of rope to vertical						Angle of rope to vertical			
Nominal rope diameter, in.	Nominal weight per 100 ft, lb	Minimum breaking strength, lb	Vertical hitch	Choker hitch	0°	30°	45°	60°	Vertical hitch	Choker hitch	0°	30°	45°	60°
½	4.7	3,990	650	350	1,300	1,200	950	650	1,200	600	2,400	2,100	1,700	1,200
9/16	6.1	4,845	800	400	1,600	1,400	1,100	800	1,500	780	2,900	2,500	2,100	1,500
⅝	7.5	5,890	1,000	500	2,000	1,700	1,400	1,000	1,800	900	3,500	3,100	2,500	1,800
¾	10.7	8,075	1,300	700	2,700	2,300	1,900	1,300	2,400	1,200	4,900	4,200	3,400	2,400
13/16	12.7	9,405	1,600	800	3,100	2,700	2,200	1,600	2,800	1,400	5,600	4,900	4,000	2,800
⅞	15.0	10,925	1,800	900	3,600	3,200	2,600	1,800	3,300	1,600	6,600	5,700	4,600	3,300
1	18.0	13,300	2,200	1,100	4,400	3,800	3,100	2,200	4,000	2,000	8,000	6,900	8,600	4,000
1 1/16	20.4	15,200	2,500	1,300	5,100	4,400	3,600	2,500	4,600	2,300	9,100	7,900	6,500	4,600
1⅛	23.7	17,385	2,900	1,500	5,800	5,000	4,100	2,900	5,200	2,600	10,500	9,000	7,400	5,200
1¼	27.0	19,950	3,300	1,700	6,700	5,800	4,700	3,300	6,000	3,000	12,000	10,500	8,500	6,000
1 5/16	30.5	22,325	3,700	1,900	7,400	6,400	5,300	3,700	6,700	3,400	13,500	11,500	9,500	6,700
1½	38.5	28,215	4,700	2,400	9,400	8,100	6,700	4,700	8,500	4,200	17,000	14,500	12,000	8,500
1⅝	47.5	34,200	5,700	2,900	11,500	9,900	8,100	5,700	10,500	5,100	20,500	18,000	14,500	10,500
1¾	57.0	40,850	6,800	3,400	13,500	12,000	9,600	6,800	12,500	6,100	24,500	21,000	17,500	12,500
2	69.0	49,400	3,200	4,100	16,500	14,500	11,500	8,200	15,000	7,400	29,500	25,500	21,000	15,000
2⅛	80.0	57,950	9,700	4,800	19,500	16,500	13,500	9,700	17,500	8,700	35,000	30,100	24,500	17,500
2¼	92.0	65,550	11,000	5,500	22,000	19,000	15,500	11,000	13,500	9,900	39,500	34,000	28,000	19,500
2½	107.0	76,000	12,500	6,300	25,500	22,000	18,000	12,500	23,000	11,500	45,500	39,500	32,500	23,000
2⅝	120.0	85,500	14,500	7,100	20,500	24,500	20,000	14,300	25,500	13,000	51,500	44,500	36,500	25,000

Synthetic webbing (nylon, polyester, and polypropylene)

Each synthetic web sling shall be marked or coded to show name or trademark of manufacturer, rated capacities for type of hitch, and type of material. The rated capacity of web slings shall not be exceeded.

Shackles and hooks

Safe working loads of various sizes of shackles shall be determined from OSHA Table H-19, 1926.251. Higher safe working loads are permissible, however, when recommended by the manufacturer for specific identifiable products, provided that a safety factor of not less than 5 is maintained.

Safe working loads, recommended by the manufacturer, shall be followed for various sizes and types of specific and identifiable hooks. All hooks for which no applicable manufacturer's recommendations are available shall be tested to twice the intended safe working load before they are initially put into use.

The employer shall maintain a record of the dates and results of such tests.

Slings (1910.184)

Alloy steel chain slings

A thorough inspection of alloy steel chain slings in use shall be made on a regular basis, to be determined by frequency of sling use, severity of service conditions, nature of lifts being made, and experience gained on the service life of slings used in similar circumstances. Such inspections shall in no event be at intervals greater than every 12 months.

The employer shall make and maintain a record of the most recent month in which each alloy steel chain sling was thoroughly inspected, and shall make such record available for examination.

Wire rope slings

Cable-laid and 6 × 19 and 6 × 37 slings shall have a minimum clear length of wire rope 10 times the component rope diameter between splices, sleeves, or end fittings.

Braided slings shall have a minimum clear length of wire rope 40 times the component rope diameter between the loops or end fittings.

Cable-laid grommets, strand-laid grommets, and endless slings shall have a minimum circumferential length of 96 times their body diameter.

Fiber core wire rope slings of all grades shall be permanently removed from service if they are exposed to temperatures in excess of 200°F. When nonfiber core wire rope slings of any grade are used at temperatures above 400°F or below −60°F, recommendations of the sling manufacturer regarding use at that temperature shall be followed.

Welding of end attachments, except covers to thimbles, shall be performed prior to assembly of the sling. All welded-end attachments shall not be used unless proof-tested by the manufacturer or equivalent entity at twice their rated capacity prior to initial use. The employer shall retain a certificate of the proof test and make it available for examination.

Fiber rope slings (natural and synthetic)

Except if wet-frozen, fiber rope slings may be used in a temperature range from −20°F to 180°F without decreasing the working load limit. For operations outside this temperature range, and for wet-frozen slings, the sling manufacturer's recommendations shall be followed.

Spliced fiber rope slings shall not be used unless they have been spliced in accordance with the following minimum requirements and with any additional recommendations of the manufacturer:

The minimum clear length of rope between eye splices must be equal to 10 times the rope diameter.

Clamps not designed specifically for fiber ropes shall not be used for splicing.

End attachments in contact with the rope shall not have sharp edges or projections.

Natural- and synthetic-fiber rope slings shall be immediately removed from service if the rope shows abnormal wear, powdered fiber between strands, broken or cut fibers, variations in the size or roundness of strands, discoloration or rotting, or distortion of hardware in the sling.

Synthetic web slings

Synthetic webbing shall be of uniform thickness and width; selvage edges shall not be split from the webbing's width.

All fittings shall be of a minimum breaking strength equal to that of the sling, and free of all sharp edges that could in any way damage the webbing.

Stitching shall be the only method used to attach end fittings to webbing and to form eyes. The thread shall be in an even pattern and contain a sufficient number of stitches to develop the full breaking strength of the sling.

When synthetic web slings are used, certain precautions should be taken:

Nylon web slings shall not be used where fumes, vapors, sprays, mists, or liquids of acids or phenolics are present.

Polyester and polypropylene web slings shall not be used where fumes, vapors, sprays, mists, or liquids of caustics are present.

Web slings with aluminum fittings shall not be used where fumes, vapors, sprays, mists, or liquids of caustics are present.

Synthetic web slings of polyester and nylon shall not be used at temperatures in excess of 180°F; slings of polypropylene shall not be used at temperatures in excess of 200°F.

Synthetic web slings shall be immediately removed from service if there exists any of the following: acid or caustic burns, melting or charring of any part of the sling surface, snags, punctures, tears or cuts, broken or worn stitches, or distortion of fittings.

Safety Lifelines, Belts, Lanyards (1926.104)

Lifelines, safety belts, and lanyards shall be used only for employee safeguard. Any such equipment subjected to in-service loading, as distinguished from static-load testing, shall be immediately removed from service and shall not be used again for employee safeguarding.

Lifelines

Lifelines shall be a minimum of ¾-in.-diameter manila rope or equivalent, with a minimum breaking strength of 5,400 lb, except that lifelines used in areas where they may be subjected to cutting or abrasion shall be a minimum of ⅞-in.-diameter wire core manila rope. All lifelines shall be secured above the point of operation to an anchorage or structural member capable of supporting a minimum dead weight of 5,400 lb.

Safety belt lanyard

The lanyard shall be a minimum of ½-in.-diameter nylon rope, or equivalent, with a maximum length to provide for a fall of no greater than 6 ft. The rope shall have a nominal breaking strength of 5,400 lb.

All safety belt and lanyard hardware shall be drop-forged or pressed steel, cadmium plated in accordance with type 1, Class B plating specified in Federal Specification QQ-P-416. Hardware surfaces shall be smooth and free of sharp edges. Except for rivets, such hardware shall be capable of withstanding a tensile loading of 4,000 lb without cracking, breaking, or taking a permanent deformation.

Safety Nets (1926.105)

Safety nets shall be provided when workplaces are more than 25 ft above the ground or water surface, or other surfaces where the use of ladders, scaffolds, catch platforms, temporary floors, safety lines, or safety belts is impractical. Where safety net protection is required, operations shall not be undertaken until the net is in place and has been tested.

Nets shall extend 8 ft beyond the edge of the work surface where employees are exposed and shall be installed as close under the work surface as practical, but in no case more than 25 ft below such work surface. Nets shall be hung with sufficient clearance to prevent the user's contact with the surfaces or structures below. Such clearances shall be determined by impact load testing. Only one level of nets is required for bridge construction.

The mesh size of nets shall not exceed 6 by 6 in. All new nets shall meet accepted performance standards of 17,500-ft-lb minimum impact resistance as determined and certified by the manufacturers, and shall have a label of proof test. Edge ropes shall provide a minimum breaking strength of 5,400 lb.

Forged steel safety hooks or shackles shall be used to fasten the net to its supports. Connections between net panels shall develop the full strength of the net.

Employees working over or near water, where the danger of drowning exists, shall be provided with U.S. Coast Guard approved life jackets or buoyant work vests. Prior to and after each use, the buoyant work vests or life preservers shall be inspected for defects that would alter their strength or buoyancy. Defective units shall be removed from service immediately.

Ring buoys with at least 90 ft of line shall be provided and be readily available for emergency rescue operations. The distance between ring buoys shall not exceed 200 ft.

At least one lifesaving skiff shall be immediately available at locations where employees are working over or adjacent to water.

Chapter

25

OSHA: Cranes, Derricks, Winches, and Hoists; Safety Practices for Helicopters

> The use of cableways, cranes, derricks, hoists, hooks, jacks, and slings is subject to certain hazards that cannot be met by mechanical means, but only by the exercise of intelligence, care, and common sense. It is therefore essential to have competent and careful personnel involved in the use and operation of the equipment, physically and mentally qualified, trained in the safe operation of the equipment and the handling of the loads.—*ANSI/ASME Standard.*

A brief summary of the major points of the OSHA safety practices for rigging equipment and accessories follows.

Cranes and Derricks (1926.550)

The employer shall comply with the manufacturer's specifications and limitations applicable to the operation of any and all cranes and derricks. Where manufacturers' specifications are not available, the limitations assigned to the equipment shall be based on the determinations of a qualified engineer competent in this field, and such determinations will be appropriately documented and recorded.

Rate load capacities, recommended operating speeds, and special hazard warnings or instructions shall be posted conspicuously on all equipment. Instructions or warnings shall be visible to the operator while at the control station.

Attachments used with cranes shall not exceed the capacity, rating, or scope recommended by the manufacturer.

Hand signals to crane and derrick operators shall be those prescribed by ANSI for the type of crane in use:

ANSI B30.17-1980, "Cranes (Top Running Bridge, Single Girder, Underhung Hoist), Overhead and Gantry"

ANSI/ASME B30.6-1984, "Derricks"

ANSI/ASME B30.8-1982, "Floating Cranes and Floating Derricks"

ANSI/ASME B30.3-1984, "Hammerhead Tower Cranes"

ANSI/ASME B30.5-1982, "Mobile and Locomotive Truck Cranes"

ANSI B30.11-1980, "Monorails and Underhung Cranes"

ANSI/ASME B30.2-1983, "Overhead and Gantry Cranes (Top Running Bridge, Single or Multiple Girder, Top Running Trolley Hoist)"

An illustration of the signals shall be posted at the job site (see Fig. 25.1).

Accessible areas within the swing radius of the rear of the rotating superstructure of a crane, either permanently or temporarily mounted, shall be barricaded in such a manner as to prevent an employee from being struck or crushed by the crane.

Inspections

The employer shall designate a competent person to inspect all machinery and equipment prior to each use and during use, to make sure it is in safe operating condition. Any deficiencies shall be repaired, or defective parts replaced, before continuing to operate the machinery or equipment.

A thorough, annual inspection of the hoisting machinery shall be made by a competent person, or by a government or private agency recognized by the U.S. Department of Labor. The employer shall maintain a record of the dates and results of such inspections for each hoisting machine and piece of equipment.

No modifications or additions which affect the capacity or safe operation of the equipment shall be made by the employer without the manufacturer's written approval. If such modifications are made, the capacity, operation, and maintenance instruction plates, tags, or decals shall be changed accordingly. In no case shall the original safety factor of the equipment be reduced.

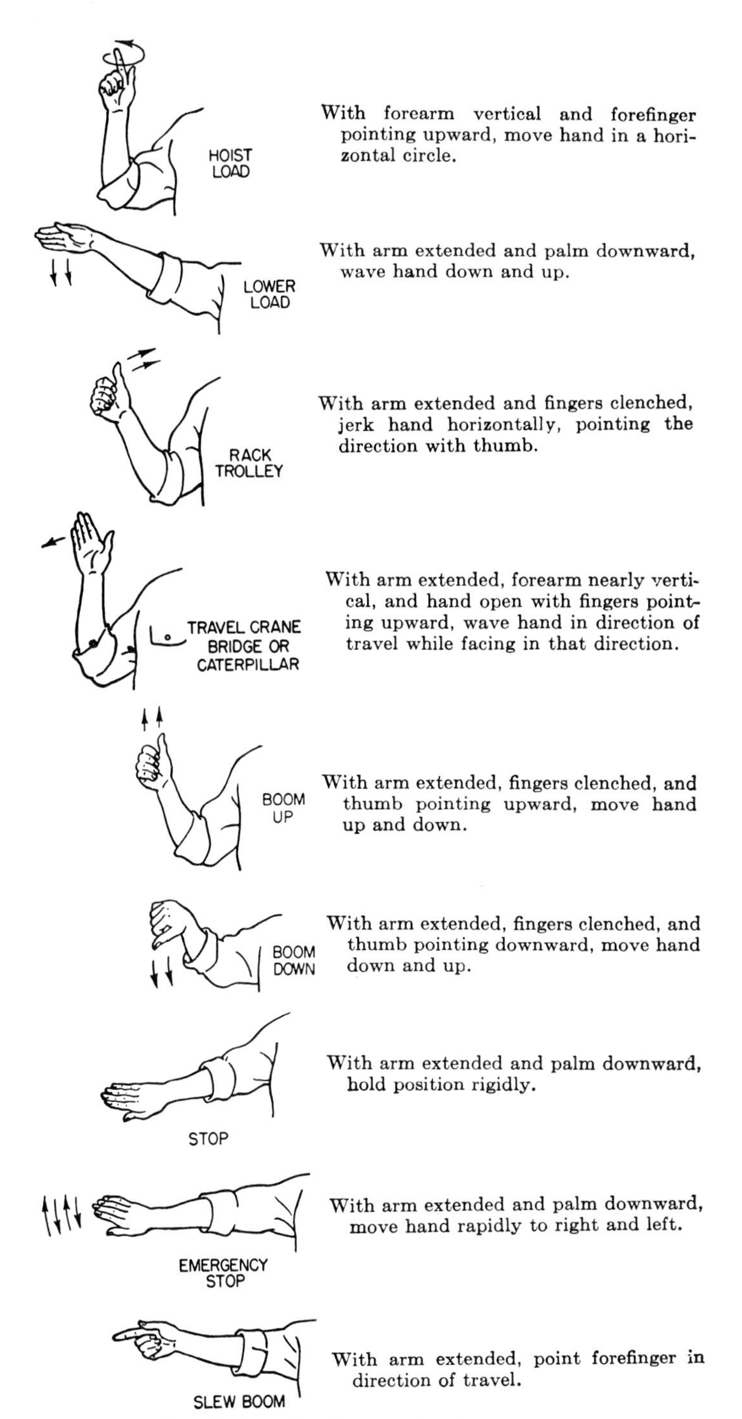

Figure 25.1 Standard one-hand crane signals.

The employer shall comply with "Power Crane and Shovel Association Mobile Hydraulic Crane Standard No. 2."

Wire rope

Wire rope used with cranes or derricks shall be removed from service when:

In running ropes, there are six randomly distributed broken wires in one lay, or three broken wires in one strand in one lay

One-third the original diameter of outside individual wires is worn

Kinking, crushing, birdcaging, or any other damage results in distortion of the rope structure

Any heat damage from any cause is evident

The nominal diameter of the wire is reduced more than:

1/64 in. for diameters up to and including 5/16 in.

1/32 in. for diameters 3/8 to 1/2 in. inclusive

3/64 in. for diameters 9/16 to 3/4 in. inclusive

1/16 in. for diameters 7/8 to 1 1/8 in. inclusive

3/32 in. for diameters 1 1/4 to 1 1/2 in. inclusive

In standing ropes, there are more than two broken wires in one lay in sections beyond end connections, or more than one broken wire at an end connection.

Wire rope safety factors shall be in accordance with ANSI B30.5-1982.

Operational precautions

Belts, gears, shafts, pulleys, sprockets, spindles, drums, fly wheels, chains, or other reciprocating, rotating, or moving parts of equipment shall be guarded if such parts are exposed to contact by employees, or otherwise create a hazard. Guarding shall meet the requirements of ANSI/ASME B15.1-1984, "Safety Standards for Mechanical Power Transmission Apparatus."

All windows in cabs shall be of safety glass, or equivalent, which introduces no visible distortion that will interfere with the safe operation of the machine.

Where necessary for rigging or service requirements, a ladder or steps shall be provided to give access to a cab roof. Guardrails, handholds, and steps shall be provided on cranes for easy access to the car and cab, conforming to ANSI/ASME B 30.5-1982, "Mobile and Locomotive Truck Cranes."

Platforms and walkways shall have antiskid surfaces.

All exhaust pipes shall be guarded or insulated in areas where contact by employees is possible in the performance of normal duties.

Whenever equipment, powered by internal combustion engines, exhausts in enclosed spaces, tests shall be made and recorded to see that employees are not exposed to unsafe concentrations of toxic gases or oxygen-deficient atmosphere.

Fire prevention

Fuel tank fiber pipe shall be located in such a position or protected in such manner as to not allow spill or overflow to run onto the engine, exhaust, or electric equipment of any machine being fueled.

An accessible fire extinguisher of 5BC rating or higher shall be available at all operator stations or cabs of equipment.

All fuels shall be transported, stored, and handled to meet OSHA regulations. When fuel is transported by vehicles on public highways, U.S. Department of Transportation rules contained in 49 CFR Parts 177 and 393 concerning such vehicular transportation are considered applicable.

Power line precautions

Any overhead wire shall be considered to be an energized line unless and until the person owning such a line, or the proper authorities of the local electric utility, indicate that it is not an energized line and it has been visibly grounded.

Except where electric distribution and transmission lines have been deenergized and visibly grounded at the point of work, or where insulating barriers, not a part of or an attachment to the equipment or machinery, have been erected to prevent physical contact with the lines, equipment or machines shall be operated near power lines with a minimum clearance between lines and any part of the crane or load:

10 ft for lines rated 50 kV or less

10 ft plus 0.4 in. for each 1 kV of lines rated over 50 kV; or twice the length of the line insulator (but never less than 10 ft)

Equipment or machines in transit, with no load and with boom lowered, shall maintain a minimum clearance between lines and any part of crane:

4 ft for lines rated less than 50 kV

10 ft for lines rated over 50 kV up to 345 kV inclusive

16 ft for lines rated over 345 kV up to 750 kV inclusive

A person shall be designated to observe clearance of the equipment and give timely warning for all operations where it is difficult for the operator to maintain the desired clearance by visual means.

While cage-type boom guards, insulating links, or proximity warning devices may be used on cranes, the use of such devices shall not preclude following the requirements of any other regulations set forth.

Transmitter tower precautions

Prior to working near transmitter towers where an electric charge can be induced in the equipment or materials being handled, the transmitter shall be deenergized or tests shall be made to determine whether electric charge is induced on the crane.

When necessary to dissipate induced voltages:

Equipment shall be provided with an electric ground directly to the upper rotating structure supporting the boom

Ground jumper cables shall be attached to materials being handled by boom equipment

Crews shall be provided with nonconductive poles having large alligator clips or other similar protection to attach the ground cable to the load

Combustible and flammable materials shall be removed from the immediate area prior to operations

Crawler, locomotive, truck cranes

All such cranes in use shall meet the applicable requirements for design, inspection, construction, testing, maintenance, and operation as prescribed in ANSI/ASME B30.5-1982, "Mobile and Locomotive Truck Cranes." All crane jibs shall have positive stops to prevent their movement of more than 5 in. above the straight line of the jib and boom on conventional-type crane booms. The use of cable-type belly slings does not constitute compliance with this rule.

Derricks

All derricks in use shall meet the applicable requirements for the design, construction, installation, inspection, testing, maintenance, and operation as prescribed in ANSI/ASME B30.6-1984, "Derricks."

Mobile cranes mounted on barges

When a mobile crane is mounted on a barge, the rated load of such a crane shall not exceed the original capacity specified by the manufacturer. Any such crane shall be positively secured on the barge.

Each crane shall be provided with a load-rating chart, having clearly legible letters and figures, that is securely fixed at a location easily visible to the operator. When load ratings are reduced to stay within the limits for list of the barge with a crane mounted on it, a new load-rating chart shall be provided.

Permanently mounted floating cranes and derricks

When a crane or derrick is permanently installed on a barge, the capacity and limitations of use shall be based on competent design, construction, installation, testing, maintenance, and operation as prescribed by the manufacturer.

Each crane shall be provided with a load-rating chart, having clearly legible letters and figures, that is securely fixed at a location easily visible to the operator.

The employer shall comply with the applicable requirements for protection of employees working onboard marine vessels as specified in OSHA 1926.605.

Hammerhead tower cranes

All hammerhead tower cranes in use shall meet the applicable requirements for design, construction, installation, testing, maintenance, inspection, and operation as prescribed by the manufacturer.

Adequate clearance shall be maintained between moving and rotating structures of the crane and fixed objects to allow the passage of employees without harm.

Employees required to perform duties on the horizontal boom of hammerhead tower cranes shall be protected against falling by guardrails or by safety belts and lanyards attached to lifelines in conformance with OSHA requirements.

Buffers shall be provided at both ends of travel of the trolley. Cranes mounted on rail tracks shall be equipped with limiting switches to limit the travel of the crane on the track; and with stops or buffers at each end of the tracks.

Overhead and gantry cranes

All overhead and gantry cranes in use shall meet the applicable requirements for design, construction, installation, testing, maintenance, inspection, and operation as prescribed in ANSI/ASME B30.2-1983, "Overhead and Gantry Cranes."

Hoists and Elevators (1926.552)

The employer shall comply with the manufacturers' specifications and limitations applicable to the operation of all hoists and elevators. Where manufacturers' specifications are not available, the limitations assigned to the equipment shall be based on the determinations of a professional engineer competent in the field.

Rated load capacities, recommended operating speeds, and special hazard warnings or instructions shall be posted on cars and platforms.

Hoisting ropes shall be installed in accordance with the wire rope manufacturer's recommendations.

Wire rope used in hoists and elevators shall be removed from service when:

In hoisting ropes, there are six randomly distributed broken wires in one rope lay or three broken wires in one strand in one rope lay

Abrasion, scrubbing, flattening, or peening causes loss of more than one-third of the original diameter of the outside wires

Any heat damage from any cause (such as torch or contact with electric wires) is evident

The nominal diameter of the wire is reduced more than:

3/64 in. for diameters up to and including 3/4 in.

1/16 in. for diameters 7/8 to 1 1/8 in. inclusive

3/32 in. for diameters 1 1/4 to 1 1/2 in. inclusive

Installation of live booms on hoists is prohibited. Using endless-belt-type personnel lifts for construction is prohibited.

Material hoists

All material hoists shall conform to the requirements of ANSI A10.5-1981, "Safety Requirements for Material Hoists," and shall be designed by a licensed professional engineer.

Operating rules shall be established and posted at the operator's station of the hoist. Such rules shall include the signal system and allowable line speed for various loads. Rules and notices shall be posted on the car frame or crosshead in a conspicuous location, and shall include the statement "NO RIDERS ALLOWED."

No person shall be allowed to ride on a material hoist except for the purposes of inspection and maintenance.

All entrances of the hoist way shall be protected by substantial gates or bars, which shall guard the full width of the landing entrance. All hoist way entrances, bars, and gates shall be painted with diagonal contrasting colors, such as black and yellow stripes.

Bars shall not be less than 2-by-4-in. wooden members or the equivalent, located 2 ft from the hoist way line and not less than 36 in., nor more than 42 in., above the floor.

Gates or bars protecting the entrances to hoist ways shall be equipped with a latching device.

Overhead protective covering of 2-in. planking, ¾-in. plywood, or other solid material of equivalent strength shall be provided on the top of every material hoist cage or platform. The operator's station for a hoisting machine shall be provided with overhead protection equivalent to tight planking not less than 2 in. thick. The support for the overhead protection shall be of equal strength.

Hoist towers may be used with or without an enclosure on all sides. However:

> When a hoist tower is enclosed, it shall be enclosed on all sides for its entire height with a screen enclosure of ½-in. mesh, No. 18 U.S. gage wire or equivalent, except for landing access.
>
> When a hoist tower is not enclosed, the hoist platform or car shall be totally enclosed (caged) on all sides for the full height between the floor and the overhead protective covering with ½-in. mesh of No. 14 U.S. gage wire or equivalent. The hoist platform enclosure shall include the required gates for loading and unloading. A 6-ft-high enclosure shall be provided on the unused side of the hoist tower at ground level.

Car arresting devices shall be installed to function in case of rope failure.

Personnel hoists

All personnel hoists used by employees shall be constructed of materials and components that meet the specifications for materials, construction, safety devices, assembly, and structural integrity as prescribed by ANSI A10.4-1981, "Safety Requirements for Personnel Hoists."

Following assembly and erection of hoists, and before being put into service, an inspection and test of all functions and safety devices shall be made under the supervision of a competent person. A similar inspection and test are required following major alteration of an existing installation. All hoists shall be inspected and tested at not more than three-month intervals. Records shall be maintained and kept on file for the duration of the job.

Hoist towers outside the structure shall be enclosed for the full height of the structure on the side or sides used for entrance and exit. At the lowest landing, the enclosure shall be enclosed to a height of at least 10 ft. Other sides of the tower adjacent to floors or scaffold platforms shall be enclosed to a height of 10 ft above the level of such floors or scaffolds.

Towers located inside the structure shall be enclosed on all four sides throughout the full height of the structure.

All towers shall be anchored to the structure at intervals not exceeding 25 ft, and in addition to tie-ins, a series of guys shall be installed. Where tie-ins are not practical, the tower shall be anchored by means of guys made of wire rope at least ½ in. in diameter securely fastened to anchorage to ensure stability.

Hoist way doors or gates shall be not less than 6 ft 6 in. high and shall be provided with mechanical locks that cannot be operated from the landing side and are accessible only to persons on the car.

Cars shall be permanently enclosed on all sides and the top, except sides used for entrance and exit that have car gates or doors. A door or gate shall be provided at each entrance to the car, protecting the full width and height of the car entrance opening. Doors or gates shall be provided with electric contacts, which do not allow movement of the hoist when a door or gate is open.

Overhead protective covering of 2-in. planking, ¾-in. plywood, or other solid material of equivalent strength shall be provided on the top of every personnel hoist.

Safety mechanisms shall be capable of stopping and holding the car and rated load, when traveling at governor-tripping speed.

Cars shall be provided with a capacity and data plate secured in a conspicuous place on the car or crosshead.

Internal combustion engines shall not be permitted for direct drive.

Normal and final terminal stopping devices shall be provided. An emergency stop switch shall be provided in the car and shall be marked "STOP."

The minimum of hoisting ropes used shall be three for traction hoists and two for drum-type hoists. The minimum diameter of hoisting and counterweight wire ropes shall be ½ in. Minimum safety factors for suspension wire ropes are listed in Table 25.1.

Personnel hoists used in bridge tower construction shall be approved by a registered professional engineer and erected under the supervision of a qualified engineer competent in this field.

TABLE 25.1 Minimum Factors of Safety for Suspension Wire Ropes

Rope speed, ft/min	Minimum factor of safety
50	7.60
75	7.75
100	7.95
125	8.10
150	8.25
175	8.40
200	8.60
225	8.75
250	8.90
300	9.20
350	9.50
400	9.75
450	10.00
500	10.25
550	10.45
600	10.70

When a hoist tower is not enclosed, the hoist platform or car shall be totally enclosed (caged) on all sides for the full height between the floor and the overhead protective covering with ¾-in. mesh of No. 14 U.S. gage wire or equivalent. The hoist platform enclosure shall include the required gates for loading and unloading.

These hoists shall be inspected and maintained on a weekly basis. Whenever the hoisting equipment is exposed to winds exceeding 35 mi/h, it shall be inspected and put in operable condition before reuse.

Wire rope shall be taken out of service when:

In running ropes, there are six randomly distributed broken wires in one lay or three broken wires in one strand in one lay

One-third the original diameter of outside individual wires is worn

Kinking, crushing, birdcaging, or any other damage results in distortion of the rope structure

Any heat damage from any cause is evident when the nominal diameter of the wire is reduced more than:

3/64 in. for diameters up to and including ¾ in.

1/16 in. for diameters ⅞ to 1⅛ in. inclusive

3/32 in. for diameters 1¼ to 1½ in. inclusive

In standing ropes there are more than two broken wires in one lay in the section beyond end connections or more than one broken wire at an end connection.

Permanent elevators under the care and custody of the employer, and used by employees for work, shall comply with the requirements of ANSI A17.1-1984 and be inspected in accordance with A17.2-1985.

Base-Mounted Drum Hoists (1926.553)

All base-mounted drum hoists in use shall meet the applicable requirements for design, construction, installation, testing, inspection, maintenance, and operations, as prescribed by the manufacturer.

Exposed moving parts, such as gears, projecting screws, setscrews, chain, cables, chain sprockets, and reciprocating or rotating parts which constitute a hazard, shall be guarded.

All controls used during the normal operation cycle shall be located within easy reach of the operator's station.

Electric-motor-operated hoists shall be provided with:

A device to disconnect all motors from the line upon power failure and not permit any motor to be restarted until the controller handle is brought to the OFF position.

An overspeed preventive device where applicable

A means whereby remotely operated hoists stop when any control is ineffective

Overhead hoists

All overhead hoists in use shall meet the applicable requirements for construction, design, installation, testing, inspection, maintenance, and operation, as prescribed by the manufacturer.

The safe working load of the overhead hoist, as determined by the manufacturer, shall be indicated on the hoist, and this safe working load shall not be exceeded.

The supporting structure to which the hoist is attached shall have a safe working load equal to that of the hoist.

The support shall be arranged so as to provide for free movement of the hoist and shall not restrict the hoist from lining itself up with the load.

The hoist shall be installed only in locations that will permit the operator to stand clear of the load at all times.

Air hoists shall be connected to an air supply of sufficient capacity and pressure to safely operate the hoist. All air hoses supplying air shall be positively connected to prevent their becoming disconnected during use.

Aerial Lifts (1926.556)

Aerial lifts shall be designed and constructed in conformance with applicable requirements of ANSI A92.2-1979, "Vehicle Mounted Elevating and Rotating Aerial Devices."

Aerial lifts include various types of vehicle-mounted aerial devices used to elevate personnel to job sites above ground, such as extendable boom platforms, aerial ladders, articulating boom platforms, vertical towers, and a combination of any of these devices.

Aerial equipment may be made of metal, wood, fiberglass reinforced plastic (FRP), or other material; may be powered or manually operated; and are deemed to be aerial lifts whether or not they are capable of rotating about a substantially vertical axis.

Aerial lifts may be field-modified for uses other than those intended by the manufacturer, provided the modifications have been certified in writing by the manufacturer or by any other equivalent entity (such as a nationally recognized testing laboratory) to be in conformity with all applicable provisions of ANSI A92.2-1979; and to be at least as safe as the equipment was before modification.

An aerial lift truck shall not be moved when the boom is elevated in a working position with workers in the basket, except for equipment that is specifically designed for this type of operation in accordance with the provisions of OSHA.

Before moving an aerial lift for travel, the boom shall be inspected to see that it is properly cradled and outriggers are in stowed position.

Ladder trucks and tower trucks

Aerial ladders shall be secured in the lower traveling position by the locking device on top of the truck cab, and the manually operated device at the base of the ladder before the truck is moved for highway travel.

Extendable and articulating boom platforms

Lift controls shall be tested each day prior to use to determine that such controls are in safe working condition.

Articulated boom and extendable boom platforms, primarily designed as personnel carriers, shall have both upper (platform) and lower controls. Upper controls shall be in or beside the platform within easy reach of the operator. Lower controls shall provide for overriding the upper controls. Controls shall be plainly marked as to their function. Lower level controls shall not be operated unless permission has been obtained from the employee in the lift, except in case of an emergency.

Only authorized persons shall operate an aerial lift.

Employees shall always wear a body belt and a lanyard attached to the boom or basket when working from an aerial lift; and shall stand firmly on the floor of the basket, and not sit or climb on the edge of the basket or use planks, ladders, or other devices for a work position.

Climbers shall not be worn while performing work from an aerial lift.

Boom and basket load limits specified by the manufacturer shall not be exceeded.

The brakes shall be set when using the aerial lift; and when outriggers are used, they shall be positioned on pads or on a solid surface. Wheel chocks shall be positioned on pads or on a solid surface. Wheel chocks shall be installed before using an aerial lift on an incline, provided they can be installed safely.

The insulated portion of an aerial lift shall not be altered in any manner that might reduce its insulating value.

Electrical tests

All electrical tests shall conform to the requirements of ANSI A92.2-1979, Section 5.

Bursting safety factor

The provisions of ANSI A92.2-1979, Section 4.9, "Bursting Safety Factor," shall apply to all critical hydraulic and pneumatic components. Critical components are those in which a failure would result in a free fall or free rotation of the boom. All noncritical components shall have a bursting safety factor of at least 2:1.

Welding standards

All welding shall conform to the following standards as applicable:

AWS B3.0-41, "Standard Qualification Procedure"

AWS D8.4-61, "Recommended Practices for Automated Welding Design"

AWS D10.9-69, "Standard Qualification of Welding Procedures and Welders for Piping and Tubing"

AWS D2.0-69, "Specifications for Welding Highway and Railway Bridges"

Helicopters (1925.551)

Helicopter cranes shall be expected to comply with any applicable requirements of the FAA.

Prior to each day's operation, a briefing shall be conducted to set forth the plan of operation for the pilot and ground crew.

Sufficient ground personnel shall be provided when required for safe helicopter loading and unloading operations.

Good housekeeping shall be maintained in all helicopter loading and unloading areas. Open fires shall not be permitted in an area that could result in such fires being spread by the rotor downwash.

Every practical precaution shall be taken to provide for the protection of the employees from flying objects in the rotor downwash. All loose gear within 100 ft of the place of lifting the load, depositing the load, and all other areas susceptible to rotor downwash shall be secured or removed.

Personnel protective equipment for employees receiving the load shall consist of complete eye protection and hard hats, secured by chinstraps. Loose-fitting clothing, likely to flap in the downwash and thus be snagged on hoist lines, shall not be worn.

Whenever approaching or leaving a helicopter with blades rotating, all employees shall remain in full view of the pilot and keep in a crouched position. Employees shall avoid the area from the cockpit or cabin rearward unless authorized by the helicopter operator to work there.

When employees are required to perform work under hovering craft, a safe means of access shall be provided for employees to reach the hoist hook and engage or disengage cargo slings. Employees shall not perform work under hovering craft except when necessary to hook or unhook loads.

The weight of an external load shall not exceed the helicopter manufacturer's rating, with the helicopter operator solely responsible for the size, weight, and manner in which loads are connected to the helicopter. If, for any reason, the helicopter operator believes the lift cannot be made safely, the lift shall not be made.

All loads shall be properly slung, according to ANSI B30.12-1975.

All electrically operated cargo hooks shall have the electric activating device so designed and installed as to prevent inadvertent operation. In addition, such cargo hooks shall be equipped with an emergency mechanical control for releasing the load. The hooks shall be tested prior to each day's operation to determine that the release functions properly, both electrically and mechanically.

Pressed sleeve, sedged eyes, or equivalent means shall be used for all freely suspended loads to prevent band splices from spinning open, or cable clamps from loosening. Tag lines for controlling loads during lifting and depositing shall be of a length that will not permit their being drawn up into the craft's rotors.

Static charge on the suspended load shall be dissipated with a grounding device before ground crews touch the suspended load. If not, then ground crews should wear protective rubber gloves when touching a suspended load.

Hoist wires or other gear, except for pulling lines or conductors that are allowed to "pay out" from a container or roll off a reel, shall not be attached to any fixed ground structure or allowed to foul on any fixed structure.

Signal systems between aircrew and ground crew shall be understood and checked in advance of hoisting the load. This applies to either radio or hand signal systems. Hand signals are illustrated in Chapter 14.

There shall be constant reliable communication between the pilot and a designated member of the ground crew who acts as a signal person during the period of loading and unloading. This signal person shall be distinctly recognizable from other ground crew members.

No unauthorized person shall be allowed to approach within 50 ft of the helicopter when the rotor blades are turning.

Chapter

26

Safe Practices for Scaffolds and Portable Ladders

Although certain equipment and material standards should be considered in discussing the safe use of scaffolds and portable ladders, most of those standards are incorporated in quality equipment and materials purchased from reputable vendors.

Stationary Scaffolds

Construction jobs of all types require the use of scaffolds for work that cannot be done safely from the ground, or for solid construction, where the use of ladders is not appropriate.

The larger the job, the bigger the scaffold requirement, and the more workers who entrust their lives to it. For this reason it is essential that all scaffolds be erected, used, and dismantled under close supervision of persons who have been properly trained for these tasks.

The safe use of any scaffold starts with proper maintenance and regular inspection of a scaffold's components. All scaffold material, frames, and accessories should be maintained in good repair, and any defect or unsafe condition should be corrected before the scaffolding is used further. Proper maintenance includes painting.

Systematic care and proper storage of component parts of scaffolding and work platforms will keep them in good shape. Return all parts to a common storage area. Take inventory to make sure no parts are missing, and to check the soundness of components.

Always inspect scaffolding material and equipment before every use. Broken, bent, excessively rusted, altered, or otherwise structurally damaged components should not be used until repaired or replaced.

Remove from service and repair any scaffold damaged or weakened from any cause. *Do not* put such units back into service until repairs are completed. Never store defective or damaged equipment without first tagging it "DANGER—DO NOT USE."

Scaffolds should be designed, built, and inspected by experienced and competent persons.

When erecting scaffolding, be sure to provide firm footings or anchorages that are sound, rigid, and capable of carrying the maximum intended load without settlement or displacement. Never use unstable objects such as barrels, boxes, loose bricks, or concrete blocks to support scaffolds or planks. Proper footings along with a safe tie-in to the structure will provide a rigid and secure scaffold.

Make sure poles, legs, or uprights of scaffolds are plumb and securely and rigidly braced. Securely lash all scaffolds to permanent structures using anchor bolts, reveal bolts, or other equivalent means to prevent scaffolds from swaying. Never use window cleaner anchor bolts.

The extra time that it takes to secure scaffolding to the structure is important in preventing accidents and provides a rigidity to the scaffolding that cannot be achieved in any other way. A rigid scaffold gives confidence to the workers it supports, and confidence in the equipment being used translates into greater work efficiency.

Scaffolds erected over 25 ft high should be tied to the structure at a height of three times the scaffold's narrowest width, and at every two sections thereafter as a minimum—at intervals not greater than 25 ft horizontally.

Every scaffold work platform 10 ft high or more should have guardrails, midrails, and toeboards on all open sides and the ends. Guardrails should be made of 2 × 4 lumber, installed not less than 36 in. nor more than 42 in. high. When a midrail is required, it should be constructed of 1 × 4 lumber. Vertical supports should be spaced no further than 10 ft apart. Toeboards should be at least 4 in. high.

Both guardrails and midrails should be able to withstand a 200-lb load. As a rule of thumb, scaffolds should be capable of supporting, without failure, at least four times the maximum intended load.

Tubular steel scaffolds have all but replaced wood framed scaffolding for most jobs, because they are more durable than wood structures, reduce fire exposure, and usually can be erected faster. Such systems should be erected, used, and disassembled according to the manufacturer's recommendations. Always use the correct devices for proper seating and locking of all connections.

When wood framed scaffolding is used, all load-carrying timber members should be a minimum of 1,500 lb (stress grade) construction grade lumber.

Each scaffold should be designed for the loads that will be carried during construction, including workers, building materials, and the weight of the scaffold structure itself. A factor of safety of at least 4 should be provided in the design.

When single-pole scaffolds are used, heavier ledgers are recommended. More numerous connections should be made to ensure against the scaffold swinging away from the building or structure. Single-pole scaffolds should be cross-braced in both directions, along the face of the building (as for independent scaffolds), and at right angles to the building face at every third or fourth upright.

Make sure wood pole scaffolds are within the reach of effective fire-fighting equipment.

Do not use outrigger scaffolds where any other type can be used. However, if this type is used, it should be constructed of only first-grade straight-grained timber, with a minimum dimension of 3 × 10 in. set on edge. The inner end of the outriggers must be securely and rigidly anchored to the structure.

Cantilevered or braced outrigger beams or thrust outs must be secured against overturning by the use of blocking or bracing. Thrust outs should not be built into the wall of the structure and left with no other support.

Ladders or equivalent safe access should be provided for every stationary scaffold.

Platform Planks

All planking should be scaffold grade lumber, constructed of 2 × 10 or wider lumber, for which the maximum permissible spans are given in Table 26.1.

The maximum permissible span for full-thickness planks, 1¼ × 9 in. or wider, is 4 ft with a medium-duty loading of 50 lb/ft. Never load scaffolds in excess of the design working load.

Platform planks should be laid with their edges close together so that there are no spaces in the platform through which tools or materials can fall. They should overlap at least 12 in. or be secured by wire or proper cleating to prevent slippage from both sudden shifts and creep. When commercial cleats are not available, planks can be secured

TABLE 26.1 Maximum Permissible Spans for Platform Planks

	Full-thickness undressed lumber			Nominal-thickness lumber*	
Working load, lb/ft^2	25	50	75	25	50
Permissible span, ft	10	8	6	8	6

* Not recommended for heavy-duty use.

quickly and efficiently with No. 9 wire. Scaffold planks should extend over their end supports not less than 6 in. nor more than 18 in.

Never bridge two or more scaffolds being used on a building or structure. Instead, maintain them at an even height with platforms butting closely.

Mobile Scaffolds

Workers should never ride mobile or rolling scaffolds, nor should they attempt to move such scaffolds by pulling on overhead pipes or structures.

Never alter or move scaffolds horizontally while they are in use or occupied. Always remove workers, material, and equipment from the platform. Stabilize the tower during scaffold movement. Apply force to the scaffold as near or as close as possible to its base. Make sure floors are level and free of obstructions and openings. Watch out for overhead obstructions.

The working platform height of rolling scaffolds should not be higher than four times the smallest base dimension, unless guyed or otherwise stabilized. Be sure to consider the overturning effect of brackets used on rolling scaffolds. Use horizontal, diagonal bracing on all rolling scaffolds, near top, bottom, and at intermediate levels of 30 ft.

When rolling scaffolds are not being moved, be sure to apply the caster brakes. Do not extend leg-adjusting screws more than 12 in. to level scaffolding.

Suspended Scaffolds

To ensure safe use, swinging stage scaffolds must be supported by hooks that are correctly shaped to engage the eave, cornice, or parapet wall of the structure, and securely anchored by rope or cable where necessary. Hooks should be fabricated of wrought iron or mild steel of ⅞-in.-diameter or equivalent minimum area.

Test all swinging scaffolds by loading at ground level with four times the load to be carried, then raise the scaffold a foot or two above the ground.

Carefully and frequently inspect suspender ropes or cables, and replace damaged or frayed ones when necessary.

If the platform is raised and lowered with block and tackle, make sure all blocks and sheaves are of the proper size, recommended for the rope or cable being used. The hook of the lower block should pass through the eye or loop provided in the stirrup. When the platform is in position, the free end of the suspending rope should be securely fastened to the point of the hook with a nonslip hitch.

Mechanical devices for lowering and raising scaffolds are recommended in preference to block and tackle. Both hand-crank and electric-motor-driven devices should incorporate a positive locking feature. Make sure the lock and winch mechanisms are maintained in first-rate operating condition at all times and inspected frequently. Above all, follow the manufacturer's recommendations pertaining to installation, use, maintenance, storage, and loading.

Cantilevered thrust outs, used to support suspended scaffolds, should be securely fastened to the structural frame by means of U bolts, beam clamps, or other suitable connections. Never use rope tie-downs, bent reinforcing bars, or weights to hold down the ends of thrust outs.

Thrust-outs should project at least 12 in. beyond the suspended scaffold.

Wire or fiber ropes used to suspend scaffolds should be capable of supporting at least six times the intended load. Make sure fiber rope is treated or protected for any work involving the use of corrosive substances or chemicals. When acid solutions are used for cleaning buildings 50 ft or higher, use wire rope for supporting the scaffold.

Take special precautions to protect all scaffold members, including wire or fiber ropes, when using heat-producing equipment.

Regularly service suspended scaffold hoisting machines, cables, and equipment. Inspect after each installation and every 30 days thereafter. Renew hoisting wire rope every 12 months.

Worker Protection

One very important way to protect workers on scaffolding is to keep the work platform clean. Do not permit the storage of tools, air lines, or other materials on the platform when such items are not needed. A cluttered platform presents a tripping hazard.

Another way to protect workers on scaffolds is to hoist materials properly, using a tag line as required by OSHA regulations for all hoisting onto scaffolds.

Be sure to provide overhead protection for persons required to work or pass under scaffolding, as well as for workers on a scaffold exposed to overhead hazards. Install a No. 18 gage U.S. standard ½-in. mesh screen (or equivalent) between toeboard and guardrail extending along the entire opening. Provide a wire mesh or heavy lumber canopy for overhead protection on the platform itself.

Always barricade or rope off the area immediately adjacent to scaffolding to keep unauthorized persons away. Use rope with brightly colored pennants or high-visibility tape. Make sure that there are adequate warning signs to prevent workers and operators of equipment from bumping into the scaffold accidentally.

Keep workers off all scaffolds during storms or high winds, and do not allow them to work on any scaffold covered with ice or snow. Remove all ice or snow, and sand the planking to prevent slipping.

Allow no more than two workers at a time on suspension scaffolds designed for a working load of 500 lbs, no more than three workers on suspension scaffolds having a working load of 750 lb.

Make sure each worker uses an approved safety line belt attached to a lifeline. The lifeline should be securely attached to substantial members of the structure (not the scaffold), or to securely rigged lines that will safely suspend a worker in case of a fall.

Checklist: Scaffold Safety

General

Inspect all scaffolding equipment each time before using.

Use only equipment in good repair and safe condition.

Inspect all planking to ensure that it is of sound quality, straight-grained, free from through knots, and graded for scaffold use.

Erect, move, and dismantle scaffold only under the supervision of competent and experienced persons.

Install and use all scaffolding accessories according to the manufacturer's recommended procedures.

Make no changes of any kind in scaffolds without proper engineering approval.

Protect scaffold structures from trucks and other vehicles operating in the work area.

Make sure the working platform is free of ice, snow, oil, or any other hazardous materials before allowing workers to use it.

Do not permit open fires on or near scaffolds.

Conspicuously display, and make sure workers observe, notices regarding the use of scaffolds.

Stationary scaffolds

Provide adequate sills for scaffold legs and use base plates.

Compensate for any unevenness of ground by using adjusting screws.

Never use unstable objects, such as barrels, boxes, bricks, or concrete blocks, to support scaffolding or planks.

Plumb and level all scaffolds as erection proceeds, and maintain during use.

Brace each scaffold leg securely.

Provide proper access to scaffold platforms—*do not* climb scaffold bracing.

Anchor running scaffold securely to structure wall at least every 30 ft of length and 25 ft of height.

Always consider the effect of wind loading on scaffold design.

Guy or brace all free-standing towers every 25 ft of elevation.

Use unit lock arms, rivet, and hairpins to lock frames together where uplift may occur.

Use horizontal diagonal bracing when necessary to prevent racking of the scaffold.

Do not use ladders or makeshift devices on top of the scaffold to increase work height.

Never place scaffolding in the proximity of power lines, unless the power company is notified and special precautions are taken to protect workers from hazards.

Platforms

Equip all planked or staged areas of scaffold with proper guardrail and intermediate railing on all open sides and ends of scaffolds over 10 ft high.

Use toeboards on all scaffolds over 10 ft high.

Use wire mesh screening between the toeboard and guardrailing whenever persons are required to work or pass under the scaffold.

Provide overhead protection for workers on scaffolds exposed to overhead hazards.

All platform planking should have at least 12 in. overlap. It should extend 6 in. beyond the center of support or be cleated at both ends.

Do not allow unsupported ends of planks to extend more than 18 in. beyond the support.

Make sure scaffold platforms are fully planked between guardrails.

Secure planks to scaffold when necessary to prevent uplift or displacement.

Do not overload scaffolds.

Rolling scaffolds

Level and plumb all rolling towers at all times.

Rolling towers should be no higher than 4 times the minimum base dimension.

Apply caster brakes at all times when scaffold is not in actual motion.

Do not extend adjusting screws more than 12 in.

Fully brace rolling towers on both sides.

Use horizontal diagonal bracing on the bottom and at intermediate levels of 20 ft.

Do not use brackets on rolling towers.

Do not ride rolling scaffolds or towers.

Remove or secure all material and equipment on the platform before moving scaffold.

Do not attempt to move a rolling tower without sufficient help.

Always apply force to move the tower as close to the base as possible.

Watch for holes or other obstructions or irregularities in the floor or overhead areas when moving a rolling tower.

Suspended scaffolds

Inspect wire ropes frequently. Make sure adequate lubrication is maintained.

Inspect and test each time a scaffold is rerigged.

Make sure that the roof or supporting structure is capable of safely supporting loads to be imposed by the scaffold.

Make sure the weight of people, material, and components on the scaffold does not exceed the manufacturer's rated load.

Do not combine two or more scaffolds by overlapping platforms on one stirrup, or by bridging between them, except for multipoint suspension scaffolding.

Overhang of outrigger beams must not exceed the distance specified by the manufacturer.

Use the correct number of counterweights as specified, and be sure that they are capable of safely supporting the maximum total loads, with a minimum safety factor of 4.

Connect wire rope to the rigging with proper fittings designed for the purpose.

Maintain two-point suspension hoisting machines at the same center distance as the overhead attachment points.

Roof irons, hooks, thrust-outs, and outrigger beams should be capable of sustaining the applied load and must be securely installed and anchored.

Counterweights for inner ends of thrust outs or outrigger beams should be securely fastened to the outrigger beam and tiebacks must be provided.

Handle all equipment with care and prevent wire ropes from becoming kinked.

Always tie suspended scaffolds into building or structure from which it hangs, except while scaffold is being raised or lowered.

Protect wire ropes when using welding equipment.

Appendix

American National Standards for Safety and Health—Construction (1987)

Accident Prevention

Accident Prevention Signs, Specifications for, ANSI Z35.1-1972, $5.00
Accident Prevention Tags, Specifications for, ANSI Z35.2-1968 (R1974), $5.00
Informational Signs Complementary to ANSI Z35.1-1972, Accident Prevention Signs, Specifications for, ANSI Z35.4-1973, $5.00

Acoustics

Designation of Sound Power Emitted by Machinery and Equipment, Method for the, ANSI S1.23-1976 (R1983), $6.25
Measurement of Sound from Pneumatic Equipment, Test Code for the ANSI S5.1-1971, $5.00
Noise Abatement and Control, Assessment of and Recommendations for Standards for (A special report—not an American National Standard), ANSI SR22, $8.00
Sound Level Meters, Specification for, ANSI S1.4-1983, $35.00
Sound Power Levels of Noise Sources, Survey Methods for the Determination of, ANSI S1.36-1979 (R1985), $25.00
Sound Pressure Levels, Methods for the Measurement of, ANSI S1.13-1971 (R1986), $9.00

Color Coding

Marking Physical Hazards, Safety Color Code for, ANSI Z53.1-1979, $5.00
Marking Physical Hazards, Swatches for Highway Colors Mentioned in ANSI Z53.1-1979, ANSI Z53.1 Set B, $20.00
Marking Physical Hazards, Swatches for Safety Colors Mentioned in ANSI Z53.1-1979, ANSI Z53.1 Set A, $30.00

Compressors

Compressors for Process Industries Safety Standard for, ANSI/ASME B19.3-1986, $25.00

Construction

Floor and Wall Openings, Flat Roofs, Stairs, Railings, and Toeboards, Construction Safety Requirements for Temporary, ANSI A10.18-1983, $6.00
Hoists, Safety Requirements for Rope-Guided and Non-Guided Workmen's, ANSI A10.22-1977, $5.00
Material Hoists, Safety Requirements for, ANSI A10.5-1981, $6.00
Personnel Hoists, Safety Requirements for, ANSI A10.4-1981, $10.00
Safeguarding Construction Alteration and Demolition Operations, ANSI/NFPA 241-1986, $11.50
Safety Belts, Harnesses, Lanyards, Lifelines, and Drop Lines for Construction and Industrial Use, Requirements for, ANSI A10.14-1975, $5.00
Safety Nets Used During Construction, Repair, and Demolition Operations, ANSI A10.11-1979, $5.00
Scaffolding, Safety Requirements for, ANSI A10.8-1977, $7.00
Temporary and Portable Space Heating Devices and Equipment Used in the Construction Industry, Safety Requirements for, ANSI A10.10-1981, $6.00
Work Platforms Suspended from Cranes or Derricks, Safety Requirements for, ANSI 10.28-1983, $6.00

Exhaust Systems

Blower and Exhaust Systems, ANSI/NFPA 91-1983, $11.50
Design and Operation of Local Exhaust Systems, Fundamentals Governing the, ANSI Z9.2-1979, $14.00

Eye and Face Protection

Occupational and Educational Eye and Face Protection, Practice for, ANSI Z87.1-1979, $9.00

Firefighting Equipment

Fire Extinguishers, Rating and Fire Testing of, Safety Standard for, ANSI/UL 711-1984, $12.00
Portable Fire Extinguishers, ANSI/NFPA 10-1981, $12.00

Fire Protection

Life Safety Code, ANSI/NFPA 101-1985, $19.50

Floor and Wall Openings

Floor and Wall Openings, Railings, and Toeboards, Safety Requirements for, ANSI A12.1-1973, $5.00

Foot Protection

Safety-Toe Footwear (includes Men's and Women's Safety-Toe Footwear), ANSI Z41.1-1983, $7.00

Gases, Compressed

Marking Portable Compressed Gas Cylinders to Identify the Material Contained, Method of, ANSI/CGA C-4-1978, $2.50

Head Protection

Industrial Workers, Protective Headware for ANSI Z89.1-1986, $7.00
Vehicular Users, Specifications for Protective Headgear for (includes supplements ANSI Z90.1a-1973 and Z90.1b-1978), ANSI Z90.1-1971, $7.00
Vehicular Users, Specifications for Protective Headgear for (supplement to ANSI Z90.1-1971), ANSI Z90.1b-1979, $3.00
Vehicular Users, Specifications for Protective Headgear for (supplement to ANSI Z90.1-1971), ANSI Z90.1a-1973, $3.00

Ladders

Fixed Ladders, Safety Requirements for, ANSI A14.3-1984, $7.00
Job-Made Ladders, Safety Requirements for, ANSI A14.4-1979 (R1984), $6.00
Metal Ladders, Safety Requirements for Portable, ANSI A14.2-1982, $10.00
Plastic Ladders, Safety Requirements for Portable Reinforced, ANSI A14.5-1982, $12.00
Wood Ladders, Safety Requirements for Portable, ANSI A14.1-1982, $10.00

Lifting Devices

Base Mounted Drum Hoists, ANSI/ASME B30.7-1984, $30.00
Belt Manlifts, ANSI/ASME A90.1-1985, $30.00
Cranes, Overhead and Gantry (Top Running Bridge, Single Girder, Underhung Hoist), ANSI/ASME B30.17-1980, $32.00
Derricks, ANSI B30.6-1984, $30.00
Hammerhead Tower Cranes, ANSI/ASME B30.3-1984, $30.00
Handling Loads Suspended from Rotorcraft, ANSI B30.12-1986, $32.00
Hoists, Performance Standard for Electric Chain, ANSI/ASME HST-1M-1982, $10.00
Hooks, ANSI B30.10-1982, $6.00
Jacks, ANSI B30.1-1986, $30.00
Mobile and Locomotive Truck Cranes, ANSI/ASME B30.5-1982, $35.00
Monorails and Underhung Cranes, ANSI B30.11-1980, $8.00
Overhead and Gantry Cranes (Top Running Bridge, Single or Multiple Girder, Top-Running Trollery Hoist), ANSI/ASME B30.2-1983, $30.00
Overhead Hoists (Underhung), ANSI B30.16-1981, $8.00
Portal, Tower, and Pillar Cranes, ANSI B30.4-1981, $10.00
Slings, ANSI/ASME B30.9-1984, $45.00

Lighting

Emergency Lighting Equipment and Power Equipment, Safety Standard for, ANSI/UL 924-1983, $18.75

Personnel Protection

Lock Out/Tag Out of Energy Sources, Safety Requirements for the, ANSI Z244.1-1982, $7.00

Respiratory Protection

Identification of Air-Purifying Respirator Canisters and Cartridges, ANSI K13.1-1973, $5.00
Physical Qualifications for Respirator Use, ANSI Z88.6-1984, $6.00
Practices for Respiratory Protection, ANSI Z88.2-1980, $9.00
Respiratory Protection for the Fire Service, Practices for, ANSI Z88.5-1981, $6.00

Scaffolds and Platforms

Boom-Supported Elevated Work Platforms, ANSI A92.5-1980, $5.00
Elevating Work Platforms, Manually Propelled, ANSI A92.3-1980, $5.00
Mobile Ladder Stands and Scaffolds (Towers), Manually Propelled, ANSI A92.1-1977, $5.00
Powered Platforms for Exterior Building Maintenance, Safety Code for, ANSI A120.1-1970, $7.00
Vehicle Mounted Elevating and Rotating Aerial Devices, ANSI A92.2-1979, $6.00
Work Platforms, Self-Propelled Elevating, ANSI A92.6-1979, $5.00

Tools, Electric

Machine-Tool Wires and Cables, Safety Standard for, ANSI/UL 1063-1986
Stationary and Fixed Electric Tools, Safety Standard for, ANSI/UL 987-1982, $25.00

Tools, Hand

Chisels, Safety Requirements for Ripping and Flooring/Electricians', ANSI B209.7-1982, $5.00
Hammers, Riveting, Scaling and Tinners, Safety Requirements for, ANSI B173.8-1982, $5.00
Hammers, Safety Requirement for Ball Pein, ANSI/HTI B173.2-1985, $6.00
Hatchets, Safety Requirements for, ANSI B173.7-1980, $5.00
Strinking Tools, Safety Requirements for Heavy, ANSI HTI B173.3-1985, $7.00

Tools, Power

Brushing Tools Constructed with Wood, Plastic, or Composition Hubs or Cores, Safety Requirements for the Design, Care, and Use of Power Driven, ANSI B165.2 1982, $8.00
Cutting-Off Machines, Safety Code for the Construction, Use and Care of Gasoline-Powered, Hand-Held, Portable Abrasive, ANSI B7.5-1983, $10.00
Power Driven Brushing Tools, Safety Requirements for the Design, Care, and Use of, ANSI B165.1-1985, $15.00

Vehicles, Surface—Sound Level

Bystander Sound Level Measurement Procedure for Small Engine Powered Equipment, ANSI/SAE J1175, $8.00

Construction and Industrial Machinery, Lighting and Marking of, ANSI/SAE J1029-MAR 86, $8.00
Engine Sound Level Measurement Procedure, ANSI/SAE J1074, $8.00
Operator Ear Sound Level Measurement Procedure for Small Engine Powered Equipment, ANSI/SAE J1174, $8.00
Operator, Safety Considerations for the, ANSI/SAE J153, $8.00
Safety Signs, ANSI/SAE J115-SEP79, $8.00
Starting or Movement of Machines, Unauthorized, ANSI/SAE J1083-JUL85, $8.00
Trucks and Buses, Exterior Sound Level for Heavy, ANSI/SAE J366-NOV84, $8.00

Ventilation

Ventilation Control of Grinding, Polishing, and Buffing Operations, ANSI Z43.1-1966, $7.00
Working in Tanks and Other Confined Spaces, Safety Requirements for, ANSI Z117.1-1977, $6.00

Welding and Cutting

Cutting and Welding Processes, ANSI/NFPA 51B-1984, $11.50
Sampling Airborne Particulates Generated by Welding and Allied Processes, Method for, ANSI/AWS F1.1-85, $12.00
Transformer-Type Arc-Welding Machines, Safety Standard for, ANSI/UL 551-1986
Welding and Cutting, Safety in, ANSI Z49.1-1983, $20.00

Index

About the Authors

W. E. Rossnagel (deceased) was a consulting and fire-protection engineer and was a safety engineer with the Consolidated Edison Company of New York.

Lindley R. Higgins has been a writer and consultant in the construction industry for more than 25 years. During his career, he has lectured at the University of Wisconsin on construction technology. In addition to his industrial activities, he is editor in chief of *Handbook of Construction Equipment and Maintenance*, and *Maintenance Engineering Handbook*.

Joseph A. MacDonald is editor in chief of *Imaging and Display Report*, published by his own company, Meta Data, Inc. In prior years, he was an editor with the Society for Information Display, Pequot Publishing, and the Kajima Corporation. From 1962 through 1978, he held several editorial positions with McGraw-Hill's *Construction Methods and Equipment* and *Engineering News Record* publications.